Electronic Warfare in the Information Age

DISCLAIMER OR WARRANTY

For a complete listing of the *Artech House Radar Library*, turn to the back of this book.

Electronic Warfare in the Information Age

D. Curtis Schleher

Software to accompany this book is available for download at:
http://www.artechhouse.com/static/schleher526.html

Artech House
Boston • London

Library of Congress Cataloging-in-Publication Data
Schleher, D. Curtis, 1932–
Electronic warfare in the information age / D. Curtis Schleher.
p. cm. — (Artech House radar library)
Includes bibliographical references and index.
ISBN 0-89006-526-8 (alk. paper)
1. Electronics in military engineering. 2. Information warfare.
I. Title. II. Series.
UG485.S3497 1999 99-27424
623'.043--dc21 CIP

British Library Cataloguing in Publication Data
Schleher, D. Curtis
Electronic warfare in the information age. — (The Artech House radar library)
1. Electronics in military engineering
1. Title
623'.043

ISBN 0-89006-526-8
Cover design by Lynda Fishbourne

685 Canton Street
Norwood, MA 02062

International Standard Book Number: 0-89006-526-8
Cataloging-In-Publication: 99-27424

10 9 8 7 6 5 4 3 2 1

Contents

Preface

Electronic Warfare in the Information Age is a modern treatment of a number of advanced topics in the EW field. It captures the major change of emphasis in the EW field where EW, previously viewed as a defensive action, is now viewed as both an attacking as well as a defensive action. One result of this is a change in EW definitions. The new definitions—electronic attack (EA), electronic support (ES) and electronic protection (EP) replace the old ECM, ESM, and ECCM nomenclature.

EW is basically a battle for control of the electromagnetic (EM) spectrum. However, it is now recognized that a primary function of the EM spectrum is as a carrier or provider of information which is essential to any military operation. Thus, EW is an essential component of what is termed information warfare (IW). The function of IW is to deny the enemy use of critical information while preserving one's own information resources.

EW fits well in the current revolution in military affairs. Attrition is no longer the major objective, being replaced by surgical strikes and an objective to separate a commander from his forces. This action is termed command and control warfare (C2W) in which EW is a critical function. Situational awareness is another EW function receiving increased emphasis. This function, primarily involving ES receivers, provides complete battlefield information to all military echelons.

Chapter 1 describes these new and many continuing functions of EW. Included is a discussion of the jamming of communications nets since this is a function receiving increased emphasis in EW support to information warfare.

Chapter 2 describes several of the advanced radar techniques whose evolution has stressed older EW systems. Also described is the low intensity EW threat mounted by regional military forces. One of the challenges faced by EW system designers is to incorporate sufficient capability to engage both modern and older threat systems.

Chapter 3 discusses the architecture and capabilities of modern EA systems. The trend towards off-board systems to cope with the monopulse radar threat is completely discussed. Also, current issues in support jamming which must contend with the development of radars with ultra sidelobe responses are described.

Chapter 4 focuses on EA actions directed against modern radars with advanced capabilities. These include EA techniques against pulse compression, pulsed Doppler, monopulse, and sidelobe cancellation threat systems.

Chapter 5 provides a complete description of digital radio frequency memories (DRFMs) which are considered the "silver bullet" of EW systems. These devices from an essential component in the design of modern EA systems which must contend with pulse Doppler and pulse compression radars. The DRFM allows the signature of the victim emitter to be stored and manipulated, which in turn, allows the EA system to penetrate the signal processor of the victim radar.

Chapter 6 discusses advanced electronic warfare support (ES) systems, including advanced receiver and processing systems. The main problem facing these advanced receiver and processing systems is the large number of pulses (e.g., 10^7 pps) which they must handle. This situation is further exacerbated by the large number of pulse Doppler radars currently being fielded. The solution is to provide a receiver which can separate the received signals into individual channels where they can be analyzed and processed.

The ultimate receiver of this type, which is under current intensive research, is the digital receiver. This receiver, using the FFT or similar algorithm, separates the signal set into individual channels where each signal can be further processed and analyzed. The key component in this receiver is the A/D converter which is placed as close to the system's antenna as possible. Current A/D converters have bandwidths in the gigahertz range while planned A/D converters will operate with tens of gigahertz bandwidth. The dynamic range of this receiver is determined by the number of bits in the A/D converter. A/D converters with ten or more bits will be required to provide dynamic ranges of 60 dB or more which is comparable with those of analog receivers.

In Chapter 7, we cover expendables and decoy systems. This form of EW is becoming increasingly popular. The difficulty of jamming monopulse radars is the main reason why off-board decoys have become the solution of choice. In addition, this form of EW is more compatible with the philosophy of low observable countermeasures than are on-board systems. In this chapter, we also discuss infrared missile attack which is a difficult problem affecting the survivability of modern aircraft.

In the last chapter, two relatively new applications are discussed. Directed energy weapons (DEWs) have great potential for destroying or damaging military targets. Advantages include propagation at the speed-of-light and an unlimited munitions supply. However, propagation phenomena limit their ultimate capability.

Signature control is recognized as an important element of EW. The ultimate capability is to make a military target invisible, and this is called stealth. The reality is that this may prove impossible to achieve at a feasible cost and active EW may be required for the survivability of a low observable target.

The current thrust in low observable technology is to reduce the radar signature of targets such that current radars are under powered to detect these types of targets. The radar equation of these radars will inevitably be rebalanced to diminish the advantage of this form of signature control. The operation of radars to detect stealth targets operating in clutter is still an open question requiring further research to determine if radars can be stabilized to meet this requirement.

In *Electronic Warfare in the Information Age,* a set of problems at the end of each chapter has been included. These problems, in many cases, are intended to extend the material presented in the chapters.

Also included are 50 MATLAB programs which document the many examples interspersed throughout the text. In the spirit of MATLAB, the reader can edit these programs to adapt them to a particular use. The programs contain numerous comment statements to allow the reader to follow the development of each program.

The material in this book resulted from several short courses given in the United States and Europe, as well as from my teaching at the Naval Postgraduate School. A current version of this short course is given for Johns Hopkins University.

The units in this text use both the English and metric systems. This stems from my belief that practitioners in this field from different countries must be fluent in both systems if they are to communicate with other workers in this field. Also, communication with users requires specification in units with which they are familiar.

As is appropriate, the author acknowledges a number of people who provided indispensable help in preparing the manuscript for publication. My wife Carol, typed the entire manuscript and its many revisions, including the detailed equations that appear throughout the book. David K. Barton constructively reviewed the entire manuscript and suggested many useful revisions. Also, a number of students made useful contributions to the MATLAB programs included in the text. Lts. Rohrer and Harris helped translate the

RGJMAT program from its Quick Basic origins to MATLAB,while Lt. Cardenas provided the gray-code conversion in the DIFM program.

D. Curtis Schleher
June 1999

1

Electronic Warfare—Threats, Requirements, and Principles

Electronic warfare (EW) is a military action whose objective is to control the *electromagnetic* (EM) spectrum. To accomplish this objective, both offensive *electronic attack* (EA) and defensive *electronic protection* (EP) actions are required. In addition, *electronic warfare support* (ES) actions are necessary to supply the intelligence and threat recognition that allow implementation of both EA and EP.

Figure 1.1 depicts the formal military terminology and definitions associated with EW. These broadened definitions have evolved from former definitions where functions were called *electronic countermeasures* (ECM), *electronic counter-countermeasures* (ECCM), and *electronic warfare support measures* (ESM) [1]. The new definitions envision an increased EA function including the use of directed energy weapons (lasers, microwave radiation, particle beams), *antiradiation missiles* (ARMs) and electromagnetic pulses (nuclear weapons destruction of electronics) to destroy enemy electronic equipment. The use of EP is also expanded to include not only protection of individual electronic equipment (ECCM), but use of such measures as *electromagnetic control* (EMCON), electromagnetic hardening, EW frequency deconfliction and *communications security* (COMSEC).

ES focuses on action taken for the purpose of near real-time threat recognition in support of immediate decisions involving EA, EP, weapon avoidance, targeting, or other tactical employment of forces. The key actions taken here are intercepting, identifying, analyzing, and locating enemy radiations. EM radiations are generally intercepted using sensitive receivers (i.e., ES receivers) that cover those frequency bands associated with significant threats.

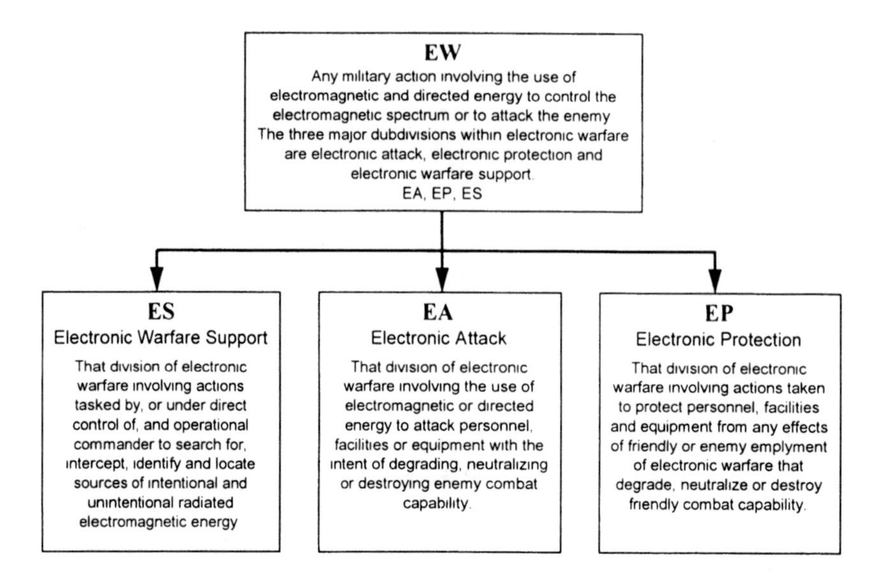

Figure 1.1 EW terminology.

Identification involves comparing the intercepted data against threat signatures stored in a threat library. Location is generally accomplished through a combination of spatially dispersed interceptions.

EA focuses on the offensive use of the electromagnetic spectrum or directed energy to directly attack enemy combat capability. It combines nondestructive actions (called "soft kill") involving electronic jamming and deception to degrade or neutralize enemy weapons with the destructive capabilities (called "hard kill") of ARMs and directed energy weapons. Jamming is defined as the deliberate radiation, reradiation, or reflection of EM energy for the purpose of destroying, damaging, or disrupting enemy use of the EM spectrum; while deception is defined as the deliberate radiation, reradiation, alteration, suppression, absorption, denial, enhancement, or reflection of EM energy in a manner intended to convey misleading information and to deny valid information to enemy electronics-dependent weapons. In a tactical environment, jamming radiations are subject to enemy actions (i.e., ARM attack) and must be carefully controlled. Examples of tactical deception are the use of ARM decoys and the radiation of false emissions to deceive an enemy's *signal intelligence* (SIGINT) system.

The destructive capability of EA is a relatively new concept in EW. ARMs have been effective in recent conflicts and are the most advanced method of employing destructive EA. Directed energy weapons use lasers or high-powered microwave transmitters to either destroy or disable electronic equipment.

EP focuses on defensive EW for the protection of friendly forces against enemy use of the EM spectrum and also against any unintentional radiations from friendly emitters. Electronic masking, emission control, the use of *wartime reserve modes* (WARM), electronic hardening, and the integration of EW systems into overall spectrum management are examples of EP actions.

An integral part of EP involves measures taken to imbed various techniques (ECCM) into electronic equipment to make them less vulnerable to EA. These actions reflect the continuing battle between EA and EP designers whereby each side attempts to gain the upper hand over the other. In essence, this is a battle of resources, with the advantage going to the side that invests the most resources.

1.1 Information Warfare

Information warfare (IW) is a broad concept, embraced by the military, whose objective is to control the management and use of information to provide military advantage. Information-based warfare is both offensive and defensive in nature—ranging from measures that prohibit adversaries from exploiting information to corresponding measures to assure the integrity, availability, and interoperability of friendly information assets. Information-based warfare is also waged in political, economic and social arenas and is applicable over the entire national security spectrum under both peacetime and wartime conditions.

From the military viewpoint, IW (sometimes called information operations) is difficult to define and suffers from a paucity of conceptual and descriptive models, theory, and accompanying doctrine. As such, IW is a term that has come to represent an integrated strategy to recognize the importance of information in the command, control, and execution of military forces and in the implementation of national policy. A definition proposed for IW is: "Actions taken to achieve information superiority in support of national military strategy by affecting adversary information and information systems while leveraging and protecting our own information and information systems" [2,3].

The current revolution in military affairs is largely rooted in the explosive advances which occurred in information technology. Vast amounts of data can be assimilated, processed, and made available to military users, allowing precision weapons to be directed against long-range targets. Protecting the integrity of these data is a major objective of current IW activity.

The global information infrastructure is a worldwide interconnection of communication networks, computers, data bases, and electronic equipment that make vast amounts of information available to users. It encompasses a wide range of equipment, including computers, satellites, fiber-optic transmission lines, microwave links, nets, scanners, television sets, displays, cable, video and audio tapes, fax machines, and telephone lines. This equipment forms information systems that collect, process, transmit, and disseminate information.

IW capitalizes on the growing sophistication of, connectivity to, and reliance on information technology. It targets information and information systems in order to affect the information dependent process [4]. Such information-dependent processes range from energy, finance, health, logistics, maintenance, transportation, personnel, control systems (e.g., air, sea, rail, road, river, pipeline, and canal transport systems), intelligence, command and control, and communications. All depend upon an assured availability of correct information at the time needed. Destroy or degrade the information or information service and the function is stopped or delayed. Exploiting this dependency relationship is the basis of IW [2,3,5].

As depicted in Figure 1.2, IW deals with a broad set of potential actions occurring in a timeline that involves both competitive (the introductory phase to military conflict) and conflict situations [6]. In the competitive phase, IW allows national level objectives to be achieved without combat. It involves covert actions that occur invisibly to most observers. As the combat timeline

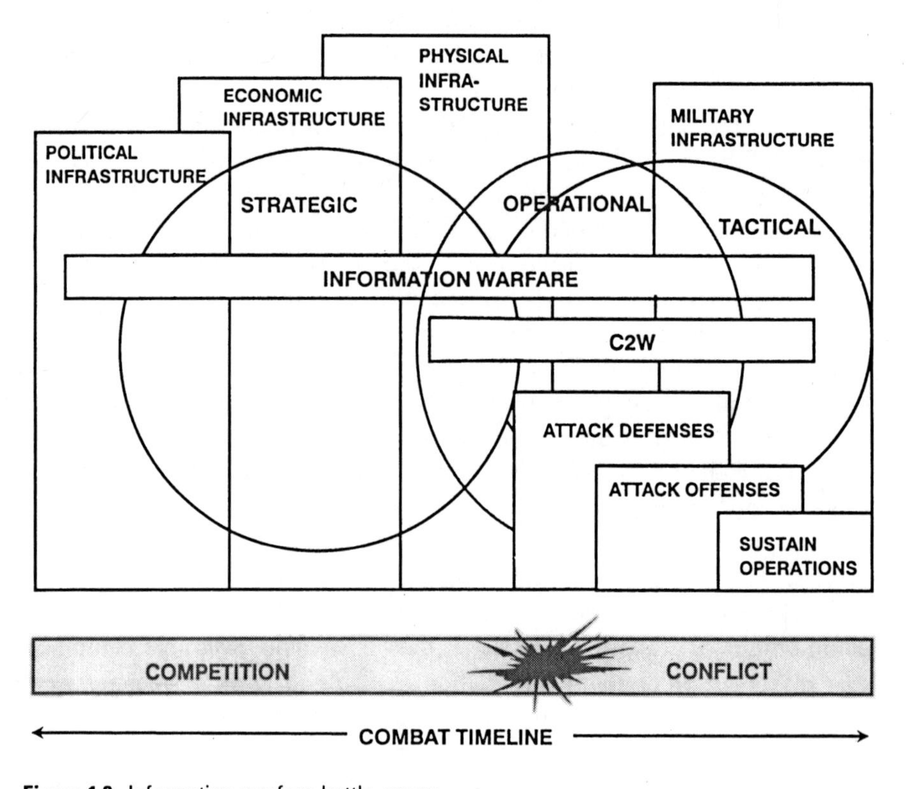

Figure 1.2 Information warfare battle space.

moves into conflict situations, IW is implemented using *command and control warfare* (C2W). C2W is fought at the military level against military decision makers, command and control systems, and combat systems. Its objective is to avoid or limit combat warning to the lowest possible level. It is fought visibly with military systems using overt actions.

1.1.1 Command and Control Warfare

Modern military commanders are closely coupled to their battlefield forces through *command and control* (C2) systems. Command decisions are made on the basis of information and intelligence that flow through these systems. These decisions result in military actions that are communicated to battlefield forces through the C2 system structure.

Command and control systems are the link through which leaders or commanders exercise authority over their assets. Break this connectivity and the leadership is effectively decoupled from its forces. The strategy that implements this excision is called C2W. The successful implementation of this strategy has the effect of depriving both the military commander and his forces from the information necessary for them to accomplish their missions. In this way, C2W can be viewed as an application of IW in military operations and, hence, is a subset of IW. C2W applies across the range of military operations at all levels of conflict and involves both offensive and defensive operations.

C2W consists of the integrated use of five military actions: *operation security* (OPSEC), EW, *psychological operations* (PSYOP), military deception, and physical destruction, which are mutually supported by intelligence. These elements are integrated into both C2-protection and C2-attack actions. C2W can be used throughout the range of military operations from routine peaceful competition, where OPSEC and deception are most effective, to war, where all five elements are essential. It is the integration of all five elements that makes C2W a powerful and important strategy. The objective of C2W is to obtain C2 superiority, which leads to success in military operations.

To understand C2W, we must examine the architecture of C2 systems. A C2 system is composed of a number of subsystems consisting of a network of (1) sensors, (2) navigation components, (3) command and fusion centers, (4) communications links, and (5) decision components (i.e., computers). These subsystems come together to form nodes. A critical node is an element, position, or communications entity whose disruption or destruction immediately degrades the ability of a force to command, control, or effectively conduct combat operations. Breaking a critical node disables the C2 system; hence, C2-attack focuses on this objective, while C2-protect functions to prevent the enemy from severing a critical node of one's own C2 system.

▶ *Example 1.1*

Figure 1.3 depicts the communications net of an air defense system [6]. Radar sensors on the E-3A, E-3D, and E-2C provide *airborne early warning* (AEW) surveillance for the various weapon systems whose function is to destroy incoming aircraft and missiles. The information generated by the AEW sensors is linked to the weapon launching systems through a series of redundant data links.

The vulnerable nodes of this C2 system include:

1. The radar sensors in the E-2C, E-3A, and E-3D;
2. The JTIDS/TADIL-J data links;
3. The TADIL-A data link;
4. The *Tactical Information Broadcast System* (TIBS).

Possible C2W actions might involve jamming or destroying the AEW radar, jamming or deception of the data links, jamming the data link or destroying the *airborne command post* (ABCCC), and jamming the satellite and data links to the surface naval assets. The basic concept of C2W is to sever a critical node that, for example, might involve deception of the JTIDS data link rather than the brute force jamming of the E-3A AWACS radar.

The five elements of C2W act synergistically to achieve C2-attack and C2-protect. The functions of these elements are [7] OPSEC, EW, PSYOP, deception, and physical destruction.

Operational Security

OPSEC is defined as the process of denying adversaries' information about friendly capabilities and intentions by identifying, controlling, and protecting indicators associated with planning and conducting military operations. OPSEC is not a stand-alone process and must be coupled with a well-developed deception plan. A good OPSEC C2-attack plan starves the enemy intelligence system, limiting the enemy commander's ability to control his forces effectively. The objective of OPSEC C2-protection is to hide friendly C2 information using deception to feed false information to the enemy [8].

Electronic Warfare

As previously described, EW consists of ES, EA, and EP. Within the C2W framework, EA primarily supports C2W-attack, while EP primarily supports

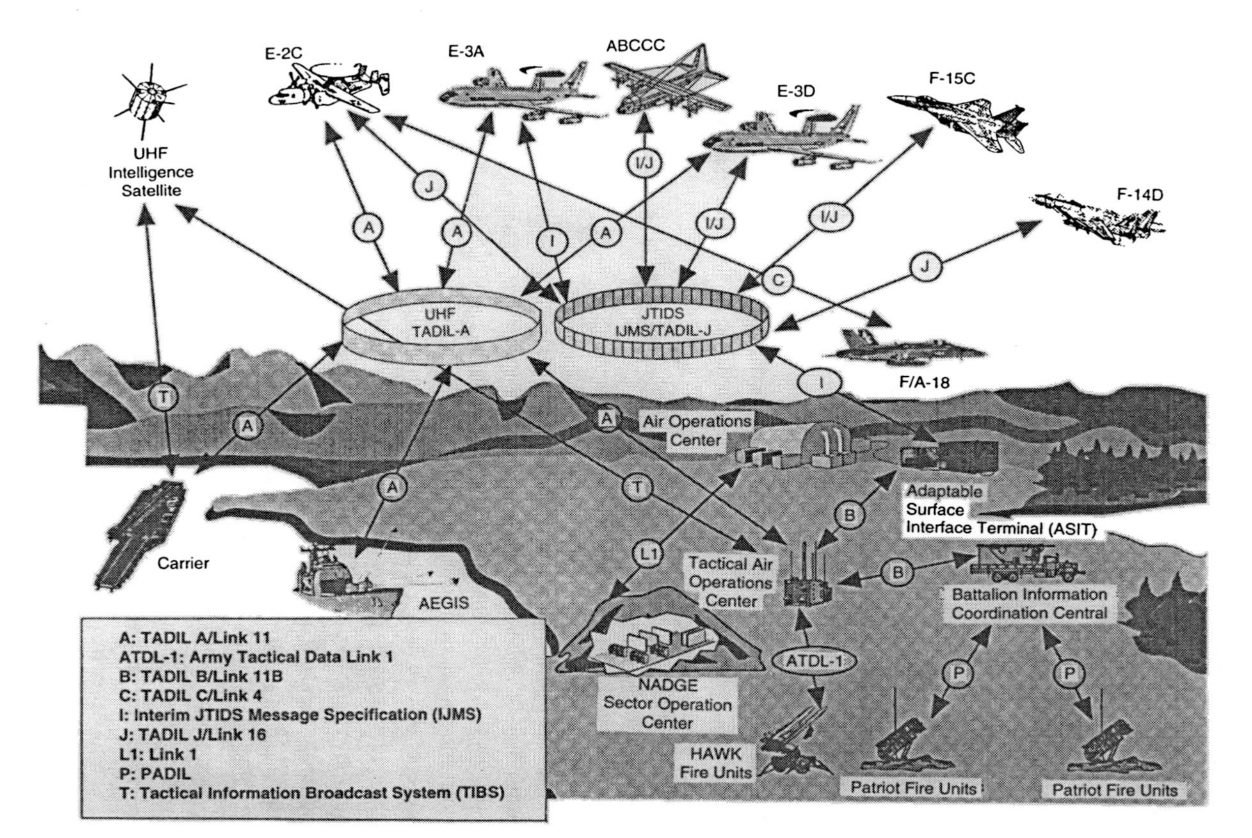

Figure 1.3 Communications net of air defense system. (*Source:* [6].)

C2W-protection, and ES both provides information into the intelligence cycle and supports EA and EP. ES is used to search for, intercept, identify, and locate sources of radiated electromagnetic energy for the purpose of near real-time threat recognition in support of immediate decisions involving EA, EP, avoidance, targeting, and other tactical employment of forces. ES data is also used to produce SIGINT (both COMINT and ELINT), which, after processing, becomes part of the intelligence base. This updated information can be used to plan C2-attack operations and provide battle damage assessment and feedback on the effectiveness of the overall C2W plan.

EA—whether jamming, electromagnetic deception, or destruction of C2 nodes with *directed-energy* (DE) weapons or ARMs—has a major role to play in almost all C2-attack operations in a combat environment. It can also be used to defend a friendly force from adversary C2-attack [3]. EP is used in C2-protect to safeguard friendly forces from exploitation by adversary ES-SIGINT operations. Equipment and procedures designed to prevent adversary disruption or exploitation of the EM spectrum are the best means that friendly forces have to ensure their own uninterrupted use of the EM spectrum during C2-attack operations.

Psychological Operations

PSYOP have been used throughout history to influence adversary groups and leaders. Modern PSYOP is enhanced by the expansion of mass communication capabilities. The effectiveness of this communication on the target audience depends on the perception of the communicator's credibility and capability to carry out the threatened actions. Military PSYOP constitutes a planned, systematic process of conveying messages to and influencing selected enemy groups. The messages conveyed by military PSYOP are intended to promote particular themes that result in desired enemy attitudes and behaviors. Therefore, PSYOP is a valuable element in both C2-protection and counter C2.

PSYOP is a formidable weapon that focuses largely on the control side of C2. The essence of PSYOP is to convince an enemy commander to either do or not do something, whereby it influences the decision cycle of C2. To accomplish this, the commander is confronted with evident reality that either creates or reinforces his perceptions. Effective PSYOP is closely integrated with OPSEC and deception in that all three seek, through perception management, to portray a picture of reality in a way beneficial to achieving some desired effect. EW can be a media for conveying the PSYOP message [9].

Deception

Deception attempts to mislead the adversary by manipulating, distorting, or falsifying friendly dispositions and capabilities to induce one to act or fail to act in a manner prejudicial to his own interests and exploitable by friendly forces. Deception operations degrade the accuracy of hostile *reconnaissance, surveillance, target acquisition* (RSTA) and intelligence gathering resources. It also seeks to lead an adversary to draw the wrong conclusion from what he does see. Deception causes the adversary commander to see the battlefield incorrectly and act in a predicted manner. To accomplish deception, it is necessary to allow the enemy's intelligence channel to function so as to use it as a conduit to feed the deceptions.

Deception and OPSEC are synergistic actions used in C2-attack and C2-protect operations. Both attempt to manage perception indicators. OPSEC seeks to limit an adversary's ability to detect or derive useful information from his observation of friendly activities. Deception seeks to create or increase the likelihood of detection of certain indicators to cause an adversary to derive an incorrect conclusion. Deception can be used as an OPSEC measure.

EW supports deception with electromagnetic deception as both a form of EA and a technical means of deception. EW attacks on intelligence collection and radar systems can be used to shape and control an adversary's ability to see certain activities [10].

Physical Destruction

In C2W a major strategy consists of breaking critical nodes of the enemy's C2 system. Destruction is just one method by which this can be accomplished. However, timing is critical since destructive actions are effective for only a limited time. In general, the adversary is able to recover from destruction, given sufficient time and resources. From a military viewpoint, it is imperative to employ destruction as an element just before it is desired to shut down the adversary's C2 function [7].

Against the command function of C2W, destruction is applied against the command headquarters; while against control, the attack focuses on critical nodes of the C2 system's communication, computer, or sensor nets. Since the destruction process involves the use of weapons, both targeting and *battle damage assessment* (BDA) are important parts of the process. BDA is an intelligence function that relies heavily on overhead imagery, both electronic and optical. Against radiating targets, a method commonly used is to monitor the transmissions before and after the attack. If a node was transmitting prior to being targeted and stopped transmitting following the attack, it is expedient to assume at least temporary success. The destruction process is aided by the development of precision guided weapons that allow the surgical removal of the various elements of the enemy's C2 system.

EA is an important element of the destruction process. ARMs have been developed to a high level of sophistication. ARMs are guided to their targets by following the radiations of critical sensors or communication links back to their source [1]. This inhibits sensors and communication links from operating when ARM attacks are likely. Moreover, many emitters utilize remote antennas and decoys as EP measures against ARM attack.

Directed-energy weapons (DEW) are another form of destructive EA [7]. These weapons can employ lasers, or charged particle or microwave/RF beams and are attractive because they attack at the speed of light. At present, the power generated by these weapons generally limits their tactical use to the disruption or burnout of electronic equipment [11]. For laser and particle beam DEW, propagation losses can be substantial. For microwave DEW, focusing the beam to concentrate sufficient energy on the target area is a concern. DEW attack can be either directly into the reception mechanism of the target (i.e., antenna), which is called "front door" attack, or through power lines, equipment enclosures, connector cables, or any other leakage path into the equipment (called "back door" attack).

1.2 Intelligence

Every element of EW is based upon accurate, timely, and focused intelligence. EA and EP are intimately dependent on knowledge of the adversary's systems, operational tactics, and capabilities. Since intelligence is perishable, its timeline is important.

Critical to operational success is gaining intelligence dominance of the battle space. All sides will attempt to determine adversary capabilities, objectives, and operational concepts. All sides will deploy their collection and analysis capabilities and endeavor to conduct successful deceptions in attempts to gain the advantage of surprise and, hence, provide operational security. Gaining and maintaining this intelligence dominance is the key element in successful EW operations.

Intelligence sources are the means or systems used to observe, sense, and record or convey information of conditions, situations, and events. There are eight primary source types: *imagery intelligence* (IMINT), *human intelligence* (HUMINT), *signal intelligence* (SIGINT), *measurement and signature intelligence* (MASINT), *open-source intelligence* (OSINT), *technical intelligence* (TECHINT), *counter intelligence* (CI), and *unintentional radiation intelligence* (RANT). These forms of intelligence and their major subdivisions are listed in Table 1.1 [7].

EW depends upon all-source, timely intelligence. SIGINT (both communications intelligence and electronics intelligence-derived intelligence products, particularly data bases) can be especially useful. Primary intelligence support

Table 1.1
Intelligence Sources

Imagery intelligence (IMINT)
Photo intelligence (PHOTINT)
Signal intelligence (SIGINT)
Communications intelligence (COMINT)
Electronic intelligence (ELINT)
Foreign instrumentation signals intelligence (FISINT)
Telemetry intelligence (TELINT)
Radar intelligence (RADINT)
Human intelligence (HUMINT)
Measurement and signature intelligence (MASINT)
Acoustical intelligence (ACINT)
Optical intelligence (OPINT)
Infrared intelligence (IRINT)
Laser intelligence (LASINT)
Nuclear intelligence (NUCINT)
Open source intelligence (OSINT)
Technical intelligence (TECHINT)
Counter intelligence (CI)
Unintentional radiation intelligence (RANT)

for EW comes from the *electronic order of battle* (EOB) and signal data bases such as the *Electronic Warfare Integrated Reprogramming Data Base* (EWIR) and the *ELINT parameter limits* (EPL).

EW planners derive EW targeting information from the EOB. This information and data bases are used by mission planners to load threat libraries utilized in all ES and EA equipment; accurate threat libraries are essential to the identification of emitters and weapon systems obtained by matching signatures obtained against the data base.

Figure 1.4 depicts the SIGINT threat environment as viewed from an airborne platform through the millimeter-wave frequency band. These threat emitters include communications systems (both voice and digital), data links, satellite communications, cellular telephones, and radar systems generally associated with some form of weapon system.

The SIGINT system functions to collect, analyze, identify, and locate emitter signals throughout the entire frequency band and is generally composed of both a COMINT system, which functions against communication emitters, and an ELINT system, which primarily looks for radar signals.

The COMINT system is concerned with activity detection and collection, classification and identification of emitters, and direction finding. When precision emitter location is desired, the hyperbolic *time difference of arrival* (TDOA)

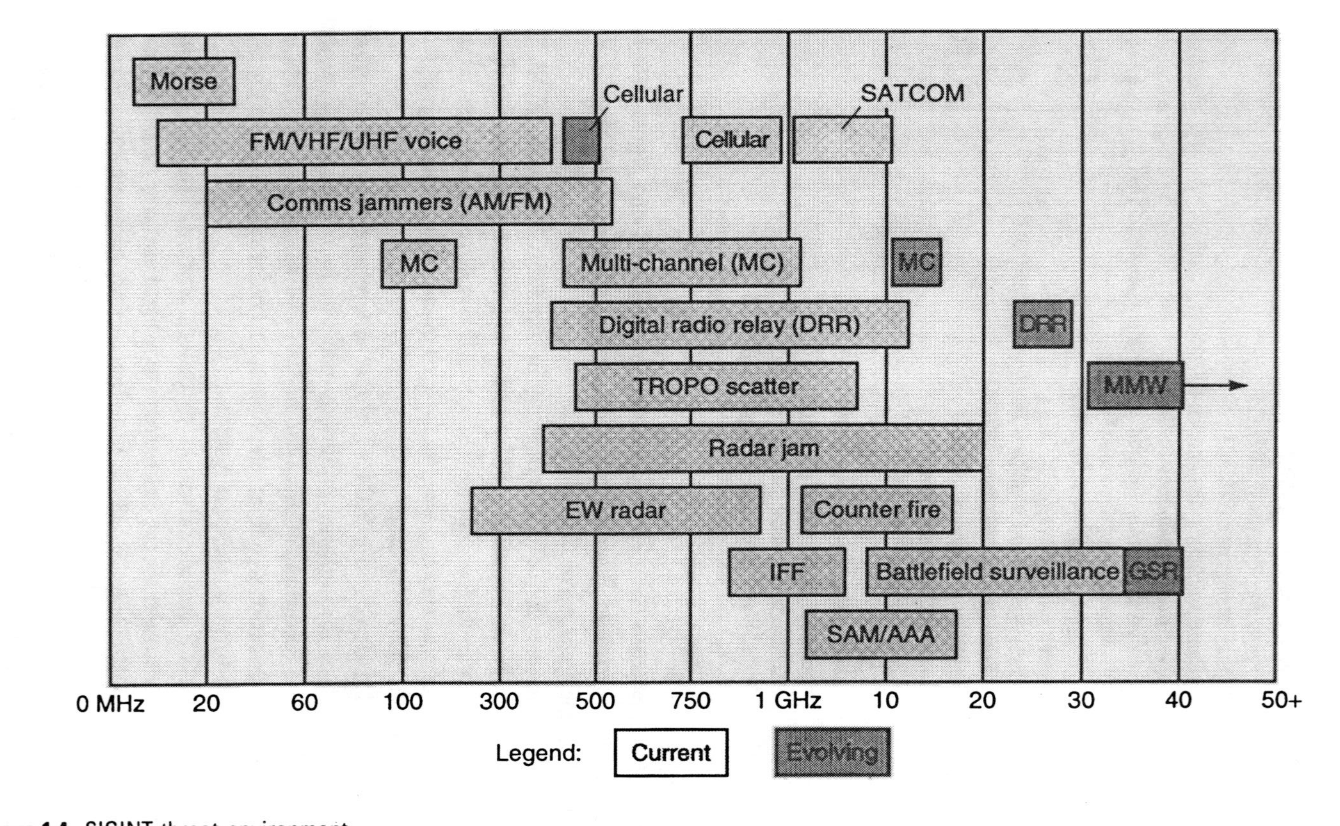

Figure 1.4 SIGINT threat environment.

and *differential Doppler* (DD) systems are used from multiple aircraft. These techniques are suitable for real-time targeting.

ELINT systems generally measure both *direction* and *time of arrival* (DOA, TOA) along with signal signature parameters (e.g., pulsewidth, frequency, pulse repetition interval, and pulse type). These collected data, after processing, are then compared with the stored threat parameter data base generated from the EOB and EPL to allow emitter identification. Emitter location is accomplished through triangulation or TDOA techniques. Weapon identification is determined through the association of intercepted emitters.

1.3 EA Effect on Radar

Figure 1.5 depicts the effect of random pulse jamming in combination with chaff as displayed on a search radar. The random pulse jamming is synchronized with the antenna's scan pattern using an inverse gain function to provide a uniform jamming pattern throughout the jammed sector. The jamming is also synchronized with the radar pulse transmissions to prevent detection of target patterns (i.e., radar blips) in the jammed sectors. It is apparent from the illustration that it would be very difficult to detect targets in the jammed areas without extensively processing the jamming waveform.

Figure 1.5 illustrates two common types of EA employed against search radar. The first involves active self-protection jamming whose purpose is to deny range, velocity, and angle information about the protected target. To prevent jammer strobing (and hence target location through triangulation), the jammer technique illustrated must be able to penetrate the radar antenna's sidelobe response. The second type of EA illustrated has the same objective as the first but is passive in nature. Chaff consists of many elemental radar reflectors that are injected into a volume that then provides a large radar reflection (relative to the shielded target's reflection) throughout the covered region, thus preventing target detection. The chaff effects may persist for many hours and, coupled with its economy, make this type of EA a favorite technique employed by EA designers.

A third type of EA commonly employed is stand-off or support jamming, which generally operates in the radar antenna's sidelobe region. Figure 1.6 depicts PPI displays of three types of jamming waveforms that are commonly utilized in EA support jammers.

The first (Figure 1.6(a)) consists of a random noiselike waveform, usually produced using FM techniques, that resembles receiver thermal noise [1]. The quality of this waveform is such that it is less efficient than ideal Gaussian noise (i.e., more power must be generated to produce an equivalent effect to

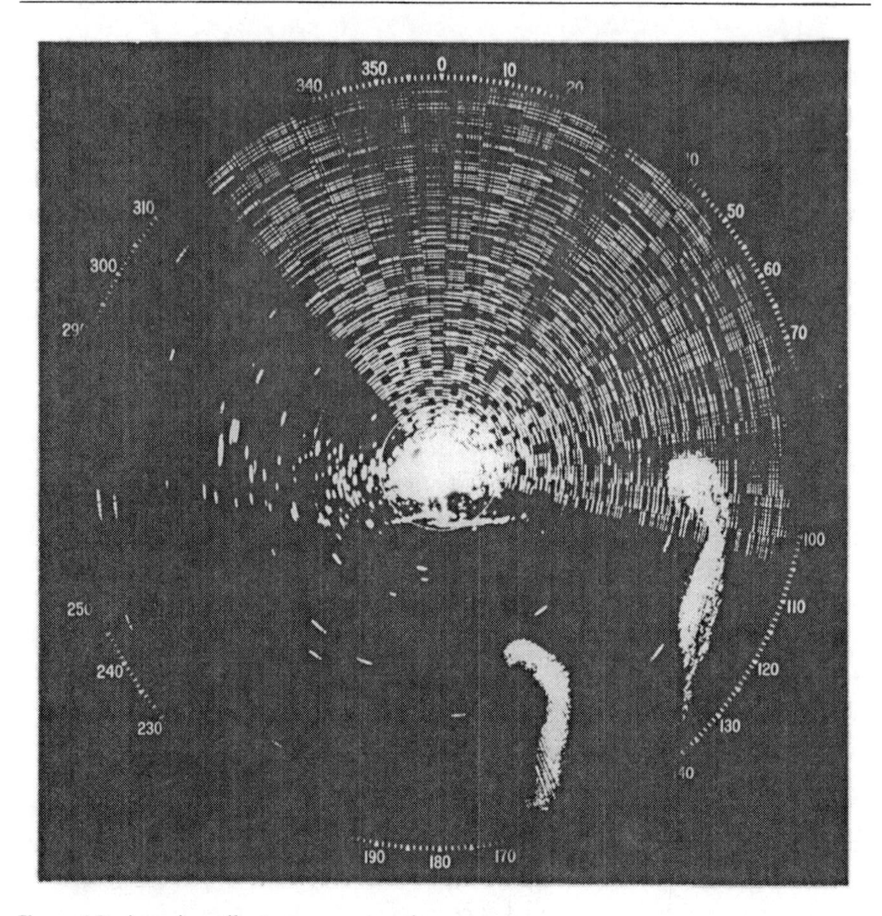

Figure 1.5 Jamming effects on search radar.

Gaussian noise). In addition, the radar signal processor is designed to discriminate against such a waveform (much as it discriminates against receiver noise), further diluting the efficiency of this type of jamming. The discriminate employed by the radar signal processor is that bona fide targets will generate a stream of returns as the radar's main beam antenna response sweeps over the target, each synchronized to the radar's transmission. This mechanism produces the radar blips displayed in Figure 1.5. On the other hand, random noise waveforms are completely random with respect to the radar transmissions. Thus, when target waveforms are integrated (summed either coherently or noncoherently) using electronic circuitry or on the radar's display, a significant processing gain results for the synchronized target signals as contrasted to the random noiselike jamming signals.

The second waveform (Figure 1.6(b)) attempts to exploit the radar's signal processor by generating a random false targetlike jamming waveform

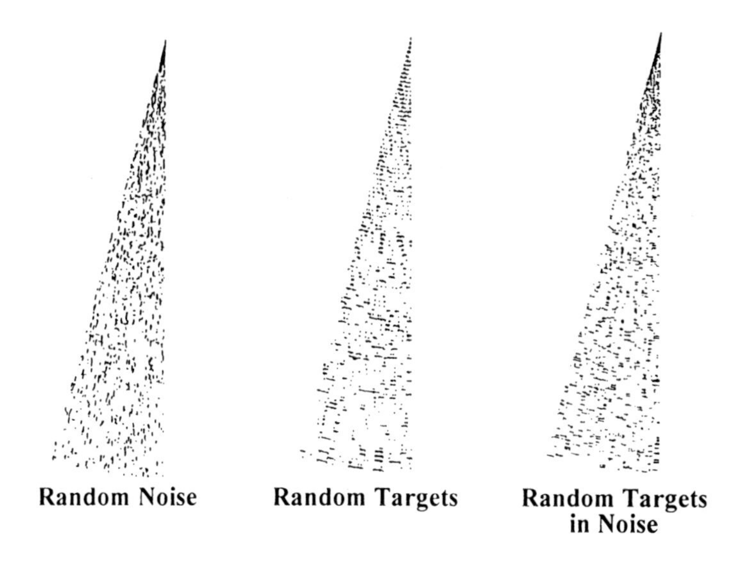

Figure 1.6 Support jamming waveform effects.

pattern that will be accepted by the radar's signal processor as a bone fide target signal. When successful, the signal processing gain will be equal for both the target and jamming waveforms, resulting in greater efficiency for this type of jamming waveform as compared to the noise-type jamming waveform of Figure 1.6(a). Comparing Figures 1.6(a) and (b), it is apparent that it would be very difficult to detect a true target return in the false target jamming response, while with the noiselike waveform even a weak target response would be detectable.

Both false target and noise jamming are combined in the third jamming waveform depicted in Figure 1.6(c). This jamming waveform induces further randomness into the jamming pattern, compounding the problem faced by the signal processor in extracting the true target from the jamming background. The high duty cycle jamming waveform combining both random pulses and noiselike responses is designed to provide resistance to rejection by sidelobe blankers and cancelers. If the sidelobe blanker responded to all false target responses, then excessive blanking of the mainbeam target response would occur. On the other hand, sidelobe cancelers are designed with time constants that do not respond to pulses and hence generally have little capability to reject pulse-type sidelobe signals.

This discussion illustrates some of the common EA threats that are encountered by military radar. A general radar jamming scenario of this type is illustrated in Figure 1.7. This scenario is not intended to be all inclusive but rather to illustrate the important point that military radar is generally subjected to multiple jamming sources. Hence, the design of radar EP is not

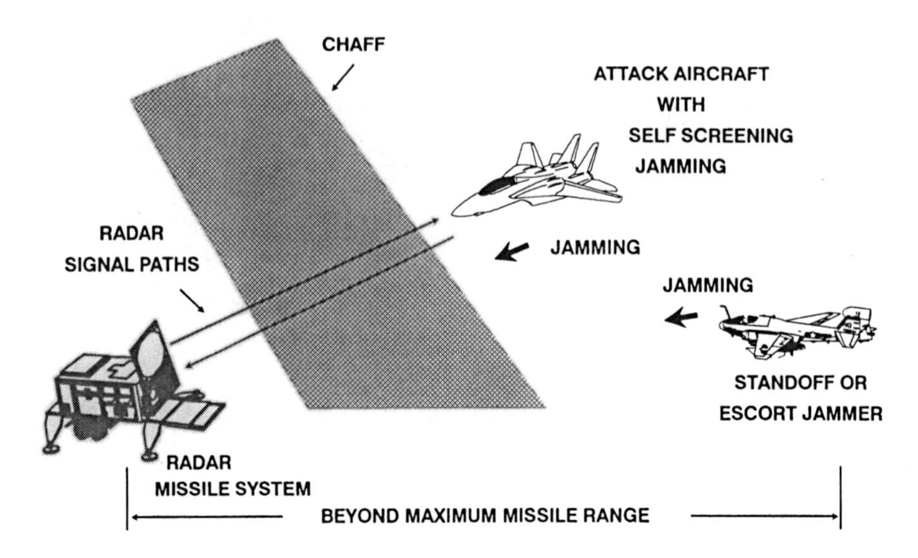

Figure 1.7 General radar jamming scenario.

a simple one-on-one problem; very often an EP that is effective against one type of EA is actually counterproductive against another common type of EA, and its employment will actually increase the vulnerability of the radar.

To understand the philosophy of EA design, it is necessary to consider the effects of EA on the victim radar. These effects and the deployment of EA are different for surveillance and tracking radar, so it is natural to discuss these radar types separately.

1.3.1 The Effect of EA on Surveillance Radar

Surveillance or search radars are by far the most common type of radar used in both military and civil applications [1]. Their function is to search a large volume of space and locate the position of targets within the search coverage. The target reports generated by a surveillance radar can be processed to form target tracks. The target-tracking vectors formed by these track-while-scan radars allow the absolute motion characteristics of the targets to be determined. The motion characteristics of the target can then be used to ascertain the relative threat posed by a target and to predict the future position of the target.

The radar data-processing function associated with the surveillance radar generally determines the data rate at which the radar must sample the target. A single radar data processor can be used with several spatially dispersed radars or to combine radar data with that of other complementary sensors (e.g., electro-optical, IR). This netting process provides accurate target tracks, mitigates shadowing and glint problems associated with surface-based radar and, as we shall discuss, allows noise jammers to be located by triangulation on the azimuth strobes from multiple radars.

Usually, military surveillance radar must provide three-dimensional information for an air search and two-dimensional data for a surface search or when a dedicated height-finding radar is coordinated with the air search radar. Radar data are naturally in spherical coordinates (e.g., range, azimuth, and elevation). Conversion to Cartesian coordinates is sometimes advantageous when multiple radars or sensors are netted.

Considering the scenario depicted in Figure 1.7, there are generally three mechanisms employed by EA to degrade the performance of a surveillance radar.

First, noise jamming from support jammers (i.e., spatially displaced from the protected target) and self-screening jammers (i.e., located on the protected aircraft) can be used to deny a measurement of the range of the penetrating aircraft [1]. Support jammers can be further divided into stand-off jammers and escort jammers (Figure 1.8).

Stand-off EA missions are those that are conducted outside the lethal zones of hostile weapon-control systems to provide EA support for friendly forces subject to hostile fire. Stand-off EA systems employ high-power noise

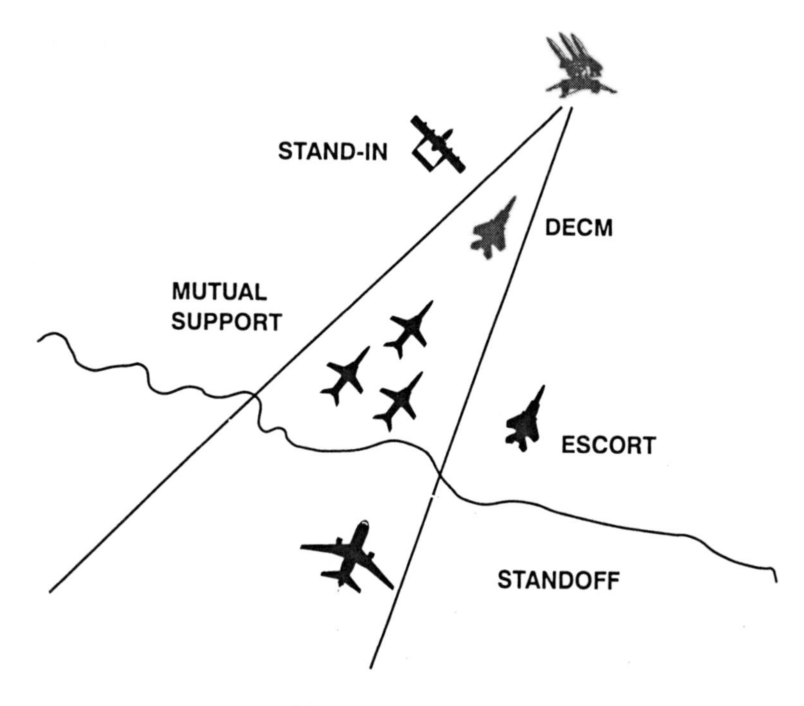

Figure 1.8 EA missions. (*Source:* [1].)

jamming (1- to 2-kW average jamming power per band) that must penetrate through the antenna and receiving systems at long ranges.

Escort EA is conducted by EA elements assigned to accompany and support the combat elements. In airborne escort jamming, the jamming aircraft accompanies the strike aircraft into the combat zone, making it vulnerable to enemy weapons such as radiation homing missiles designed to suppress EA. The effectiveness of escort jammers would be expected to be greater than stand-off jammers because of their closer proximity to the victim systems and their penetration into the radar's main beam or high-order sidelobes. Escort jamming is generally used when the penetrating aircraft do not have enough available space or payload to protect themselves or when the protected aircraft does not want to radiate (i.e., quiet target).

Using noise jamming in either support or self-screening modes attempts to conceal both the magnitude and direction of a raid. When deployed from a stand-off jammer with a quiet attack aircraft, it is possible to completely conceal the raid's direction until a late stage of the operation.

Use of escort or self-screening jammers can have a similar effect in concealing the magnitude of the raid, but a general indication of the possible attack direction can be obtained using jamming strobes. Confusion can be created by deploying widely spaced escort jammers, thereby creating strobes from many directions, only one of which supports the penetrating aircraft. The resulting confusion will have a significant effect on the reaction time of the defenses. For example, an *early warning* radar will require collateral information from alternate sensors or, alternately, an investigation of each strobe direction. *Surface-to-air missile* (SAM) systems may attempt to obtain collateral data by slewing the radar (or associated electro-optical or IR tracker) to examine each direction in turn. In each of these cases, the noise jammer, if successful, will extract a significant price in performance by lengthening the reaction time of the weapon system.

The second EA mechanism that degrades the radar's performance involves the use of *deceptive electronic countermeasures* (DECM) techniques, whose signals attempt to emulate those from bona fide target or clutter returns except that they are introduced into the radar at erroneous ranges or directions. This type of EA can be initiated from either on-board or off-board equipment and its primary characteristic is that it is synchronized with the radar transmissions. Transponder jammers are synchronized temporally with the victim radar, while repeater jammers employ some form of microwave memory (e.g., *digital RF memory* (DRFM)) to provide both phase and time synchronization.

Against surveillance radar, a principal tactic is to generate a sufficient number of false targets to overload the radar's data processing system. The jammer, by generating deception pulses whose amplitude is inversely related to the incident radar signal and sufficiently penetrates the radar's sidelobe structure, can deny the radar the use of received jamming signal strength as an indicator of the direction of the jamming aircraft. This inverse gain jamming when used on-board the protected aircraft can confuse the direction of the attack and its intensity. The use of staggered *pulse repetition frequency* (PRF) on the surveillance radar causes all false targets to occur at ranges greater than the jamming aircraft, but sufficient confusion can generally be generated by this form of jamming to deny an accurate estimate of raid size.

A more subtle type of deceptive jamming is termed camouflage or surgical countermeasures [1]. The objective of this form of EA is to confuse the radar's signal processor into suppressing the actual target return. As applied to surveillance radar, any radar signal processing function that suppresses target returns in the presence of clutter or interference is a prime candidate for this form of EA. One example of this form of EA is the radar's *constant false-alarm rate* (CFAR) function, which desensitizes the radar's detection threshold in the presence of clutter residues. The introduction of synthetic clutter into the radar's signal processor (e.g., weather clutter) has the potential for suppressing target returns as a self-protection measure. An example of deceptive support jamming is the introduction into the radar's sidelobe of a jamming pulse that covers the target return, causing the radar's sidelobe blanker to remove the target return.

A third form of EA is through the deployment of corridors of chaff to conceal the presence and magnitude of a raid. If many corridors can be laid, then the true direction of attack can be denied to the defense system. The best antichaff measures generally employ Doppler processing techniques, which rely on a high degree of correlation between radar transmissions during the *coherent processing interval* (CPI) of the radar's signal processor. On the other hand, the best antijamming measures generally rely on a high degree of randomness between radar transmissions. Together, chaff and active EA can present a formidable challenge to radar.

A fourth type of EA not explicitly depicted in Figure 1.7 is the reduction of *radar cross section* (RCS) to escape detection. This falls into a general class of techniques called low-observable or stealth technology. Low-observable technology and other EA techniques (i.e., both active and passive) are generally synergistic methods that complement each other in reducing the detectability of targets.

When the target's RCS is reduced, several effects are produced. First, the free space detection range of conventional radar is reduced in proportion to $\sigma_t^{1/4}$, where σ_t is the radar's RCS. Further, active EA, which has a figure of merit equal to $P_j G_j / \sigma_t$, where P_j is the jammer's transmitted power and G_j is the jammer's antenna gain, becomes more effective for a given jammer *effective radiated power* (ERP) (P_jG_j). Ultimately, active EA is not required for target self-protection, but support jamming becomes increasingly effective. Detection of stealth targets in various forms of clutter (i.e., ground, sea, weather, chaff) becomes more difficult. Extraneous angle radar targets (e.g., birds, insects, anomalies) whose RCSs are of the same order of magnitude as the stealth targets become difficult to eliminate as radar sensitivity is increased to allow stealth target detection.

A fundamental design modification to surveillance radar that must detect reduced RCS targets is to increase their power-aperture product, thereby rebalancing the radar equation so that they are sensitive enough to detect these targets at maximum range. However, the increased power aperture also increases the response of clutter and angel targets in direct proportion to that of the stealth target, so more innovative radar solutions are required for this situation [12]. One result of this increased radar sensitivity is that active EA capabilities must also be rebalanced, resulting in ERP requirements similar to that required for conventional targets against a conventional radar.

1.3.2 The Effect of EA on Tracking Radar

The classic form of tracking radar is dedicated to following the path of a single target and measuring its position in a spherical coordinate system. The target position is tracked in angle using servo control loops that move the radar antenna in azimuth and elevation. The target is tracked in range using a servo-controlled early-gate/late-gate range tracker. When the tracking radar is on a moving platform (e.g., ship, aircraft, or missile), gyro stabilization loops are also supplied in azimuth and elevation to remove the effects of vehicle motion. In many cases, the tracking radar supplies velocity data obtained by differentiating the measured position data. Velocity data are used to predict the future position of the target, which is useful in pointing a weapon with the proper lead angle.

Self-protection ECM systems to counter tracking radar tend to use deceptive EA (called DECM) types of techniques. This reflects a strategy of attempting to cause the tracking radar to break lock rather than have the penetrator mask its own position through noise jamming. Noise jamming is somewhat suspect against tracking radar because the potential for tracking the EA radiations in angle (track-on-jam) always exists. In addition, angle tracking is sufficient, to a first order, for a missile system that uses semiactive guidance. Also, multiple angle-tracking radars have the potential for computing range using triangulation techniques.

Another reason for employing DECM techniques for self-protection against tracking radar is that they are highly efficient in their use of average transmitter power, thereby providing the capacity for countering multiple threats. DECM against tracking radar is generally achieved through the radar's main beam, which further reduces the required jammer ERP.

Target range can be denied to the radar via a *range gate walk off* (RGWO) in which a false echo generated by the jammer is progressively delayed to seduce the target tracking gate away from the true target echo so that, in effect, an infinite *jam-to-signal* (J/S) ratio is obtained for angle deception purposes. Noise jamming might be used against a target-tracking radar that does not have a track-on-jam (also called home-on-jam in missile seekers) mode of operation. In some situations, its effect may be to create uncertainty as to whether the target is within the weapon system coverage. Short-range air defense antiaircraft systems generally rely on laser range finders to provide immunity to noise jammers.

Leading edge range trackers in combination with PRF jitter can be an effective EP measure against repeater type DECMs. The function of PRF jitter is to prevent the jammer from anticipating the radar's transmission, thus denying any possibility of forward pull-off. This guarantees that the true target return will always lead the deceptive signal and allows the leading edge tracker to select the true signal.

Repeater jammers must have very short turn-around times (e.g., 50 ns to 100 ns) to minimize the probability of leading edge range trackers rejecting deceptive pulses. One form of leading edge range tracker gives preferential weighting to the forward gate signal, which can result in an advantage of the order of 10 dB to the true signal. This leads to the use of high J/S in repeater jammers as one method that tends to negate the effects of leading edge range tracking.

PRF jitter is feasible in most types of tracking radar. However, in pulse Doppler radars that form multiple Doppler filters (i.e., medium- and high-PRF pulsed Doppler radars), the PRF must be constant over the CPI used to form the target filter. Hence, in these types of radar, a block-to-block PRF stagger is generally used to increase their susceptibility to anticipating deceptive jamming techniques.

Transponder-type jammers generally use a cover pulse waveform that extends over the target's return echo. This can be effective in situations where repeater jammers might be ineffective (i.e., with PRF jitter and leading edge range trackers). The jammer's cover pulse is queued from the previously intercepted radar transmission with width sufficient to encompass the expected arrival time of the target's return echo, which is uncertain due to any PRF jitter employed by the radar. It is also possible to simulate inbound targets in this manner. Pulse-to-pulse frequency agility in conjunction with PRF jitter (i.e., the agile-agile radar emitter) is an effective counter to transponder jammers.

However, when a radar uses Doppler processing, it must maintain a constant transmitter frequency over the coherent processing interval. Hence, a frequency agile radar waveform is generally not appropriate for Doppler-type radar.

Many radars use coherent modes of operation that allow them to distinguish targets from background clutter on the basis of their relative Doppler shifts [12]. Generic types include the pulsed Doppler radar, which is used for detecting low-flying targets in both ground and airborne situations, and the CW radar, which is used in semiactive missiles and low-flying target acquisition radar.

From an EA viewpoint, the significant characteristic of a Doppler radar is that the final detection process involves the use of a narrowband filter (e.g., 50- to 500-Hz bandwidth) that is matched to the Doppler-shifted target spectrum. This provides the Doppler radar with a natural protection against noise jamming, which must spread its energy over the expected range of possible target returns. This leads to a general preference for DECM against Doppler radar. The DECM is able to concentrate its energy into the narrow Doppler filter and, hence, makes optimum use of the available jamming power.

Velocity deception is employed in most DECM systems by repeating a frequency-shifted replica of the received radar signal. The frequency shift is initially programmed so that the repeated signal is within the passband of the Doppler filter containing the target return (e.g., within 50 Hz to 1000 Hz of the target frequency). This is designed to allow the jammer to capture the Doppler filter containing the target through the radar's *automatic gain control* (AGC) action. The repeated jammer signal is then swept in frequency until the maximum expected Doppler frequency of the radar is reached. The repeated signal is then switched off, forcing the victim radar to reacquire the target. This form of velocity deception is called *velocity gate pull-off* (VGPO) and, as the name suggests, is most effective against those Doppler radars that employ velocity trackers (automatic Doppler frequency-control tracking loops). This technique is fully compatible with *range deception* (RGPO) and angle deception jamming.

By using VGPO in combination with RGPO (and possible angle deception), the DECM ensures that the victim-pulsed Doppler radar is not able to reject the jammer signal through Doppler filtering. VGPO is particularly important against tracking radars (pulsed or continuous wave) that employ Doppler filters (called speed gates) in their angle-tracking loops. The speed gates ensure that angle error information is only derived from that part of the signal that contains the target of interest. This type of operation is typical of tracking radars used in semiactive missile guidance systems and in airborne and ground-based tracking radars used against low-flying aircraft in heavy clutter environments. Alternately, if the victim radar does not contain a Doppler mode of operation, little is lost by employing velocity deception because the radar simply ignores the frequency modulation induced on the jamming signal.

The frequency modulation waveform used to induce the false target information onto the repeated jammer signal is a critical parameter of a velocity deception repeater. First, the initial frequency offset must lie within the passband of the Doppler filter containing the target. It must remain within the target Doppler filter for a sufficient duration to allow the AGC attack time to take effect. Also, the lower the initial offset, the longer it takes to develop the repeater waveform (e.g., a 1-Hz offset would take 1 sec to develop the false Doppler signal). Because of this, the initial offset should be as high as possible, with 25% or less of the Doppler resolution capability of the radar suggested as an appropriate value. If the Doppler resolution is not known, then 20 Hz is considered a good design value. Maximum frequency offsets of 5 kHz to 50 kHz are typical, as are pull-off periods of from 1 sec to 10 sec [1].

DECM utilized against Doppler radars generally employ parabolic or modified exponential velocity pull-off characteristics. This prevents jumps in acceleration from occurring as the jamming signal replaces the target signal in the velocity-tracking loops. Also, DECM systems using both RGPO and VGPO generally ensure that the related velocities simulated by the false jammer target are consistent in both the temporal (range) and frequency (Doppler) domains. Since RGPO is more easily accomplished when simulating an outbound target (delay in range), this leads to a preference for decreasing Doppler frequency in the compatible VGPO mode.

Generally, an effective DECM against a tracking radar employs both range (and possible Doppler) and angle deception techniques. Range deception techniques are somewhat independent of radar implementation as compared to angle deception techniques, which must be tailored to a specific radar implementation. For this reason, range deception is almost universally employed in deception jammers, but its effect is primarily limited to introducing false range information into the victim radar. While the false range information is being absorbed by the radar, it still provides accurate angle information. It is only when the radar's range gate has been captured and the DECM is turned off that angle information is denied to the victim radar. Reacquisition in the range dimension may be rapid (i.e., on the order of milliseconds) if the radar is pointed in the direction of the target. For this reason, it is appropriate to introduce false angle information into the radar at the same time RGPO is being attempted. If the radar is forced to search in both angle and range during reacquisition, then this cycle is appreciably lengthened and the radar is rendered ineffective during this period.

Angle deception must be directed against the specific angle measurement technique employed in the victim radar. Many older type tracking radars (e.g., conical scan) employ a sequential lobing-type angle-tracking mechanization that induces amplitude modulation onto the target's return. When the amplitude modulation induced onto the target is nulled, then the radar antenna's boresight is pointed exactly at the target. This type of system is easily confused by amplitude modulated type jamming. A particularly effective type of jamming induces amplitude modulation onto a repeated signal whose rate corresponds to that of the sequential lobing but whose phase is 180 degrees with respect to the phase of the modulation induced onto the target. This type of jamming is achieved by a technique called inverse gain jamming where, in general, the necessary information required for effective jamming can be easily derived from the intercepted radar signal. For this reason, many radars of this type scan on receive-only (e.g., COSRO, LORO) while employing a fixed beam on transmit. However, although inverse-gain jamming is most effective when the jamming waveform is synchronized to the radar antenna's scan pattern, it can also be effective in a sweeping mode that searches for the most suitable modulation for perturbing the radar's angle-tracking circuits [1].

The ease with which sequential lobing radars are confused by amplitude modulated jamming has resulted in a preference for monopulse (simultaneous lobing) angle-tracking techniques in most modern radar. In a monopulse tracking radar system, an angular error estimate is made on each return pulse, thereby rendering the system insensitive to amplitude fluctuations in the data. This makes this type of radar highly resistant to angle deception modulation from a single jamming source.

There are several techniques that are applicable to jamming of monopulse systems. These can be divided into several categories. The most fundamental category involves multiple source techniques whose objective is to distort the EM wave's angle of arrival at the monopulse antenna such that either the monopulse tracker is prompted to point away from the target's angular direction or spurious modulation, introduced into the tracker's servo system causing the tracker to break lock. These techniques against monopulse radar include blinking, cooperative jamming, cross-eye (artificial glint), and terrain bounce (ground reflections). Another category relies on imperfections in the monopulse designer implementation that are exploited by the jammer. The most prominent of such techniques is cross-polarization jamming, which reverses the sense of the angle error signal developed by the monopulse angle discriminator. Other techniques in this category are image frequency jamming, skirt frequency jamming, and countdown, which attacks the radar's AGC. A third category involves the use of an active expendable jammer or chaff launched in controlled bursts, which attempts to lure the tracker's pointing angle away from the bona fide target. Needless to say, monopulse radars are difficult to jam, and most of these techniques are technically and operationally difficult to implement.

1.3.3 Defense Suppression

EA against military surveillance and tracking radars generally involves lethal actions as well as electronic suppression techniques. These lethal actions have as a primary objective the physical destruction of a radiating emitter, which usually is a component of an enemy defense system. The most prominent lethal defense suppression system is the ARM.

ARMs are primarily directed against radar-type emitters, although any radiation source is potentially vulnerable. The *high-speed ARM* (HARM) is typical of current ARM technology. This high-performance, air-to-ground tactical missile utilizes a broadband RF monopulse sensor to seek-out, home-on, and destroy an enemy radar [1].

ARMs are often launched from dedicated defense suppression aircraft, which have the ES facilities for detection, identification, and location of threat systems such as SAM batteries. These systems include an onboard digital processing capability coupled with extended frequency range receivers, which provide a comprehensive prioritized display of the immediate threat [1].

Another form of lethal defense suppression uses precision emitter location techniques. The objective of this system is to provide an accurate location of the position of an emitter. The system uses the difference in the TOA from an emitter pulse, as measured in two or more cooperating aircrafts separated by precisely known distances, to determine the target's position in a common coordinate grid. A weapon using *distance-measuring equipment* (DME) can be guided to the emitter location, even though the emitter has stopped radiating. A major advantage of this technique is that it is potentially effective against any emitter despite the fact that it only comes on the air occasionally. Also, the attacking aircraft can operate in a stand-off mode using its DME guided weapon [1].

Lethal defense suppression techniques are generally part of an overall EA strategy. To understand this strategy, we must digress and describe the structure of an air defense system. The classic air defense system depicted in Figure 1.9 uses two types of weapons against the penetrating bomber. The first weapon system is a manned interceptor that primarily serves as a launch platform for *air-to-air missiles* (AAMs). The second weapon system is a SAM that is placed in the path that the penetrating bomber must traverse in order to reach a high-value target.

The first element of the defense system encountered by the penetrating bomber or cruise missile is the early warning radar net, whose function is to provide initial detection, thereby alerting the defense system. The surface radars employed are generally low-resolution, low-frequency, two-dimensional search radars with free-space detection ranges of the order of 300 nm on a 1-m^2

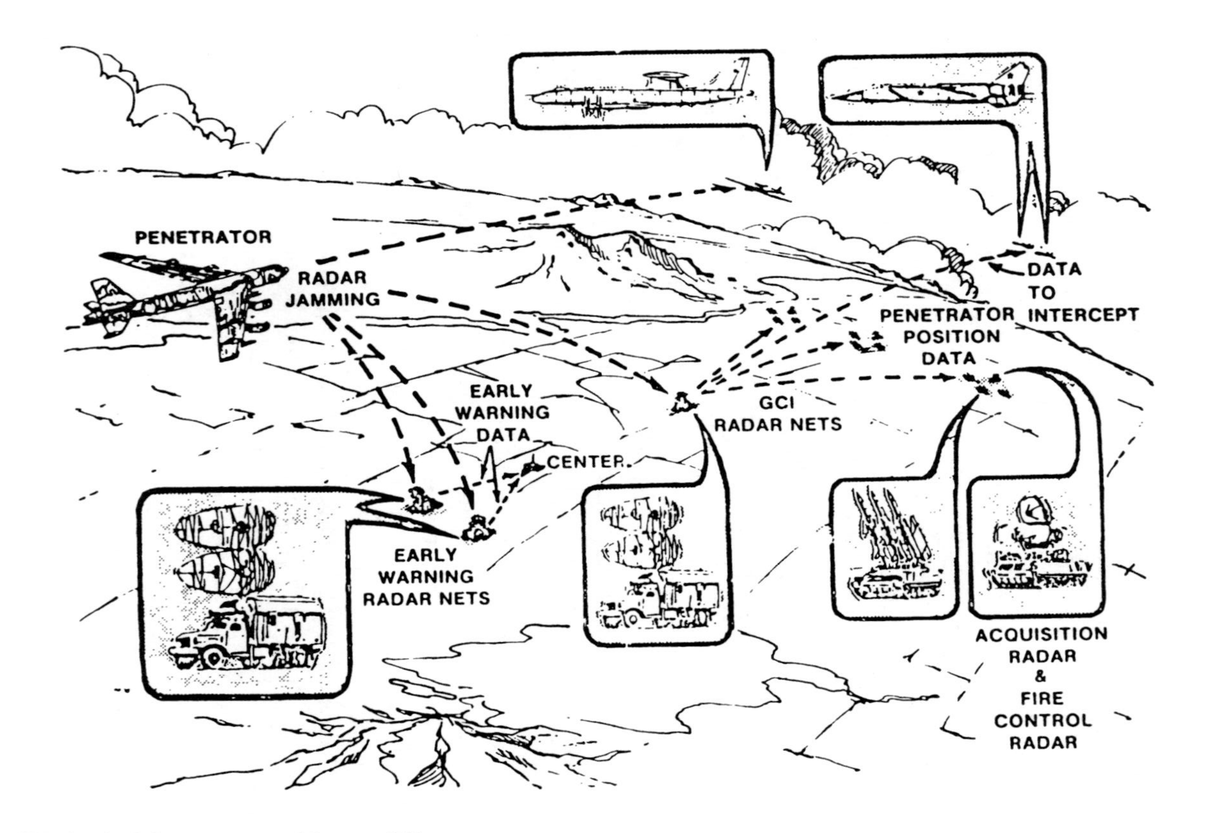

Figure 1.9 Classic air defense system. (*Source:* [1].)

target. Detection ranges on low-altitude penetrating targets are limited by radar line-of-sight to the target and are considerably reduced from the maximum free space range.

The next elements encountered are *airborne early warning* (AEW) radars (i.e., AWACS), which take advantage of the increased line-of-sight to low-altitude targets afforded by a medium- to high-altitude platform. This solves the terrain shielding problem associated with surface-based early warning radar but introduces another significant problem. The AEW radar must now "look down" into the surface (ground or water) at high grazing angles, which causes a radar clutter return of high magnitude that competes with the normal target return. Moving targets can be separated from the clutter using Doppler processing techniques. Operation over water has been feasible for many years, but over-land operation requires subclutter visibilities of the order of 60 dB to 80 dB, which have only recently become practical [12].

Operation of AEW radar within the integrated air defense system is more complex than for surface early warning radar. In addition to its function of detecting hostile aircraft and missiles, it also provides a control function to guide interceptors to their targets. The control function is partially a self-defense function, but primarily the AEW aircraft provides an excellent platform for communications links to all elements of the air defense system.

The early warning reports of the air defense system are generally passed onto a centralized air defense operations center, or filter center. The function of the operations center is to control the assets of the air defense system to manage the air battle effectively.

In general, the early warning report contains the information that a potentially hostile target has been detected and provides a low accuracy indication of its location. The next action of the air defense system is to form an accurate track on this target, allowing an engagement strategy to be formed. Accurate tracks are formed by *ground control intercept* (GCI) radar or by their equivalent function in airborne AEW radar.

GCI radars generally provide higher accuracy data than early warning radar. In addition, they must provide height information that is essential to allow *airborne interceptors* (AI) or SAM acquisition radar to lock onto the hostile target. In addition to forming tracks on hostile targets, GCI radars also track friendly aircraft, which are identified through an *identification from friend or foe* (IFF) system.

The GCI radar net employs either two- or three-dimensional surveillance radar types. The two-dimensional fan beam types generate range and azimuth information, which is translated into a target's plan position in geographical coordinates. Associated with the two-dimensional radar is a nodding height-finding radar directed at the azimuth of selected targets and measures target elevation and range, which is converted into height information. The three-dimensional radar simultaneously measures the target's azimuth, elevation, and range. The measurements are then converted into plan position and height.

Each radar of the GCI net generates target reports on each scan of its antenna through the coverage volume (typically, 5- to 10-sec frame time). These reports containing two- to three-dimensional target data are transmitted to the operations center through data links (see Figure 1.9). These data are then combined to form target plots, which is a process known as plot extraction. Successive plots are then combined using a tracking algorithm (e.g., Kalman filter) to form target velocity vectors, which are smoothed estimates of the trajectories of both the penetrating aircraft or missiles and any interceptor aircraft.

The GCI radar data are displayed on a computer-driven display system in the operations center, which functions to speed the transfer of data, filter the extraneous target reports, calculate the threat, and advise the operating personnel on the best defensive measures. The two basic options that the air defense commander has are either to engage the hostile target by launching an airborne interceptor or to assign SAMs or antiaircraft point defenses to the target.

If an interceptor is launched, it must be guided to the vicinity of the target so that its AI weapon control radar can acquire the target. Control is exercised via voice command or data link (see Figure 1.9) from the air defense operations center or, alternatively, from an *airborne warning and contract system* (AWACS) platform. Once the AI radar locks onto the target, it can launch its *air-to-air missile* (AAM). AAM guidance can be either infrared or radar.

The terminal element of an air defense system is generally the SAM system. Its radar assets may consist of an acquisition radar, IFF system, target tracking radar, illuminator for semiactive radar guidance, missile tracking radar, command guidance link, and missile fuse sensor. These components can be highly integrated as in the Patriot missile system or functionally implemented as in several older missile systems (e.g., Hawk Missile System).

The acquisition radar is critical to the functioning of the SAM system. It provides the target detection, identification, and evaluation necessary to allow the target tracking radar to acquire the target. If command guidance is used, it requires at least two tracking radars, one to track the target and another to track and send guidance commands to the missile. Guidance computations are performed by a computer on the ground, which takes target and missile radar tracking data and converts the information into missile guidance commands. The target-tracking radar is sometimes supplemented with a range-only radar to perform the critical target-tracking range measurement. This guards against the system's vulnerability to simple noise jamming, which would deny the range measurement necessary for successful functioning of the missile's guidance system.

Semiactive guidance functionally needs only one radar for tracking and illuminating the target and a data link, which provides a stable reference signal for the missile. When a single radar is used, it is usually of the pulsed Doppler type [12]. In some cases, a separate *continuous wave* (CW) illuminator radar is slaved to the tracking radar. In the semiactive guidance system, the missile seeker bistatically tracks the reflections from the target using a narrowband filter, which accepts the Doppler shifted target signals, while rejecting direct signals from the illuminator and reflected signals from radar clutter, which are at different Doppler frequencies.

In effect, the semiactive seeker is the same as having a complete target tracking radar aboard the missile, which generates its own navigation commands as the missile "homes" onto the target without the weight and cost penalty of carrying a heavy transmitter. In contrast to command guidance, a semiactive system becomes increasingly accurate as the target-to-missile range decreases, leading to a more effective missile system whose size and warhead can be minimized. The EA vulnerability of semiactive missiles is low due to the passive nature of the missile seeker and the use of very narrow receiver bandwidths in the Doppler tracking circuitry. EA techniques such as terrain bounce, active and passive decoys, and VGPO are generally attempted against semiactive guidance systems.

Several missile systems (e.g., AMRAAM, Harpoon) employ active radar guided missile seekers, eliminating the need for target illumination employed in semiactive guidance systems. The advantage of active radar guidance is that it provides a "launch and leave" capability, whereby multiple missiles can be in the air simultaneously, leaving the launching platform (i.e., aircraft) free to maneuver in order to avoid counter missiles.

The active radar-guided missiles' range is limited by the relatively small antenna aperture/average transmitter power product that can be physically achieved in a small-diameter missile. This generally restricts the active radar-guided missile to terminal applications of the order of 10 nm or less. When longer range operation is desired, then midcourse guidance can be supplied by semiactive (e.g., Phoenix) or command guidance techniques.

The *suppression of an enemy air defense* (SEAD) system generally requires the coordinated effort of a number of EW assets. The strategy employed during a SEAD mission is generally as follows.

First, a number of high-powered wide-area support jammers (e.g., EA-6B/EF-111A) are employed to degrade early warning, GCI, and target acquisition radar. If they are successful, then the air defense system does not have the necessary information to launch either interceptor or missile strikes against penetrating aircraft within regions covered by the jamming. However, in general, the support jamming aircraft are vulnerable to triangulation from netted radars within the air defense system that form multiple strobes on the jamming radiations, thereby locating the jammer's position. This requires that support jammers performing escort missions also carry self-protection EA systems.

The next step in the air defense suppression strategy is the jamming of communications links used by the command and control (C^3I) system. This is accomplished by dedicated communication jamming aircraft (e.g., Compass Call). When successful, this paralyzes the ability of the air defense net to utilize the radar and other sensor data obtained by various individual elements of the air defense system. In addition, the effective jamming of the C^3I communication links makes it impossible for the force commander to exercise authority over his assets. Thus, each element of the air defense system is basically on its own without the support of any other element in fulfilling its mission.

Further, within this strategy, each penetrating aircraft is equipped with its own self-protection and threat warning system. The radar portion of these systems generally consists of a *radar warning receiver* (RWR) and complementary DECM system. The function of these systems is to protect the aircraft against SAMs and AI launched AAMs. The objective of the RWR is to alert the crew of a missile attack and give information as to the position of the various missile launchers threatening the aircraft. The objective of the DECMs is to jam the terminal threat radars that provide guidance information to the missiles. Since these functions are complementary, it is possible to integrate the RWR function into the DECM (e.g., ALQ-135/ALR-56C, INEWS).

A complementary system employed in some aircraft adds an active *missile attack warning* (MAW) system in addition to the DECM. The MAW system generally consists of either a pulsed Doppler radar or infrared or ultraviolet sensor that locates the presence, range, and direction of an attacking missile. Infrared flares and radar chaff are then dispensed to lure the missile away from the protected aircraft.

Complementary threat warning and self-protection jammers against electro-optical threats are currently under development. Laser threat warning systems warn against laser range finders and designators. Infrared jammers are used to negate *forward-looking infrared* (FLIR) and missiles that home in on the infrared signature of the protected aircraft.

Within the overall defense suppression strategy is the use of ARMs, which attempt to destroy a smaller set of early warning, GCI, and SAM target acquisition radars that cannot be jammed with confidence. Defense suppression aircraft use sophisticated ES direction finding and location receivers to target potential emitters for HARM launches. An ancillary principle of this form of defense suppression is that an enemy will be inhibited in the full use of his electronic systems by the presence of a potentially lethal destroyer of radiation sources.

A general radar strategy employed when confronted with an ARM attack is to stop radiating, thereby causing the ARM to lose its guidance information, and then turning on a spatially remote decoy transmitter to lure the ARM away from the radar site. The timing of this strategy depends upon the detection of the ARM as it approaches the radar site. The ARM trajectory is usually selected to attack the radar through the zenith hole region above the radar where its detection capability is minimal. Thus, a successful implementation of this strategy requires a supplemental radar, which provides a high probability of detection in the zenith hole region.

The potential susceptibility of ARM-type systems to decoys and radar shutdown have led to the development of precision emitter location and strike systems. These systems, using TDOA measurements, are also potentially insensitive to EP techniques such as frequency agility and *pulse repetition interval* (PRI) agility and short "on" time radar techniques, since they can operate on a single pulse. With these systems, a weapon using a DME transponder can be guided to the emitter's location, even though it has stopped radiating, using the system's common coordinate system formed by the original measurement. Although developed, this system has not been operationally deployed because of its high cost.

1.4 EA Effect on Communications

Communication systems provide a vital role in modern military strategy. First, they provide a pipeline by which a commander exerts control over his forces. Further, they allow transmission of battlefield information to all echelons, allowing timely and effective decisions relative to force deployment and movement. The latter function is an essential component of an area generally referred to as "situational awareness." This is defined as detecting information in the environment, processing the information with relevant knowledge to create a composite picture of the current situation and acting on this picture to make a decision or to perform further exploration [13].

EA in a battlefield environment poses several problems. First, communication jamming equipment is generally large, cumbersome, and difficult to handle and operate. Jamming in a crowded electromagnetic environment poses a major C2 problem since proper coordination is needed to avoid jamming friendly transmissions. A high-power jammer also acts as a beacon to enemy ARMs.

The future communications services envisioned to be required by tactical forces are depicted in Figure 1.10 [2]. Tactical C2 nets are intended to provide connectivity among the distributed ground, sea and air mobile tactical networks used for low data-rate information exchange, and voice connectivity among mobile tactical units. This net is characterized by small, mobile terminals aimed at supporting low-throughput requirements. The low data rates (5 kHz to 25 kHz per channel) make these links unsuitable for moving large amounts of information or high data demand products, such as digitized imagery, except under emergency circumstances.

These military tactical networks include the *Single Channel Ground and Air Radio System* (SINCGARS), *Joint Tactical Information Distribution System* (JTIDS), *Mobile Subscriber Equipment* (MSE), and data links associated with *Cooperative Engagement Capabilities* (CEC). The tactical networks generally connect force structures that are highly mobile and require connectivity via satellite communications. A large portion of these military satellites operate at UHF (250 MHz to 400 MHz) via the naval *Fleet SATCOM* (FLTSATCOM), follow-on UHF (UFO) satellites, and the *Air Force satellite communications network* (AFSATCOM). The UHF band, because of its resulting large antenna beam widths, does not offer any protection from EA and can be easily disturbed using unsophisticated techniques. This renders the critical UHF satellite connectivity among tactical terrestrial command and control networks vulnerable to ground mobile jammers. For these reasons, *extremely high frequency* (EHF) connectivity among tactical networks is currently planned or deployed within the *Military Strategic Relay* (MILSTAR) and portions of the UFO satellites operating at 44-GHz uplinks and 20-GHz downlinks. The EHF allows employment of narrow spot antenna beams, which force sidelobe jamming, thereby providing a degree of antijam protection (EP).

The high-capacity system illustrated in Figure 1.10 provides two-way, point-to-point connectivity between upper echelons of the military and connectivity to support activities requiring a high data rate. This connectivity involves high data rate C2 and distributed data base transfer. These functions are currently implemented via SATCOM using the *Defense Satellite Communications System* (DSCS) operating at 7 GHz to 8 GHz with a 1.5-Mb/s data rate and commercial SATCOM and fiber optic systems [14]. The DSCS system provides relative insensitivity to jamming interference when spot beams and large antennas are used at the higher echelons of command, since jammers are unlikely to be deployed within the beams servicing the upper echelons of command. Commercial systems can provide the connectivity and bandwidth required, but availability in locations where forces are deployed cannot be guaranteed.

The *Direct Broadcast System* (DBS) depicted in Figure 1.10 involves a wideband link to small, mobile terminals, servicing all echelons of the military

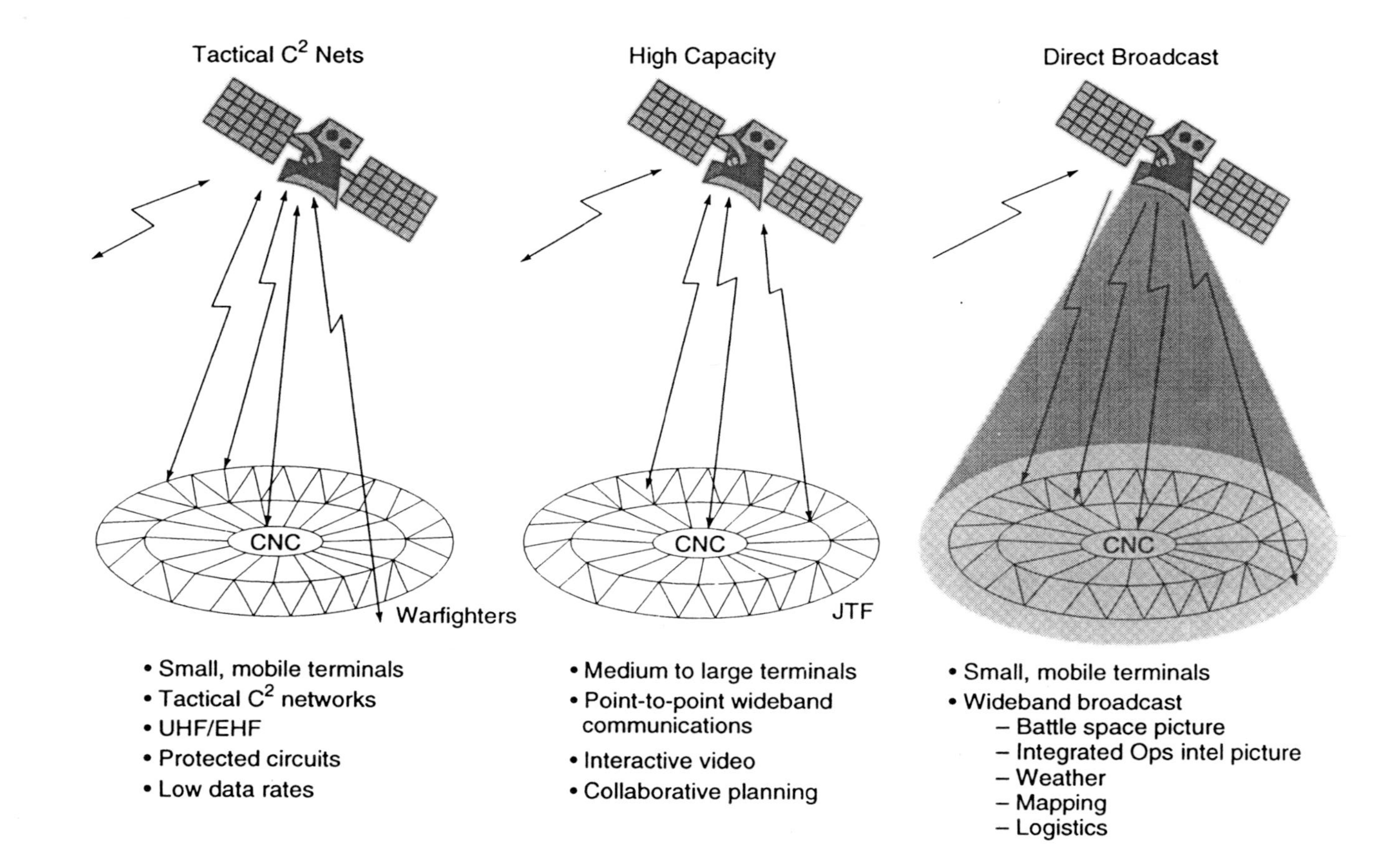

Figure 1.10 Communications system requirements. (*Source:* [2].)

[15]. It is patterned after the digital television broadcast concept and can provide both low- and high-bandwidth demand digital imagery. As currently configured, GBS satellite communications traffic is carried on MILSTAR military satellites as well as Inmarsat transponders for noncritical information that does not require encryption. Future implementations envision a dedicated DBS satellite.

The DBS is intended to provide integrated situational awareness and critical support information to tactical users at all levels of command. This category of service provides subscribers with quick, efficient, and simultaneous access to broadband information via small, mobile, and inexpensive receiver-only terminals. The user can employ filters to select broadcast information. A satellite broadcast system can be made insensitive to ground mobile EA threats expected in the future in that these threats cannot easily attack the downlink broadcast information. Only an airborne or space-based jamming threat can attack the downlink, requiring a high level of sophistication not expected in many future operations. A DBS could transmit the joint battlespace picture, vital intelligence data, weather, maps, logistics, and other vital information. The use of GPS time and positional accuracy to label DBS military communications would allow the achievement of precise global situational awareness in both time and space. The ability of operational commanders to shift a high percentage of the information dissemination needs to the DBS mode is a major advantage of this approach.

The *Global Information Infrastructure* (GII) that interacts with or supports military operations is a vast complex set of information systems supported by commercial grids and infrastructure (see Table 1.2). The advent of a variety of low-cost commercial information services in space-based commercial communications, cellular telephony, navigation, imagery, and environment services supports a trend for the military to use these commercial networks for economic reasons. The protection of critical segments of the GII is a concern requiring that information be protected commensurate with its intended use. In general, the commercial elements of the GII are highly susceptible to attacks that either disrupt information flow (availability) or corrupt the data (integrity) within the infrastructure.

The integrity of information over communication links can generally be protected using encryption devices. EA of commercial communication systems to deny availability is a delicate issue complicated by legal and treaty implications. However, it is permissible to deny military users the use of commercial communication links in times of conflict. For example, EA of the downlink from a commercial satellite to break a communication mode associated with a C^3 system would be an expected military action, as would jamming of cellular telephones if they carried military traffic. This discussion indicates the continued

Table 1.2
Global Information Infrastructure Supports Military Operations

Media and Infrastructure
U.S. public-switched networks
Commercial communications satellite systems—United States and International
Intelsat, Inmarsat, Panamsat, Iridium, Odyssey, and Globalstar
Navigation systems
Transoceanic Cable System
Global Positioning System
International telephone and telegraph
Data bases
Internet
DoD Milsatcom
Milstar, DSCS, and UHF
Tactical networks and C2
Supporting infrastructure
Power grid, commercial system support, spares, maintenance, and transport
Cellular communications

need for dedicated military satellites, particularly in the EHF region where commercial satellites are unavailable [14].

1.4.1 Communications Jamming Principles

The use of EA against communication networks in C2W operations is axiomatic. The primary objectives in military communications jamming are to separate a force commander from his forces and to limit situational awareness by restricting information transfer among the various military echelons. Communications EA can be effected by either disrupting enemy communication or injecting false or deceptive information into the information net. Disruption of communication links is best accomplished by breaking nodes in the communications network [1].

The principles of EP are embedded in most modern communications systems. These include (1) extensive use of fiber optic channels to inhibit the interception and injection of stray radiation into the channel; (2) use of highly directive antennas (generally operating at high frequencies) with low sidelobes to inhibit detection and jamming; (3) use of antijam waveforms in the time, frequency, and coding domains to inhibit detection and restrict injection of deceptive or disrupting signals; and (4) use of encryption to encode sensitive messages so that they cannot be compromised. In addition, most communications systems make extensive use of digital technologies that further inhibit spoofing, which entails the substitution of deceptive messages for valid ones [16].

A counter philosophy to communication jamming contends that it is advantageous to intercept an enemy's communication traffic for intelligence purposes rather than to employ jamming. As discussed, the widespread use of modern digital and encryption techniques has generally made the interception of enemy messages difficult, if not impossible. This leads to the conclusion that if you cannot use an enemy's communication for your own advantage, you must deny him its use.

In the following, we will discuss EA against three different types of military communication systems: tactical data links, combat net radios, and military satellite communications. Tactical data links are used in C^3 systems to transfer data from one element to another (see Figure 1.3). The data are generally in the form of digital messages. The digital format allows data transmission using error-correcting codes [1]. Combat net radios provide short- and medium-range voice communications over typical distances of up to 50 km, generally using VHF in the range of 30 MHz to 88 MHz. Military satellite communications provide worldwide connectivity between users allowing high data rate data transmission.

1.4.2 EP of Communications Systems

Military communications systems generally use a number of EP techniques that are listed in Figure 1.11 [17]. A technique common to most systems is the use of spread-spectrum communications; whereas COMSEC is the encryption/decryption of the transmitted information, spread spectrum can be thought of as the encryption/decryption of the RF signal carrying this information. In spread-spectrum communications, the signal is spread over a wide RF band in accordance with a pseudorandom code. Despreading takes place through the correlation with a stored replica of the pseudorandom code at the receiver. The correlation process has the property that it attenuates noncorrelated interference by the ratio of the spread-to-information bandwidth, which can be the order of 20 dB to 60 dB.

Spread-spectrum signals provide strong resistance to interference and jamming, including a significant reduction of the effects of multipath. In addition, the use of *code division multiple access* (CDMA) allows for processing of multiple spread-spectrum signals through satellite communications nets without cointerference, and also without requiring overall network timing control. This combination provides both antijam and *low probability of intercept* (LPI) characteristics, which are especially attractive to military users. Also, the same principle allows for the multiplexing of several information signals onto

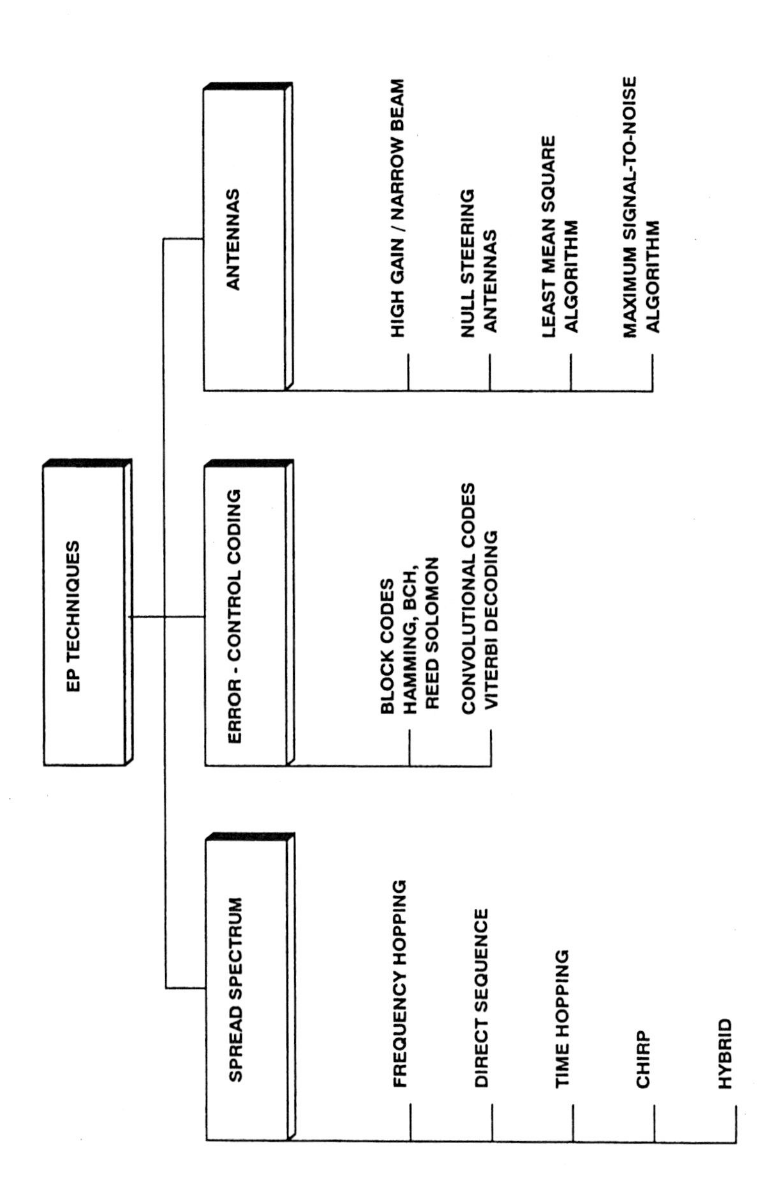

Figure 1.11 Communications EP techniques. (*Source:* [17].)

one transmission. In addition, the use of spread-spectrum signals allows for highly accurate range measurements (e.g., GPS).

The two main types of spread-spectrum signals are called direct sequence and frequency hopping. With direct-sequence spreading, a random code carrier is used to multiply the information-carrying signal. The code carrier is a sequence of code chips occurring at a symbol rate r_C. If the message signal occurs at the rate r_S, then the spread-spectrum signal occupies a bandwidth proportional to $r_C + r_S$. The ratio $(r_C + r_S)/r_S$ ($\approx r_C/r_S$) is called the processing gain and provides an accurate measure of the benefits provided by direct sequence spreading. For example, if the processing gain is 30 dB, an enemy jammer would have to increase its jamming power by 30 dB in order to offset the added protection of the spreading. Similarly, an enemy ES receiver would have a 30-dB disadvantage relative to a friendly receiver in being able to detect the signal.

With direct sequence spreading, the instantaneous bandwidth is the entire spread bandwidth. The processing gain available is limited by practical code rates and is reduced as information waveform rates are increased. Due to the typically short chip interval, direct sequencing provides good resistance to follow jammers. Usually this technique is used with coherent-message waveforms, such as phase shift keying, which provides additional detection improvement relative to noncoherent methods [17].

A second popular technique is frequency hopping. With frequency hopping, the transmitted frequency is pseudorandomly changed at the hopping rate, with the total available bandwidth being the spread bandwidth. The faster the hopping rate, the less time a follower jammer has to disrupt communications. Also, a higher hopping rate reduces the effects of multipath. Hopping rates can be either higher, lower, or equal to the information waveform rate. If the jammer spreads its energy over the full frequency range of the communication signal, then the processing gain equals β_n/β_m, where β_n is the frequency band over which the hopping occurs and β_m is the message bandwidth. However, often a partial-band jamming strategy is employed whereby the jamming energy is concentrated in a portion of the band. In this case, the processing gain can be considerably less than that promised by the ratio of the hopping to information bandwidths. Error-correcting coding becomes essential in this case to allow the actual gain to approximate the available processing gain [17].

With frequency hopping, the instantaneous bandwidth is less than the hopping frequency range. Process gain is limited by the available bandwidth, but the hopping rate is limited by the ability of frequency synthesizers to hop over a wideband of frequencies in a pseudorandom manner. With long hop intervals, frequency hopping is vulnerable to EA by a follower jammer. Frequency hopping is usually used with noncoherent modulation techniques such as *frequency shift keying* (FSK) and has a lower overall gain relative to coherent methods. Use of channeled receivers makes interception of frequency hopping signals easier than direct sequence signals. Frequency hopping techniques are particularly well suited when combined with compatible null-steering array antenna techniques.

Two other types of spread spectrum are time-hopping and chirp. With time hopping, the available time for an information symbol is divided into several shorter time slots, with only one of these slots actually used. From frame-to-frame, the choice of the time slot is controlled by the time hopping code in a similar way to frequency hopping.

Frequency hopping spread-spectrum signals can be generated by mixing linear FM chirp waveforms [18]. With this method, two impulses, separated in time, are applied to a *surface acoustic wave* (SAW) dispersive delay line. Each impulse generates a linear chirp waveform, but the chirps are separated in time and, hence, in frequency. The two chirp waveforms are then mixed to generate a fixed frequency of magnitude proportional to the time difference between impulses. Through control of the time difference between impulses, using a psuedorandom code, a frequency hopping waveform is generated. This method provides the advantage of rapid hopping and wide bandwidth under control of an external pseudorandom code. In practice, these other types of spread-spectrum systems have been used less extensively than direct sequence or frequency hopping.

Error-correcting coding is an important complement to frequency hopping spread spectrum. It basically involves adding parity check symbols to the message, which allows errors to be automatically detected and corrected unless the number of errors exceeds the built-in power of the code. Frequently, error-correcting coding follows a source-encoding step involving data compression, which removes redundancy from the message, which is then replaced by the efficient redundancy associated with error-correcting coding [17].

Error-correcting coding can be divided into block and convolutional coding. With block codes, r parity check bits, formed by linear operations on the k data bits, are appended to each block of k bits. These codes can be used to both detect and correct errors in transmission. Block coding is appropriate (1) in forward correcting schemes whereby error detection in a received message triggers the transmitter to repeat the message and (2) when errors tend to occur in bursts instead of being randomly distributed over time [19].

Examples of well-known block codes are: Hamming, *Bose-Chaudhuri-Hocquenghem* (BCH), and Reed-Solomon. Hamming codes are simple and provide for the correction of only one error in a block. BCH codes are the generalization of Hamming codes, providing for the correction of many errors. In addition, BCH codes allow for greater flexibility in choosing the best way to optimize data rate, bandwidth, energy, error probability, and complexity. Hamming codes are special cases of general BCH codes, as are Reed-Solomon codes, which are excellent for burst error cases and make use of nonbinary symbols.

Another powerful type of error-correcting code is the convolutional code group. In these codes, a sliding sequence of past data bits is used to generate several code bits. Distinct blocks are no longer sent and successive transmitted bits contain the history of a sequence of data bits. Redundancy added in this manner permits significant improvement in error performance. The best known technique for decoder correction is known as the Viterbi algorithm, which makes use of a path-tracing concept.

Antennas can supply EP using their spatial filtering ability. In the microwave frequency region, highly directive antennas can be designed that are useful for point-to-point communication systems. These systems coupled with low-sidelobe technology can provide a high degree of protection against jammers located at angles displaced from the axis of the main antenna beams. This principle is applied in satellite systems operating in the SHF or EHF regions where a raster of skewed spot beams is generated to allow coverage of a large surface area. The beams are then selectively energized to cover desired spot locations on the surface while attenuating any jamming signals that originate from any of the desensitized regions.

Sidelobe cancellation systems are also used with directive antennas to provide additional attenuation of sidelobe jammers. These systems utilize auxiliary or guard antennas to adaptively generate antenna patterns that provide nulls in the direction of jammers. In theory, a number of sidelobe jammers equal to the number of auxiliary antennas can be nulled. The potential for sidelobe cancellation is increased when a phased array antenna is used. The number of degrees of freedom (independent nulls that can be formed) is equal to the number of elements minus one. In addition, for satellite systems the adaptive processor can be either located on the satellite or in the ground central station.

1.4.3 EA Waveforms and Strategy

In an EA against a digital communications link, a jammer conceptually may choose between jamming the information recovery circuit or the synchronization circuitry. The system may involve several levels of synchronization: phase, bit, word, frame, and spreading code. Prevention of synchronization or causing loss of synchronization may be feasible and more effective than causing errors in information symbols. Both must be considered when investigating vulnerability. Another strategic choice involves the temporal and frequency characteristics of the jamming signal. Partial-band jamming may be preferred against certain types of frequency hopping spread-spectrum systems. In some cases, keyed time transmissions are effective against certain types of communications systems.

A basic EA strategy is the injection of noise into the communications receiver. The effect of this in digital systems is measured by the *bit error rate* (BER). When the BER exceeds 20%, the intelligibility of the message degrades substantially [20,21]. The J/S ratio for noise jamming, ignoring propagation effects, is

$$\frac{J}{S} = \frac{P_j G_{jr} G_{rj} R_{tr}^2 L_r B_r}{P_t G_{tr} G_{rt} R_{jr}^2 L_j B_j} \tag{1.1}$$

where

P_j = jammer power
P_t = communication transmitter power
G_{jr} = antenna gain, jam → com recvr.
G_{rj} = antenna gain, com recvr. → jam
G_{rt} = antenna gain, com recvr. → com xmtr
G_{tr} = antenna gain, com xtmr → com recvr.
B_r = communications receiver bandwidth
B_j = jamming transmitter bandwidth
R_{tr} = range between communications transmitter and receiver
R_{jr} = range between jammer and communication receiver
L_j = jammer signal loss (including polarization mismatch)
L_r = communication signal loss

In general, J/S ratios of 5 dB to 10 dB are sufficient to provide significant interference. The equation can also be used to solve for the jamming power (or *effective radiated power* (ERP)) required in a particular situation or the maximum range at which the jamming is effective.

▶ *Example 1.2*

A 4-W VHF frequency hopping radio is used to provide a 25-kHz bandwidth communications channel over a 10-km range. Hopping is accomplished at 100 hops/s over a frequency range of 6.4 MHz. The radios use omnidirectional antennas. A broadband stand-off noise jammer operates at 15 km. What jammer ERP is required to effectively jam the radio (SJR = 5 dB)? If the jammer can follow the hopping, what jammer ERP is required?

```
% Communication Jamming
% --------------------------
% comjam.m

clear;clc;clf;

% Define Jammer Erp from Erpmin to Erpmax

Erpmin=100;Erpmax=10000;   % watts
Erp=Erpmin:Erpmax;

% Communications System & Jammmer Parameters

Jsrdbx=5;     % JSR for Effective Jamming - db
Pt=4;         % Comm Transmit Pwr - watts
Gtr=0;        % Antenna Gain Comm Xmit-Comm Rcv - dbi
Grj=0;        % Antenna Gain Comm Rcv-Jam - dbi
Grt=0;        % Antenna Gain Comm Rcv-Comm Xmit - dbi
Br=.025;      % Comm Receiver Bandwidth -Mhz
Bj=6.4;       % Jammer Transmitter Bandwidth- Mhz
Rtr=10;       % Range between Comm Xmit & Rcv - km
Rjr=15;       % Range between Jam & Comm Rcv - km
Lj=0;         % Jammer Loss(includes Pol)- dB
Lr=0;         % Comm Loss- dB

% Convert dB to ratio

Gtr=10^(Gtr/10);Jsrx=10^(Jsrdbx/10);
Grj=10^(Grj/10);Grt=10^(Grt/10);
Lj=10^(Lj/10);Lr=10^(Lr/10);

% Compute Jam-to-Signal Ratio for Frequency Hopping

x=Grj*Rtr^2*Lr*Br;
y=Pt*Gtr*Grt*Rjr^2*Lj*Bj;
JSR=Erp.*x/y;
Jsrdb=10*log10(JSR);

% Find Jammer ERP for Jsrdbx

Erpj=Jsrx*y/x;

% Plot JSR vs Jammer Erp
```

```
Jsrdb0=Jsrdbx*ones(size(Erp));
plot(Erp,Jsrdb,Erp,Jsrdb0);grid;
xlabel('Jammer ERP - watts');
ylabel('Jam-to-Signal Ratio -dB');
title(['Communication Jamming ERP     ' ...
'(ERPj =',int2str(Erpj),'  watts)']);
```

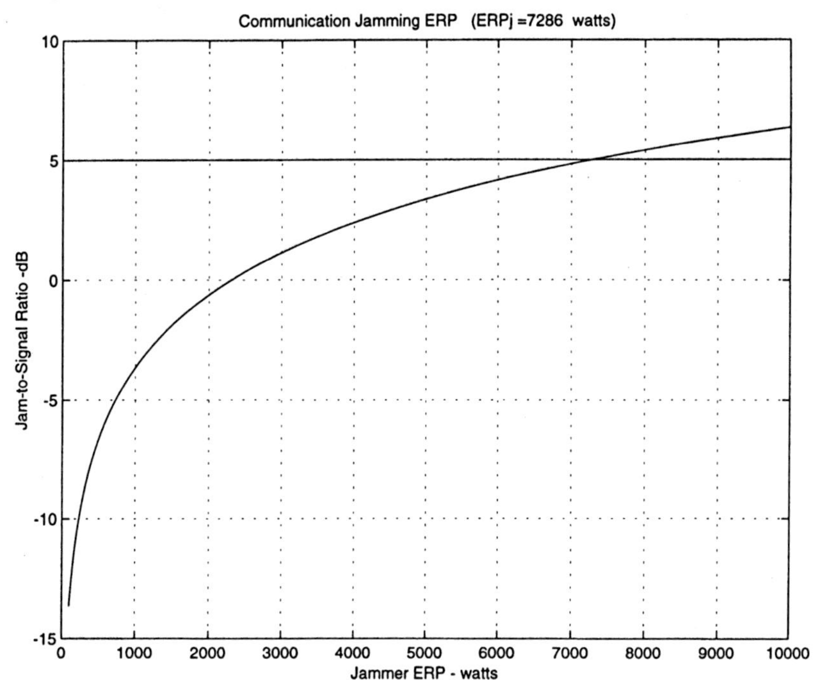

An examination of the equation indicates that for maximum jamming effectiveness, the jammer ERP, which is the product of jammer power output and jammer antenna gain, should be high. This indicates the use of a directional jamming antenna whose polarization is matched to that of the communications receiver antenna. The jammer should be located to optimize propagation to the communications receiver, and a jamming waveform should be selected to minimize the required J/S ratio.

While band-limited noise can be useful as a jamming waveform, there are other waveforms that are useful for jamming communication channels. These include:

- Noise-modulated AM;
- Noise-modulated FM;

- Noise burst (wideband and narrowband);
- Amplitude-keyed signals;
- Swept FM signals;
- Phase-shift-keyed signals;
- Garbled speech;
- CW tones;
- Frequency-shift keyed signals;
- Single-sideband waveforms.

For susceptibility evaluations, it is usually sufficient to consider the following EA waveforms [17]:

- Narrowband noise (spot jamming);
- CW tone (spot jamming);
- Wideband noise (barrage jamming);
- Swept-FM (swept-spot jamming).

When jamming voice channels, modulation that has spectral energy in the 5- to 7-Hz region tends to be most effective. For digital communication links, modulation components at the chip rate tend to insert false bits into the message, thereby changing the code. Deceptive jamming that uses recorded messages repeated-back at random times to garble the message can be utilized. EA waveforms and strategy are summarized in Figure 1.12.

1.4.4 EA Against Military Communications Systems

In the following, we will discuss EA against three different types of military communication systems—tactical data links, combat net radios, and military satellite communications. Tactical data links are used in C^3 systems to transfer data from one element to another (see Figure 1.3). The data is generally in the form of digital messages. The digital format allows data transmission using error-correcting codes [1]. Combat net radios provide short- and medium-range voice communications over typical distances of up to 50 km, generally using VHF in the range of 30 MHz to 88 MHz. Military satellite communications provide worldwide connectivity between users allowing high data-rate data transmission.

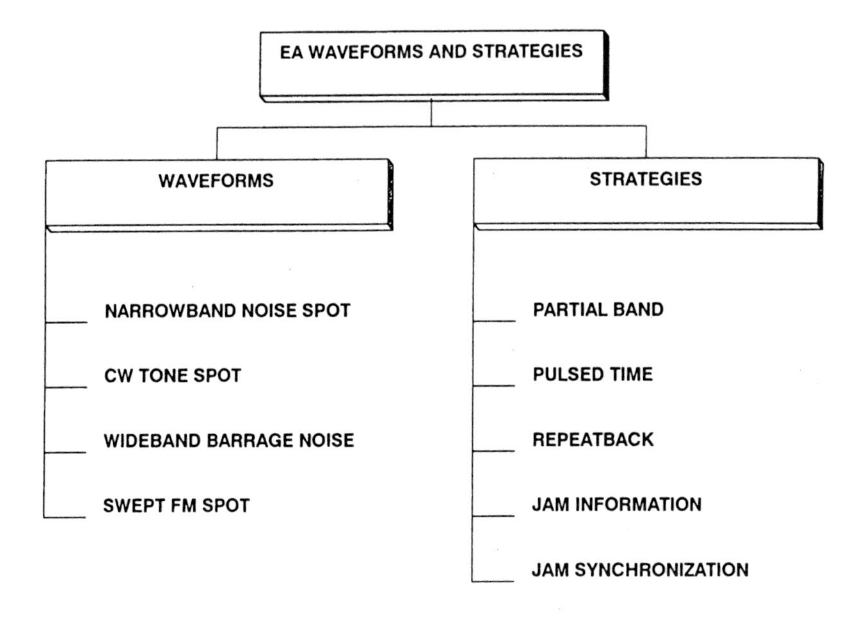

Figure 1.12 EA waveforms and strategies. (*Source:* [17].)

1.4.4.1 Tactical Data Links

The basic purpose of a tactical data link is to transfer wide bandwidth information from one element of the battle force to all other elements. A primary characteristic of Tactical Data Links is that they must be nodeless, meaning that no communications center is needed since all users transmit to everyone else. If one element of the net is destroyed or simply not transmitting, then all other members will still be able to communicate.

In the following, we will concentrate on the characteristics of the *Joint Tactical Information Distribution System* (or JTIDS; TADIL-J) data link. The JTIDS data link was developed to provide a high-capacity, reliable, jam-protected, secure digital information distribution system allowing a high degree of interoperability between data collection elements, combat elements, and command control centers within a military theater of operations. The first practical implementation of JTIDS was in the AWACS (see Figure 1.3) to control Air Force fighters [22].

The L-band (960 MHz to 1215 MHz) JTIDS allows for the transfer of digital data between any properly equipped users within line-of-sight. It uses a *time-division multiple access* (TDMA) format, whereby all users transmit on a common channel. With the "party line" TDMA approach, all the information from each user is available in real time to anyone who needs the data.

With the TDMA format, each user is assigned one or more time slots, which are repeated at regular 12.8-min cycles. Each 12.8-min cycle is divided into 64 twelve-second frames. Each twelve-second frame is further divided into 1,536 time slots, each of 7.8125-ms duration. Thus, JTIDS has a capacity of 98,304 message-exchange time slots, each of 7.8125-ms duration.

Each message-exchange slot starts with a synchronizing preamble, which allows the receiver to synchronize to the pseudonoise digital code that is preprogrammed for that user. An identification band that allows the receiver to determine whether he is interested in the sender is next. Then follow 109 information packets that carry the basic message data. Each of the 109 information packets is 6.4 μs long and contains 32 bits of data, resulting in a basic chip rate of 200 ns. The RF bandwidth necessary to support this data rate is 10 MHz. A guard-band follows the message-band packet transmission to allow propagation to line-of-sight before the next transmission.

The pseudonoise or spread-spectrum coding utilized in JTIDS provides communications security, since the nonsynchronized intercept receiver has no knowledge of the code for a particular user, and hence cannot synchronize to a particular message packet. However, the 10-MHz bandwidth spread-spectrum transmission utilizes only a portion of the available 255-MHz overall bandwidth, which would result in a susceptibility to spot jamming. To correct this situation, 51 separate transmission frequencies are used to cover the band of interest. The JTIDS transmission frequency then hops in accordance with a preprogrammed code (that is available to all users) in random sequence with minimum jumps on the order of 30 MHz. This forces a jammer to spread its energy across the entire band, thereby providing a significant processing gain at the JTIDS receiver.

Another feature that increases the reliability of JTIDS transmissions is the use of Reed-Solomon codes, which allow reconstruction of messages even though part of the message is missing or in error. Average error rates of 0.6% were experienced during AWACS testing, but error rates from 1% to 12% were noted during high bank angle turns [22].

The unique feature of JTIDS is that all users transmit on a common channel, thereby making real-time exchange of data possible. Time slots for 98,304 users are theoretically available, but a practical number is between several hundred to a maximum of 2000 users. The AWACS application uses up to approximately 10% of the time slots. A fighter aircraft would be expected to use only a few time slots. Transmitter power on the order of 1 kW is required for AWACS, while 100W to 500W is needed for fighter aircraft. The development of a *distributed time-division multiple access* (DTDMA) architecture allows for simultaneous transmission and reception, which increases capacity and eliminates the need for guard intervals.

JTIDS, through a combination of direct sequence and frequency hopping, achieves a spread-spectrum bandwidth of 255 MHz with an information data rate of 57.6 kHz [23]. This represents a 36.5-dB processing gain against a broadband noise jammer. In addition, a forward error-correcting Reed-Solomon code permits the reconstruction of the information content of the messages in which up to eight errors occur in the 31-bit code [22]. These features make JTIDS highly resistant to conventional noise jamming.

Tone jamming is one possible mode of EA. With this method, a jammer picks some of the frequency channels, jams them, and lets the signal pseudorandomly jump into the jamming band. If five out of the 51 channels are blocked, a significant 10% error rate would be incurred [22].

Another EA possibility is to compromise the pseudorandom code that synchronizes the spread-spectrum characteristics of the system. This would allow the jammer to follow the frequency and time excursions of the data link system, which would negate the spread-spectrum characteristics of the system. In this regard, the use of the more secure CDMA, which requires synchronization of the code sequences employed for user identification, would result in higher jam resistance.

1.4.4.2 Tactical Communication Radio Nets

The command and control of tactical forces is exercised primarily using combat net radio [20,21]. *Combat net radio* (CNR) provides short- and medium-range communication over typical distances of up to 50 km throughout the main forces deployed in battle areas, generally using VHF in the range of 30 MHz to 88 MHz.

Operation is normally based on the net principle, with one net used for simplex communication between a group of users who are synchronized on a particular channel of the net. The message transmitted on the first net is received by a rebroadcast station, which then communicates the message to a second channel of the net, usually at a different carrier frequency. This sequence is repeated until all users of the net are interconnected.

There has long been a desire to encrypt all military communications, but in the past this was not practical because cryptographic equipment was large, consumed a lot of power, and synchronization was difficult. It is now possible to build security features into every radio, even man-packs, due to advances in digital microprocessor technology. If every radio has communication security features with complex codes, an unfriendly listener must have considerably more equipment than is practical to deploy in a battlefield situation. Hence, future C2W against CNR will most probably consist of direction finding, jamming, and, if necessary, destruction of the links [24].

Another technological advance that makes eavesdropping on enemy signals difficult is spread-spectrum communication. Whereas COMSEC is the encryption/decryption of the transmitted information, spread spectrum can be thought of as the encryption/decryption of the RF signal carrying this information. In spread-spectrum communication, the signal is spread over a wide RF band in accordance with a pseudorandom code. Despreading takes place through correlation with a stored replica of the pseudorandom code.

Several types of spread-spectrum techniques exist, but the type most commonly used in CNR is that of frequency hopping. The technique apparently provides a better near and far performance and can be implemented with less complex circuitry than the other available techniques [20,21]. Frequency-hopping spectrum spreading is used in the VHF single-channel ground and air radio system (SINCGARS-V) utilized by the U.S. Army and the Jamming Guarded Radio (Jaguar-V) man-pack radio used by certain elements of NATO. The Jaguar-V will soon be replaced by the Panther 2000-V [25].

It is relatively simple for a listener, or eavesdropper, to despread a single spread-spectrum link when no other such links are operating in the same region. However, this simple single-link situation will not normally exist on the battlefield, since there are likely to be many nets using spread-spectrum techniques, all operating asynchronously, with their RF signals spread in accordance with different pseudorandom codes. Despreading of a single net operating in the presence of other nets is practically impossible [26]. The trend toward frequency-hopping encrypted radios, which is being pursued actively by many manufacturers, promises to make battlefield communication highly secure. This will limit the intelligence that can be obtained by listening to radio traffic and place a premium on communication jamming in a battlefield situation to disrupt the enemy's forward-area C^3 systems.

A complete description of frequency-hopping CNR is beyond the scope of our discussion. Instead, we will give a brief description of the tactical jamming situation and the selection of frequency-hopping rates.

Against non-spread-spectrum communication links, and more particularly against non-frequency-hopping nets, it is generally adequate if the target signal can be intercepted and jamming started in 2 sec to 3 sec. Intercepting and attempting to jam a frequency-hopping link is a totally different proposition [24].

With a few exceptions, frequency-hopping communications are digital. Even with voice links, the signals are digitized. Many of the currently available radios used for such communications are so-called slow-hoppers, changing frequency at 50 to 500 hops per second. Consequently, a jammer that requires 2 sec or 3 sec to intercept and begin jamming cannot be effective, even against a slow-hopping radio.

To jam a digital signal effectively, a minimum of 20% of the bits should be in error [24]. A radio that hops at 100 hops per second will dwell on each specific frequency for approximately 10 ms. The responsive jammer, therefore, must intercept and begin jamming in 8 ms at the most. In practice, a responsive jammer must be even faster to compensate for the greater distance that the jamming signal must travel as compared to the main radio-link path. Therefore, against frequency-hopping communications, jammer response becomes critical. If the jammer elects to use broadband jamming to cover the whole communications band, then a sizable increase in jammer ERP (e.g., 24 dB for the Jaguar-V CNR) is required [26].

Frequency-hopping radio is implemented using identical stable frequency synthesizers in both the spatially remote radio transmitter and receiver link, which provide a number of selective frequencies. The radio's output frequency sequence is generated using a pseudorandom base code, which is available at both the transmitter and receiver ends of the link. When the transmitter and receiver base-code sequences are synchronized, the link is locked for secure communication. A critical parameter is the time the CNR stays at a particular frequency that determines the frequency-hopping rate.

Protection against intercept and direction finding tends to increase with hop rate. Complete protection against a sophisticated fast-follower jammer would require a hop rate of the order of 10,000 hops per second. The cost of implementing a frequency hop CNR tends to increase above 1,000 hops per second. Hop rates above 1,000 hops per second tend to ring IF circuitry, causing internal interference. Switching the transmitter from frequency to frequency tends to reduce average power at rates higher than 500 hops per second [26].

The Jaguar-V is an example of an advanced frequency-hopping VHF radio available in man-pack and vehicular configurations. It transmits approximately 50W (3W to 4W for the man-pack version), has a receiver sensitivity of −113 dBm, and an antenna gain of 13 dB. The signal hops at medium speed between 50 to 500 hops per second. The Jaguar-V system is designed to hop over a bandwidth of 6.4-MHz, corresponding to 256 channels at 25-kHz spacing. Up to nine 6.4 MHz bands can be preprogrammed, providing the capability of covering the overall band in 25 kHz synthesized steps. The system also has a *steerable null antenna processor* (SNAP), which minimizes the effects of jamming by tuning to a null in the jamming signal [24,26].

Recent developments in CNRs involve *direct digital synthesizers* (DDS), which permit fast hopping rates of over 1,000 hops per second [27]. These devices are also used in EA equipment (see Section 5.4) and operate using a digital phase accumulator that drives a waveform map (phase input to sine wave look-up). DDS provide the capability for synthesizing any waveform within their bandwidth capability under digital control. This provides the capability for producing complex modulation pseudonoise waveforms that enhance EP of these systems.

A basic EA method against CNR is to employ broadband noise jamming. Typical CNRs provide a processing gain of the order of 20 dB to 30 dB, while man-pack versions transmit powers of 100 mW to 4W. Thus, jamming powers in the range of 10W to 4000W are required to defeat this type of CNR. Fixed versions use the order of 50-W power and, hence, are beyond the capability of most tactical communications jammers.

Some of the older types of CNRs that use slow hopping (order of 10 to 100 hops/s) are susceptible to follower-type jammers. In this hopping range, jammers have from 10 ms to 100 ms to tune to the CNR frequency. This time must include the propagation time between the CNR transmitter-to-the-jammer and then the jammer-to-the-CNR receiver.

Another technique used against CNRs is partial band jamming, whereby the jammer's energy is concentrated in a particular band covered by the CNR. When the frequency hopping causes the CNR to enter the jammed band, the messages are interrupted. As the message is interrupted, the BER increases and, if it reaches the order of 20%, the CNR is jammed.

With CNRs, a synchronization waveform must be stored in the receiver. The methods vary as to how this is achieved [25]. Some require accurate measurement of absolute time, others require a "wrist watch" approach of modest accuracy, while others require relative time rather than absolute time. In the future, most radios will be synchronized to the GPS clock. Compromise of the synchronization mechanization may provide a fruitful area for EA, but the specifics depend upon the particular CNR involved.

1.4.4.3 Satellite Communications

Military Satellite Communication (MILSATCOM) networks operate at UHF (250 MHz to 400 MHz), SHF (7 GHz to 8 GHz) and EHF (20 GHz and 44 GHz). Also, extensive use is made of commercial capabilities. The UHF capability has the advantage of using simple antennas that are compatible with mobile platforms, such as aircraft, small ships, and small-land vehicles. However, EP is limited at UHF, motivating utilization of the higher frequency bands with their wider bandwidths and steerable directive narrow antenna beams [1].

Major military requirements for communications satellites are a high data rate, signal security against eavesdropping (low-intercept probability), protection of satellite vulnerability to attack by nuclear blast and *electromagnetic pulse* (EMP), lasers, kinetic energy as well as EP against EA, and other countermeasures. These requirements have generally been met by a dedicated set of military satellites. However, trends in commercial satellite communications indicate that a major portion of future military needs could be satisfied using these systems. The basic task of any communications system is to transmit as much data as possible. Commercial capability in this area has already outstripped military satellite systems, and the commercial sector will continue to advance their capabilities at a dramatic pace.

The main requirements for a commercial system to fulfill military requirements in time of crisis are [30]:

- Polar as well as equatorial LEO orbits;
- Interface with handsets and other nodes (i.e., telephone satellite stations and mobile telephone systems), so that messages could be routed either completely via satellite or partially over terrestrial systems;
- Intersatellite communication to route messages—here, the development of laser communication, with its modest size and energy requirements, could enormously speed up data transfers; store-and-dump systems for large data packets, which should be incorporated into civilian systems, more than is the case at present; the movement of very large data packets, in particular an integral part of many internet applications, could be accomplished more effectively and economically;
- High frequencies, between 40 GHz to 150 GHz, to counter disturbances not only from nuclear blast, but also those caused by infrequently occurring powerful solar storms;
- More hardened and reliable electronics, solar cells, and other adaptations to cope with natural occurrences like the Van Allen radiation belts (VAB), solar storms, and space junk; these will not meet current military standards, but the high degree of redundancy of a civil satellite system would go some way to compensate for this, enabling it to fulfill the military mission just as well as a purely military system.

Falling military budgets will probably lead in the direction of commercial satellites. Presently, considerable military traffic uses the INMARSAT commercial satellite system for voice traffic (2.4 kbits/s).

MILSTAR is a primary example of a current military satellite system. The MILSTAR communications package provides uplinks at EHF (44 GHz) and UHF (300 MHz), downlinks at SHF (20 GHz) and UHF (250 MHz), and cross links between satellites at 60 GHz. Operation at EHF results in low probabilities of intercept and detection. A typical 1-degree uplink beam is less than 250-m wide at a height of 40-k feet, and a terminal can typically operate within 4.8 km of an enemy listening post. The block 1 design provides 100 channels for traffic at rates from 75 bits/s to 2.4 kbits/s. The signals are encrypted and hop several thousand times a second over a 2-GHz bandwidth. The block 2 design provides a TI link at a rate of 1.544 Mbits/s that accommodates at least 600 users simultaneously per satellite. This design includes two spot antennas that can null jammers.

MILSTAR is a *geostationary* (GEO at 36,000-km altitude) system of only three satellites with no coverage at the poles, while the trend in commercial systems is to *low Earth orbit* (LEO), using much smaller satellites. Coverage by a LEO system necessitates many more satellites than by a GEO, but there are advantages in smaller antenna size and lower energy requirements, which can outweigh the apparent disadvantage of higher satellite numbers. GEO satellites, because of their high altitude above the Earth's surface, tend to introduce appreciable time delay in communications that is confusing and inefficient for interactive communications. As a result, GEO satellites are preferred for video transmissions while LEO satellites are used for voice communications.

LEO satellites generally operate below the Van Allen radiation belts at altitudes from 700 km to 1,400 km. The coverage area of a single satellite is low, so many satellites are needed to provide uninterrupted coverage. In addition, because the space crafts are so close to the Earth's surface, they pass overhead more rapidly than higher orbiting satellites. In order to maintain continuity throughout a call, LEO satellites must either employ complex satellite cross links or hundreds of surface cross links spanning the globe.

Examples of mobile communications commercial LEO systems are Iridium and Teledesic, which orbit, respectively, at 780 km and 695 km to 700 km. Iridium consists of a 66-satellite constellation, while Teledesic uses 840 satellites. These systems are intended to provide worldwide telephone service. Data rates for commercial systems are generally lower than the military MILSTAR, but each is capable of supporting 2,000 to 3,000 simultaneous links; so that if they were used together, data rates of 1.5 Mbits/s would be easily obtained.

The large number of satellites used in LEO systems has a number of important consequences for military applications, including the high resilience to laser and kinetic attack provided by the small size and high number of satellites used in systems like Teledesic. This follows due to the small size of the satellite, which makes it harder to hit, and the sheer number of targets to be simultaneously engaged. For any individual satellite disabled, a replacement is always just over the horizon. The same principles hold for resistance to jamming since the effort to continuously block out a sector requires multiple high-powered jammers with rapid slewing ability. The main threat to any satellite system would be from detonation of a nuclear device in space.

The *Global Broadcast Service* (GBS) GEO satellite is intended to provide integrated situational awareness and critical support information to all tactical users in a theater of operations. It is patterned after direct broadcast TV satellites. Approved military recipients can receive GBS simultaneously broadcast information (video, imagery, data) via an 18-in diameter antenna. Service is worldwide.

GBS is intended to augment current MILSATCOM systems, relieving them of much of the one-way traffic they currently carry. Further, GBS will incorporate a battlefield-responsive broadcast management structure that transmits data from military C^3 centers while also allowing in-theater direct injection of data. Two modes of operation are envisioned using either wide area coverage or steerable spot beams via the satellite's phased array antenna.

The initial service is created using the last three Navy UFO satellites configured with GBS transponders. GBS provides two steerable 500-nmi diameter coverage spot beams with a 24-Mbits/s data rate and one wide area 2000-nmi-diameter coverage operating with a 1.544-Mbits/s data rate. Characteristics of the GBS satellite transponders are given in Table 1.3 [31].

EA against MILSATCOM can be on either the uplink or downlink segments or both. Brute-force jamming operates on the premise that if enough jamming power is radiated into a receiver, the user signal will be indistinguishable from the jamming signal and the system will collapse. Against spread-spectrum signals, the jammer may distribute its energy over the whole or part of the spreading bandwidth, which corresponds, respectively, to full-band or partial-band jamming. Another strategy may consist of distributing the jamming power into a series of tones that occupy part or all of the spreading bandwidth. As a special case, jamming power may be contained in a single tone, operating in pulsed or CW modes.

Partial-band jamming is effective when the jammer energy is not sufficient to cause high enough BERs in the channel by jamming the whole bandwidth.

Table 1.3
GBS Transponder Characteristics [31]

Parameters	Per Spacecraft	Global
Number of spot beams	2	6
Transponders	4	12
Total data rate (Mbps)	96	288
Transponder redundancy	5 for 4	
Downlink EIRP (dBW)	54.5	
Spot beam antenna size (in)	22	
Total TWT power (W)	120 (47% eff)	
Receive antenna size (in)	18	
Uplink frequency (EHF)	30 to 31 GHz	
Downlink frequency (SHF)	20.2 to 21.2 GHz	

Under these conditions, the jammer trades the jamming bandwidth by concentrating jamming energy in portions of the band to be more effective. A percentage of the frequency bandwidth that can be jammed increases with increased jamming energy and full-band jamming becomes feasible when it is sufficiently high. Using forward error-correction coding and increasing the spread-spectrum processing gain constitute the principal EP measures against such an attack [32].

A pulsed jammer can exploit the time scale by jamming at duty cycles less than or equal to unity. Pulse jamming can also be used to jam several satellites by time sharing.

Uplink EA is a preferred strategic jamming strategy since its effects are net-wide and affect all the users. The position of a GEO satellite is easily located for this purpose, while LEO satellites may present more difficulty in this regard. Different transponders of a satellite can be jammed by time sharing if the jammer bandwidth is narrower than the satellite bandwidth. A jammer may jam a single satellite or could jam multiple satellites by beam switching.

Satellite transponders saturated by an uplink jammer operate at a maximum downlink power level that is proportional to their received uplink power level. The net effect of a jamming signal is then to steal power from the wanted signals. When the jammer power dominates the total uplink power received by the jammed transponder, the power in the wanted signal becomes too low to be detected by the user satellite ground terminal. In some cases, satellite transponders employ hard-limiting receivers so that jammer signals that saturate this receiver can cause up to 6-dB suppression of the wanted signal by this effect [33]. Spread-spectrum signals are as vulnerable to this attack as conventional signals. This is a specific weakness of hard-limiting transponders that can be countered by the use of on-board processing [32].

A jammer can exploit the nonlinear characteristic of a satellite transponder by sweeping its frequency through regions that affect the carrier tracking loop and communications demodulator. With high-jamming ERP, the satellite transponder becomes saturated and the downlink signals undergo AM modulation that degrades automatic gain control/tracking loop responses. Pulse-jamming signals can produce the same effect by judiciously choosing the duty cycle and pulse repetition frequency to produce maximum effect on the tracking loops [32].

Under normal operating conditions, the total capacity available from a satellite is allocated among various ground terminals, and each ground terminal has a satellite resource budget (power and bandwidth) according to its traffic requirements. A number of ground terminals may compete with each other for a share of the satellite capacity. Under jamming conditions, the capacity available for a ground terminal is set mainly by the jamming level and to some extent by the presence or absence of other ground terminals accessing the same transponder.

The maximum data rate (R in bits/s) that can be supported by a hard-limiting transponder using binary digital signals against uplink jamming can be approximated by [32]

$$R = \frac{\alpha \cdot W_s}{(1 - \alpha) \cdot (E_s/N_0) \cdot M_L} \tag{1.2}$$

where

W_s = spreading bandwidth (Hz)

α = $S'/(L_s \cdot P_o)$, percentage of transponder ERP used by the desired signal

E_s/N_0 = Energy per bit to noise spectral density to achieve specified BER

M_L = link margin

L_s = limiter signal suppression

P_o = total power receiver by transponder

S' = desired signal power received by satellite

An examination of the equation indicates that the magnitude of the spreading bandwidth, the percentage of the desired signal retained in the presence of jamming, and the required S/N to achieve the BER determining the capacity of the jammed communications channel.

▶ *Example 1.3*

A hard-limited satellite transponder is illuminated by a satellite ground terminal with a 95-dBW EIRP. The 40-MHz bandwidth frequency-hopping link requires a 10-dB signal-to-noise ratio with a 6-dB link margin to achieve an acceptable BER. The transponder is jammed from a surface-based terminal within the satellite main beam coverage. Propagation and atmospheric losses are identical for both the jammer and the satellite ground terminal. Determine the maximum communications data rate for jammer EIRPs from 90 dBW to 130 dBW. What is the effect of 10 additional ground terminals of equal EIRP on the data rate?

```
% Traffic Capacity with Uplink Jamming
% Hard Limited Satellite Transponder
% ------------------------------------
% jamchan.m

clear;clc;clf;

Erpmin=90;Erpmax=130;

% Define Jammer Erp(dbw) from Erpmin to Erpmax

Erpj=Erpmin:.1:Erpmax;

% Enter Link Parameters

W=40;    % Bandwidth - Mhz
x=10;    % Es/No - dB
Ml=6;    % Link Margin - dB
Erpsgt=95; % SGT Erp - dbw

W=W*1e6;x=10^(x/10);Ml=10^(Ml/10);

% Compute ratio - Signal-to-Total Transponder Power

S=10^(Erpsgt/10);J=10.^(Erpj/10);

% Find Signal Suppression Loss - Ls
% Jones/IT Trans/ Jan 1963

jsr=J/S;
snr=x;snr=10^(snr/10);
jnr=jsr*snr;

   f(1,:)=[0,1,1.11,1.33,2,5,10,100];
   f(2,:)=[10,1,.88,.7,.5,.33,.31,.29];
   f(3,:)=[20,1,.8,.62,.43,.32,.30,.28];
   f(4,:)=[30,1,.79,.6,.41,.31,.29,.27];
   f(5,:)=[40,1,.77,.58,.4,.3,.28,.26];
   f(6,:)=[50,1,.75,.57,.4,.3,.27,.25];
   f(7,:)=[60,1,.74,.56,.4,.3,.27,.25];
   f(8,:)=[70,1,.74,.56,.4,.3,.27,.25];
   f(9,:)=[80,1,.73,.55,.4,.3,.27,.25];
   f(10,:)=[90,1,.73,.55,.4,.3,.27,.25];
   f(11,:)=[100,1,.726,.547,.394,.294,.27,.25];

for j=1:length(Erpj);
if jsr(j)<1;Ls(j)=1;end;
if jsr(j)>100 | jnr(j)>100;Ls(j)=.25;end;
if  jsr(j)>=1 & jnr(j)<=100;
 Ls(j)=table2(f,jnr(j),jsr(j));end;
end;
```

```
% Compute Bit Data Rate

fac=Ls./(1+jsr+(1/snr));
R=W*fac./((1-fac)*x*Ml);

% Plot Channel Rate vs Erpj

semilogy(Erpj,R);grid;
xlabel('Jammer Erp - dBw');
ylabel('Channel Capacity - bits/s');
title(['Channel Capacity of Jammed Transponder']);
```

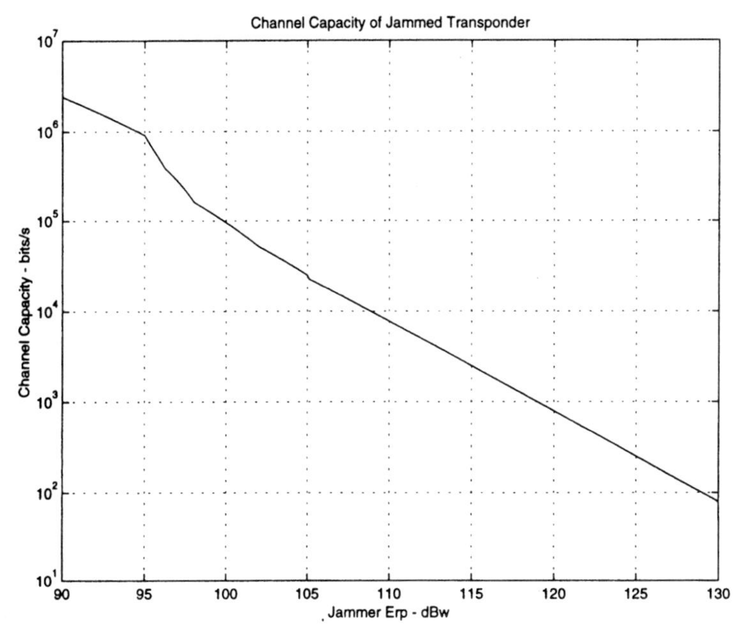

Downlink jamming differs from uplink jamming in that it generally attacks an operationally significant airborne, land-based, ship-borne, or submarine-based platform. In contrast, uplink jamming concentrates on the entire SATCOM network.

Downlink jamming is best accomplished from an airborne platform that accomplishes both interception and jamming. Jamming can be directed at satellite beacon and telemetry signals as well as communications channels. The jammer may attack to distort the synchronization information provided by the satellite beacon in order to increase acquisition time, disturb satellite tracking, or cause the loss of telemetry data. The beacon signal received at a user satellite ground terminal may be downlinked-jammed either using a stronger pseudobeacon signal simulating the actual beacon or by a separate jamming signal.

The jammer ERP required for downlink jamming can be determined from (1.2). Satellite terminals generally employ large-diameter (e.g., 20 ft to 40 ft) antennas that provide high gains and narrow beams resulting in a sidelobe jamming situation for an airborne jammer. For example, a 42-ft-diameter ground terminal antenna has a gain of 58 dBi at 7.5 GHz. Antenna sidelobes are −10 dBi at greater than 48 degrees off boresight and approximately $G(\theta) = 32 - 25\log(\theta)$ for angles from 1 degree to 48 degrees [32]. This provides the downlink signal with an advantage as great as 68 dB. A further advantage can be obtained using adaptive sidelobe cancelers or ultra low sidelobe antennas. On the other hand, free-space propagation losses are proportional to the square of the ratio of the distance between the ground terminal and the satellite and the ground terminal to the airborne communications jammer. This provides an advantage to the airborne jammer. For EHF (e.g., 20 GHz), downlinks atmospheric and rain attenuation must be considered in the jammer calculations, which generally result in an advantage to the satellite link.

Repeat-back jamming is a threat mainly against slow frequency hoppers. It consists of interpreting, processing, and retransmitting the signals received from the user transmitter to the user receiver before the user system hops to a new set of frequency channels. To be effective, the time delay corresponding to the differential path delay between the repeater jammer and the user receiver should be less than the difference between the frequency hopping dwell time and the time required to process the intercepted signals. Thus the higher the hopping rate, the more protected is the communications link and the closer the repeater platform must be to the ground terminal. Repeater jamming applies to both up- and downlink EA. In contrast to an uplink repeater, a downlink repeater can intercept all downlink channels within the coverage area of the satellite.

1.5 Problems

1. Describe the relationship of Electronic Warfare to Information Warfare.

2. What are the five pillars of C2W and how do they interact?

3. Shannon's Noisy Channel Coding Theorem states that codes can be found that allow information transmission at arbitrarily small error rates with channel capacity in bit/s of $C = B\log_2(1 + S/N)$, where B (Hz) is the channel bandwidth and S/N is the signal-to-noise power ratio. Plot the channel capacity for a 20-MHz bandwidth channel and various S/N ratios. For a 500 °K receiver and a 20-dB S/N ratio, what jamming power is required to reduce the channel capacity to 10% of its unjammed value? How is the channel capacity changed if a 2-bit A/D converter is used to digitize the signals? (*Hint*: consider quantization S/N ratio).

4. A UHF FM communications transmitter radiates 500W from an omnidirectional antenna. The airborne receiver 30 nmi away also uses an omnidirectional antenna with unity gain. How much noise jamming power is required in an airborne jamming antenna with 3-dB gain to produce a jam-to-signal ratio of 5 dB when the jammer-to-receiver distance is 1 nmi, 10 nmi, and 50 nmi? The jammer bandwidth is 2 MHz, while the receiver bandwidth is 75 kHz.

5. If a repeater jammer is used instead of a noise jammer in Problem 4, what is the effect on the required jammer power?

6. Figure 1.9 depicts a multilayered air defense system. The terminal weapons that engage the airborne target are the SAM and the AAM carried by the interceptor aircraft. What are the critical nodes for each of these systems?

7. The flux density on the Earth of satellite communication systems must be limited so that they do not interfere with terrestrial radio systems that operate in the same frequency bands. The limit in the band from 8.025 GHz to 11.7 GHz is -140 dBW/m^2 within any 4-kHz band at angles of arrival from 25 degrees to 90 degrees. What satellite transmitter ERP produces this flux density for GEO orbit (19,322 nmi) and for LEO (420 nmi)? What ERP from an airborne noise jammer operating at 20,000 ft is required to produce this same flux density?

8. Computer hacking is one form of IW. Explain the terms computer viruses, worms, Trojan horses, logic bombs, trapdoors, chipping, nano machines and microbes, HERF guns, and EMP bombs as they relate to computer attack.

9. The *Global Positioning System* (GPS) receiver uses signals from at least four satellites to solve for the user's three-dimensional position at a particular time. The GPS operates at frequencies of 1575.42 MHz and 1227.6 MHz. The signal power at the receiver through an omnidirectional antenna is -160 dBW. Using MATLAB, plot the jam-to-signal ratio at the GPS receiver versus the jammer range for jammer ERP of 10 mW, 100 mW, 1W, 10W, and 1000W. On this plot, superimpose jam-to-signal ratio lines of 24 dB required to jam the acquisition/low accuracy code link, 54 dB required to jam the precision code link, and 43 dB required to jam the carrier link (used for ultrahigh precision measurement).

10. Assume that the GPS system described in Problem 9 is equipped with an adaptive array type antenna whose function is to attenuate jamming signals. If the adaptive array can attenuate the jamming signal by 20 dB, what ERP is required to jam the GPS system from a range of 20 km?

11. A ground-to-synchronous satellite link operates with a data rate of 1 kbit/s while the ground antenna uses a 60-ft-diameter antenna. Antijam protection is provided using a 10-mbits/s direct sequence spread-spectrum code. The jammer uses a 150-ft-diameter antenna with a transmitter power of 400 kW. Assume equal space and propagation losses. How much power is required at the Earth station to achieve signal-to-jam ratio of 16 dB at the satellite receiver? Assume receiver noise is negligible [34].

12. A communications data link uses frequency hopping at a hop rate of 10,000 hops/s to avoid the threat of repeat back jamming. Ignoring Earth curvature and assuming transmission to a satellite in geosynchronous altitude (36,000 km) that is directly overhead, compute the radius of vulnerability, which is the radius outside of which the data link is unconditionally safe from repeat back jamming by a ground based jammer [34].

13. Repeat Problem 12 under the assumption that the jammer requires a minimum of 10 s to identify the transmission frequency and tune the jammer output [34].

14. Spread-spectrum techniques can be used to meet regulations regarding power density radiating the surface of the Earth. If a satellite at synchronous altitude (36,000 km) transmits 4-kbit/s data using 100W of ERP, what spreading bandwidth is required to maintain a flux density on the Earth's surface no greater than -151 dBW/m^2 in any 4-kHz band [34]?

References

[1] Schleher, D.C., *Introduction to Electronic Warfare,* Norwood, MA: Artech House, 1986.

[2] Report of the Defense Science Board Summer Study Task Force on "Information Architecture for the Battlefield," Oct. 1994.

[3] Joint Doctrine for Command and Control Warfare (C2W), Feb. 1996.

[4] Libicki, M., *What is Information Warfare?,* National Defense University, Aug. 1995.

[5] Adams, J., "Warfare in the Information Age," *IEEE Spectrum,* Sept. 1991.

[6] Carney, J., and C. Corsetti, "Operational Effectiveness Through Interoperability," First International Symposium on Command Control Research Technology, National Defense University, June 1995.

[7] Joint Command and Control Warfare Staff Officers Course, Armed Forces Staff College, Jan. 1995.

[8] Joint Doctrine for Operations Security, Joint Pub 3-54, 15 April 1994.

[9] Doctrine for Joint Psychological Operations, Joint Pub 3-53, 30 July 1993.

[10] Joint Doctrine for Military Deception, Joint Pub 3-58, 6 June 1994.

[11] Taylor, C., and N. Younan, "Effects from High Power Microwave Illumination," *Microwave J.*, June 1992.

[12] Schleher, D. C., *MTI and Pulsed Doppler Radar,* Norwood, MA: Artech House, 1991.

[13] Garner, K., and T. Assenmacher, "Improving Airborne Tactical Situational Awareness," *J. Electronic Defense,* Nov. 1996.

[14] Hardy, S., "Catching Falling Stars," *J. Electronic Defense,* Jan. 1995.

[15] Cooper, P., "Global Broadcast Service to Widen Battlefield View," *Defense News,* Sept. 9–15, 1996.

[16] Herskovitz, D., "Come Whisper in My Ear," *J. Electronic Defense,* June 1995.

[17] Pettit, R., "Communications Jamming in Today's Electronic Warfare," *International Countermeasures Handbook,* 1985.

[18] Darby, B., and J. Hannah, "Programmable Frequency-Hop Synthesizer Based on Chirp Mixing," *IEEE Trans. on Microwave Theory and Techniques,* Vol. MTT-29, No. 5, May 1981.

[19] Schwartz, M., *Information, Transmission, Modulation and Noise,* Fourth Ed., New York: McGraw-Hill, 1990.

[20] Van, P., "New Concepts in Battlefield Communications. Part 1 Fast Frequency Hopping," *International Defense Review,* March 1983.

[21] Sundaram, G., "New Concepts in Battlefield Communications. Part 2 Slow/Medium Frequency Hopping," *International Defense Review,* May 1982.

[22] "JTIDS/TIES Consolidate Tactical Communications," *EW,* Sept./Oct. 1977.

[23] Stiglitz, M., "The Joint Tactical Information Distribution System," *Microwave J.*, Oct. 1987.

[24] "The Future of Communications Jamming," *International Defense Review,* May 1983.

[25] Pengelley, R., "Casting the Net," *International Defense Review,* April 1996.

[26] Munday, P., and M. Pinches, "Jaguar-V Frequency Hopping Radio System," *IEE Proc.*, Vol. 129, Pt F, No. 3, June 1982.

[27] Boyle, D., and G. Sundaram, "Future Combat Radios, the Key Lies in Digitation," *International Defense Review,* May 1991.

[28] Hardy, S., "Catching Falling Stars," *J. Electronic Defense,* Jan. 1995.

[29] Knoth, A., "Space-Based Comms in the 21st Century," *International Defense Review,* May 1995.

[30] Wilson, J., "A Commanding View," *International Defense Review,* May 1995.

[31] Murates, J., "Tactical DMS: A Global Broadcast Service Option," Naval Postgraduate School Thesis, June 1996.

[32] Sakat, M., "An Overview of ECCM Factors in MILSATCOM Systems," *AGARD Conference Proc., ECCM for Avionic Sensors and Communications Systems,* AGARD-CP-488, Feb. 1991.

[33] Jones, J., "Hard-Limiting of Two Signals in Random Noise," *IEEE Trans. on Information Theory,* Jan. 1963.

[34] Sklar, B., *Digital Communications—Fundamentals and Applications,* Englewood Cliffs, NJ: Prentice-Hall, 1988.

2

Advanced Radar Threat

Radar is a key sensor in most modern weapon systems. Its ability to function in an all-weather environment at long ranges is unmatched by any other available sensor. Also, it is one of the few sensors capable of providing accurate range information.

Land-based radars are used for a variety of military tasks ranging upward in size and complexity from man-portable vehicle and personnel locators to ballistic-missile-tracking phased arrays. Between these extremes are mortar and artillery location radar, artillery fire control radar, short- to long-range air surveillance radar, and SAM target and illumination radar.

Naval applications include air surveillance, surface search, surface-to-air, and surface-to-surface weapon control and target acquisition radar. Specialized applications include radar for aircraft carrier landing, air traffic control, collision avoidance, close-in weapon control, and target illumination for semiactive missile seekers.

Airborne applications for radar are more diverse than either ground- or naval-based systems. Radar located on aircraft provide multiple functions of target location and weapon delivery, navigation, terrain following and avoidance, weather detection, missile detection and warning, altimetry, and surface mapping. The challenge in airborne radar is to provide the multiple air-to-air and air-to-surface functions, many of which are vital to aircraft survival, within the limited antenna-aperture space available. The solution to this problem for fighters and bomber aircraft has evolved from multiple radars that shared the available aperture to the present multifunction array radars that share a common aperture to the future where conformal arrays will be built into the aircraft structure, thereby enlarging the available aperture. An alternative approach to the limited aircraft antenna-aperture problem, which is used in present AEW radars, is to use a large rotodome located above the aircraft. The rotodome limits the aerodynamic performance of the aircraft and, hence, can be applied only to a specialized class of aircraft.

Missile radars are primarily applied for weapon guidance and navigation. Active and semiactive radars are the principal sensor used for target tracking in all-weather missile guidance systems. The semiactive radar used in missile systems is the only operational example of a successful bistatic radar. In airborne applications, active radar is the current trend in missile guidance because it eliminates the illumination function of the aircraft AI radar, thereby allowing the aircraft to maneuver while the missile is in flight, and to have the capability of firing multiple missiles simultaneously.

Radar navigation of missiles is used in very long range *air-launched cruise missiles* (ALCM), where an altimeter provides *terrain contour matching* (TERCOM), and in the terminal correction mode of an intermediate range surface-to-surface tactical missile, which applies radar map matching against a stored target scene.

Space-based and high-flying manned and UAV radars provide the potential for greatly increased surveillance coverage in terms of both the instantaneous field-of-view and total region surveyed. Other advantages include line-of-sight access to most places on Earth and relative immunity to weather (since the maximum path through the atmosphere is generally short, approximately 10 nmi). The possibility of bistatic or multistatic operation has been studied, with the transmitter in space (presumably a safe sanctuary) and passive receivers located on airborne platforms or on the ground. The advantage of passive receivers (which can be numerous, small, and inexpensive) is that they are covert, thereby reducing the probability of detecting the weapon system by ES or ELINT while also providing an immunity to antiradiation missiles.

The high cost and large amounts of data that can be generated by space-based radar have limited their applications primarily to experimental use, with the possible exception of a Soviet ocean surveillance space-based radar. The experimental space-based radars (Seasat and SIR-A) have been used for remote sensing of the sea and land to determine ocean and geological characteristics.

Imagery generated using Synthetic Aperture Radar [1] carried on *high-altitude long-endurance* (HALE) platforms is used by the military to provide situational awareness of strategic areas. HALE platforms include the manned SR-71 Blackbird and U-2S aircraft, which carry the *Advanced Synthetic Aperture Radar* (ASAR) in a pod beneath the aircraft. The ASAR operates at X-Band, has an electronically scanned antenna, and includes wide-area and spot MTI modes [2,3]. The U-2R (predecessor of the U-2S) operates at 70,000 ft and has an operating range that exceeds 3,400 nmi.

The Predator UAV is classified as a medium altitude endurance platform and operates at 25,000 ft, remaining on station for up to 40 hrs at a distance of 500 nmi. from base. The Ku-band SAR surveys an 800-m-wide strip with a maximum resolution of 0.3m. The radar beam can be steered through 150 degrees in azimuth and 40 degrees in elevation [4]. The predator will be replaced by the long-endurance Global Hawk UAV.

The Global Hawk UAV cruises at 65,000 ft, has a range of over 3,000 nmi, and can loiter for 24 hrs. The SAR aboard the Global Hawk will be the first to employ onboard image processing, whereby the radar image is formed onboard the platform rather than on the ground. The SAR radar has multiple operating modes: SAR strip map with 1-m resolution, SAR spot mode with 0.3-m resolution, and ground moving target indication with speeds down to 4 knots. All modes cover slant ranges from 20 km to 200 km. During a 24-hr mission, the UAV can provide sensor coverage of about 40,000 nmi^2 in the wide-area mode or 1900 spot measurements (2 km × 2 km) with 0.3-m resolution [5].

Another stealth UAV called the Dark Star carries the same sensors as the Global Hawk. The Dark Star is intended to loiter undetected over a target for 8 hrs, at a height of 45,000 ft, with a range of 500 nmi.

The characteristics of radar employed in modern weapon systems are covered in several references [6–8]. An examination of these radars indicates a wide variety of techniques and characteristics. This is a consequence of the design cycle of radar, which typically takes five to ten years to develop and then approaches technical obsolescence within two years after being operationally deployed. Yet, simple economics dictate at least ten years of operational deployment. The result is a mix of operational radars, whose designs reflect the technological eras in which they were designed rather than the operational requirement for which they were designed.

The classic air defense system depicted in Figure 2.1, which is primarily composed of radar threats, has served as the prototype against which many of the current EW systems were designed. The success of these EW systems is indicated by the *Joint Tactical Air Electronic Warfare Study* (JTAEWS), which reported that, in recent conflicts, only 5% of aircraft combat losses were due to RF guided weapons [9]. The fact that 90% of the losses were due to IR guided missiles points to the need for better IR countermeasures [9]. However, the study should not be interpreted to indicate that RF guided weapons are no longer a major threat to aircraft survivability. Also, long-range stand-off air-to-ground weapons and beyond visual range air-to-air missiles (e.g., AMRAAM) put the engagement at ranges where radar will remain the principal aircraft threat for the foreseeable future. Recent advances in radar design and

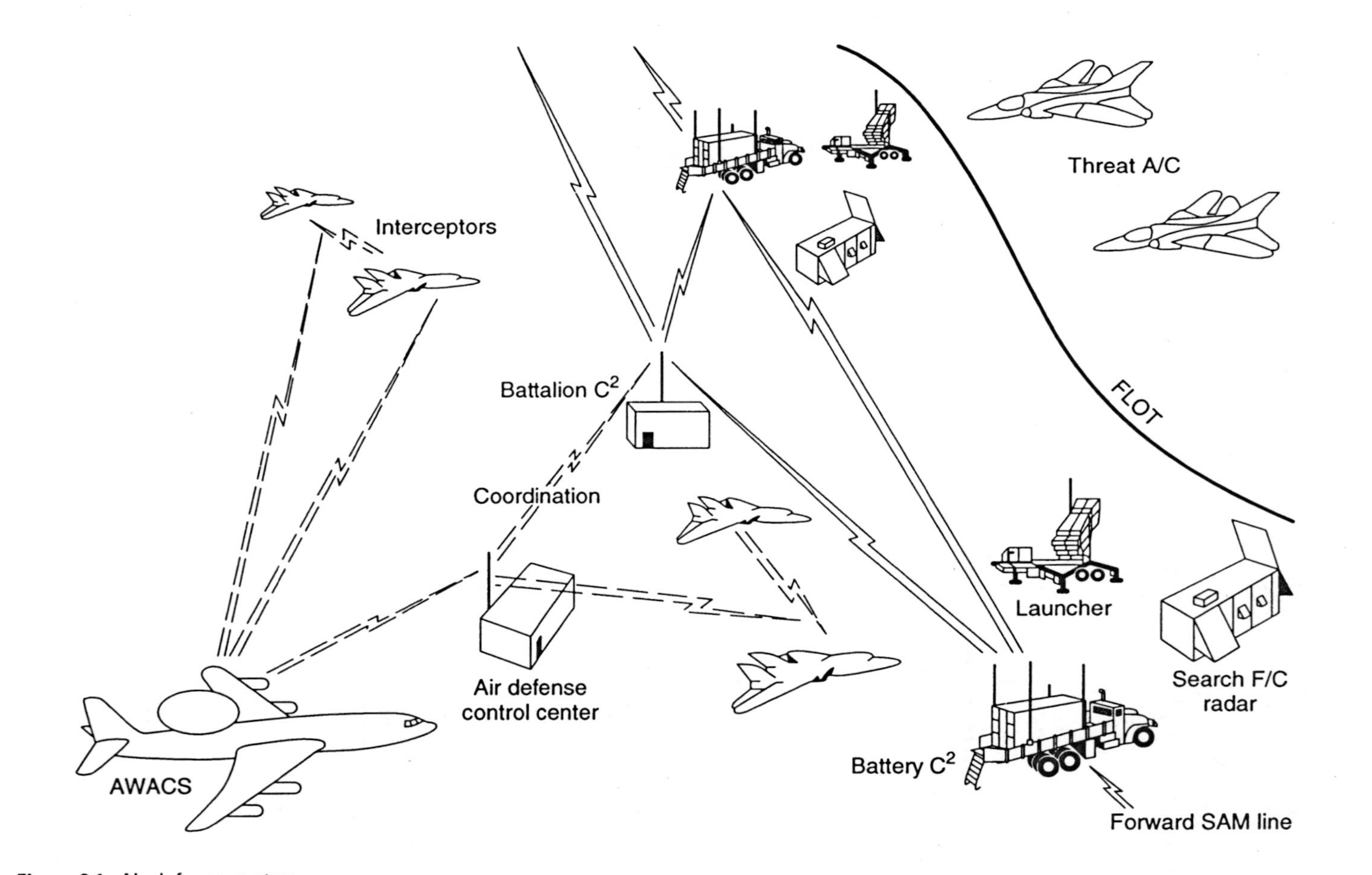

Figure 2.1 Air defense system.

weapon system componentry have advanced the state-of-the-art in these systems to the point where the capabilities of current EW systems are suspect against these advanced type weapon systems. In the following, we describe several of the advances and trends in modern radar systems that present the advanced threat due to RF guided weapons.

Recent advances in radar and missile technology are listed in Table 2.1 along with their projected affect on EW systems. The results of this technology are apparent today in the many systems available on the international weapons market. Although affordability is not listed in the table, it is a significant factor in the proliferation of high-technology weapons. The result is that a combination of older and newer type systems is employed to accommodate the economic situation of the particular user involved.

Several trends in Table 2.1 are fueled by advancements in component technology. The rapid advancement in digital integrated circuits, which supports the computer industry, has allowed the development of advanced signal processors and computers. These allow coherent radar systems using Doppler and pulse compression techniques to be implemented in a practical manner. Computers are important in accomplishing fusion in multisensor systems and in phased array systems where multiple agile antenna beams must be formed and scheduled. The precise control of the spectral, temporal, and spatial domains of the radar using digital techniques makes it difficult for a jammer to penetrate through the radar's protective shell to supplant the true target's signature.

Monopulse tracking is a strong trend in most modern tracking radar and missile seekers. The monopulse action rejects amplitude modulated jamming signals while providing a strong home-on-jam capability against noise jammers. This makes monopulse tracking radars and missile seekers very difficult and expensive to jam using onboard techniques. The present trend in EA against monopulse systems is to use towed decoys against monopulse missile seekers, which function to lure the missile away from the protected aircraft or ship.

Another strong radar trend is the use of ultra low sidelobe antennas that are supplemented in some systems by coherent sidelobe cancellers and sidelobe blankers. When these techniques are employed, sidelobe jamming becomes very difficult, particularly using noise-jamming techniques. This has resulted in the upgrade of support jammers to add coherent (DRFM) or quasi-coherent *direct digital synthesizers* (DDS) threat generators that synthesize matched jamming waveforms. These matched jamming waveforms introduced through the radar's sidelobes allow full processing gain to be achieved in the radar's signal processor.

Phased array radars are gaining broader acceptance, but their use is generally confined to the larger, more expensive weapon systems (e.g., Patriot, Aegis, Pave Paws, SA-10, SA-12). They are also used in counter battery radar

Table 2.1
Advanced Radar and Missile Techniques

Advanced Technology	Effect on EW System
Coherent radar systems	Reduces or eliminates effect of noise jamming in surveillance radar
Pulsed Doppler	Requires DRFM or DDS in jammer to provide matched jamming waveform
Pulse compression	Produces mismatch in ES and RWRs which are tuned to conventional threats
Multichannel receivers	Reduces or eliminates jamming of tracking radar through onboard techniques
Monopulse Tracking radar Bistatic missile seekers	Requires off-board jamming technique such as towed decoys against tracking radar
Sidelobe cancellers Sidelobe blankers	Reduces or eliminates sidelobe jamming of surveillance radar
Multiple sensor systems	Jamming of RF system is ineffective
RF/IRST RF/EO RF/Laser	Requires coordinated multispectral jammers to negate threat
Ultralow sidelobes	Reduces or eliminates effect of support jammers on surveillance radar
Multiple simultaneous threats Higher power Frequency and parameter agility Netting	Requires increased jamming energy in amplitude, spectral and spatial domains for effective jamming
Stealth/low observables	Eliminates need for active onboard radar jammer
Reduced aircraft, missile, and ship signatures	Accentuates need for ES system to determine effectiveness Increases effectiveness of support jamming
LPI Spread spectrum Power management Random pulse compression	Eliminates detection by ES and RWR receivers tuned to conventional threats
Multifunction radar Phased arrays	Reduces effect of support jammers on search function through low sidelobe, burn-through, and adaptive array features
Search-track-illuminate for semiactive seekers	Agile beam control allows rapid reacquisition of jammed targets
Adaptive array	Increased opportunity for EA to affect time allocation of phased array
High-speed missiles	Decreased EA reaction time
Mach 3+ velocity	Missile may acquire alternate target if jammed
High –g maneuvers Improved onboard processing	Soft-kill may be preferred over hard-kill due to faster reaction time

(e.g., TPQ-36, TPQ-37), which locate artillery and mortar launch sites, and in the B-1B radar. Affordability is definitely a factor and, in general, phased arrays are only used when more conventional type radar are not practical. The development of phased array modules using *Monolithic Microwave Integrated Circuits* (MMIC) is a thrust intended to produce a low-cost active aperture phased array. The objective is to develop modules that cost approximately $100, but the realities are that the cost will be about $1,000 per module [10].

EW against phased array radar is not as well developed as that for conventional threat targets. The fundamentals of EW are generally the same for phased arrays as they are for conventional radar. However, the beam agility of phased arrays requires a number of adjustments to conventional EW approaches, while the time sensitivity of phased array beam control offers an EA opportunity not present in conventional radar.

The use of multiple sensors in weapon systems is a natural consequence of the increased performance of EO/IR and laser systems, coupled with the natural synergism of active radar, which provides all-weather performance, and the passive high-angular precision provided by EO/IR systems, which suffer degradation in conditions other than clear air. The onboard fusion of multiple sensors is made possible by the rapid growth of fast, miniature digital computers and signal processors. A particularly effective combination in combat air-to-air weapon systems is to detect the target using a passive FLIR or IRST and then measure its range using a laser range finder. An RF or IR missile is then launched against the target, which has no warning that an attack is in progress. The radar that is aligned with the IRST or FLIR angular position radiates during the end-game to provide semiactive radar guidance when RF guided missiles are used.

A system similar to that described previously is used in the ground-based *Air Defense Anti-Tank system* (ADATS) [11]. In this system, targets are initially located using the X-band LPD-20 surveillance radar. A FLIR supplemented with a contrast tracker provides high-accuracy (better than 0.1 mrad) angle data on the target. A laser beam operating at 10.2 μm is aligned with the target angle. The target range is confirmed using a laser range finder. A Mach 3 missile is launched and rides the laser beam to engage the target. This system is effective against low-flying aircraft and helicopters, but, like all EO/IR systems, operates only under fair-weather conditions.

Another multisensor system is used with *Anti-Shipping Cruise Missiles* (ASCM), which rely primarily on radar guidance to attack ships. Because of their lethality, most military ships are outfitted with extensive EW systems to counter this threat. When the radar is jammed or decoyed, several types of ASCM seekers employ a supplementary IR tracker that provides guidance against the generally large IR signature of the ship. This is an increasing trend in ASCM design.

The increasing use of low-observable techniques can severely stress the capability of current radars that were designed to detect much larger targets. When land clutter is involved, the primary factor is the radar's ability to detect targets determined by its subclutter visibility, which can range from 30 dB for older systems to 90 dB for upgraded or newly designed systems [11,12]. In stealth platforms, the function of the classical onboard ECM system is replaced by the stealth characteristics. For low-observable platforms (i.e., the RCS is reduced by a factor on the order of 10), onboard ECM systems are still required, but their ERP is reduced and support jamming becomes more effective. In general, ESM systems are required by stealth and low-observable platforms to provide situational awareness.

Several operational LPI radars have entered the marketplace. Since it is not possible to alter the physics of radar detection, these radar must radiate sufficient ERP to accomplish target detection. However, they employ waveforms that are mismatched to those for which ESM sets are tuned [6]. Generally, these LPI radars attempt to provide detection of targets at much longer ranges than an ESM system on the target can accomplish detection of the radar. Retuning of the ESM to detect LPI radar is not a simple process, requiring significant modification of current ESM systems, which would cost billions of dollars to implement.

The increasing velocity of attack missiles to Mach 3 and greater reduces the survivability of aircraft and ships, whose defensive systems do not have the necessary reaction time to cope with such a system [13]. Requirements caused by this trend include greater detection time, more accurate missile location, and automatic engagement of the defensive system since there is not sufficient time for operator reaction in a manual or semiautomatic mode of operation.

2.1 Low-Intensity Threat

Advanced radar and missile systems described in Table 2.1 can be fielded only by countries that have a high technology or economic base. More typical systems, mainly directed at air defense, are those exported by the former Soviet Union that provide capabilities for low- to mid-intensity conflicts. These air defense systems are patterned after Soviet air defense doctrine, which stresses density, diversification, and large numbers of early warning, GCI, and target-acquisition radars that are netted through C^3 systems. They represent two general principles: forward defense and defense in depth [6]. The forward defense doctrine is to destroy the attacker as far out as possible, preferably before it leaves its launch pad or airfield. The doctrine of defense in depth dictates zone-after-zone of defensive means from the extremely long-range position back to the point to be defended. These systems are interesting to study since many existing EW systems were developed against this threat.

The heart of the air defense system is a mix of SAMs. A typical set attributed to the Iraqi air defense system is listed in Table 2.2 [14]. The SA-2 (Guideline) strategic command-guided missile system provides high-altitude coverage up to 82,000 ft using a missile with a maximum speed of Mach 3.5 and a range of 31 nmi. The SA-3 Goa strategic command guided missile provides low- to medium-altitude coverage, a range of 16 nmi, and uses a Mach 2 missile. This missile system has a television guidance option and advanced ECCM options. The SA-6 (Gainful) is a mobile tactical low- to medium-altitude missile. The kill zone of this missile is from 300 ft to 43,000 ft with a range of 18.5 nmi. This missile uses terminal semiactive guidance with midcourse command guidance and an optional infrared homing guidance. The mobile SA-8 (Gecko) tactical low- to medium-range missile has a kill zone between 150- to 20,000-ft altitude. The SA-7 (Grail) is a shoulder-held (Manpads) IR heat-seeking missile with a range of 5 to 6 nmi and 5000- to 10,000-ft-altitude coverage. It is not effective against high-speed maneuvering aircraft. The mobile SA-9 (Gaskin) is a tactical low-altitude IR guided missile. It has a maximum range of 16,500 ft.

The radar associated with this mix of SAMs are listed in Table 2.3. The frequency diversity of this radar suite is apparent. In addition to frequency diversity, these radars also provide overlapping coverage, which imposes a difficult problem for any defense suppression system. The Bar Lock is an E/F-band early warning and GCI radar used with the SA-2 missile system. This radar has six stacked beams. Other radars associated with the SA-2 include the B-band Tall King, which is effective against high-altitude targets out to a range of 300 nmi to 350 nmi; the E-band Side Net height-finding radar (normally colocated with Bar Lock and Tall King radar) having a range of 60 nmi to 70 nmi; and the D/E/F Fan Song missile control radar. The Fan Song operates in a track-while-scan mode, scanning designated sectors with two orthogonal fan beams (2 degrees by 10 degrees). This allows the SA-2 to track up to six targets simultaneously while guiding three missiles. The advanced versions of this radar have a range of about 40 nmi.

The Spoon Rest is an A-band early warning radar, with a range of 175 nmi, associated with the SA-3 missile system. This radar evolved from the Knife Edge radar. Other radars associated with the SA-3 missile system are the C-band Squat Eye and Flat Face radars that are used in target acquisition. The Squat Eye is used when low-angle coverage is required and has a range of about 125 nmi. The Flat Face is also used with the SA-6 and SA-8 missile systems. It uses two stacked parabolic reflector antennas and has a peak power of 900 kW providing a range of 70 nmi to 80 nmi. The E-band Rock Cake

Table 2.2
Surface-to-Air Missile Performance

NATO Designation	Guidance	Maximum Speed	Maximum/Minimum Range	Maximum/Minimum Altitude	Lethal Burst Radius
SA-2 Guideline	RC	Mach 3–3.5	35–50/7–9.3 km	28/1.5–4.5 km	
SA-3 Goa	RC	Mach 3.5	18–29/2.4–6 km	12.2–18.3/1.5 km	12.5m*
SA-6 Gainful	RC/SAH	Mach 2.8	30+/4 km	18 km/30m	5m*
SA-7 Grail	IR	Mach 1.5–1.95	3.5–5.6 km/45m	1.5–4.3 km/23–150m	
SA-8 Gecko	RC	Mach 3.0	12/1.6–3 km	13–19 km/45m	5m*
SA-9 Gaskin	IR	Mach 2.0	6.5–8/0.56–0.8 km	5–6.1 km/15–20m	

RC, radio command; IR, infrared; SAH, semiactive homing; *at low altitude against an F-4-sized target.

Table 2.3
Air Defense Radar Systems

NATO Designation	Role	Frequency	Peak Power	Scan	Range	PRF
Bar Lock	EW/GCI	2695–3125 MHz*	650 kW per beam		390 km (E)	375 pps
Fan Song A	T/A	2965–2990 MHz	600 kW		60–120 km (M)	
Fan Song B	T/A	3025–3050 MHz	600 kW		60–120 km (M)	
Fan Song D	T/A	4910–4990 MHz	1.5 MW		70–145 km (M)	
Fan Song F	T/A	5010–5090 MHz	1.5 MW		70–145 km (M)	
Flat Face	EW/A	810–850/880–905 MHz	900 kW		210 km (e)	200–800/600–800 pps
Gun Dish	FC	J-band	100–135 kW		20 km (M)	
Knife Rest A	EW	70–73 MHz	70–75 kW		150–250 km (E)	
Knife Rest B	EW	79–93 MHz	50–100 kW	1–6 rpm	180–280 km (E)	50–100 pps
Land Roll	S/T	S: 4–8 GHz T: 13–15 GHz			S: 30 km (M) T: 20–25 km (M)	
Low Blow	LAFC	9000–9460 MHz	250 kW		85 km (M)	1750–3500 pps
Rock Cake	HF	2 GHz	2 MW		300 km (E)	
Side Net	HF	2.56–2.71 GHz		5–30**	175 km (E)	
Spoon Rest A	EW	147–161 MHz	180–350 kW	2–6 rpm	200–275 km (E)	310–400 pps
Spoon Rest B	EW	147–161 MHz	180–350kW	1–3 rpm	200–275 km (E)	310–400 pps
Squat Eye	LAS	As Flat Face	As Flat Face		As Flat Face	As Flat Face
Straight Flush	A/TI	A: 5–6 GHz TI: 8–10 GHz CW			60–90 km	
Tall King	EW	150–180 MHz		2–4 rpm	500–600 km (E)	
Thin Skin	HF	6 GHz			240 km (E)	

T, tracking; A, acquisition; EW, early warning; FC, fire control; S, search; LAFC, low-altitude fire control; LAS, low-altitude surveillance; GCI, ground controlled interception; TI, target illumination; HF, height finding; CW, continuous wave; M, maximum; E, effective; *, six beams; **, nodding cycles per second.

provides height information for the SA-3. This radar has a large peel-shaped nodding antenna that produces a 1- by 5-degree beam that scans 30 to 40 times a second through the coverage area. The range of this radar is 50 nmi. The I-band Low Blow is the missile control radar for the SA-3. It uses two scanning parabolic reflector antennas, one set above the other at an angle of 45 degrees from the horizontal. It has a range of 20 nmi to 28 nmi.

The Straight Flush is the acquisition and illumination radar for the SA-6 missile system. The antenna has an upper tracking antenna that rotates independently of the lower antenna, with both mounted on a rotating turntable. The lower antenna produces a low angle coverage beam and several high-coverage beams that are generated at different frequencies. Command signals are in J-band and tracking in the G/H band. The range is about 23 nmi to 29 nmi.

The J-band Gun Dish radar is used with the SA-9 and provides a range of 3.5 nmi to 4.5 nmi. It is also used with the ZSU-23-4 quadmounted cannon air defense vehicle, where it rotates with the gun turret.

Newer missiles of the type described in Table 2.1 include the SA-10 (Grumble), SA-11 (Gadfly), SA-12 (Gladiator), SA-13 (Gopher), SA-15 (Gauntlet), and the SA-17 (Grizzly). These missiles represent a considerable advancement over the previous versions.

The SA-10, for example, uses a Mach 6 missile with active radar terminal guidance. It is designed to engage both low-flying cruise missiles and high-flying theater ballistic missiles. It has a range of 54 nmi and engages targets flying between 100 ft and 100,000 ft.

The SA-10 system uses three ground-based radar systems. The E/F-band Big Bird radar provides three-dimensional coverage using a phase-scanned planar array. The Clam Shell horizon search radar provides low-angle coverage. The I/J-band Flap-Lid phased array radar can track up to six targets simultaneously. It uses a space-fed lens phased array antenna with 10,000 elements [16].

2.2 Air Defense Radar

Practically every developed country in the world has some form of air defense system, the extent of which varies from several surveillance radars to the highly complex and sophisticated systems developed by the United States and former Soviet Union [17]. Modern systems generally consist of powerful long-range three-dimensional radars (with associated IFF systems); medium- and short-range two-dimensional radars that detect low-flying aircraft, helicopters, and cruise missiles; and over-the-horizon and AEW radars [6]. Passive electro-optical and *infrared search and track* (IRST) systems are also used in air defense but, since their range is limited to the visual horizon, generally act as very short-range backup on the battlefield and aboard naval vessels against low-flying aircraft and missiles. Air defense systems are often interconnected with the civil air traffic network. This occurs in many countries with the air traffic system either integrated within the military system or the civil system acting as a backup during national emergencies [17].

Modern air defense radars are facing a wider range of threats that includes inherently small targets such as *unmanned airborne vehicles* (UAVs), cruise missiles, and low-observable targets with RCSs as low as 0.01 m^2 [18]. Also, in some applications, they must detect theater ballistic missiles at altitudes as high as 500,000 ft. Radar survivability is an issue that dictates mobility and reduction of the radar signature and/or the addition of decoys to protect it against ARM attack. Naval air defense radars, operating in littoral waters, have the problem of detecting small RCS slow-flying cruise missiles attacking over land from coastal batteries. Many existing naval radars designed for "blue water" operation do not have the necessary subclutter visibility to cope with this threat.

AEW radar take advantage of the increased line-of-sight to low-altitude targets afforded by a medium- to high-altitude platform. This solves the terrain-shielding problem associated with surface-based early warning radar but introduces another significant problem. The AEW radar must now "look down" into the surface (ground or water) at high grazing angles, which causes a radar clutter return of high magnitude that competes with the normal target return. Moving targets can be separated from the clutter using Doppler processing techniques. Operation over water has been feasible for many years, but over-land operation requires subclutter visibilities on the order of 50 dB to 70 dB, which are now practical.

Operation of AEW radar within the integrated air defense system is more complex than for surface early warning radar. In addition to its function of detecting hostile aircraft and missiles, it also provides a control function to guide interceptors to their targets. The control function is partially a self-defense function, but primarily the AEW aircraft provides an excellent platform for communication links to all elements of the air defense system.

The *command, control, and communications* (C^3) system associated with the use of AWACS by NATO is illustrated in Figure 1.3. Note that the AWACS must communicate with all elements of the air defense net. This is accomplished through a variety of communication systems that operate at HF, UHF, and L-band frequencies. The TADIL and JTIDS links, which provide guidance and control information to the interceptors, as well as the AEW radar are prime targets for EA. The ground elements of the air defense net use land lines (LINK-1) to increase their resistance to EA.

Currently, the most widely used air defense radars are the conventional two-dimensional azimuth-scanning surveillance types. Basically, these radars scan a fan beam in elevation through 360 degrees in azimuth. The scanning is accomplished relatively slowly (4- to 12-sec frame time) by mechanically rotating a horn-fed reflector through the surveillance sector. The key design factors generally result in an operating frequency in the UHF, L-band, or S-band regions.

The two-dimensional radar had been the standard type used for surveillance service for many years. In military applications this type of radar is generally being replaced by newer three-dimensional radar types. In some cases, the older two-dimensional types are retained to form a surveillance net of two- and three-dimensional radars. Civilian air traffic control applications still utilize the two-dimensional radar type, except in those situations where a combined air traffic control and air defense system is required.

The move toward three-dimensional radar in military applications is driven by the need to provide height data in high traffic-density situations. The alternative to the three-dimensional radar is a two-dimensional radar with a number of associated height-finder radars. A typical height-finder radar supplies on the order of 20 heights per minute, while some 200 to 400 aircraft might be involved in a dense traffic situation. Reaction time for a military-type system or conflict-type civilian system might be as short as 1 min to 2 min. The resulting number of height-finder radars would be uneconomically large under the conditions described previously. Alternatively, a three-dimensional radar can supply an almost unlimited number of height reports in its 5- to 10-sec frame time.

In civilian-only air traffic control systems, a radar-beacon *secondary surveillance radar* (SSR) is colocated with the primary two-dimensional radar system. The SSR has an altitude reporting mode incorporated into the airborne transponder that supplies the needed target height data. The function of the two-dimensional radar is to provide data on small aircraft, which are not required to carry a transponder, and on any other aircraft whose transponder is not operating.

A primary function of both two- and three-dimensional surveillance radars is the estimation of target ground-track velocity vectors, which is usually accomplished using a *track-while-scan* (TWS) system operating on the radar data. The track data is used in military systems for (1) threat identification, (2) threat evaluation, (3) weapon assignment, (4) predicting target position, and (5) kill evaluation; while in civilian air traffic control systems it is used for (1) traffic control, (2) conflict alert, and (3) approach control. The requirements for military systems are more stringent than for civilian systems due to the higher accelerations (10 to 50 m/s^2) found in military systems. Accelerations are lower in civilian systems (2 to 3 m/s^2) because of path regularity and pilot collaboration associated with civilian air traffic control operation.

Modern three-dimensional radar used in military air defense networks must provide long-range detection to 300 nmi, on small RCS targets (0.1 m^2), with altitude coverage to 100 kft, providing a height accuracy of ±2500 ft in a frame time of 10 sec to 12 sec. This requires high-power operation with transmitter outputs of 4-MW peak/10-kW average [19].

Three-dimensional surveillance radar measures the target's elevation angle in addition to its azimuth angle and slant range. The elevation angle and slant range are used to compute the target's height. As previously discussed, three-dimensional radars are rapidly replacing two-dimensional radar in military applications to satisfy the requirement of providing height data in a high traffic-density environment.

Three-dimensional coverage is provided by mechanically scanning an antenna through 360-degree azimuth rotation while covering the elevation angle with multiple beams or a scanning pencil beam. Most modern designs use flat planar-array antennas, which provide the low sidelobes (−5 dBi to −20 dBi) necessary for protection against ECM and ARMs. Frame times of 10 sec to 12 sec are typical to allow for processing of the multiple elevation beams in clutter. *Moving target indication* (MTI) processing of those beams intercepting ground or weather clutter is generally used rather than *moving target detector* (MTD) processing, which requires more pulses per azimuth beam position (16 *versus* two or three), but this MTI is of limited usefulness, a major limitation of three-dimensional radar.

The second critical issue with three-dimensional radar is the technique employed to provide the elevation-angle measurement. There are two fundamental types with variations on each type. One fundamental type uses a single pencil beam that is scanned electronically up and down through the elevation coverage angle. Variations on this type involve scanning the pencil beam using either phase or frequency beam steering. The second type uses a stack of multiple beams throughout the elevation coverage. Variations of this type involve forming the multiple beams at RF, IF, or digitally. A fundamental difference between the basic types is that the beams are formed sequentially in the scanning beam type and simultaneously in the stacked beam type. A variation of the scanning beam type is to scan multiple beams through the elevation coverage angle (e.g., Series 320). Another variation is to scan the whole elevation coverage angle within a radar pulse width.

The advantages and disadvantages of each type of three-dimensional radar are itemized in Table 2.4. The fundamental types and variations have become associated with particular manufacturers, whose products have been selected for various requirements, resulting in a wide variety of types in current operation.

Table 2.4
Three-Dimensional Radars—Performance Characteristics

Types	Examples	Advantages	Disadvantages
Stacked-Beam	TPS-43E TPS-70, Martello ARSR-4	High-target data rate Compatible MTD processing High-elevation accuracy	Poor elevation sidelobes Partial aperture use Clutter in higher beams Expensive receivers
Phase Scan	TPS-59 FPS-117 TRS-2215	Full aperture use Good elevation sidelobes No MTI in upper beams Solid-state transmitter Simple receiver	Poor MTI performance Complex transmitter waveforms Data rate limited Complex beam management
Frequency Scan	TPS-32D SPS-48 Series 320	Full aperture use Good elevation sidelobes No MTI in upper beams Simplified antenna	Poor MTI performance Complex transmitter waveforms Data rate limited Complex beam management Susceptible to spot jamming

In the following, each of the types are described with their associated advantages and disadvantages.

The stacked beam three-dimensional radar generally radiates a $\csc^2$ fan-shaped beam on transmission that floodlights the elevation coverage angle (20 degrees to 30 degrees). The three-dimensional resolution is provided on reception by a vertical stack of pencil beams (typically six to twelve beams) that are formed from a single aperture. Antenna designs for stacked beams have evolved from the double-curved reflector fed by an array of horns (e.g., AN/TPS-43E) toward flat planar arrays where the multiple beams are formed at intermediate frequencies (e.g., Martello). The future trend is toward the use of digital beam forming using in-phase and quadrature channels whereby beam shape and sidelobe jamming can be controlled using adaptive array techniques. The elevation measurement is generally made using amplitude comparison between adjacent beams in the stack.

There are several advantages to the stacked beam approach. A high target data rate can be achieved because all the elevation angles are covered simultaneously in the time required for the azimuth beam to sweep over the target. The technique is compatible with MTD processing, which provides nearly optimal performance in clutter. MTD or MTI processing can be optimized in each beam against the expected clutter environment (e.g., ground clutter in lower beams and weather clutter in higher beams). Elevation accuracy is not degraded by target scintillations because the measurement is made with simultaneous beams rather than sequential beams as in the scanning beam case.

On the debit side, stacked beam systems suffer a number of disadvantages. Several stem from the fact that it is difficult to achieve ideal beam patterns with the complex antenna required to generate multiple beams from a single aperture. The stacked beam antenna tends to have poor sidelobes, which is a serious disadvantage insofar as ground clutter is illuminated by the peak of the transmit beam and thus, targets at higher elevation angles are only decoupled from the clutter by the relatively high one-way receiving elevation beam sidelobes. Also, good height accuracy requires matched receivers with accurate control of beam shape and stability for off-axis target measurement, both of which are difficult to achieve. Further disadvantages are the cost of the multiple matched receivers and the loss of efficiency in the power-aperture product due to the use of a shaped floodlight beam on transmit rather than a pencil beam.

In the scanning beam approach, the elevation beam is rapidly scanned or stepped through the entire coverage in the time that the antenna rotates through the azimuth beamwidth. Antenna elevation beam control can be accomplished by either phase (e.g., AN/TPS-59) or frequency (e.g., AN/TPS-32) scanning. The basic height measurement in this type of system is made by estimating the elevation angle at which the peak of the scanning elevation beam illuminated the target. A characteristic of scanning beam three-dimensional radar is that the dwell time on target must be of the order of two or three PRIs in order to restrict the beam shape loss and provide reasonable MTI operation. This generally limits the overall data rate that can be provided by the radar due to the large number of individual beam positions within the overall azimuth and elevation angular coverage. Also, the basic elevation measurement accuracy is somewhat limited by the small number of data samples per measurement and the effect of target scintillation during the scan, which is particularly troublesome in frequency scanned radar.

There are several advantages of the scanning beam three-dimensional radar technique. The full antenna aperture is used on both transmission and reception and thus yields the maximum single-hit signal-to-noise ratio for a given power aperture product. In addition, ground clutter in the upper elevation beams is decoupled by the two-way antenna sidelobe pattern (50 dB to 60 dB), which considerably simplifies the MTI requirement, excepting the lowest elevation beam position. (However, MTI is not generally used in the upper beams because of time limitations.) Also, the receiver requirements are considerably simplified with respect to the stacked beam approach, resulting in a reduced cost and complexity. Furthermore, the elevation angle measurement is made on boresight, which leads to higher accuracy.

A perceived disadvantage of the frequency-scanning approach is that an enemy can precisely determine the radar's transmission frequency from a knowledge of their relative locations. This leads to a spot-jamming ECM situation, which favors the jammer over the radar. This, in turn, has led to a preference for the phase-scanning approach (which is compatible with frequency agility) over the frequency-scanning approach in several of the more modern designs.

Another disadvantage of the scanning beam approach is due to the sequential nature of the elevation beam measurement, which is susceptible to target scintillation errors. This may be aggravated in the frequency-scanning approach because frequency changes are known to induce target scintillation.

A typical frequency scan three-dimensional radar, such as the TPS-32D, uses a flat planar-array antenna that is driven by a serpentine microwave-feed assembly. The serpentine feed has the characteristic of providing a differential phase shift between rows of the array, which is a function of the applied frequency. Thus, at each particular applied carrier frequency, a unique elevation beam position is obtained. Elevation beam steering is accomplished by changing frequency. Also, a group of elevation beam positions can be generated by driving the antenna with a group of pulses each at a different frequency (e.g., Series 320, AN/SPS-48).

An example of the complex beam energy management employed in a scanning beam three-dimensional radar is illustrated in Figure 2.2 for the E/F-band TPS-32D frequency scanned radar. This radar uses long- and short-range waveforms in normal operation to search seven elevation beam positions within a 50-ms dwell time. The lower beam, pointed just above the radar horizon, provides an MTI waveform with 60-kW peak power that searches the clutter horizon (order of 10–15 nmi) and a long-range waveform with 2.2-MW peak power that detects targets beyond the clutter horizon to 300 nmi. The next two beam positions (positions 2 and 3) use the long-range 2.2 MW peak power waveform to detect targets up to 100,000-ft in altitude. These beam positions do not have to contend with ground clutter that is over the horizon at 160-nmi range. The peak power is reduced to 665 kW in beam positions four and five since the maximum range of targets at 100,000-ft altitude is 85 nmi to 110 nmi. At the maximum elevation beam positions (positions 6 and 7), 60-kW peak power is transmitted, enabling the detection of targets at 80 nmi to 85 nmi. The maximum elevation scan angle of the TPS-32 is 18 degrees, the scan time is 10 seconds, the azimuth beamwidth is 2.15 degrees, the elevation beamwidth is 0.80 degrees, and the pulsewidth is 30 μs, while the PRF varies from 265 pps to 917 pps [6].

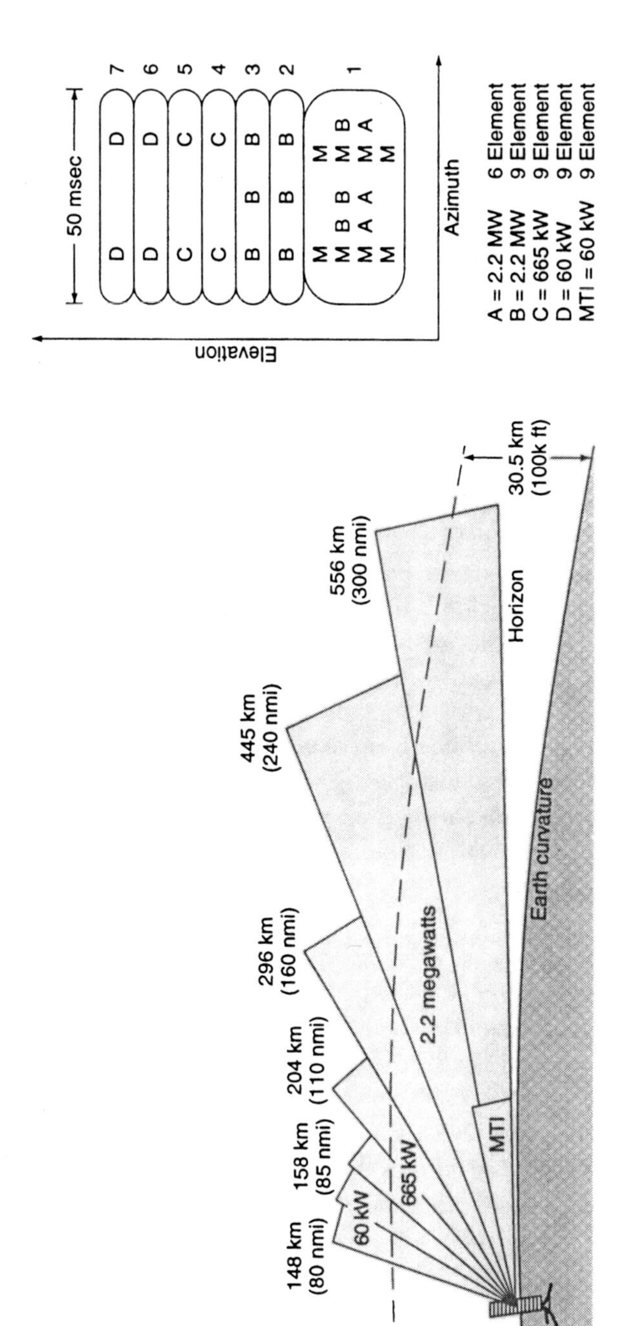

Figure 2.2 Beam management for three-dimensional radar.

2.2.1 EP for Air Defense Radar

The major EW threats to an air defense radar involve: (1) noise jamming, (2) deception jamming, (3) chaff, (4) decoys and expendables, and (5) ARMs. These major threats against radar represent three possible actions on the part of the enemy: (1) radiation of energy to confuse the radar, (2) injection of spurious targets into the radar's surveillance volume, and (3) destruction of the radar. The first two actions are sometimes referred to as "soft kills," while the last is called a "hard kill."

The most common type of jamming is noise jamming. For this discussion, we will assume that noise jamming refers to band-limited noise whose spectral density is uniform over the band of interest. When the band of interest is much wider than the radar's acceptance bandwidth, the noise is referred to as "white noise."

In the air defense situation, main-lobe noise jamming can originate from the target aircraft as a self-protection action or from a support or escort jammer that is strategically positioned with respect to the radar and the target. Alternatively, a stand-off jammer can be employed that attempts to screen targets of interest through the radar's sidelobes. Also, sidelobe stand-in jamming is under current study using UAVs and other platforms.

The noise jammer situation is basically an energy battle between the radar and the jammer. In the main-lobe noise-jamming situation, the advantage is with the jammer because the radar experiences a two-way propagation loss of its energy as contrasted with the one-way propagation loss between the radar and the jammer. With sidelobe jamming, the radar designer can equalize the jammer's advantage by low-sidelobe design coupled with the use of sidelobe-cancellation techniques.

With main-lobe noise jamming, the radar design principles are clear. First, the radar must maximize the energy received from the target with respect to that received from the jammer. It can maximize the received target energy by transmitting more average power, dwelling longer on the target, or increasing antenna gain. If the radar's data rate is fixed and a uniform angular search rate is dictated by mechanical or search strategy, then the only option for the radar is to increase its average transmitter power. The next option is to reduce the data-rate requirement, thereby allowing a longer dwell time on the target. This is sometimes referred to as a "burn-through mode" and reflects the philosophy that it is desirable to have some target data in a heavy EW environment rather than no data at all. The ability to vary the data rate in an optimal manner is one of the principal advantages of phased array radar. Adaptive scan rate and "look back" scanning are also available in a limited number of advanced surveillance radar.

The second principle of EP design for surveillance radar in a mainbeam noise-jamming environment is to minimize the amount of jamming energy accepted by the radar. This is accomplished by spreading the transmitted frequency range of the radar over as wide a band available while maintaining a radar bandwidth consistent with the radar range resolution requirement. If, for example, a 150- to 300-MHz transmitting frequency range is available at E/F-band for a 1-MHz radar resolution bandwidth, then the potential for a 150 to 300 dilution of jamming energy is possible. A jammer that works against the 1-MHz radar bandwidth is referred to as a "spot" jammer, while a jammer that works over the full 150- to 300-MHz radar RF bandwidth is called a "barrage" jammer. The EP objective is to force the jammer into a barrage-jamming mode of operation.

The operation of radar over a wider bandwidth than that dictated by range resolution requirements can be accomplished in several ways. Some radar incorporate a spectrum analyzer that provides an advance look at the interference environment, hence allowing the radar frequency to be tuned to that part of the environment that contains minimum jamming energy. This can be defeated, however, if the noise jammer has a "look-through" mode and can follow the radar frequency changes.

Frequency agility refers to the radar's ability to change frequency rapidly after a time period that corresponds to the radar's Doppler processing time. For an MTI radar, this period may be as short as every two transmitted pulses and typically every three or four transmitted pulses. For pulsed Doppler and MTD radar, a "block" processing interval may typically be eight to sixteen transmitted pulses. Frequency agility usually forces a noise jammer into a barrage-jamming mode.

Frequency diversity refers to the use of several complementary radar transmissions at different frequencies, either from a single or several radars. The diversity is usually limited by practical considerations to a finite number of frequencies (five to ten). Examples of this mode of operation are a two-dimensional surveillance radar coupled to a height-finder radar at a different frequency or a number of spatially dispersed radars in a netted configuration at different frequencies.

Another method that is employed to reduce the effect of main-beam noise jamming raises the transmitter frequency in order to narrow the antenna's beamwidth. This restricts the sector that is blanked by mainlobe noise jamming and provides a strobe in the direction of the jammer. Strobes from two spatially separated radars pinpoint the jammer's location. However, with multiple jammers ghosting can be a problem.

Increasing the radar frequency does not affect the signal-to-interference energy ratio within a radar resolution cell when the antenna aperture and radar data rate are held constant. The increased frequency, in turn, increases both the antenna gain and the number of radar resolution cells that must be searched by equivalent amounts. The net effect is that the target return power is increased by the same amount that the target dwell time is decreased, thereby holding the target-to-jamming energy constant.

The EP design principles for mainlobe noise jamming also apply to sidelobe noise jamming, with the exception that the sidelobe response in the direction of the jammer must be minimized. Ultralow sidelobes, which are of the order of 50 dB below the antenna's mainlobe peak response, are feasible using currently available advanced technology. These ultralow sidelobe antennas have an average sidelobe level of the order of –20 dBi (with respect to an isotropic radiator), which is another way of saying that only 1% of the radiated power is in the sidelobes. Key to this antenna technology, which was operationally introduced in the AWACS radar, are tight antenna tolerance and aperture illumination function control; control of systematic, random, and spill-over errors in the antenna's design; and a physical antenna length encompassing many radar wavelengths. The last criterion can be directly related to frequency in situations where random errors dominate. The random errors cause a fixed average sidelobe level, which is independent of frequency, while the antenna's gain or directivity is proportional to the frequency squared. Thus, increasing frequency improves the antenna's sidelobes relative to the main lobe's peak response in a manner that is proportional to the frequency squared. Definitions of antenna sidelobe levels are given in Table 2.5.

The control of sidelobe noise jamming using ultralow sidelobe antennas is not usually sufficient to completely protect an air defense radar against sidelobe jamming. In addition, most operational radars do not use ultralow (greater than –40 dB) or low (–30 to –40 dB) sidelobe antennas and have antenna sidelobes in the –13- to –30-dB region with average sidelobes of 0 dBi to –5 dBi. *Sidelobe cancellation* (SLC) is a coherent processing technique that has the potential of reducing noise jamming through the antenna sidelobes and is employed in a number of operational radars for this purpose. Present

Table 2.5
Definition of Antenna Sidelobe Levels

Performance	Relative Level	Average Level
Ordinary	–13 dB to –30 dB	0 dBi to –5 dBi
Low	–30 dB to –40 dB	–5 dBi to –20 dBi
Ultralow	Below –40 dB	Below –20 dBi

SLCs have the capability of reducing sidelobe noise jamming by 20 dB to 30 dB, but their theoretical performance is potentially much higher.

The operation of a sidelobe canceller is best understood by relating its operation to that of an adaptive antenna array processor. In an adaptive array, a weighting factor (phase and amplitude) is applied to each of the antenna elements, which are then combined in a summing network. By controlling the weighting factors through an adaptive-loop antenna pattern, nulls can be generated in any direction due to the interferometric action between any two sets of elements. Thus, the number of nulls that can be formed is equal to the number of elements minus one, which is sometimes referred to as the number of degrees of freedom of the array. The same adaptive array action can be applied to groups of elements combined into subarrays, which reduces the processing requirements for a phased array to a reasonable size. Thus, an adaptive antenna array has the potential of placing antenna pattern nulls in the direction of sidelobe jammers while still maintaining the mainlobe pattern, thereby reducing the effects of the jammer at the output port of the antenna.

While adaptive arrays are applicable to phased arrays, they are not appropriate for conventional single-element antennas. However, by adding auxiliary antennas to a conventional radar, an adaptive array type of action can be formed between the main antenna and the added antennas. This configuration is an SLC, and the only requirement is that the auxiliary antennas must have a greater response in the direction of the jammer than the sidelobe response of the main antenna in that direction.

One of the limitations of the SLC is that the number of degrees of freedom is usually low, since only a small number of auxiliary antennas can be practically added to the main antenna (e.g., Patriot, five SLC arrays; AEGIS, six SLC arrays). Because the maximum number of sidelobe jammers that can be handled is equal to the number of auxiliary antennas, the cancellation system is easily saturated. This problem is compounded by any jammer multipath from objects in the proximity of the radar that add an additional degree of freedom for each multipath signal that has an angular direction significantly different than that of the main jammer. Another complication occurs for antennas whose cross-polarization response is significantly different than its main polarization response. This causes two orthogonally polarized auxiliary antennas to be added for each degree of freedom of the SLC system.

Sidelobe cancellation loops are best suited to cancel narrowband jamming signals. When wideband jamming signals impinge on the antenna array (main and auxiliary antennas) at angles other than normal to the array, the signal appears to span an arc in angle. Thus, a single null is not adequate to cancel a wideband jammer and closely spaced multiple nulls are required in this case.

Typically then, the number of sidelobe cancellation loops might be specified at up to three times the number of expected jamming sources. The implementation of more than three or four cancellation loops presents a difficult operational design problem.

The other major type of active EA is deceptive jamming (DECM), which attempts to introduce false targetlike information into the radar. Deceptive jamming usually has an objective of overloading the radar data-processing subsystem, which, in turn, prevents the radar from developing accurate target tracks. A different form of deceptive jamming called range bin masking attempts to cover the target's skin return with a wide pulse in order to confuse the radar's signal-processing circuitry into suppressing the actual target return [6]. The advantage of deceptive jamming is a concentration of the available jamming energy into the radar's acceptance bandwidth, a technique that becomes significant in an airborne EA situation when many radar must be jammed simultaneously and jamming energy is at a premium.

There are basically two types of radar deception jammers: transponders and repeaters. Transponders generate noncoherent returns that emulate the temporal characteristics of the actual radar return. Repeaters generate coherent returns that attempt to emulate the amplitude, frequency, and temporal characteristics of the actual radar return. Repeaters usually require some form of RF memory to allow anticipatory returns to be generated.

Deception jammers have a number of specific characteristics that can be used by radar to identify their presence. The most prominent is that the false target returns must usually follow the return from the jammer-carrying target and, in any event, must all lie in the same direction within a radar PRI. If the deception jammer uses a delay that is greater than a PRI period to generate an anticipatory false target return, then pulse-to-pulse frequency agility eliminates the false target and pulse-to-pulse PRI jitter identifies the false target returns.

Generation of false targets in a different direction other than that of the jammer-carrying aircraft requires injecting jamming signals into the radar's sidelobes. Many radars employ sidelobe blankers to defeat this type of EA. A sidelobe blanker compares the signal detected in its main channel against a signal detected in an auxiliary antenna channel. The response of the auxiliary antenna channel is arranged to be greater than any main antenna sidelobe response in that direction but less than the antenna's mainlobe response. A signal is blanked if its auxiliary antenna response is greater than the main antenna response, which signifies that a sidelobe jammer is present. The orthogonal polarization response of the auxiliary antenna must be consistent with that of the main antenna's cross-polarization response, or two orthogonal auxiliary antennas must be employed to allow sidelobe blanking to work against cross-polarization jammers.

An effective sidelobe blanker confines deception returns to the same azimuth direction as the jammer-carrying target. If TWS target tracking is employed to determine target trajectories, then all tracked targets with only radial trajectories are suspect. Also, if two or more radars are employed, then only true targets will have the same location on all radars while false targets will appear at different positions.

True target returns tend to fluctuate from scan to scan with fixed-frequency radar and pulse to pulse with frequency-agility radar. Transponder jammers generally send the same amplitude reply to all signals they receive above a threshold and hence do not simulate actual target fluctuation responses; in addition, they usually appear wider in azimuth than real targets due to the two-way pattern effect of the scanning antenna's response on real target echoes. Repeater jammers can be made to simulate the actual amplitude response of real targets and, hence, are preferred over transponder-type jammers.

Another countermeasure that frequently confronts radar consists of clouds of electrically conducting dipoles, called chaff, that are injected into the radar's coverage volume. The chaff dipoles are cut to approximately a half-wavelength of the hostile radar frequency. Chaff is usually packaged in cartridges that contain a broad range of dipole lengths designed to be effective over a wide frequency band. For use against low-frequency radar at VHF and UHF, the chaff (called "rope") must be on the order of hundreds of feet in length. Chaff is often dispensed for aircraft self-protection, and its use is coordinated with the information obtained from an onboard radar warning receiver.

The primary advantage of chaff is its simplicity and the fact that very little knowledge about the victim radar is required for its effective use. When chaff is dispensed from an aircraft, the drag on an individual dipole is so great compared to its mass that it rapidly comes to rest in the air. It is further spread (blooming) by air turbulence and falls at a slow rate of 1/3 to 2 m/s, thereby causing it to hang in the air for a long time. Chaff is often launched by a rocket in the direction of a hostile radar to allow a penetrating aircraft to fly through the chaff cloud and escape detection.

From a radar viewpoint, the properties of chaff are very similar to those of weather clutter, except that its broadband frequency capability extends down to VHF. The rapid deceleration of dispensed chaff causes its Doppler signature to be different from that of an actual target. Its steady-state mean Doppler frequency (0 to 30 m/s) is determined by the mean wind velocity, while its spectrum spread (on order of 1 m/s for thin layers of chaff) is determined by wind turbulence and a shearing effect, which is due to different wind velocities as a function of altitude.

Doppler processing in the form of MTI or pulsed Doppler processing is used by a radar to extract targets from chaff. With MTI signal processing, a notch or null response is required at a different frequency than that required for ground or sea clutter. Because the notch must be adaptive to allow for different chaff mean velocities, a difficult MTI design problem is generated. With pulsed Doppler processing, a number of velocity or Doppler cells are spread throughout the expected range of Doppler frequencies in each range cell. By employing a CFAR processor that examines range cells surrounding the test cell, it is possible to blank those cells containing chaff returns while passing targets contained in other Doppler cells. A variation of MTI and pulsed Doppler processing called an MTD processor is also effective in processing chaff returns. Pulse compression is effective in reducing the size of the radar resolution cell. This, in turn, reduces the effective RCS of the chaff that competes with the target return in any range cell. Also, it should be noted that the use of chaff, in forcing MTI or Doppler processing, can deny the radar the ability to use pulse-to-pulse frequency agility and that extension of chaff beyond the unambiguous range can deny the use of PRF stagger, both factors opening the way for more effective deception jamming.

Decoys refer to small radar targets whose RCSs are generally enhanced using corner reflectors or Luneburg lenses to simulate fighter or bomber aircraft. The objective of decoys is to cause a dilution of the assets of the defensive system, thereby increasing the survivability of the penetrating aircraft.

Both decoys and chaff are subsets of a larger class of EA devices known as expendables. The term expendable connotes that the EA device is used up in its employment. Expendables include both active and passive devices. Their primary objective is to saturate the data-processing subsystem of the radar net. In some cases, they are also used in a self-protection mode to divert enemy weapons away from their intended targets.

A problem with expendables is that their period of operation must coincide with the activity that they support. This implies careful timing of their use because, once expended, they can no longer provide the function for which they were designed. The practical implication is that expendables must always be used in conjunction with some other form of EA.

From a radar viewpoint, expendables that have the general attributes of real targets are very difficult to identify as false targets. A simple method sometimes employed is to test the scintillation characteristics of detected targets to determine that they follow those of real targets. Expendables that tend to be designed under stringent economic constraints often return only a steady signal to the radar.

A more sophisticated method is to look for returns from rotating components of the target that any form of a powered target must possess. Examples of this are jet engine or propeller modulation returns associated with aircraft targets. Other methods include high-resolution techniques to classify targets and correlation with other sensors such as ES systems.

ARMs pose a serious threat to most surveillance radars. A general radar strategy employed when confronted with an ARM attack is to stop radiating, thereby causing the ARM to lose its guidance information, and then to turn on a spatially remote decoy transmitter to lure the ARM away from the radar site. The timing of this strategy depends upon the detection of the ARM as it approaches the radar site. The ARM trajectory is usually selected to attack the radar through the zenith hole region above the radar where its detection capability is minimal. Thus, a successful implementation of this strategy requires a supplemental alerting radar that provides a high probability of detection in the zenith hole region.

Also, the possibility of a multisensor ARM must be considered that contains an IR sensor in addition to its RF sensor. When confronted with this threat, it is important to have a cool antenna that contains no components (active or otherwise) that generate heat. This possibility is often cited as an argument against active aperture phased array radar that generate considerable antenna heat.

Another approach often suggested for ARM protection is the use of multistatic or bistatic radar where the transmitters and receivers are spatially separated. The transmitters in this approach are still vulnerable to ARM attack, but the receivers do not provide any homing information for the missile. The many problems involving synchronization, range determination, blind zones, and Doppler processing difficulties have precluded general operational deployment of this type of radar.

There are certain advantages in choosing a low transmitting-frequency (UHF or VHF) for the supplemental alerting radar. First, the RCS of the ARM becomes greater as the dimensions of the missile approach the wavelength of the radar, causing a resonance effect. Also, it is easier to build a radar with good MTI capability at the lower frequencies. A low-frequency radar is somewhat less vulnerable to an ARM attack due to the difficulty of implementing a low-frequency antenna with the limited aperture available in the missile.

As a general principle, surveillance radar should be designed with low average sidelobes to provide maximum resistance to ARM attacks. This follows since most ARMs are designed to attack through the radar's sidelobes where the signal is continuously available. Also, anti-ARM procedures using decoy transmitters and blinking within a radar net are more effective when low average sidelobe radar are employed.

► *Example 2.1*

The characteristics of the E/F-band TPS-70 stacked beam three-dimensional radar are given in this example. This radar uses an ultralow sidelobe antenna with azimuth sidelobes specified as –48 dB referred to the mainbeam antenna gain. It also employs selectable frequency agility with a 200-MHz total spread. The TPS-70 is an upgrade of the TPS-43E, which provides ordinary sidelobes (−1 dBi), but the rest of its parameters are similar to the TPS-70.

Using RGJMAT (see Appendix A) determine the detection range of the TPS-70 on a 1-m^2 target in the clear and in the presence of five stand-off jammers positioned at a range of 50 nmi (20,000-ft altitude) with jammer parameters listed as follow.

Radar Parameters	Jammer Parameters
P_t = 3.0 MW	P_j = 2000W
F = 3 GHz	G_j = 5 dB
PRF = 250 pps	B_j = 200 MHz
G_t = 36 dB	R_j = 50 n mi
G_r = 40 dB	L_j = 7 dB
θ_{AZ} = 1.5°	H_j = 20,000 ft
$\mathring{\theta}_{AZ}$ = 36°/s	Number = 5
NF = 4.5 dB	
PW = 6.5 μs	
L = 12 dB	
ϕ_t = 0.4°	
RCS = 1 m^2	
Length = 10 m	
FA = 200 MHz	
G_{SL} = −8 dB	

Next, find the detection probabilities of the TPS-43E for the same situation. Add three sidelobe cancellation loops that attenuate the jammer by 25 dB and recalculate the detection probabilities.

```
% Range Calculations in a Jamming Environment
% -------------------------------------------
% rjx.m
%
% This program calculates radar ranges in a jamming environment.
% It works with both Stand-off jamming and self-screening jamming
% for steady and Swerling type targets with frequency agility,
% coherent integration and standard atmosphere/rain attenuation
```

```
%
% Code translated from BASIC "RGJMAT.BAS"
%
% Last edited: 28 August 1997.

%%%%%%%%%%%%%%%%%%%%%%%%%%%%%%%%%%%%%%%%%%%%%%%%%%%%%%%%%%%%%%
%%%%%%%%%%%%%%%%%%%%%%%%%
%
clear
clc
echo off
%
% Set user variables.  The user should change the following
% variables as needed for the simulation they are running.
% Descriptions of the variables follow.
%
% Radar related:
%
trans_pwr_radar = 4000;  % Transmitter power of radar - kw
freq_radar = 3000;       % Radar's frequency - MHz
pulse_width = 6.5;       % Radar's pulse width - microsecs
GTDB_radar = 36;         % Transmitter gain of radar - dB
GRDB_radar = 40;         % Receiving gain for radar - dB
GRDB_sl_radar =-5;       % Side-lobe of radar - dBi
noise_fig_radar = 4.5;   % Noise figure of receiver - dB
loss_radar_dB = 12;      % Radar losses - dB
PRF = 250;               % Pulse repetition freq - pps
prob_det = 0.90;         % Probability of detection  < 1
prob_false_alarm = 1e-6; % Probability of false alarm < 1
BW_dop = 250;            % Doppler filter BW - Hz
BW_fa = 200;             % Frequency agility BW  - MHz
Azimuth_bw = 1.5;        % Azimuth beamwidth - degrees
az_rate = 36;            % Azimuth scan rate - degrees/sec

% Target related:
%
RCS=1;                   % Radar cross section - m^2
el_tgt_deg = 0.4;        % Elevation of target - degrees
target_length = 10;      % Length of target - m

% Select jamming type (enter 1 for yes, 0 for no; can select
% none, one, or both):
%
SSJ = 0;                 % Selfscreening Jamming? - boolean
SOJ = 1;                 % Stand-off Jamming? - boolean

% Enter characteristics for selfscreening Jamming.  If not
% selected, values won't be used.
%
SSJ_pwr_jam = 10;        % SSJ power of jammer - w
SSJ_gain_dB = 0;         % Gain of SS jammer in dB - dB
```

```
SSJ_bw = 20;              % Bandwidth of SSJ - MHz
SSJ_loss_dB = 7;          % Losses with SSJ - dB

% Enter characteristics for stand-off Jamming.  If not, selected
% values won't be used.
%
% Each array must be equal and there must
% be a value for each stand-off jammer you
% wish to simulate
%
SOJ_pwr_jam = [2000 2000 2000 2000 2000];       % SOJ power - W
SOJ_gain_dB = [5 5 5 5 5];                      % SOJ Gain - dB
SOJ_bw = [200 200 200 200 200];                 % SOJ Bandwidth - MHz
jam_range = [50 50 50 50 50];                   % Range of jammer - Nmi
jam_height = [20000 20000 20000 20000 20000]; % Height of jammer - ft
SOJ_loss_dB = [7 7 7 7 7];                      % SOJ Losses  - dB

% Weather
% rain_fall = 0;          % Rain fall rate - mm/hr

% Print radar, target , jamming and weather parameters

fprintf('Radar Parameters \n\n');
par1=[trans_pwr_radar;freq_radar;pulse_width;PRF;GTDB_radar;
RDB_radar;noise_fig_radar; ...
     GRDB_sl_radar];
fprintf(' Pt(kw)          f(Mhz)      PW(1e-6*s)   PRF(pps)    Gt(dB)
Gr(dB)      NF(dB)');
fprintf('         Gsl(dB)   \n ');
fprintf('%-14.1f',par1);fprintf('\n\n');
par2=[loss_radar_dB;BW_dop;BW_fa;Azimuth_bw;az_rate];par3=
[prob_det;prob_false_alarm];
fprintf('Lr(dB)        Dop  Bw(Hz)    FA  Bw(MHz)    Az  Bw(deg)    Az
Rate(deg/s)    Pd              Pfa\n');
fprintf('%-17.2f',par2);fprintf('%-14.1g',par3);fprintf('\n\n');
fprintf('Target and Weather Parameters\n\n');
par4=[RCS;el_tgt_deg;target_length;rain_fall];
fprintf(' RCS(m^2)  tar el(deg)  tar lgth(m)  rain(mm/hr)\n');
fprintf('%-18.1f',par4); fprintf('\n\n');
if SSJ==1;fprintf('Self Screening Jamming Parameters\n\n');
   par5=[SSJ_pwr_jam;SSJ_gain_dB;SSJ_bw;SSJ_loss_dB];
   fprintf(' Pj(w)       Gj(dB)     Bwj(MHz)    Lj(dB)\n');
   fprintf('%-14.1f',par5);fprintf('\n\n');
end;
if SOJ==1;fprintf('Stand-Off Jamming
Parameters\n\n');num=length(SOJ_pwr_jam);
   fprintf('Number of Jammers = %4.0f\n',num);
   fprintf('Pj(w) =          ');fprintf('%-
10.0f',SOJ_pwr_jam);fprintf('\n');
   fprintf('Gj(dB) =       ');fprintf('%-
12.1f',SOJ_gain_dB);fprintf('\n');
```

```
   fprintf('BWj(Mhz) = ');fprintf('%-
11.0f',SOJ_bw);fprintf('\n');
   fprintf('Rj(n mi) =     ');fprintf('%-
11.1f',jam_range);fprintf('\n');
   fprintf('Hj(ft) =       ');fprintf('%-
9.0f',jam_height);fprintf('\n');
   fprintf('Lj(dB) =       ');fprintf('%-
12.1f',SOJ_loss_dB);fprintf('\n\n');
end;

%%%%%%%%%%%%%%%%%%%%%%%%%%%%%%%%%%%%%%%%%%%%%%%%%%%%%%%%%%%%%%%%
%%%%%%%%%%%%%%%%%%%%%%
% STOP EDITING.  ONLY PROGRAM FOLLOWS
%%%%%%%%%%%%%%%%%%%%%%%%%%%%%%%%%%%%%%%%%%%%%%%%%%%%%%%%%%%%%%

%%%%%%%%%%%%%%%%%%%%%%%%%%%
%
% Begin Main
%

% Convert radar gain, loss, noise to ratios
%
loss_radar = db_ratio(loss_radar_dB);
GT_radar = db_ratio(GTDB_radar);
GR_radar = db_ratio(GRDB_radar);
noise_factor_radar = db_ratio(noise_fig_radar);

% Calculate system input noise temp
%
sys_noise_temp = 290 * (noise_factor_radar - 1) + 150;  % in K

% Calculate range without jamming, FA, etc (S/N = 1)
%
range_index = 129.2 * ((trans_pwr_radar * pulse_width * GT_radar
* GR_radar * RCS) / ...
(freq_radar^2 * sys_noise_temp * loss_radar))^.25;

% Calculate number of pulses
%

non_coh_pulses = (Azimuth_bw/az_rate) * BW_dop;
coh_pulses = PRF/BW_dop;

% If target length is greater than zero,computes number of
% independent pulses for FA radars
%
if target_length > 0
  N_exp = 1 + (BW_fa/(150 / target_length));
  else N_exp = non_coh_pulses;
end

% Determine detectability factors
```

```
%
det_fact;

% Calculate ranges for the different target types, clear, no
% atmospheric or weather attenuation accounted for here.
%

swrlg_case= ['      0       1         2         3         4
1fa        3fa'];

ranges_clear = range_index * 10.^((-det_facts)/40);

% echo on
% Now, the atmospheric attenuation and rain attenuation is
% calcuated.  First, the Swerling cases being calculated are
% listed. Then under ar_atten_ans, row 1 is the atmospheric
% attenuation (dB)and row 2 is the rain attenuation (dB).  The
% 3rd row contains the  affected ranges (NMI).Underafter_jam_ans,
% the meaning of the rows are the same as for the attenuation
% matrix.
% echo off
ar_atten;

% Effects of Jamming...
%
if SSJ==1
  pwrj_density = SSJ_pwr_jam/SSJ_bw;
  SSJ_gain = db_ratio(SSJ_gain_dB);
  SSJ_loss = db_ratio(SSJ_loss_dB);

% Range with just self-screening jamming
%
rng_ssj_only = 0.0048116 * sqrt(trans_pwr_radar * GT_radar * RCS
* pulse_width ./ ...
(pwrj_density .* SSJ_gain .* (loss_radar/SSJ_loss)));
  if SOJ==1
     soj_eff;
  end
  jam_eff;
end
if (SOJ==1)&(SSJ==0)
  soj_eff;
  rng_ssj_only = rs00;
  jam_eff;
end
```

» rjx

Radar Parameters

Pt(kw)	f(MHz)	PW(1e-6*s)	PRF(pps)	Gt(dB)	Gr(dB)	NF(dB)	Gsl(dB)
3000.0	3000.0	6.5	250.0	36.0	40.0	4.5	−8.0

Lr(dB)	Dop Bw(Hz)	FA Bw(MHz)	Az Bw(deg)	Az Rate(deg/s)	Pd	Pfa
12.00	250.00	200.00	1.50	36.00	0.9	le-006

Target and Weather Parameters

RCS(m^2)	tar el(deg)	tar lgth(m)	rain(mm/hr)
1.0	0.4	10.0	0.0

Stand-Off Jamming Parameters

Number of Jammers = 5

Pj(w) =	2000	2000	2000	2000	2000
Gj(dB) =	5.0	5.0	5.0	5.0	5.0
BWj(Mhz) =	200	200	200	200	200
Rj(n mi) =	50.0	50.0	50.0	50.0	50.0
Hj(ft) =	20000	20000	20000	20000	20000
Lj(dB) =	7.0	7.0	7.0	7.0	7.0

Detection Range with Atmospheric and Weather Attenuation

Swerling Case	0	1	2	3	4	1FA	3FA
Atmospheric Atten (dB)	2.28	1.77	2.23	2.02	2.26	2.25	2.26
Weather Atten (dB)	0.00	0.00	0.00	0.00	0.00	0.00	0.00
Detection Range (n mi)	144.21	92.69	138.21	115.61	141.18	139.83	142.00

Detection Range with Jamming

Swerling Case	0	1	2	3	4	1FA	3FA
Atmospheric Atten (dB)	1.97	1.45	1.91	1.70	1.94	1.93	1.95
Weather Atten (dB)	0.00	0.00	0.00	0.00	0.00	0.00	0.00
Jamming Range (n mi)	109.91	69.27	105.14	87.26	107.49	106.42	108.15

»

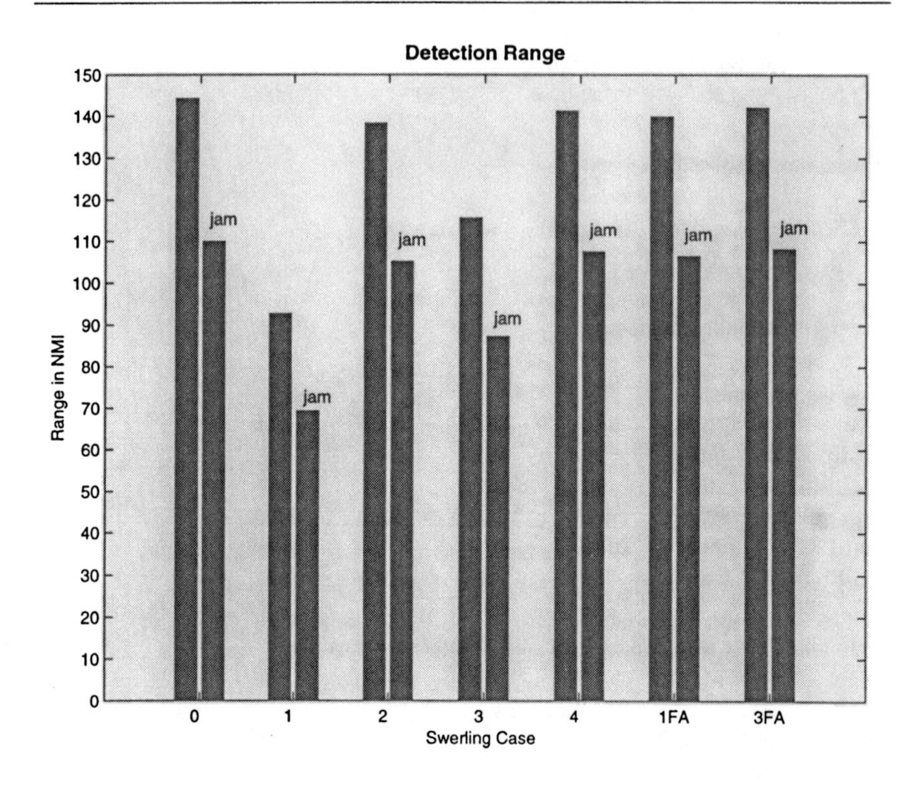

» rjx
Radar Parameters

Pt(kw)	f(MHz)	PW(1e-6*s)	PRF(pps)	Gt(dB)	Gr(dB)	NF(dB)	Gsl(dB)
3000.0	3000.0	6.5	250.0	36.0	40.0	4.5	−1.0

Lr(dB)	Dop Bw(Hz)	FA Bw(MHz)	Az Bw(deg)	Az Rate(deg/s)	Pd	Pfa
12.00	250.00	200.00	1.50	36.00	0.9	le-006

Target and Weather Parameters

RCS(m^2)	tar el(deg)	tar lgth(m)	rain(mm/hr)
1.0	0.4	10.0	0.0

Stand-Off Jamming Parameters

Number of Jammers = 5

Pj(w) =	2000	2000	2000	2000	2000
Gj(dB) =	5.0	5.0	5.0	5.0	5.0
BWj(Mhz) =	200	200	200	200	200
Rj(n mi) =	50.0	50.0	50.0	50.0	50.0

Hj(ft) =	20000	20000	20000	20000	20000
Lj(dB) =	7.0	7.0	7.0	7.0	7.0

Detection Range with Atmospheric and Weather Attenuation

Swerling Case	0	1	2	3	4	1FA	3FA
Atmospheric Atten (dB)	2.28	1.77	2.23	2.02	2.26	2.25	2.26
Weather Atten (dB)	0.00	0.00	0.00	0.00	0.00	0.00	0.00
Detection Range (n mi)	144.21	92.69	138.21	115.61	141.18	139.83	142.00

Detection Range with Jamming

Swerling Case	0	1	2	3	4	1FA	3FA
Atmospheric Atten (dB)	1.59	1.12	1.54	1.34	1.56	1.55	1.57
Weather Atten (dB)	0.00	0.00	0.00	0.00	0.00	0.00	0.00
Jamming Range (n mi)	79.11	49.50	75.62	62.55	77.35	76.56	77.82

»

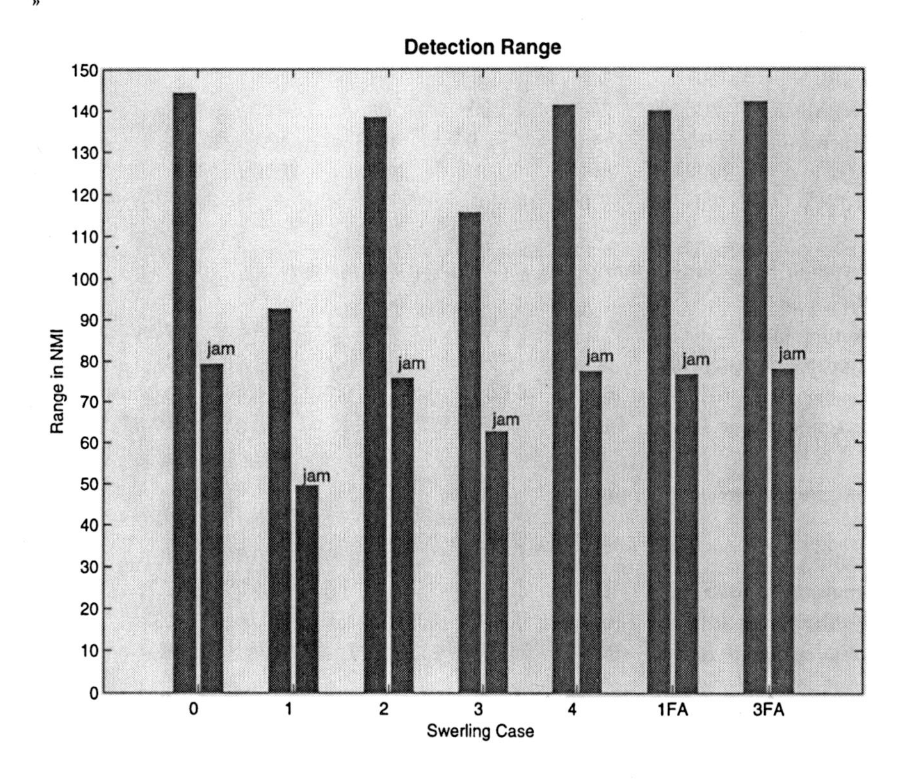

» rjx

Radar Parameters

Pt(kw)	f(MHz)	PW(1e-6*s)	PRF(pps)	Gt(dB)	Gr(dB)	NF(dB)	Gsl(dB)
3000.0	3000.0	6.5	250.0	36.0	40.0	4.5	−1.0

Lr(dB)	Dop Bw(Hz)	FA Bw(MHz)	Az Bw(deg)	Az Rate(deg/s)	Pd	Pfa
12.00	250.00	200.00	1.50	36.00	0.9	1e-006

Target and Weather Parameters

RCS(m^2)	tar el(deg)	tar lgth(m)	rain(mm/hr)
1.0	0.4	10.0	0.0

Stand-Off Jamming Parameters

Number of Jammers = 5

Pj(w) =	6	6	6	2000	2000
Gj(dB) =	5.0	5.0	5.0	5.0	5.0
BWj(Mhz) =	200	200	200	200	200
Rj(n mi) =	50.0	50.0	50.0	50.0	50.0
Hj(ft) =	20000	20000	20000	20000	20000
Lj(dB) =	7.0	7.0	7.0	7.0	7.0

Detection Range with Atmospheric and Weather Attenuation

Swerling Case	0	1	2	3	4	1FA	3FA
Atmospheric Atten (dB)	2.28	1.77	2.23	2.02	2.26	2.25	2.26
Weather Atten (dB)	0.00	0.00	0.00	0.00	0.00	0.00	0.00
Detection Range (n mi)	144.21	92.69	138.21	115.61	141.18	139.83	142.00

Detection Range with Jamming

Swerling Case	0	1	2	3	4	IFA	3FA
Atmospheric Atten (dB)	1.81	1.31	1.76	1.55	1.79	1.78	1.79
Weather Atten (dB)	0.00	0.00	0.00	0.00	0.00	0.00	0.00
Jamming Range (n mi)	96.44	60.55	92.27	76.42	94.36	93.41	94.89

»

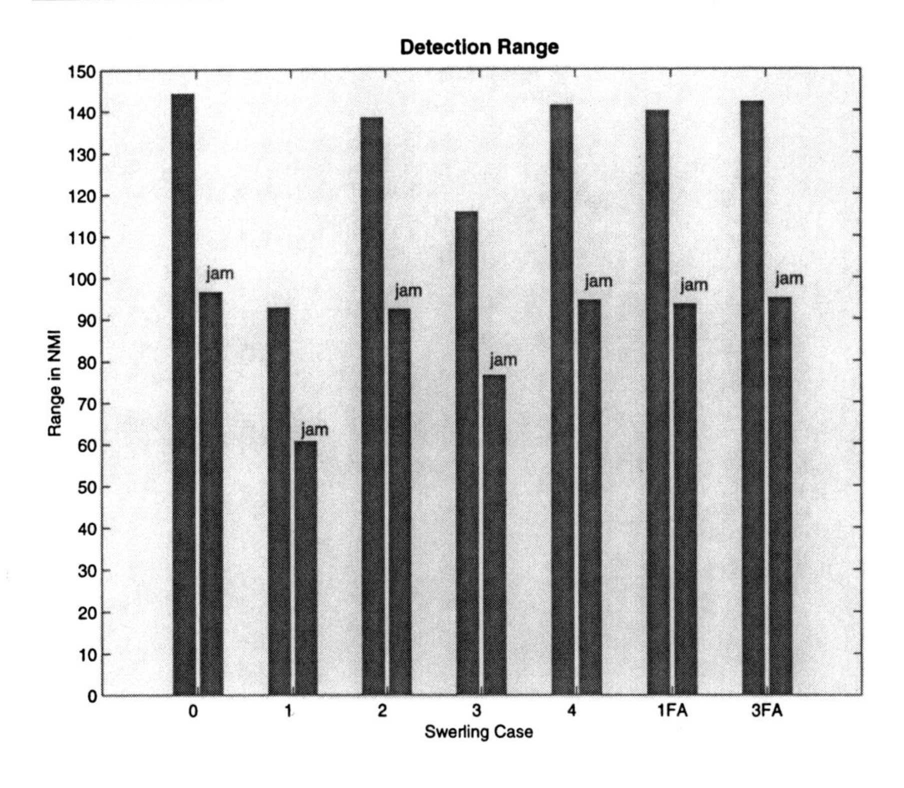

2.3 Phased Array Radars

Phased arrays are operational in basically three types of applications. Large phased arrays are used for the detection, tracking, and impact-point computation of long-range ballistic missiles (ICBMs and SLBMs) and for space tracking. Tactical-sized phased arrays perform acquisition, tracking, and target illumination functions in SAM systems and in weapon location systems (e.g., the TPQ-36 and TPQ-37 radars). Limited scan line-arrays perform elevation scanning in three-dimensional radar. Hybrid scan arrays provide small-angle scanning in both azimuth and elevation (20 degrees by 15 degrees) for precision approach-landing guidance radar (e.g., the AN/TPN-19 radar). An emerging application is on large strategic aircraft (B-1B) to perform air-to-air and air-to-ground functions from a single aperture. The characteristics of several tactical phased arrays is given in Table 2.6.

With present microwave technology, array costs are measured in hundreds of dollars per element. Monolithic technology is expected to reduce this cost to under a hundred dollars per element. Large arrays such as Cobra Dane (95-ft diameter) at L-band use 15,360 active elements and 19,049 dummy elements, while PAVE PAWS (72.5-ft diameter) at UHF uses 1,792 active

Table 2.6
Tactical Phased Arrays

System	Frequency	Power	Elements	Type	Phase Shifters	Scan
Patriot	C-Band	—	5161	Space fed	Ferrite	±60 degrees
Aegis (SPY-1)	S-Band	4–6 mW	4096 Transmit 4352 Receive 4 Faces	Subarray Steering	Ferrite	±50 degrees
TPQ-36	X-Band	26 kW	64	Frequency scan in elevation	Ferrite	±45 degrees
TPQ-37	S-Band	120 kW	359	Subarray	Diode	±45 degrees (Az) ±10 degrees (El)
B-1B (APQ-164)	X-Band	—	1526	Corporate feed	Ferrite	—
SA-10 (Flap Lid)	X-Band	—	10,000	Space fed	Faraday rotators	—
SA-12 (Grill Pan)	X-Band	—	10,000	Space fed	Faraday rotators	—

elements and 885 dummy elements. Tactical arrays such as Patriot at radar C-band use 5,161 elements, while AEGIS at S-band uses 4,096 elements on transmit and 4,362 elements on receive. Since arrays are limited to the angular coverage of approximately ±60 degrees, it is necessary to employ multiple faces in some tactical (e.g., AEGIS, four faces) and strategic (e.g., PAVE PAWS, two faces) applications, which further increases the number of elements per system [6].

Weapon system applications that require simultaneous tracking of multiple targets are candidates for a phased-array radar solution. The phased array can satisfy this requirement from a single aperture (as compared to a multiplicity of target-tracking radars) with an accuracy that is superior to a TWS tracking system. A particular advantage of the phased array radar is its energy-management ability, whereby it can concentrate maximum energy on critical targets (usually short-range targets), while dwelling for shorter times on noncritical targets (usually long-range targets) without wasting energy in sectors where there are no targets or those of limited interest. This ability optimizes the target-tracking function of the phased array and is unique when coupled with a semiactive missile guidance system, which requires a high rate of target illumination during its terminal phase.

In addition to the target-tracking function, phased array radar can also provide a target-acquisition function, a search function, and a target-illumination function for semiactive missile guidance. The principal advantage of this multifunction operation is that the large investment in the phase array antenna and the associated overhead computer and processors is spread over all the requirements. A related advantage is that the entire radar function is performed from a single aperture, which enhances mobility and leads to an antenna capable of being hardened for a battlefield environment. However, the multifunction aspects of the radar design cause a number of compromises in the performance over that which could be achieved if the phased array radar were optimized for each function individually.

The compromise in operating frequency, whereby the search function would favor a low operating frequency while the track function favors a high operating frequency, should be noted. In a heavy traffic environment, it may be necessary to schedule hundreds of search and track operations per second, providing only a few pulses per action, thereby ruling out refined Doppler processing, such as the MTD, on all but a few beams. The coverage of a single-face phased array radar is usually limited to a cone of 120 degrees of included angle, which restricts tactical operation, or necessitates the use of multiple faces, which becomes prohibitively expensive in all but the most stringent requirements. These performance characteristics and limitations are listed in Table 2.7.

Table 2.7
Phased Array Radar Characteristics

Feature	Use	Limitation
Rapid and agile scan	Faster scan with narrow beam High gain Avoids jamming Sequential detection	Doppler resolution sets dwell time; 60-degree coverage cone; limited bandwidth; large number of elements, frequency compromise
Control of sidelobe structure	Reduces jamming, clutter, multipath	Requires accurate phase and amplitude control; inefficient amplifier, efficiency
Multiple simultaneous beams	Track multiple targets, increases dwell time (Doppler resolution)	Complexity and cost of amplifier array
High power capacity of amplifier array	Extends radar range	Complexity and cost (especially) when multiple face coverage is required
Conformal arrays	Compatible with ships and aircraft; provides environmental protection	Multiple apertures may be required to provide coverage; complex feed networks

The array itself generally consists of a number of elements distributed on a planar surface (approximately 12.5 ft by 13 ft for AEGIS). Each element at least consists of a phase shifter and radiating element. The elements are grouped into sets called subarrays. Subarray steering is generally used to minimize the number of ports in the monopulse networks, and this also allows large arrays to support wide-bandwidth waveforms. Active and dummy elements can be distributed across the planar surface to control the aperture illumination function, which determines the antenna's radiation pattern and sidelobe structure.

Phased array architectures are depicted in Figure 2.3. The array aperture can be either passive or active. Most currently operational phased arrays use the passive approach.

In the passive-aperture approach, the array modules must be fed from a divider network that distributes the available power with the appropriate ampli-

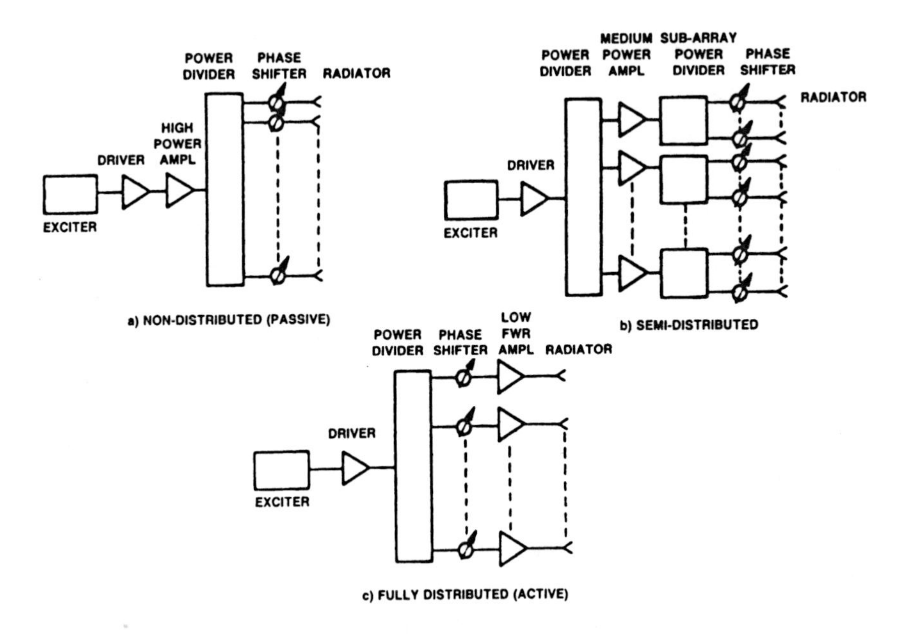

Figure 2.3 Phased array architectures. (*Source:* [12].)

tude taper among the array elements. The two available methods employ either a corporate feed network or a space feed. The corporate-feed method uses standard RF distribution components to connect the power source to the passive array elements. This approach is compatible with the use of subarrays and distributed transmitters, but the feed network becomes complex when element phase and amplitude control involves monopulse sum-and-difference radiation patterns and amplitude tapering aperture-illumination functions. The space-feed approach involves illuminating the array with a primary feed, whereby the array acts as a lens or reflector that corrects the spherical wave front and applies a linear phase shift across the aperture to steer the beam. The space-feed approach has the advantage that multiple beams (including monopulse patterns) and illumination tapers are formed in the primary feed, thereby simplifying the structure and control of the array elements.

Present development efforts tend to favor the active aperture approach [10, 20]. In the active aperture approach, each element contains a transceiver consisting of a transmitter, phase shifter, and a low-noise receiver. An advantage of this approach is that large radiated powers can be generated by the spatially coherent addition of a large number of low-power devices. *Monolithic microwave integrated circuits* (MMIC) are being developed for this purpose. A typical transceiver in G/H-band provides an output power of 3.5W; a 4-dB noise figure low-noise receiver; a 6-bit phase shifter, and a 20-dB variable gain amplifier for transmit and receiver tapers [10].

The design of an active phased array radar is a lesson in both economics and radar design. First, the detection range of the radar is proportional to the fourth root of the power-aperture product of the radar. The number of elements in the array is determined by its aperture size and frequency since the elements must be spaced approximately $\lambda/2$ apart to prevent grating lobes from entering the coverage pattern. Further, the number of elements determines the average power of the array since each element is constrained to the order of 2W to 20W average power in the microwave region due to the solid-state nature of the design. The power-aperture required to detect stealth targets (RCS = 0.001 m^2) at a range of 50 km is on the order of 150,000 W-m^2. If we assume that the aperture is limited to 5 m^2 for tactical applications, the number of required elements and power available at different frequencies in the microwave region are listed in Table 2.8.

An examination of the table indicates that the design at 6 GHz (λ = 5 cm) is the most practical from both technical and economic bases. Designs at lower frequencies do not provide the necessary power-aperture product. However, this design at the current $1,000/element phased array module cost would be very expensive. If the module cost could be reduced to the projected $100/module, then the design would be more economically viable. The preceding analysis does not take into consideration the possible RCS enhancement of stealth targets as the frequency is decreased.

Table 2.8
Power Aperture Available from Active Aperture Phased Array

λ (cm)	Elements (5 m^2)	Power/element (W)	Total Power (kW)	Power-Aperture (W-m^2)
3	22,200	2	44	2.22×10^5
5	8,000	4	32	1.60×10^5
10	2,000	8	16	8×10^4
20	500	16	8	4×10^4

► *Example 2.2*

Develop a MATLAB program that plots the detection range versus frequency available from an active aperture phased array for the parameters given in Table 2.8.

```
% Active Phased Array Detection Range
% Against Low RCS Targets
```

```
% ----------------------------------------
% phadet.m

clear;clc;clf;

% Antenna Effective Aperture

Ar=5; % sq. m

% Enter Frequency Range

fmin=1;fmax=10;
f=fmin:fmax;

% Enter Radar Parameters
% Swerling Case 0 Target

ts=1;          % Search Frame Time - s
RCS1=.1;        % Radar Cross-Section - sq. m
RCS2=.01;       % Radar Cross-Section - sq. m
RCS3=.001;      % Radar Cross-Section - sq. m
ang=1;          % Search Coverage - steradians
k=1.38e-23;     % Boltzman's Constant - w/(Hz-Ko)
NF=3;           % Receiver Noise Figure - dB
Sa=13;          % SNR for Pd=.9;Pfa=1e-6 - dB
Ls=10;          % System Losses - dB
eff=.7;         % Antenna Aperture Efficiency

F=10^(NF/10);Sa=10^(Sa/10);Ls=10^(Ls/10);

% Find System Noise Temperature

Ts=293*(F-1)+150;

% Assume Linear Increase of 2 w Module Power
% Inversely Proportional to Frequency at 10 GHz

Pm=2*10./f;    % watts

% Find Number of Modules with wl/2 Spacing

wl=.3./f;
n=sqrt(Ar)./(wl/2);

% Find Average Power

Pa=Pm.*n.^2;

% Compute Missile Detection Range

R1=(Pa*Ar*ts*RCS1/(4*pi*ang*k*Ts*Sa*Ls)).^.25;
```

```
R2=(Pa*Ar*ts*RCS2/(4*pi*ang*k*Ts*Sa*Ls)).^.25;
R3=(Pa*Ar*ts*RCS3/(4*pi*ang*k*Ts*Sa*Ls)).^.25;

% Plot Range vs Frequency

plot(f,R1/1000,f,R2/1000,f,R3/1000);grid;
ylabel('Range - km');
xlabel('Frequency - GHz');
title(['Active Aperture Phased Array Detection Range']);
text(6.5,70,'RCS = 0.001 sq. m');
text(4,110,'RCS = 0.01 sq. m');
text(1.5,160,'RCS = 0.1 sq. m');
```

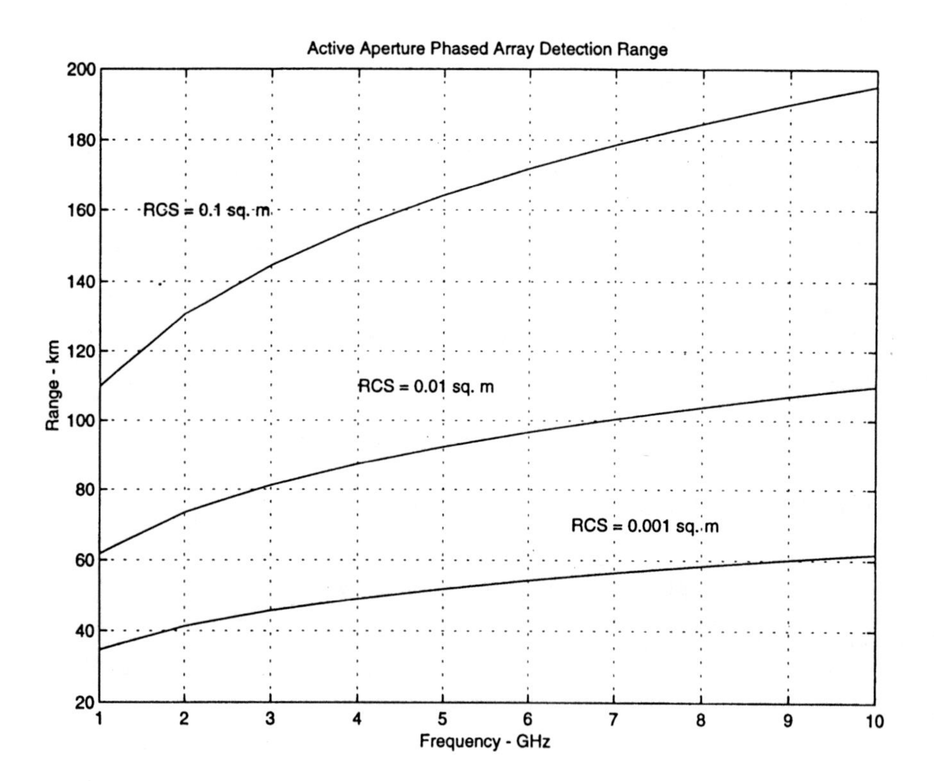

Each radiating element of the active array has its own receiver. Groups of collocated receiving elements (order of 100) are generally combined in microwave networks to form subarrays. Each subarray has a down converter and digitizer that produces an accurate version of the signal amplitude and phase. The subarrays can be summed to provide the normal antenna pattern receiving main-beam target signals, clutter, and any noise jamming entering via the sidelobes. An adaptive beamformer can be formed by weighting the output of each subarray in amplitude and phase before summing to shape the radiation pattern. The adaptive beamformer drives the nulls in the sidelobe pattern toward those direction angles containing jamming interference while maintaining the pointing direction of the main beam toward the target [20].

The element phase shifters are usually three- to five-bit digital devices, using either ferrites or PIN diodes for phase control. Diode phase shifters are basically lower power (e.g., 4-kW peak, 200-W average), compact devices with fast switching speeds (50 ns) that are most useful below 2 GHz, but they are available through Ku band. Ferrite phase shifters may be higher power devices (e.g., 150-kW peak, 400-W average) with moderate switching speeds (1 μs) that are most useful above 5 GHz, but they are available through 60 GHz. The number of phase shifter bits is a major factor in determining the average sidelobe level and the beam-pointing accuracy. Future arrays might use digital beam forming in the receiving channel by operation on each element at baseband frequencies using a *fast Fourier transform* (FFT) processor.

Wideband operation of phased arrays is desirable both from an enhanced target signature basis and for antijamming capability. Since element spacing of the order of $\lambda/2$ is necessary to avoid grating lobes, the tightest spacing is dictated by the shortest wavelength over which the array must operate. This requires development of wideband radiating elements and beam steering by variable time delay instead of phase to preserve the antenna pattern over the whole band. Present research on wideband arrays uses photonic phased array beamformers employing laser driver fiber optics delay lines to generate the antenna beams [21].

2.4 Airborne Radar

Airborne radars are generally designed to perform a variety of functions associated with weapon delivery (both air-to-air and air-to-surface), situational awareness, target identification and classification, navigation, terrain avoidance, altimetry, and imaging (both surface and targets). In many cases, all multimodes must be performed through a single aperture that is constrained to fit into a limited space. In fighter aircraft, the nose-mounted radar is normally a multimode system of the pulsed Doppler type to provide both "look-up" and "look-down" capabilities. The "look-down" mode is the most difficult since the target is likely to be heavily submerged in clutter, making the use of pulse Doppler (velocity discrimination) essential [22].

Waveform design is critical in airborne radars that generally employ low-, medium-, and high-PRF modes to accomplish their various modes [22,23]. Low-PRF designs are unambiguous in range, while high-PRF designs are unambiguous in Doppler. Medium-PRF designs are ambiguous in both range and Doppler. The various air-to-air and air-to-ground modes are listed in Table 2.9 for a typical airborne fighter radar.

Many modern airborne radars use both medium- and high-PRF designs in their "look-down" mode. High PRF designs (e.g., AWACS) are effective in detecting closing targets since this allows signal-processing Doppler filters that provide a high degree of clutter rejection to be implemented. This results due to the Doppler frequency of the clutter being separated from that of the target. However, when the target is receding from the radar, this is no longer true, resulting in the need for a medium-PRF design.

Medium-PRF operating modes used in modern air-to-air radar result in ambiguities in both range and Doppler measurement. These are resolved by switching between several PRF values during the time the antenna dwells on the target. The values of these PRFs are selected to avoid blind zone regions in range and velocity where targets cannot be detected due to eclipsing caused by radar transmissions or overlapping strong main beam surface clutter returns.

Table 2.9
Airborne Interceptor Radar Modes

	Low PRF	Medium PRF	High PRF
Air-to-air modes			
Multimode search (look-up)	✓		
Multimode search (look-down)		✓	✓
Velocity search			✓
Target track			✓
Semiactive homing (target illumination)			✓
Air-to-ground modes			
Real beam mapping*	✓		
Ground moving target indication (GMTI)	✓		
Doppler beam sharpening	✓		
Spotlight SAR	✓		
Sea surface search*	✓		
Terrain avoidance*	✓		
Terrain following*	✓		
Doppler navigation	✓		
Air-ground ranging*	✓		
Beacon ranging*	✓		

*Non-Doppler modes.

The PRF switching facilitates detection in ground clutter but tends to reduce the maximum detection range and Doppler resolution.

To achieve the desired detection performance while maintaining an acceptable false-alarm rate, medium-PRF radars often employ a "three-of-eight" double-threshold detection scheme. To determine true target range and velocity, detection is attempted in each of eight coherent dwells as the antenna scans the target. If threshold crossings are achieved at any three of the eight PRFs, which resolve to the same range and velocity value, this is accepted as a bona fide target detection [24].

Characteristics of several modern airborne pulsed Doppler radars are given in Table 2.10. A number of these radars are the result of upgrade programs, prevalent in fighter airborne radar systems, which add both improved performance and additional modes into the radar. This is generally made possible by the rapid advance in improved signal-processing capability. Most combat-type airborne radars operate in the high microwave region (X- or Ku-band) in order to provide the required angular resolution within a limited aperture size. Both AEW radars (AWACS and E-2C) operate at lower frequencies necessary for long-range operation under moderate weather conditions. They both use rotodome antennas to provide reasonable angular resolution consistent with their operating frequency band.

The pulsed Doppler radar generally has good resistance against ECM. The ability to resolve targets in Doppler frequency allows rejection of chaff and other correlated interference, which are separated from the target return

Table 2.10
Characteristics of Airborne Pulse Doppler Radar

Type	Platform	Mission	Frequency	Technique
AWG-9/APG-71	F-14	Air superiority	X-band	High/low PRF
APG-63/70	F-15	Air superiority	X-band	High/medium/low PRF
APG-66/68	F-16	Air superiority	X-band	High/medium/low PRF
APG-65/73	F-18	Air superiority	X-band	High/medium/low PRF
APQ-156	A-6A	Ground target attack	Ku-band	Low PRF
Foxhunter	Tornado	Air superiority	X-band	High PRF
APY-1	AWACS	Air surveillance	S-band	High PRF
APQ-164	B-1B	Ground target attack	X-band	
APS-125/145	E-2C	Air surveillance	UHF	TACCAR, DPCA, AMTI
APQ-181	B-2	Ground target attack	Ku-band	LPI waveform

by more than a Doppler filter bandwidth. The matched-filter aspect of pulsed Doppler operation provides the effect of coherent integration, which optimizes the target-to-jamming ratio in the presence of noise jamming. The ability of a pulsed Doppler radar to extract radial target velocity through direct Doppler measurement and by differentiating range measurements makes it necessary for a deception jammer to induce a realistic Doppler signature on the simulated target return. The large number of pulses generated by a pulsed Doppler radar provides problems for many types of EW intercept receivers that rely on deinterleaving pulse trains to identify threat radar.

The only significant disadvantage against ECM is the necessity of maintaining a stable transmitter frequency for a time equal to the inverse of the Doppler filter bandwidth. This allows a jammer the time to tune to the first transmitted pulse and then jam subsequent pulses in a spot jamming or deception mode. The vulnerability in this mode is generally to main-beam jammers, since appropriate sidelobe cancellation and blanking can be applied to reduce the effect of sidelobe jammers. Also, a sidelobe cover mode can be used, whereby multiple spurious frequencies are radiated in the sidelobe direction through an auxiliary antenna at the same time that the main radar pulse is transmitted.

2.4.1 Synthetic Aperture Radar

Synthetic-aperture radar (SAR) modes are being incorporated into many existing airborne radars as well as UAVs and other reconnaissance aircraft. SARs provide photolike high-resolution imagery of terrain and fixed surface targets at long range and in all weather. In addition, they generally provide the capability to provide *ground moving target indication* (GMTI), allowing them to detect and track slowly moving ground vehicles traveling along roads or across country.

In combat radars, the SAR mode assists in mapping, target-acquisition, and weapon delivery. Reconnaissance radars provide SAR imagery that allow targets such as mobile missile launchers as well as fixed targets to be located.

The delivery of weapons from an airborne platform against stationary and slowly moving ground targets generally requires a high-resolution image of the ground. This is difficult to achieve with a conventional airborne radar in the cross range direction due to the limited space available for antenna aperture. SAR techniques overcome this limitation through the synthesis of a large aperture by the coherent combination of a number of sequential measurements formed as the antenna moves along the aircraft's flight path. Furthermore, the rapid advance in digital signal processing and associated components has made it practical to accomplish the SAR function in an onboard processor. This mode has been incorporated into a number of currently operational airborne radars (e.g., the AN/APG-63 radar in the F-15 aircraft) [1].

The basic action of a SAR processor is to produce a higher azimuth angle resolution. As the name synthetic aperture implies, this can be viewed as the result of synthetically creating the equivalent of a large linear array antenna. Alternatively, this can be viewed as dividing the conventional antenna's azimuth cell into a number of Doppler cells, each associated with a particular azimuth segment. The two models are equivalent. The former model is generally used when explaining the operation of a synthetic aperture airborne radar (SAR), while the latter is used to explain the operation of a forward-looking radar in a Doppler beam-sharpening ground-mapping mode. An explanation of both models is given in the following, and each is used to describe some of the pertinent characteristics of SAR processors.

In an SAR, the airborne antenna generally radiates in a direction that is perpendicular to the aircraft's ground track. The antenna moves a distance that is proportional to the aircraft's velocity between each transmitted pulse. If these sequentially received samples are stored and combined, this result is identical to the signal that would have been obtained from an n-element linear array antenna that has an effective length (l_e) equal to the distance transversed during the transmission of n pulses. The maximum effective length of the synthetic array can be no longer than the angular sector subtended by the real antenna because after movement $l_e \geq R\theta_b$, the target is out of the beam.

A second effect that limits the effective array antenna length is the necessity to restrict the length so that the phase front approximates a plane wave. This is accomplished by requiring that within l_e, the two-way minimum distance from the target to the aircraft differs from the two-way maximum distance by no more than a quarter wavelength ($\lambda/4$). Applying this condition results in the effective length for an unfocused synthetic aperture of $l_e = \sqrt{R\lambda}$, which provides an effective cross-range resolution $\Delta r_c = K \cdot \sqrt{R\lambda/2}$.

It is interesting that the cross-range resolution of an unfocused SAR does not depend upon the size of the real antenna aperture but is a function of the range to the target and the wavelength. The synthetic array can be focused at range R by adding an appropriate corrective phase shift to each sequentially received signal so that when combined from range R all signals are in phase. The cross-range resolution for this focused condition is $\Delta r_c = D/2$, which is independent of both the range and wavelength but depends on the real antenna aperture (D).

The SAR mode is used primarily in reconnaissance operation [25] to produce a high-resolution strip map of an area of interest. Practical designs range from side-looking radars that use only noncoherent integration (e.g., AN/APS-94) to those using focused synthetic aperture processors (e.g., AN/UPD-8). The present trend is toward fully focused SAR using real-time digital processing. Many of the synthetic aperture designs send the unprocessed data to the ground for processing via a wide bandwidth data link (e.g., the Seasat radar).

SAR can also be used in squint and spotlight modes. In a squint mode, the radar antenna looks in the forward direction (15 degrees to 45 degrees with respect to the ground track) of aircraft motion rather than to the side. This mode is similar to Doppler beam sharpening. The spotlight mode causes the radar antenna to track the target area as the aircraft moves, thereby increasing the potential processing time and, hence, the resolution in a particular region of high interest.

Doppler beam sharpening (DBS) provides the capability for real-time ground mapping and the potential for increased subclutter visibility against slowly moving ground targets. The DBS mode can be applied to low- and medium-PRF airborne forward-looking radar and is effective in enhancing air-to-ground weapon delivery operation.

In a forward-looking airborne radar that illuminates the Earth's surface, all returns from scatterers within a basic radar resolution cell combine to form the radar return. The cell dimensions are proportional to the effective transmitter pulsewidth and the antenna azimuth beamwidth, which is relatively large in airborne applications due to the limited space available for antenna aperture.

A Doppler frequency shift is induced on the return from each scatterer that is proportional to the relative closing velocity of the aircraft to that scatterer. Because scatterers are distributed through the azimuth cell at different angles with respect to the aircraft's velocity vector, their Doppler signatures are different, depending upon the relative angular displacement with respect to the center of the cell. This creates a spectrum of Doppler return frequencies distributed about the mean frequency, which is associated with the center of the resolution cell. The spectral spreading is a function of the aircraft's velocity, antenna azimuth beamwidth, and antenna's pointing angle relative to the aircraft's pointing angle.

In the ground-mapping mode, a bank of Doppler filters, each with bandwidth inversely proportional to the scanning antenna's dwell time on the target, is used to process the return. The output of each Doppler filter within the bank then contains only those returns associated with a particular narrow angular segment, which is appreciably smaller than the antenna's azimuth beamwidth. This is the DBS effect. Also, because the returns from targets come from narrow angular regions and hence have distinct Doppler signatures, they tend to fall within a single Doppler filter in the bank. The signal-to-clutter ratio within this filter for both stationary and slowly moving targets is much larger than it would have been if the total clutter return within the azimuth beamwidth were contained within the filter. This is the subclutter visibility improvement effect of the DBS mode.

The cross-range resolution for either the squint mode of an SAR or a DBS mode of a forward-looking radar with velocity (v), angle off ground track (θ), and dwell time (T_d) is given by $\Delta r_c = \lambda R/(2vT_d \sin\theta)$. For a forward-looking radar with a uniform scan, the dwell time is a constant, and the resolution and hence the signal-to-clutter improvement factor become progressively poorer as the antenna points toward the ground track ($\theta = 0$). In an SAR with a squint mode, it is possible to increase the dwell time with decreasing θ in order to keep the resolution constant, except in a narrow angle about the ground track where the integration time becomes impractically long. When DBS is used in weapon delivery systems, it is important to bear in mind the relative deterioration about the ground track so that tactics can be planned that minimize operation in this region.

The resolution achieved by operational strip map SARs range from 3m for older systems (i.e., UPD-8) to 1m for new systems (i.e., Global Hawk) [1,2]. Resolutions of the order of 0.3m are available in spot mode SAR. This can be contrasted against the feature size necessary for feature recognition of various military significant targets listed in Table 2.11. An examination of the table indicates that current SARs allow aircraft and small military vehicles to be identified from stand-off ranges on the order of 200 km.

EA against airborne ground-mapping radar used for weapon delivery can be expected to be of high intensity. The EP considerations used for other surveillance radar are generally applicable in this case. It is to be expected that ground jammers would be located in the vicinity of high value targets whose mission is to jam weapon delivery radar.

The EA situation against SARs that are used principally for reconnaissance is not as clear. The long processing times used by these radar provide some inherent EA resistance, but they would appear to be vulnerable to a dedicated jamming effort. However, in many cases, the processing is not done in real

Table 2.11
Feature Recognition Versus Resolution

Feature	Square Resolution of Cell Size
Coastline, outline of cities, detect large discrete scatterers, outline of mountains	300m
Detect major highways (some fading), variations in fields, large airfields	40m to 50m
City street structure, reliably detect highways, large building shapes, small airfields, "roadmap" type imagery	15m to 30m
Reliably detect vehicles, shapes of houses, building	3m to 10m
Aircraft, vehicles	1/2m to 3m

time and the threat posed by these radar is of low priority. EP for these types of radar must also consider the threat against the associated wideband data link as well as the basic radar.

▶ *Example 2.3*

Using MATLAB, plot the angular resolution performance for a real beam radar, an unfocused SAR, and a focused SAR at X-band using a 10-ft-long antenna.

```
% Sythetic Aperture Radar Resolution
% ------------------------------------
% sarres.m

clear;clc;clf;

% Input Radar Parameters

f=10000;                % Frequency - Mhz
l=10;                   % Antenna length - ft

wl=(300/f)/1854;l=l/6080;

% Set Range

Rmin=1; Rmax=100;       % Range - nmi
R=Rmin:Rmax;

% Compute SAR resolution

d=.76*R*wl/l;           % Real Beam Reolution
d1=.6*sqrt(R*wl);       % Unfocussed array resolution
d2=l/2;                 % Focused array resolution

d2=d2*ones(size(R));

% Plot array resolution

loglog(R,d*6080,R,d1*6080,R,d2*6080);grid;
xlabel('Range - nmi');
ylabel('Resolution - ft');
axis([1,100,1,1000]);
title(['SAR Resolution (Real Beam, Unfocused, Focused)']);
text(2,300,'Real Beam');
text(3,50,'Unfocused');
text(4,6,'Focused');
text(30,500,['f = ', num2str(f/1000),' GHz']);
text(30,300,['l = ', num2str(l*6080),' feet']);
```

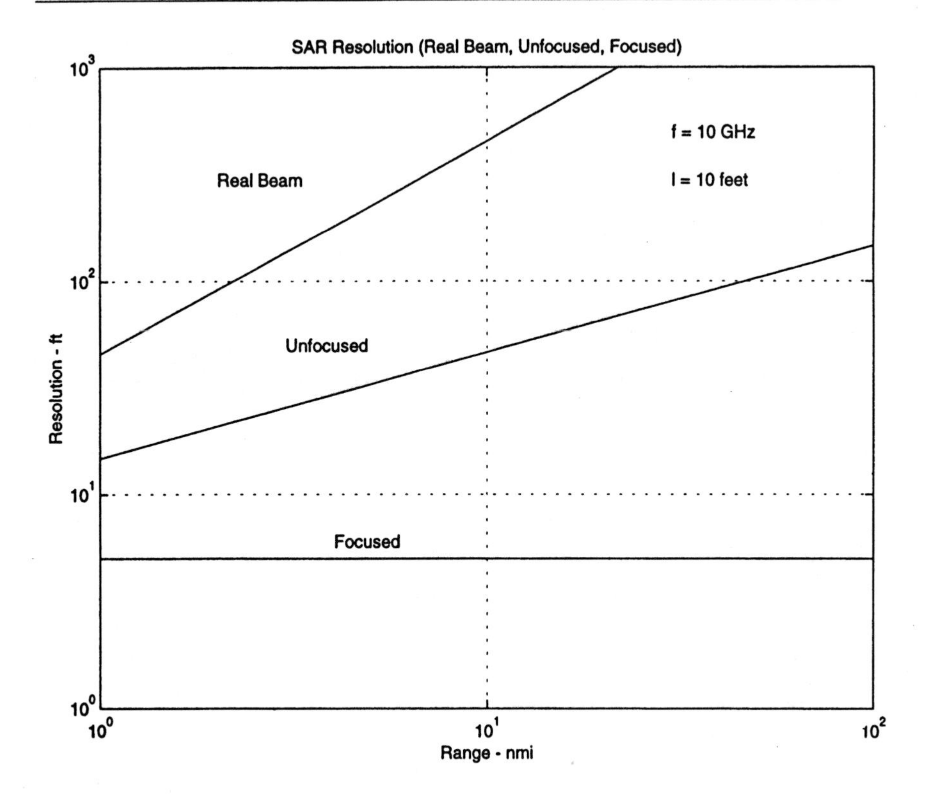

2.4.2 Inverse Synthetic Aperture Radar

An SAR achieves high resolution in the cross-range dimension by taking advantage of the motion of the vehicle carrying the radar to synthesize the effect of a large antenna aperture. The classic SAR is appropriate for providing high-resolution ground maps but ill suited for imaging targets such as ships and aircraft that have rotational motion. Unless the differential Doppler shifts, which such motions produce, are accurately predicted and compensated, they tend to defocus the array and blur the image. With slightly different algorithms, these shifts, rather than those due to the radar's relative velocity, can be used to provide the angular resolution needed for imaging. This technique is called *inverse synthetic aperture radar* (ISAR).

ISAR imaging of a ship due to its pitch motion is depicted in Figure 2.4. It can be shown that the lower limit for cross-range resolution in this mode due to migration through cross-range resolution cells (Δr_c) [24] is given by

$$\Delta r_c \geq \sqrt{\frac{\lambda D_y}{4}} \tag{2.1}$$

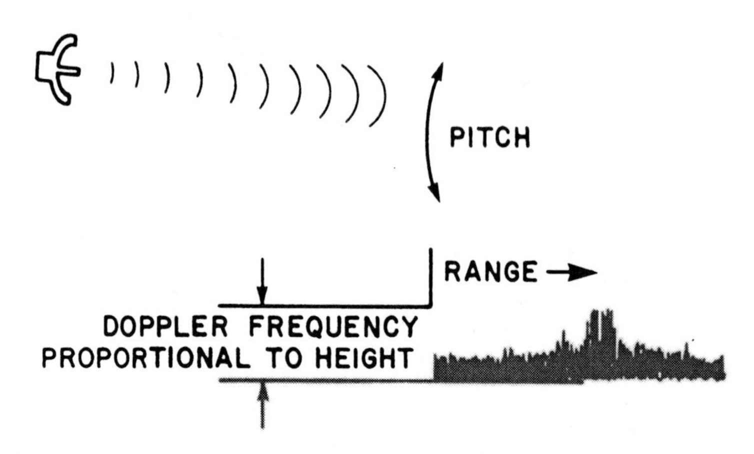

Figure 2.4 ISAR imaging of a ship. (*Source:* [13].)

where D_y is the ship's length and λ is the radar's wavelength. Thus, a cross-range resolution of 1.5m can be achieved at X-band for a ship with 200-m length. Further assuming a pitch angular rate of 1 degree/s results in an angular excursion of $\Delta\theta = \lambda/2\Delta r_c = 0.01$ rad or 0.6 degrees to achieve the limiting cross-range resolution with a processing time of $t_p = \Delta\theta/w = 0.6$ sec. ISAR image views of a ship can be produced due to a roll and yaw motion as well as pitch motion. Predicted cross-range resolutions for ship targets using an S-band radar in sea state 5 with pitch, yaw, and roll motion using a processing time of 1 sec are listed in Table 2.12 [30].

Table 2.12
Predicted Cross-Range Resolutions for Ship Targets Using an S-band Radar in Sea State 5 (t_i = 1 sec)

Ship Type	Motion Type	Average Angular Velocity (w)	Average Angular Excursion in 1 s ($\Delta\theta$)	Average Cross-Range Resolution (Δr_c)
Destroyer	pitch	1.01 degree/s	0.0177 rad	2.82m
	yaw	0.54 degree/s	0.0093 rad	5.35m
	roll	6.30 degree/s	0.1100 rad	0.46m
Carrier	pitch	0.16 degree/s	0.0028 rad	17.83m
	yaw	0.08 degree/s	0.0014 rad	35.54m
	roll	0.38 degree/s	0.0066 rad	7.56m

2.4.3 Space Time Adaptive Processing for Airborne Radar

High-resolution airborne SAR type radars provide the capability to produce a radar ground map of photographiclike image quality under all-weather conditions. This shows the precise location of bridges, buildings, airfields, and other static features together with critical time-sensitive targets such as halted convoys or even stationary military vehicles. However, it is desirable to complement this fixed target indicator with a GMTI, which shows slowly moving ground vehicles.

The critical problem for GMTIs is their ability to detect minimum velocity targets (order of 3-knots radial velocity) against a background of heavy ground clutter. In an airborne radar, this is complicated by the need to compensate for the motion effects induced on the clutter by the moving platform.

Early GMTIs used the clutter itself as a reference, thereby measuring the differential motion between the ground (at zero velocity) and the target. However, performance of this method is poor due to the interaction between the target and ground clutter components [24]. The next generation of systems used coherent processing techniques, which necessitated the use of special motion compensation techniques to produce acceptable performance.

In general, platform motion can be resolved into two components, one parallel and one perpendicular to the antenna aperture. These components have the effect of both spreading the spectrum of the clutter returns and shifting the mean of the spectrum.

With the antenna aperture perpendicular to the velocity vector, the dominant effect is the shifting of the mean of the clutter spectrum. In general, the mean of the Doppler spectrum is determined by the platform velocity, the transmitted frequency, and the angle to the clutter path under consideration $f_c = (2v_{ac}/\lambda)\cos\phi_0 \cos\psi_0$. Motion compensation of this effect is generally accomplished using the *time average coherent airborne radar* (TACCAR) technique that locks the radar's coherent oscillator to the mean clutter frequency.

The *displaced phase center antenna* (DPCA) technique compensates for the velocity component parallel to the antenna's aperture. This is accomplished by either physically or electronically displacing the phase center of the antenna in a direction opposite to the velocity component of the aircraft. A single canceller MTI can be almost perfectly compensated by DPCA provided that the beam patterns from the two displaced apertures have identical patterns. Compensation of higher order MTI cancellers is difficult, resulting in imperfect compensation for these configurations [24].

When TACCAR and DPCA motion compensation techniques are effectively employed, the overall improvement factor provided by the GMTI is limited to the ratio between the integrated sidelobe level and the mainbeam gain. This results because the antenna sidelobes essentially cover 360 degrees and the Doppler shifts received through the sidelobes range from $-v_{ac}$ (v_{ac} is the aircraft's velocity) to $+v_{ac}$. This Doppler spread can cover the entire MTI ambiguous frequency band. As a result, these sidelobe returns are not canceled and the MTI improvement factor is limited by the ratio of the integrated energy entering the system through the mainlobe to that entering through the sidelobes for the two-way antenna pattern [6].

A different way of looking at the motion compensation problem is to plot the clutter return from an airborne radar in Doppler-azimuth space, which provides the clutter ridge as illustrated in Figure 2.5. From this illustration, it

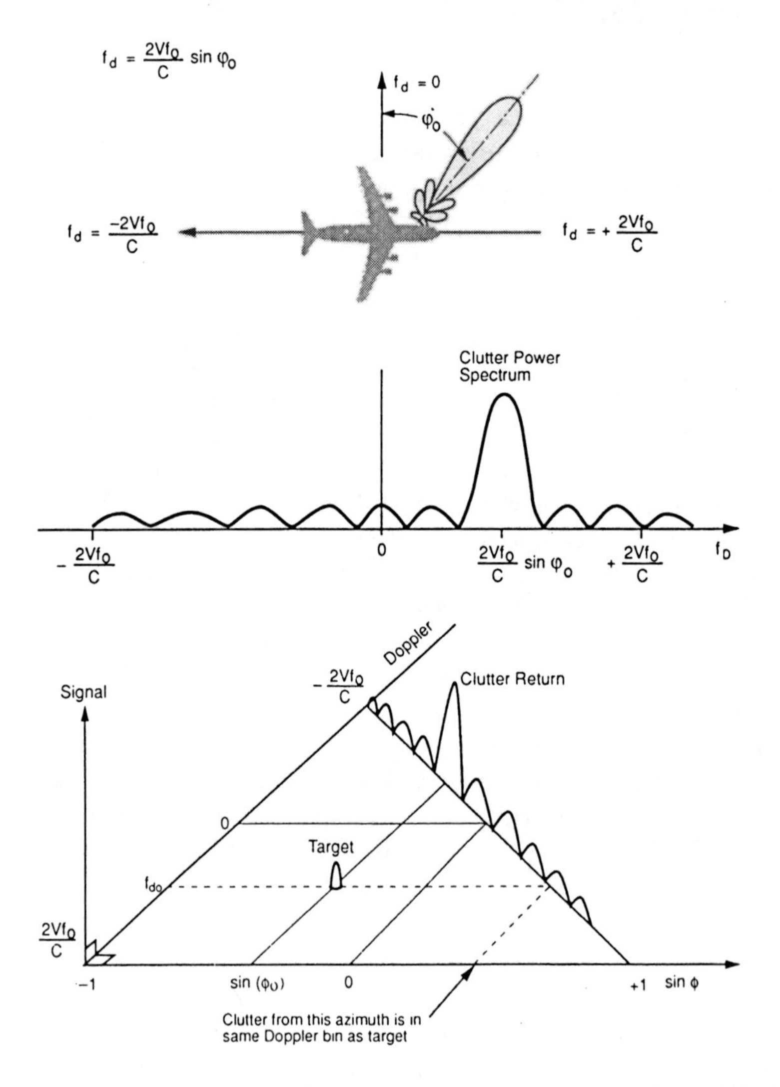

Figure 2.5 Space-time MTI filtering (Barile, E., R. Fante, and J. Torres, "Some Limitations on the Effectiveness of Airborne Adaptive Radar" *IEEE Trans. AES* © 1992, IEEE).

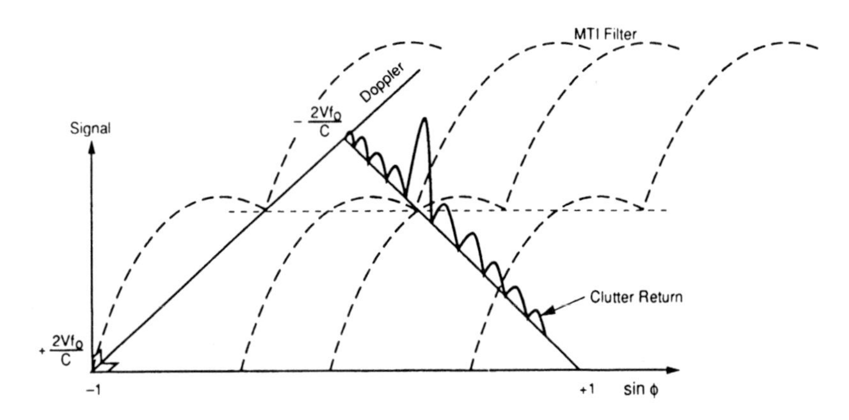

Figure 2.5 (continued).

is apparent that MTI filtering in the Doppler domain limits performance due to the presence of clutter returns through the antenna sidelobes. However, if the MTI filter's response could be adapted in the spatial (angle) domain as well as the temporal domain, then the null of the MTI filter could be placed on the clutter ridge, thereby providing greatly enhanced performance. This method, called space-time processing, is the thrust of modern GMTI systems that allow detection of slowly moving ground targets and increased performance in clutter and jamming environments [27].

A dual-port space-time processor can be synthesized by Doppler processing the signals received from each port of a dual-aperture antenna and then recombining them through a complex weighting factor that is independently determined for each Doppler filter bin to form a spatial null in the effective receive pattern in the direction corresponding to the expected location of that specific Doppler. This technique would reject ground clutter returns from their expected locations within the radar beam while passing those signals from moving targets whose Doppler frequency components are inconsistent with those anticipated from their angle of arrival. It can be shown that this processing is equivalent to DPCA motion compensation.

Adding a third port allows a monopulse angle measurement to be made on the detected moving target. This measurement provides an angular accuracy that is a fraction of an antenna beam width. Note that DBS techniques are ineffective for moving targets since their Doppler frequencies are inconsistent with the Doppler frequency of the background ground return and, hence, will be mapped at the wrong angle.

The Joint Stars radar uses the three-port space-time approach to detect slowly moving ground targets as well as to provide an SAR map. In the MTI surveillance mode, the radar can indicate ground targets on the move by day or night and in adverse weather and heavy ground clutter in areas of 300 km by 400 km. It also can pick up slow-moving helicopters. The radar display is overlaid with a digital map and symbols to label targets and gives tactical information.

Joint Stars is heavily dependent on software, requiring over 600,000 lines of code, and uses 20 vector processors [28]. The planar X-band slotted array antenna is 24-ft long and 2-ft high, with a rectilinear grid of 456 by 28 horizontally polarized slots. The antenna is electronically scanned over ±60 degrees in azimuth using 456 phase shifters [29].

Space-time adaptive processing (STAP) is a generalization of DPCA processing that uses the principles of adaptive antenna techniques. Using a phased array antenna, adaptive weights are applied to each element. The weights are adaptively computed to provide the maximum signal-to-interference (clutter plus noise plus jamming) ratio at the output of the processor. This effectively places antenna nulls over clutter returns having the same Doppler frequency as the target and over jamming signals while causing the mainlobe antenna beam to be focused on the target.

Fully adaptive STAP requires inversion of the large dimension covariance matrix formed in the space-time domain. This is generally beyond the computational capability of currently available processors. Current research in this area is focused on the development of computationally efficient algorithms that allow application of the STAP principle [27].

2.5 EP Techniques for Surveillance and Tracking Radar

Most modern radars are designed with a substantial amount of built-in EP. Older radars used for military purposes have been generally upgraded to include protection against various EA measures. In this section, we discuss the various EP techniques commonly applied to search and tracking radars.

EP against noise jamming has evolved to where it can generally be rejected in the radar antenna's sidelobe region. The principal radar EP that accomplish this are ultralow sidelobe antennas and the use of narrow bandwidth coherent Doppler radar. Noise jamming in the radar's mainlobe is generally effective, but in a search radar it produces a strobe in the direction of the jammer, while in a monopulse track radar it allows the jammer to be tracked in angle.

The limitations of noise jamming have caused the EA threat to move in the direction of deceptive jamming that attacks the radar's signal processor. A principal EA component that supports this thrust is the *digital RF memory* (DRFM), which allows the jammer to memorize the radar's waveform. Radar EP techniques are available to combat some forms of DECM. However, other forms of DECM are evolving that will require further development of radar EP to combat these new jamming threats.

Figure 2.6 depicts the EP techniques that are generally available within a search radar. Threats they combat are noise jamming, deceptive jamming, and cross-polarization jamming, both in the main antenna and in the antenna sidelobes. The general objective of the radar's EP is to restrict the EA effects to the radar's mainbeam. When this is accomplished, high-duty cycle noise and false target jamming produces strobes in the jammer's direction. DECM must be countered in the radar's signal processor by discriminating against synthetic jamming returns. EPs that prevent the jammer from anticipating the radar transmissions (i.e., frequency agility and PRF jitter) are useful in this regard.

In the radar antenna, ultralow sidelobes, *coherent sidelobe cancellers* (CSLC), and *sidelobe blankers* (SLB) are the main line of defense to restrict the radar jamming to the main antenna lobe. Another EP feature that can be deployed with low sidelobe antennas is the transmission of cover pulses through auxiliary sidelobe antennas that are synchronized to the main radar transmissions but at different random frequencies. This prevents the EA system from measuring the radar's frequency through the sidelobes, thereby forcing the jammer into a barrage mode of jamming.

The other EP action available in the radar transmit and receive antenna is to provide either a variable or an adaptive polarization. Using an adaptive polarization in the radar receiver antenna that is orthogonal to that used in the EA transmission results in cancellation (e.g., generally on the order of 20 dB) of the jamming signal. The use of a variable transmit polarization that is different from the receive polarization makes it difficult for the EA to employ cross-polarization jamming. The use of orthogonally polarized auxiliary receive antennas in the radar allows cross-polarized jamming signals to be intercepted and utilized in CSLC and SLB loops.

EP used in the transmitter has a variety of purposes that allow the radar to function against certain types of EA. The use of frequency agile transmissions with PRF agility is aimed at forcing the jammer into a noise barrage jamming mode. It is also useful against anticipatory EA techniques. Burn-through modes are aimed at increasing the energy returned from the target with respect to the jamming energy. This could involve transmitting more average power or dwelling on the target for a longer period. Pulse compression generally provides a processing gain over a noise jammer equal to the pulse compression ratio $\tau \cdot B_r$ where τ is the radar pulse width and B_r is the transmitted bandwidth. *Low probability of intercept* (LPI) waveform design attempts to prevent the jammer from intercepting the radar signal, which in turn inhibits the EA jammer from initiating a jamming response.

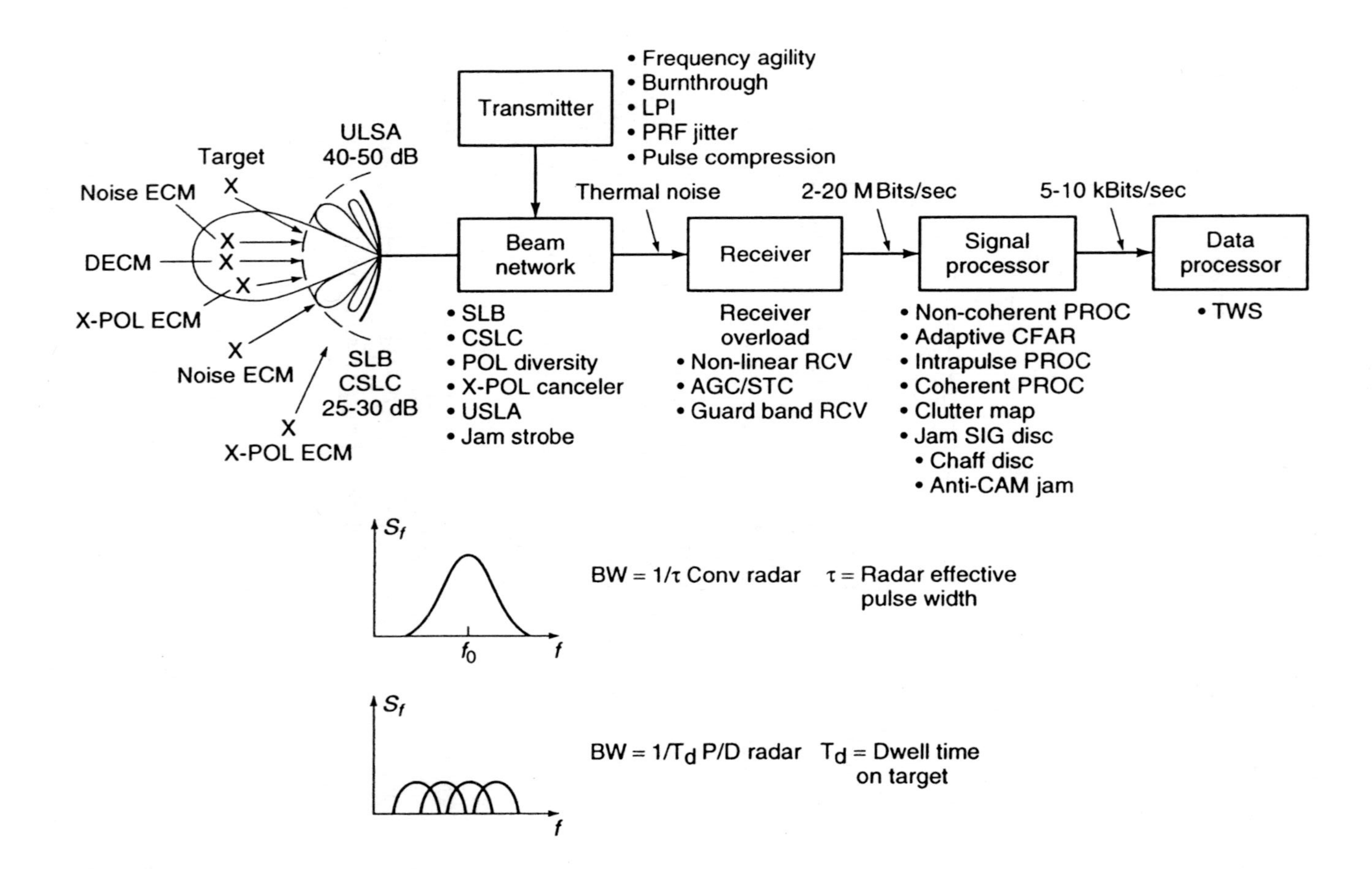

Figure 2.6 Search radar EP techniques.

EP in the radar receiver is generally aimed at preventing receiver overload that would suppress target signals. This form of EP generally involves some form of receiver nonlinear operation or AGC/STC control. Logarithmic receivers, Dicke Fix receivers, or Guard-band receivers are examples of this type of EP. Receivers that process Doppler radar signals are sometimes incompatible with nonlinear operation.

The radar signal processor generally performs significant EP operations. Its overall function is to extract target reports from the many extraneous returns due to noise, clutter, and jamming. The input throughput of 2- to 20-Mbit/s to the search radar's signal processor (see Figure 2.6) generally must be reduced to an output rate of 5 to 10 kbit/s of target reports.

Coherent Doppler processing provides a matched filter response to the target in noise jamming. The optimum Doppler bandwidth is related to the coherent dwell time on target

$$B_D = 1/T_d \tag{2.2}$$

With uniform spectral density noise jamming, the noise power entering the target Doppler filter is then reduced by the factor $F = B_j \cdot T_d$, which for the ASR-9 MTD processor (B_D = 150 Hz) and a spot noise of 10 MHz results in an improvement of 48 dB with respect to that available from a single return pulse. This coherent Doppler processing is one of the more powerful signal processing techniques against noise jamming.

Another function of the signal processor is false alarm control. This is generally accomplished using CFAR techniques, of which cell-averaging CFAR is the most prominent. The general vulnerability of the cell-averaging CFAR to a mismatch condition (e.g., variation of the noise statistics from their design values) is well known; hence, robust-type CFAR designs are required to protect against this situation [6]. Another method that provides a CFAR action in ground-based search radar is the use of a clutter map. This technique sets the threshold in a single cell on the basis of many radar scans and, hence, should be less vulnerable to mismatch than the cell-averaging CFAR.

Other functions performed in the radar signal processor involve tests that reject signals that do not have the characteristics of real radar signals. An example of this type of EP is the pulse width discriminator that rejects targets that have unreasonably long lengths.

Figure 2.7 depicts the EP that is generally available within a dedicated tracking radar to counter the active threats indicated. In the discussion that follows, it is presumed that the tracking radar has acquired the target. The search radar EP described in the previous section would apply during the acquisition mode of the tracking radar, which is generally designed to exclude

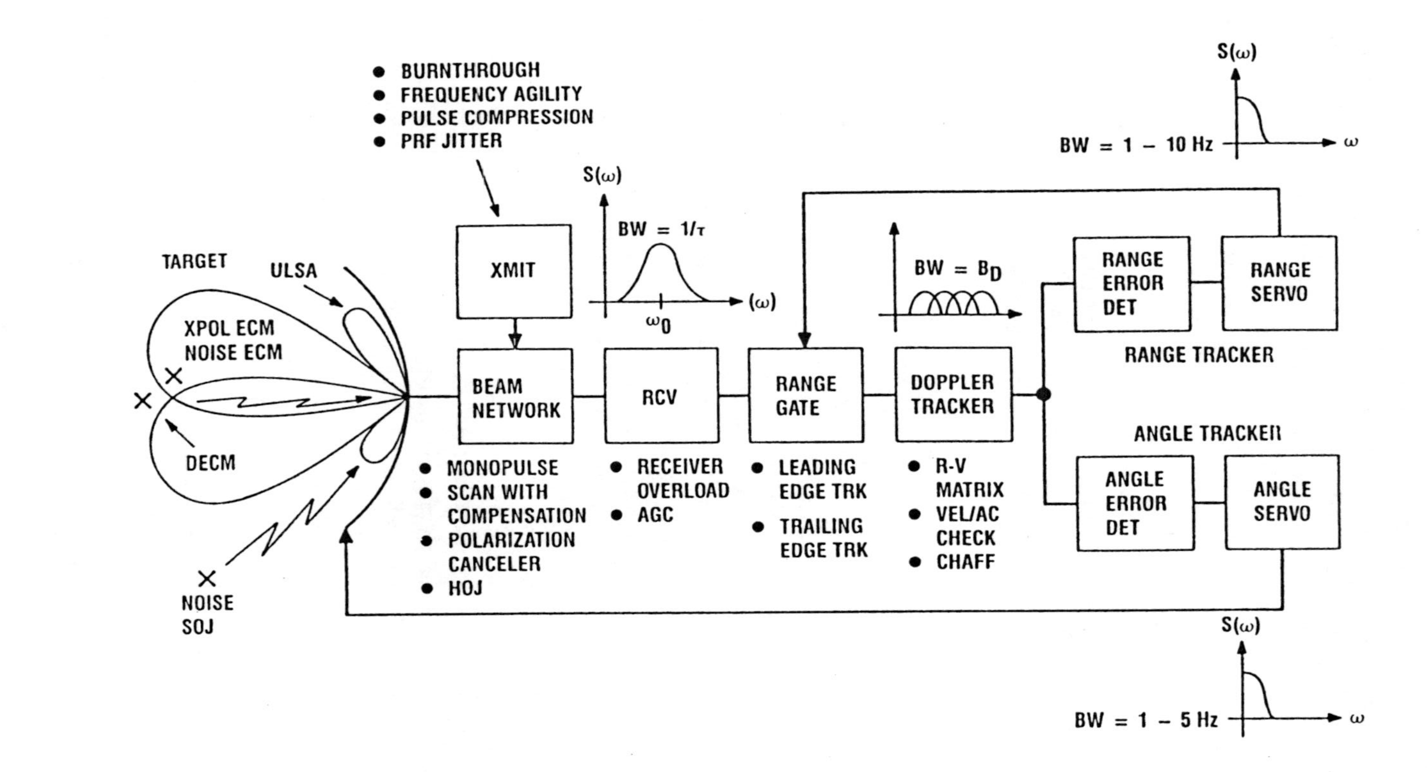

Figure 2.7 Tracking radar EP.

all signals except the signal from the target being tracked. As such, it employs a funnel-like approach whereby the track returned is first passed through a narrow range gate, then in most cases through a Doppler filter (called a speed gate), and finally the angle error information is extracted and applied to angle servos that direct the antenna beam to point at the target.

The primary self-protection EA applied against tracking radar is of the DECM variety since noise jammers are easily angle tracked by a home-on-jam mode incorporated into a monopulse tracking radar. Stand-off noise jammers have the potential of degrading the tracking radar's angular accuracy, thereby requiring low antenna sidelobes as an EP against this threat.

DECM techniques against tracking radar are initially directed toward capturing the range and speed (if appropriate) gates of the tracking radar. During this process, the radar is denied range and velocity information but still tracks the target in angle. After the range (and speed) gates are captured, angle deception is applied to cause the tracking radar to break lock.

The first EP defense against this type of jamming is to prevent the tracking radar's range gate from being captured. Against repeater-type jamming, a leading edge range tracker is employed to exploit the unavoidable delay associated with repeater jammers. Non-Doppler tracking radar are potentially vulnerable to transponder-type jammers that anticipate the radar transmissions. PRF jitter and frequency agility are effective EPs against this type of jamming.

If the DECM is unsuccessful in capturing the range tracker, then a cover pulse may be employed that overlaps the target return with an appropriate Doppler signature to penetrate the Doppler tracker. The purpose of the cover pulse is to allow the jamming waveform to penetrate into the radar's angle tracking circuits utilizing deceptive angle-jamming techniques. The type of angle deception employed depends upon the particular angle-tracking technique utilized in the tracking radar.

Conical scan-tracking radars are generally vulnerable to inverse gain amplitude modulation induced on the jamming waveform. EP techniques against inverse gain type of DECM involve using a silent scan (i.e., COSRO) or a jamming cancellation technique (i.e., scan-with-compensation or conopulse). Silent scan techniques do not remove the vulnerability but rather make it more difficult for the jammer to effectively implement the DECM [6].

Monopulse tracking radars are insensitive to amplitude modulation and hence are not susceptible to those jamming techniques generally applied against conical scan radar. Dual-source angle jamming (e.g., cross-eye) is a fundamental technique to implement. Defect-type monopulse jamming is possible by exploiting design weaknesses inherent in many monopulse tracking radars. An example of this type of jamming is cross-polarization jamming whereby the angle error signal is reversed in those monopulse implementations employing antennas that have significant cross-polarization responses. Defect-type jamming is always susceptible to design corrections that eliminate the vulnerability. Another fundamental monopulse jamming technique is the use of expendables (either active or passive) that lure the tracking radar to point away from the true target position.

2.6 Problems

1. A surface-based surveillance radar operating at a frequency of 1.3 GHz detects a 1000-ft altitude airborne target at 30 nmi. A passive IR sensor operating at a wavelength of 4 μm detects the same target at 15 nmi. Both sensors operate in clear weather conditions. What are the respective ranges of both sensors when operating in 4 mm/hr of rain? When operating in radiation fog with 100-ft visibility?

2. A surface-based monostatic radar has a range of 100 nmi on an airborne target. A surface-based receiver is used to operate in a bistatic mode and has the same receiver aperture as the radar. What is the maximum range from the radar that the target can be detected when the bistatic receiver is 50 nmi from the target?

3. Derive the bistatic radar equation for P_r from the monostatic radar equation by letting $R^2 = R_1 R_2$, where R_1 is the range between the transmitter and the target, R_2 is the range between the receiver and the target, and P_r is the received power. Plot the contours of constant power received, which are called the ovals of Cassini.

4. Derive the Doppler shift of the received signal in a bistatic radar in terms of v, the target velocity; ψ, the angle that the target's velocity vector makes with respect to the baseline connecting the receiver and transmitter; ϕ_T, the transmitter angle in the bistatic triangle; and ϕ_R, the receiver angle in the bistatic triangle. (*Hint*: The bistatic Doppler is given by $f_d = (\dot{R}_1 + \dot{R}_2)/\lambda$, where R_1 and R_2 are defined in Problem 3.) From the derived equation, show that the Doppler shift is zero for targets moving in a radial direction between the transmitter and receiver and maximum for targets moving along a line which bisects the bistatic triangle given by $\psi_{max} = \phi_R + \phi_r/2$.

5. Show that the effective length L_e of an unfocused SAR is given by $\sqrt{R\lambda}$ and the cross-range resolution by $\sqrt{R\lambda}/2$. Use the criterion that the two-way path length from the edge of the synthetic aperture is $\lambda/4$ greater than the two-way path length from the center to the target. (*Hint*: From the geometry, set up the quadratic equation $R^2 + (L_e/2)^2 = (R + \lambda/8)^2$ and by analogy to a real antenna, the beamwidth is given by $\theta = \lambda/(2L_e)$, where the factor of 2 results from the two-way propagation path.)

6. An unfocused SAR operating at a frequency of 10 GHz is on an airborne platform moving with a ground velocity of 300 knots. The range to the target is 20 km. What is the effective length L_e of the synthetic array and the cross-range resolution? If the PRF is 4000 pps, how many pulses are processed to form the array?

7. An airborne radar illuminates a target for a semiactive missile. The radar operates at 10 GHz, provides 2-kW CW power, and has a 33 dB antenna gain. The missile seeker uses a 12-in-diameter flat plate array (efficiency of 0.7), has a 500-Hz bandwidth, and a 9-dB noise figure. A 1-m^2 target is 20 nmi from the illuminator, while the missile is 10 nmi from the target. Find the S/N ratio in the missile seeker. What angular tracking accuracy is available in the missile seeker at this range assuming a monopulse tracker is used?

8. Pulsed Doppler radar can be ambiguous in both range and velocity. Show that the unambiguous range-velocity contour is given by

$$R_u v_u = \frac{2.3565 \cdot 10^4}{f_{GHz}} \tag{2.3}$$

where f_{GHz} is the radar frequency in GHz. R_u is in nmi and v_u is in knots.

9. Using the results of Problem 8, what is the maximum unambiguous range for a radar operating at 10 GHz that must unambiguously detect targets with velocities of 600 knots? Repeat for a UHF radar operating at a frequency of 425 MHz.

10. Radar using MTI filters have a blind speed region for targets whose Doppler returns correspond in frequency to multiples of the radar's PRF. What would be the first blind speed for a radar operating at 1.2 GHz using a 600-Hz PRF? How many blind speed regions would there be for targets with velocities up to 1000 knots? What is the unambiguous range of this radar?

11. The equivalent RCS (σ_c) of surface clutter can be approximated as

$$\sigma_c = \sigma^\circ \cdot \theta_{AZ} R \cdot \frac{c\tau}{2} \tag{2.4}$$

where σ degrees is the dimensionless clutter backscatter coefficient, θ_{AZ} is the azimuth beamwidth, R is the range, and τ is the radar's pulsewidth. Solve this equation for the maximum radar detection range in surface clutter in terms of the signal-to-clutter ratio (σ_t/σ_c) whose value can be determined from standard radar detection curves for a particular P_d and P_{fa}.

12. Repeat Problem 11 for volume clutter whose RCS is given by

$$\sigma_c = \eta \cdot \frac{\pi}{8} \cdot \theta_{AZ} \phi_{el} R^2 \cdot \frac{c\tau}{2} \tag{2.5}$$

where η is the clutter's backscatter coefficient, θ_{AZ} is the azimuth beamwidth, ϕ_{el} is the elevation beamwidth, R is the radar range, and τ is the radar's pulsewidth.

13. Radar clutter can be attenuated using an MTI comb filter whose rejection notches are centered over the clutter Doppler return. The MTI improvement factor (I_c) is defined as the signal-to-clutter ratio at the output of the MTI filter divided by the signal-to-clutter ratio at the input to the MTI filter. Derive the maximum detection range at the MTI filter output in terms of the parameters given in Problem 11 and the MTI improvement factor.

14. The MTI improvement factor is given by

$$I_c = \frac{\sum_{j}^{n-1} w_j^2}{\sum_{j=0}^{n-1} \sum_{k=0}^{n-1} w_j w_k \rho_c(j-k)} \tag{2.6}$$

where $\mathbf{w} = [w_0 \ w_1 \ . \ . \ . \ w_{n-1}]$ is the MTI filter's weight vector and

$$\rho_c(j-k) = \exp[-(j-k)^2\Omega^2/2] \tag{2.7}$$

and $\Omega = \sigma_f T$ with σ_f the standard deviation of the clutter spectrum in rad/s and T the radar interpulse interval. Find the MTI improvement factor for a single $\mathbf{w} = [1 \quad -1]$ and double $\mathbf{w} = [1 \quad -2 \quad 1]$ canceller under the assumption that $\sigma_f T \ll 1$.

15. Show that, in general, the MTI improvement factor for MTI cancellers that process n pulses, whose weights use a binomial coefficient, is given by

$$I_c = \frac{2^{n-1}}{(n-1)!(\sigma_f T)^{2(n-1)}} \tag{2.8}$$

where $\sigma_f T \ll 1$ and the parameters are as described in Problem 14.

16. The primary limitation in the MTI performance of a surface-based search radar is due to the spectral spreading of the clutter due to the scanning motion of the antenna beam. For a Gaussian-shaped antenna beam $\sigma_f T = 1.666/n_B$, where n_B is the number of target hits occurring within the 3-dB width of the antenna beam. Using MATLAB, plot the MTI improvement factor for single, double, and triple cancellers as a function of n_B.

17. The primary limitation in the MTI performance of an airborne search radar is due to the spectral spreading of the clutter due to platform motion. The standard deviation in Hertz of the clutter spectrum for low-altitude operation is given by

$$\sigma_{pm} = \frac{0.528 \cdot V_{ac} \sin \theta_{AZ}}{D_{AZ}} \tag{2.9}$$

where V_{ac} is the aircraft velocity, D is the aperture width in compatible units, and θ_{AZ} is the azimuth scan angle. Using MATLAB, plot the MTI improvement factor for single, double, and triple cancellers as a function of the azimuth scan angle. Assume an aircraft velocity of 300 knots, an aperture width of 3 ft, and an n_B of 20 hits per 3-dB beamwidth. Remember that standard deviations due to different effects combine as the square root of the sum of the squares.

18. Assume that a surface-based MTI radar must detect a cruise missile, whose RCS is 0.01 m^2, with a $P_d = 0.9$ and a $P_{fa} = 10^{-6}$ at a range of 15 nmi. The radar has a beamwidth of 1 degree, a pulsewidth of 1 μs, while the clutter backscattering coefficient ($\sigma°$) is −30 dB. What MTI improvement factor is needed to meet this requirement?

19. Repeat Problem 18 to find the MTI improvement factor for detection of the cruise missile in rain clutter. Assume that the MTI filter notch can be centered on the rain clutter spectrum. At the radar's frequency, 4 mm/hr of rain results in a backscattering coefficient (η) of $10^{-6.5}$/m. Assume an elevation beamwidth of 3 degrees.

20. The ARSR-4 radar has the parameters

frequency	1300 MHz
scan rate	30 degrees/s
beamwidth	1.4 degrees
PRF	216 Hz
MTI	Double canceller

What is the MTI improvement factor for this radar due to antenna scanning? Also find the first blind speed and standard deviation of the antenna scanning clutter spectrum in knots.

21. The MTI improvement factor for an N-point FFT filter is

$$I_c = \frac{N^2}{N + \sum_{i=1}^{N-1} (N - i) \cos\left(2\pi \frac{ki}{N}\right) \cdot \rho_c(i)} \tag{2.10}$$

where k (0, 1, 2, . . . , $N-1$) is the FFT filter number and $\rho_c(\cdot)$ is defined by (2.7). What is the MTI improvement factor for the $k = 2$ filter using the parameters given in Problem 20? How does this compare to that of the double canceller MTI?

22. Consider two airborne radars with parameters

Parameter	A/B #1	A/B #2
f	425 MHz	10,000 MHz
D_{AZ}	20 ft	2.5 ft
V_{ac}	300 knots	600 knots
PRF	300 Hz	4000 Hz
Scan angle	90 degrees	30 degrees

Find the relative clear region/blind region ratio for an MTI filter due to the platform motion spectrum. Assume that the blind region is defined by (2.9) while the rest of the region is clear.

23. For a pulse Doppler radar with a beamwidth of 1 degree, calculate the antenna scan rate that allows a velocity resolution of 10 ft/s. Assume a matched filter and a wavelength of 5 cm.

24. An airborne radar operating at 16.5 GHz uses a DBS mode to detect a fixed ground target. The aircraft is moving at 400 knots, has a beamwidth of 1.5 degrees, scans at 20 degrees/s, and has a PRF of 1700 Hz. The DBS processor uses an FFT filter bank that fills the region between PRF lines. How many points are required in the FFT filter to minimize the clutter output that competes with a target at 30 degrees scan angle?

25. An ISAR processor detects the Doppler shift caused by target rotation and converts this into a target image. Assume that an X-band radar ($\lambda = 3$ cm) is used to image a ship target using pitch rotational motion so that the ISAR image plane consists of the ship length for slant range and the ships height for cross range. Assume that the ship's length is $D_y = 200$m while its height is $D_x = 30$m. Find the cross-range resolution given by

$$(\Delta r_c)^2 > \Gamma \frac{\lambda D_y}{4} \tag{2.11}$$

Then solve for the maximum slant range resolution given by

$$\Delta r_c \Delta r_s > \frac{\lambda D_x}{4} \tag{2.12}$$

26. In Problem 25, assume the ship has a rate of 1 degree/s. Using the results of Problem 25, what Doppler resolution (Δf_d) is required to provide the maximum cross-range resolution? What coherent processing time is required to provide this Doppler resolution? Through what angle does the ship rotate during this coherent processing time interval (PTI)?

27. An unfocused side-looking SAR is moving at 300 knots. The azimuth beamwidth is 4 degrees, the frequency is 10 GHz, and the PRF is 4000 Hz. The slant range to the radar patch is 20 km. Determine the maximum length (l) of the synthetic aperture using the criterion that the synthetically formed antenna beam must be in the far-field region (i.e., $R > 2l^2/\lambda$). What is the processing time to form the synthetic antenna beam? How many pulses will be coherently integrated in this period? What is the resolution of the SAR?

References

[1] Sweetman, W., "Airborne Reconnaissance Radar," *International Defense Review,* Sept. 1987.

[2] Hewish, M., "Airborne Ground Surveillance," *International Defense Review,* Jan. 1996.

[3] Hewish, M., "Building a Bird's-Eye View of the Battlefield," *International Defense Review,* Feb. 1997.

[4] Knowles, J., "EW and UAVs: Payloads That Pay Off," *J. Electronic Defense,* July 1996.

[5] Entzminger, J., "Acquiring Affordable UAVs," *J. Electronic Defense,* Jan. 1995.

[6] Schleher, D. C., *Introduction to Electronic Warfare,* Norwood, MA: Artech House, 1986.

[7] Blake, B., "Jane's Radar and Electronic Warfare Systems 1996–97," Surrey CR3 2NX, United Kingdom.

[8] Brookner, E., *Aspects of Modern Radar,* Norwood, MA: Artech House, 1988.

[9] Papke, N., et al, *Electro-Optical/Infrared (EO/IR) Countermeasures Handbook,* Arlington, VA: Naval Air Systems Command, 1996.

[10] Brukiewa, T., "Active Arrays: The Key to Future Radar Systems Development," *J. Electronic Defense,* Sept. 1992.

[11] Pengelley, R. "Saturation Attacks Challenge SHORAD Capabilities," *International Defense Review,* June 1993.

[12] Hardy, W., "Land Clutter Effects on Shipboard Radar," Navy PEO (TAD) Report, 11 April 1996.

[13] Nash, T., "Close-Range Clout Defends Against Air Attack," *International Defense Review,* July 1996.

[14] Streetly, S., "Twisting the Tiger's Tail," *J. Electronic Defense,* Sept. 1990.

[15] *International Countermeasures Handbook,* 14th ed., Cardiff Publishing, 1989.

[16] Barton, D., "The 1993 Moscow Air Show," *Microwave J.,* May 1994.

[17] Blake, B., "Long-Range Air-Defense Systems," *International Defense Review,* May 1994.

[18] "Widening the Radar Net," *International Defense Review,* Sept. 1996.

[19] Bayle, D., and G. Sundaram, "Looking into the Future for High Power Radar," *International Defense Review,* June 1991.

[20] "Active Array Radar: Principles and Benefits," *J. Electronic Defense,* Jan. 1997 Suppl.

[21] Goutzoulis, A., K. Davies, J. Zomp, P. Hrycah, and A. Johnson, "A Hardware Compressive Fiber-Optic True Time Delay Steering System for Phased-Array Antennas," *Microwave J.,* Sept. 1994.

[22] Stimson, G., *Introduction to Airborne Radar,* El Segundo, CA: Hughes Aircraft Co., 1983.

[23] Richardson, D., "New Modes for Tomorrow's Fighter Radar," *International Defense Review,* July 1995.

[24] Schleher, D. C., *MTI and Pulsed Doppler Radar,* Norwood, MA: Artech House, 1991.

[25] Hewish, M., "Airborne Ground Surveillance," *International Defense Review,* Jan. 1996.

[26] Barile, E., R. Fante, and J. Torres, "Some Limitations on the Effectiveness of Airborne Adaptive Radar," *IEEE Trans. AES,* Vol. 28, No. 4, Oct. 1992.

[27] Ward, J., "Space-Time Adaptive Processing for Airborne Radar," Lincoln Laboratories Technical Report 1015, Lexington, MA, Dec. 1994.

[28] Wamstall, B., "Joint Stars Fights to Stay on Target," *Interavia,* Nov. 1988.

[29] Shnitkin, H., "Joint Stars Phased Array Antenna," *IEEE Systems Magazine,* Oct. 1994.

[30] Wehner, D., *High Resolution Radar,* Norwood, MA: Artech House, 1987.

3

Modern EA Systems—Architecture, Types, and Technology

The modern concept of *electronic attack* (EA) is to offensively use the *electromagnetic* (EM) spectrum to disable the enemy's combatability. EA involves both nondestructive (soft kill) and destructive (hard kill) actions. Jamming is an example of a soft kill action that attempts to dilute the effectiveness of an enemy weapon system through confusion, distraction, deception, or seduction. Destruction of an enemy radar by an ARM and disabling the detector in an IR missile using a high-powered laser are hard kill EA actions. In addition to EA hard kill actions, C2W encompasses the classical kinematic kill that destroys the threat weapon either by collision or by explosion.

Due to their different mechanisms, soft and hard kill have generally been separately employed; but it is apparent that both are aimed at a unitary purpose. Soft kill mechanisms are generally more economical to employ, have inherent multithreat capability; and, being nonlethal, represent a reduced political consequence in use. However, soft kill damage assessment is difficult to determine and being nonlethal may simply divert the weapon to another target. This leads to the conclusion that both soft and hard kill must be applied in a coordinated manner to optimize the desired effect.

An example that displays the interaction between soft and hard kill is the defense of a ship against an *anti-shipping cruise missile* (ASCM). Current ASCMs (i.e., Harpoon, Exocet) fly at low levels (10 ft to 50 ft) above the sea surface with speeds approaching Mach 1. They are generally radar guided, but IR (e.g., Penguin) and dual-mode (radar and IR) guided missiles are currently operational. As the ASCM appears above the radar horizon (about 15 nmi), the ship has 75 sec to 100 sec to either destroy or divert the missile. Soft kill and hard kill mechanisms are listed in Table 3.1. These defense mechanisms represent a layered defense to ship survivability [1]. The *Cooperative Engagement Capability* (CEC) or "forward pass" attempts to engage the ASCM over-the-horizon using airborne-based or ancillary sensors to locate the target and destroy it before it is seen by the ship. In addition, the ship's air defense SAM attempts to destroy the ASCM launch platform. Soft kills are generally attempted as the ASCM appears over the horizon since these techniques are most effective when the ASCM seeker is in an acquisition mode. Chaff, flares, and various passive and active decoys are used to attract the missile away from the ship. Missiles launched from the ship are aimed at the ASCM using either semiactive or active radar or IR guidance. The *Close-in-Weapon System* (CIWS's) high fire rate gun system provides the inner defense layer that is employed as the ASCM enters the zone 400m to 4 km from the ship. Research on high-powered lasers is progressing and may ultimately prove to be an effective defense mechanism. The laser solution offers almost instantaneous reaction at the speed of light, rapid re-engagement, a large and renewable magazine supply, line-of-sight accuracy, precision aim pointing, and high single shot cost effectiveness. However, in practice, various problems such as thermal blooming, atmospheric attenuation, beam focusing, power generation, and large equipment weight and volume have limited practical deployment of such a weapon [2].

Figure 3.1 depicts a functional block diagram of a generic EA system. A multispectral receiving and processing system provides situational awareness of the environment and collects threat signals in that environment. The system computer assesses the overall threat and allocates EA resources and threat techniques to counter that threat. The EA resources consist of onboard jamming transmitters in the RF, IR, and *electro-optical* (EO) (laser) spectral bands and offboard chaff and flare dispensers complemented by active and passive decoys.

The most prevalent threat against which EA is directed is the radar or IR guided missile. IR guided seekers have traditionally been vulnerable to

Table 3.1
Soft and Hard Kill ASCM Defense Systems

Soft Kill	**Hard Kill**
Chaff	Close-in-Weapon System (CIWS)
Flares	Semi-Active Homing Missile (Sea Sparrow)
Active decoy	Passive Homing Missile (RAM)
EA	DEW (High-powered laser)
Towed decoy (rubber duck)	Cooperative Engagement Capability (CEC)
IRCM/DIRCM	Surface-to-Air Missile (SAM)

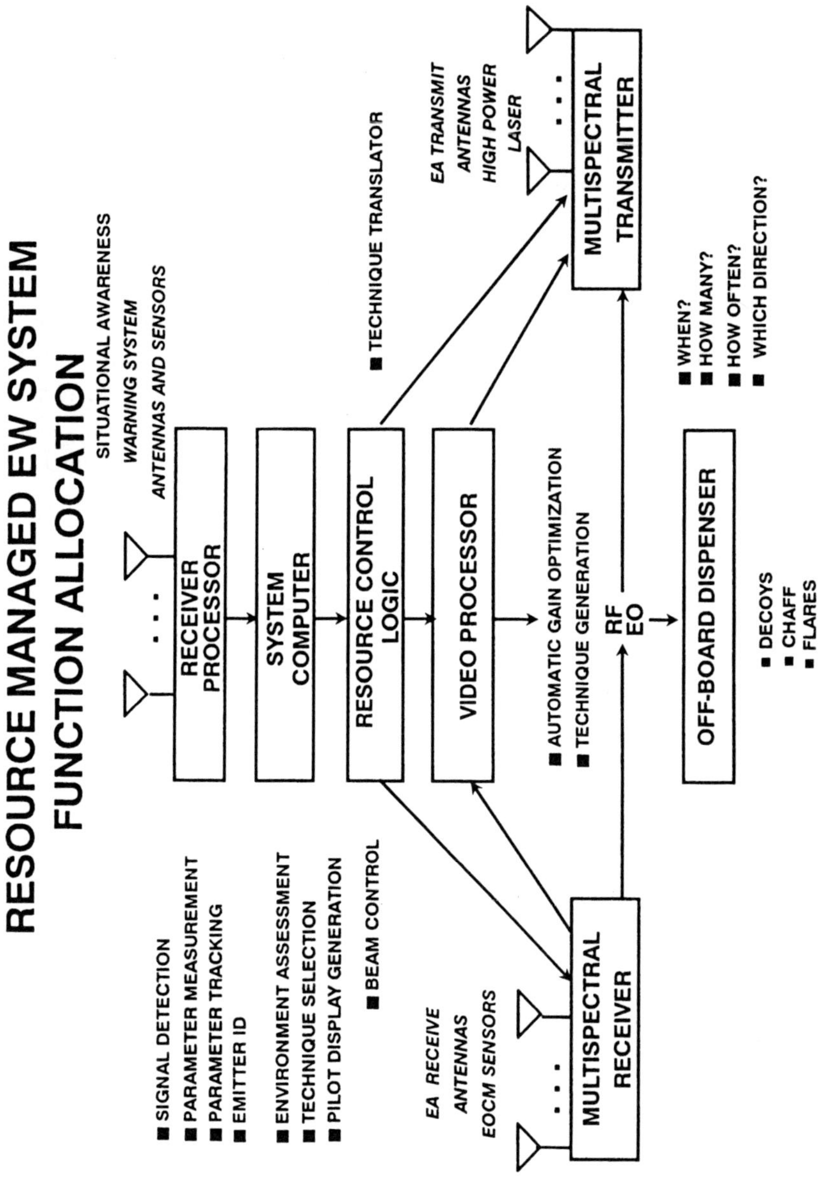

RESOURCE MANAGED EW SYSTEM
FUNCTION ALLOCATION

RECEIVER PROCESSOR
SITUATIONAL AWARENESS
WARNING SYSTEM
ANTENNAS AND SENSORS
■ SIGNAL DETECTION
■ PARAMETER MEASUREMENT
■ PARAMETER TRACKING
■ EMITTER ID

SYSTEM COMPUTER
■ ENVIRONMENT ASSESSMENT
■ TECHNIQUE SELECTION
■ PILOT DISPLAY GENERATION

RESOURCE CONTROL LOGIC
■ BEAM CONTROL
■ TECHNIQUE TRANSLATOR

VIDEO PROCESSOR
■ AUTOMATIC GAIN OPTIMIZATION
■ TECHNIQUE GENERATION

EA RECEIVE
ANTENNAS
EOCM SENSORS
MULTISPECTRAL RECEIVER

EA TRANSMIT
ANTENNAS
HIGH POWER
LASER
MULTISPECTRAL TRANSMITTER

RF
EO

OFF-BOARD DISPENSER
■ DECOYS
■ CHAFF
■ FLARES

■ WHEN?
■ HOW MANY?
■ HOW OFTEN?
■ WHICH DIRECTION?

Figure 3.1 Functional EA system.

weather, flares, and reflections off the surface that cause excessive false alarms. On the other hand, radar-guided weapons offered all-weather performance, long range, and better counter-countermeasure capability. This led to an emphasis on RF countermeasures in current EA equipment. However, the balance is changing toward IR seekers that tend to be inherently countermeasure resistant, using factors such as temperature, color, size, and shape to discriminate between target and countermeasures and using improved signal processing that makes them better at rejecting look-down clutter.

Current RF missiles have greater target range than IR guided missiles, but the differential is diminishing. However, radar seekers face the challenge of new EA techniques, such as towed decoys, terrain bounce, and the deployment of low RCS aircraft and missiles. Semiactive seekers have a particular problem where post-launch turnaway to reduce their closing rates sometimes moves the radar echo from the fighter toward the main-beam clutter [3]. In addition, steadily improving RF SAMs and AAMs demand increased self-protection capabilities to cope with new multimode, frequency agile, *pulsed Doppler* (PD), pulse compression radars using monopulse angular tracking techniques.

The net result of this enhanced SAM and AAM threat is the economic challenge of including enough EA capability onboard the platform to be protected. This has led to an interest in offboard towed decoys as an economical means for coping with these advanced threats [4].

3.1 Onboard/Offboard Architectures

Figure 3.2 depicts three EA system architectures that use onboard, offboard, or combined onboard/offboard architectures. Onboard countermeasures (Figure 3.2(a)) are typical of many current systems that are concerned primarily with the RF threat and consist of either internal or, if on an aircraft, pod-mounted equipment.

Offboard countermeasures refer to those systems that are carried by the aircraft, helicopter, or ship but are deployed when under direct attack (i.e., missile launch) by the threat. There are generally three types of offboard countermeasures: (1) dispensed expendable devices (i.e., chaff and flares) that leave the protected vehicle and are never recovered, (2) propelled devices that are intended to divert the missile from its target, and (3) towed devices that can be recovered and reused. The use of offboard countermeasures implies a *missile attack warning* (MAW) system that triggers the offboard action at the appropriate time and in the appropriate direction. In some cases, the *radar*

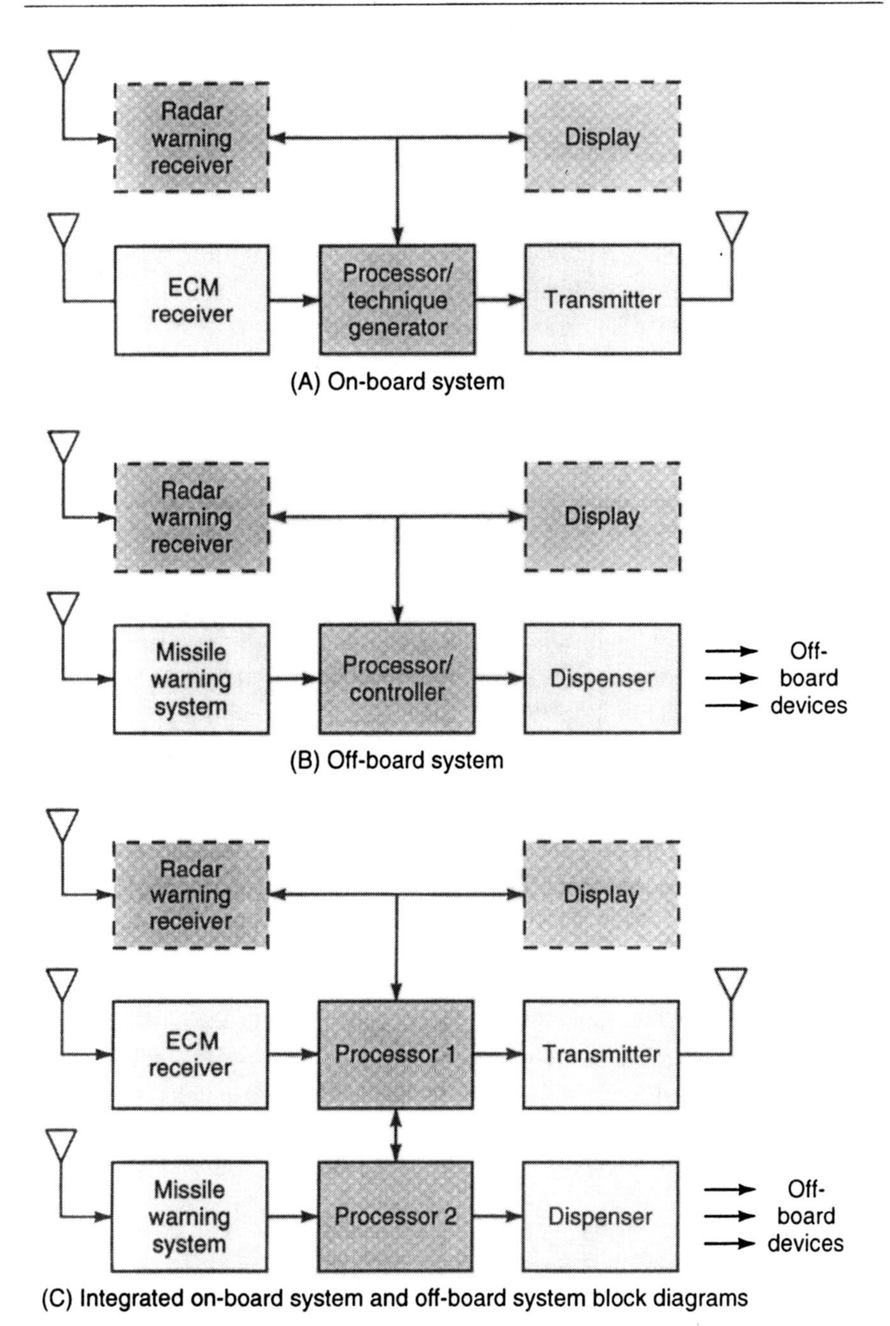

Figure 3.2 Onboard and offboard system block diagrams. (*Source:* [5].)

warning receiver (RWR) is adequate for this purpose, but generally its resolution is inadequate.

The integrated onboard/offboard system (see Figure 3.2(c)) potentially provides a higher level of survivability than either system alone. The onboard system can degrade a threat's target acquisition, target track, and missile guidance functions; while the offboard system is effective only in the endgame to decoy the missile from its intended target [5].

The *Integrated Defensive Electronic Countermeasures* (IDECM) system depicted in Figure 3.3 is an example of a combined onboard/offboard EA system [6]. The onboard (techniques generator) portion of IDECM is based upon the *Advanced Self-Protection Jammer* (ASPJ) design. The *techniques generator* (TG) is designed to apply a variety of RF techniques against pulsed, CW, and PD threats. The TG samples the RF environment and then compares it against a threat library to match a specific threat identification to the received environment. It then generates in the signal conditioning assembly the specific RF technique that counters the threat; converts the RF signals to light in a laser modulator, and transmits this signal through a *fiber optic* (FO) link to the *fiber optic towed decoy* (FOTD). The FOTD converts the laser signal to RF and transmits it using a *traveling wave tube* (TWT) amplifier.

Using an onboard jammer to generate RF techniques and an FOTD to transmit them gives the aircraft the benefits of both a smart onboard jammer and an offboard repeater. This approach provides a cost-effective method to counter RF guided missiles including those employing difficult-to-jam monopulse seekers. Note, however, that due to geometrical and deployment constraints, the towed decoy cannot guarantee protection in the front and rear sectors of attack [7].

Other possible towed-decoy configurations are depicted in Figure 3.4. The ALE-50 is a self-contained RF countermeasure system with a TG, modulator, and power amplifier contained in the decoy tow body [4]. When deployed, it independently counters a radar by repeating the threat radar signals. The ALE-50 is generally complemented with an RWR and an expendables dispenser (ALE-47) that launches chaff, flares, and seduction RF signals (GEN-X).

3.2 Operational EA Systems Architecture

The architecture of a typical self-protection airborne jammer (i.e., ALQ-165) is depicted in Figure 3.5. Typical parameters are given in Table 3.2 for this type of system [6]. Generally, self-protection jammers provide ±60 degrees fore-and-aft angular protection since the EA system cannot generate sufficient ERP to cover its broadside *radar cross section* (RCS). The frequency range is

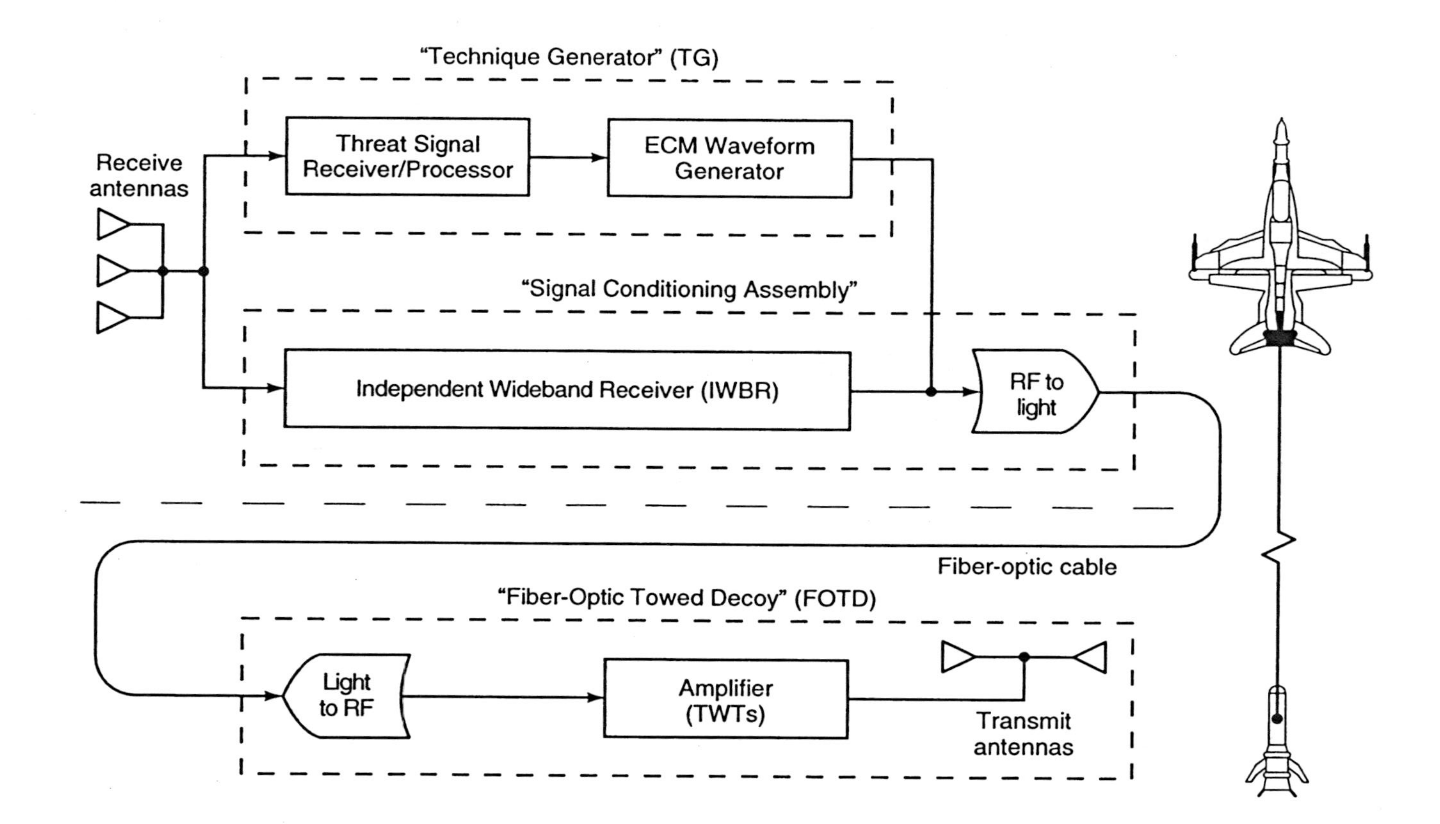

Figure 3.3 The IDECM RFCM system. (*Source:* [4].)

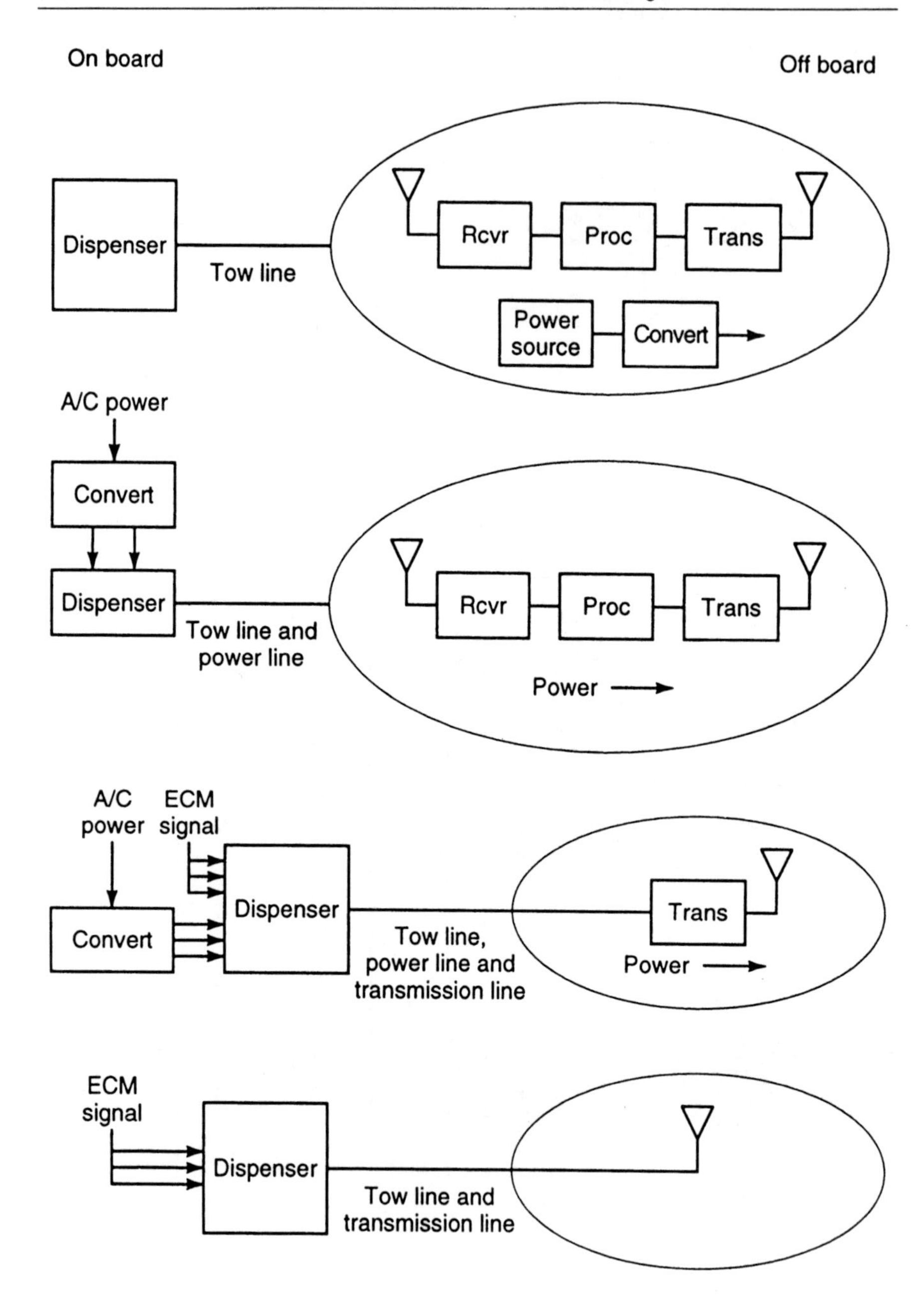

Figure 3.4 Variants of onboard/offboard towed decoys. (*Source:* [5].)

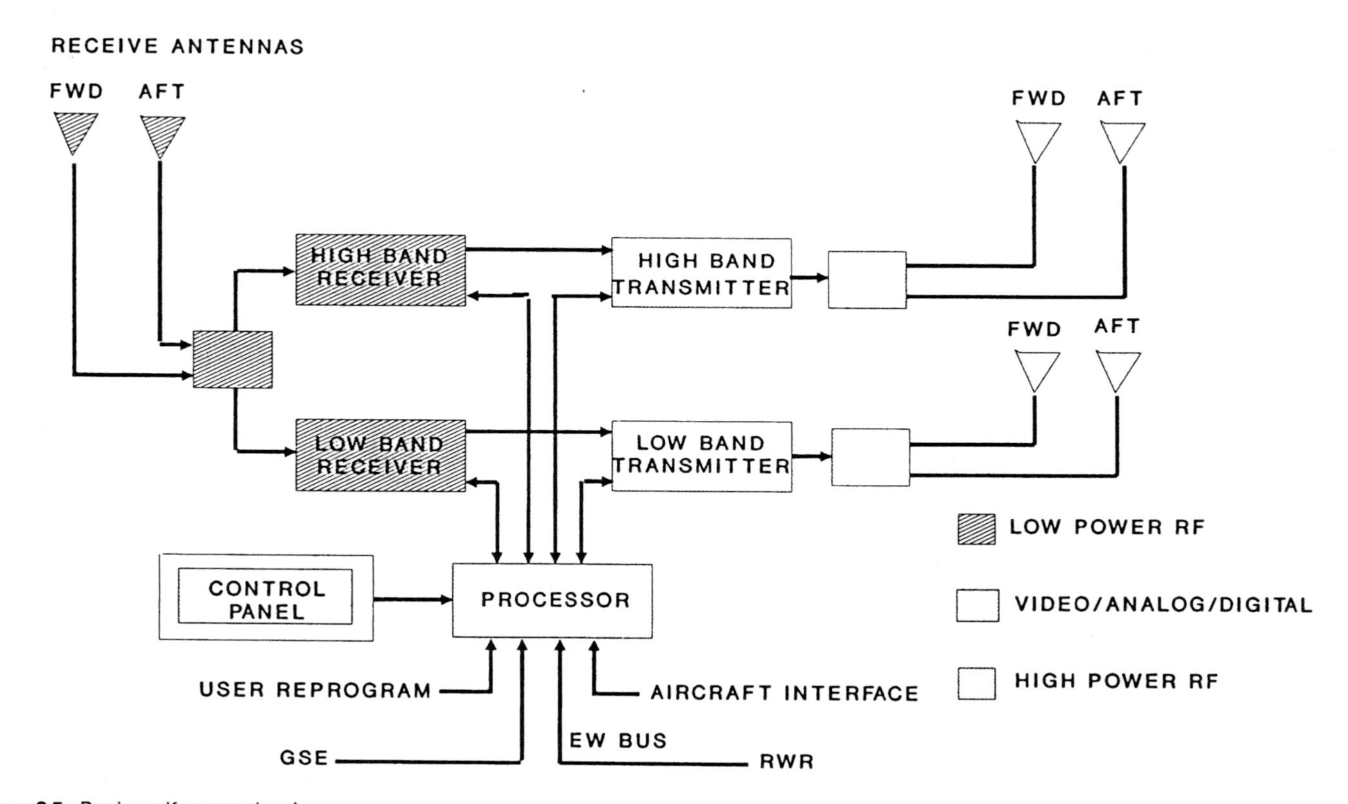

Figure 3.5 Basic self-protection jammer.

Table 3.2
Typical Advanced Self-Protection Jammer Parameters

Frequency coverage	0.7 to 18 GHz
System response time	0.1 to 0.25 sec
Receiver-processor	
Dynamic range	50 dB
Sensitivity	–71 dBm
Resolution	5 MHz
Instantaneous bandwidth	1.44 GHz
Pulsewidth	0.1 μs (min)
Input pulse amplifier	20 dBm (max)
False alarm	5/hr (max)
Signal detection capability	Pulse, CW, PD, Agile
Jammer	
Peak power	58 to 63 dBm
Pulse-up capability	5 to 7 dB
Set-on accuracy	±0.5 to ±20 MHz
Duty cycle	5% to 10%
Jamming capability	16 to 32 signals
Modes	Noise deception

covered in two bands since TWTs with adequate power are not available to cover the whole band. The processor is capable of handling a number of pulsed and *continuous wave* (CW) threats simultaneously. The CW capability requires separate additional TWTs in each band since most TWTs of the type employed cannot amplify both CW and pulsed signals simultaneously.

EW self-protection systems have become an essential part of modern tactical aircraft. This capability dictates the combination of the RWR threat warning function and an EA jamming function. These functions can be supplied separately or by an integrated system. In any event, the interaction of the jamming radiations from the EA system with the RWR must be considered. On the other hand, the stand-alone RWRs described previously do not have the capability of supplying the information required for EA operation. For example, the simple RWR does not provide: (1) the threat's frequency signature with enough resolution to permit jammer set-on, (2) threat location data with enough accuracy to permit defense suppression action, nor (3) threat priority data to allow optimal use of jammer resources for EA power management.

A *tactical EW system* (TEWS) receiver (AN/ALR-56C) combines both RWR and EA functions. This system functions to identify intercepted threats and then provides digitized parameters to the associated jammer (AN/ALQ-135). An interesting feature of this advanced RWR is that all functions are controlled by the integral computer. However, this capability is unusual in that most EA systems contain their own ES system and the RWR function is performed separately in a dedicated system (see Figure 3.5).

Another EA system architecture used to protect naval ships against air and ASCM attack is the SLQ-32V(3) depicted in Figure 3.6 [8]. An incoming signal is focused at an output port of the receiver array lens corresponding to the signal direction of arrival. In a repeater DECM mode, the signal is switched by means of the receiver switch into a repeater channel, where it is amplified, modulated, and fed by means of the transmitter switch to the selected beam port of the transmit lens. The port selected corresponds to a transmitting beam that is effectively retrodirective to the incoming signal.

In the transponder mode, the jamming signal is generated internally, within the transponder unit, by being phase-locked to the threat signal. As with the repeater mode, the direction of transmission is determined by the beam-port selection on the transmitter lens. By providing multiple repeater channels and multiple transponder RF generators, several hostile threats can be simultaneously countered by either mode or a combination of modes. The full gain of the transmit array is realized against each threat, and each threat receives an optimized countermeasure response.

An advantage of the system is that the directive properties of the transmitter and receiver lenses are matched, so beam-steering computers and complex logic circuitry are avoided. The only distinction between the receiver and transmitter lenses is that the receive arrays are optimized for minimum sidelobes and fixed beamwidths, while the transmit arrays are designed for maximum ERP. The technique used in the receive array to maintain constant beamwidth over the frequency range is to provide increasing attenuation of the outer array elements as the frequency increases, thus keeping the array aperture size constant in terms of wavelength.

The EA system depicted in Figure 3.6 uses a total of 140 medium-power (40W to 50W) TWTs to generate an ERP in excess of 1 MW. Coverage is through 360 degrees in azimuth and on the order of 24 degrees in elevation over the frequency range of E through J bands.

The key element of the lens-fed multiple beam phased array is the beamforming lens. Various lenses are used for different coverage situations. The Rotman lens is appropriate for linear arrays, the R-2R lens for a 90-degree arc array, and the R-kR lens for semicircular and circular arrays. The Rotman lens, although generally designed to feed straight linear arrays, can accommodate array curvature up to a maximum arc length of approximately 90 degrees [9].

A description of some modern EA systems is given in Table 3.3. These range from stand-off jammers (ALQ-99E) to both internal (ALQ-126B, ALQ-136, ALQ-161, ALQ-162, ALQ-165, ALQ-135, ALQ-172) and pod-type (ALQ-131, ALQ-184) self-protection systems. The most prevalent type

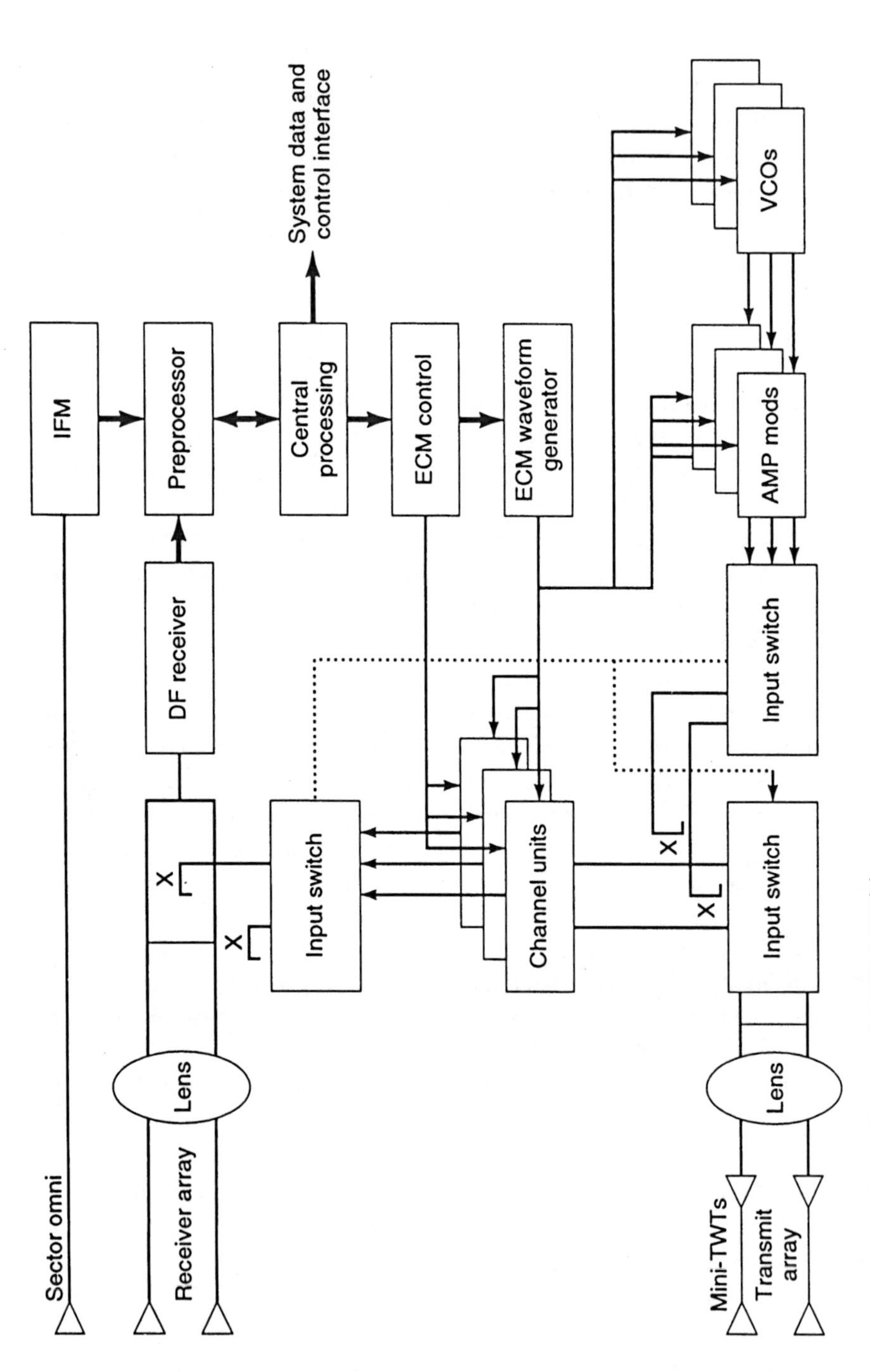

Figure 3.6 Functional block diagram of AN/SLQ-32V(3).

Table 3.3
Modern EA Systems

Type	Platform	Function	Characteristics
ALQ-99-E	EA-6B, EF-111A	Support	Search radar jamming, 10 transmitters, analysis receivers, directional antennas
ALQ-126B	A-4, A-6, A-7, F-4, F-14, F-18	Self-protection	Coverage through I/J band, 1-kW peak power, 4% to 5% duty, track radar jamming, typical 60-degree fore/aft coverage, repeater transponder modes
ALQ-136	Army helicopters	Self-protection	I/J-band jammer, AA artillery radar jammer, threat programmable
ALQ-131	F-16, F-111, A-7, A-10, F-15, F-4	Self-protection	Dual-mode pod jammer, CVR/SHR analysis receiver, phased array
ALQ-161	B-1B	Self-protection	Coverage through I band, search and track radar jamming, monopulse and Doppler radar jamming, software controllable, phased array
ALQ-162 (shadowbox)	AV-8B, F-16, F-18, RC-12D, RV-1D, helicopter	Self-protection	CW jammer, chopped-repeater lightweight, threat programmable, ±60-degree switchable coverage
ALQ-165 (ASPJ)	F-16C*, F/A-18*, F-14*, AV-8B*	Self-protection	Covered frequency range in two bands, pulse/CW capability, analysis receivers, threat programmable
ALQ-135 (TEWS)	F-15C	Self-protection	High-powered transmitters, power management, integrated with ALR-56C
ALQ-172	B-52G/H	Self-protection	Track/search radar jamming, steerable jam beams, software programmable, phased array antenna, monopulse radar jamming
ALQ-184(V)	F-4, F-15, F-16, A-7, A-10, F-111	Self-protection	Pod jammer, Rotman lens, medium-power miniature TWTs, transponder and repeater jamming, high ERP
SLQ-32V(3)	Ships	Self-protection	Lens-fed array, crystal video and IFM receivers, medium-power miniature TWTs, tactical display, transponder and repeater jamming, high ERP

*Originally planned.

of EA system is the self-protection jammer (see Table 3.3). The distribution of this type of EA is approximately 65% airborne, 25% shipboard, and 10% ground-based. The emphasis on airborne self-protection EA reflects the relative vulnerability of aircraft that must confront modern air defense systems. The importance of naval self-protection EA has recently gained impetus with the emergence of the ASCM threat.

From an operations viewpoint, the airborne and naval self-protection requirements are significantly different. Most fighter strike aircraft have low RCS, which range from under 1 m^2 to 10 m^2 for head-on aspects and from 10 to 100 m^2 for broadside aspects. Typical EA coverage is fore and aft (e.g., 60-degree cone tilted 15 degrees downward) in the regions of minimum RCS. Size and weight are limited in strike aircraft, presenting a problem in carrying EA jamming transmitters covering the full radar band (e.g., 0.5 GHz to 18 GHz). One solution is to use external jammer pods, which are specialized to the expected threats (frequency bands) to be countered on a particular mission. An alternate solution is to use an internal, power-managed EA system that covers only the terminal threat bands (e.g., E/F, G/H, I/J bands). The internal EA size and weight constraints are somewhat relaxed for bomber-sized aircraft, but the RCS ranges to be covered are also increased.

On the other hand, naval vessels have RCSs that can be as large as one million square meters, requiring very large jammer ERP to cover the target. Also, ships are relatively immobile and it is difficult to avoid an accurately targeted missile once launched. However, a significant advantage for EA on board a naval vessel is the relatively unlimited weight, size, and prime power available as contrasted with the airborne situation.

The dominant characteristic of self-protection EA is that each aircraft or ship carries sufficient EA equipment to provide for its own self-defense. In the airborne case, it is generally desirable to supplement self-protection EA with either escort or stand-off EA systems (see Figure 1.8). This type of EA is carried on escort vehicles other than the one being protected. Escort EA implies a dedicated aircraft carrying high-powered jammers that accompanies the friendly strike force and provides a protective jamming shield in support of the entire force. Stand-off EA implies a jamming vehicle that stands off at a distance beyond the effective range of target defenses.

The stand-off jammer has the advantage of carrying a large dedicated EA payload, which can be advantageously positioned for maximum jamming effectiveness. The disadvantage is that large ERP is required due to the potentially long jamming range and the need to jam into the sidelobe pattern of the victim antenna. The escort jammer has the advantage of a favorable jamming geometry, which includes the possibility of main-lobe jamming, thereby maximizing its effect. However, the escort jammer becomes a high-priority target for the enemy defense system and, when rendered inoperable, leaves the entire force vulnerable to enemy action.

Another concept that is currently under study is that of the stand-in jammer. This jammer operates in a UAV much like the stand-off jammer. However, since it is much closer, the ERP jamming requirements are reduced by the square of the operating range ratio. In this way, a small 100-lb jammer equipped with a phased array antenna can generate over 100W of ERP, which when operated in proximity to the radar can produce the equivalent of 40 kW of ERP from the stand-off jammer.

3.3 EA Radar Jamming Waveforms

There are basically two types of radar that must be jammed by EA equipment. The first is the surveillance or search radar, which functions to locate the position of a target within a large coverage volume. Conceptually, the search radar can be modeled as a rotating antenna beam whose main lobe sequentially scans the search volume while its sidelobes provide response in all directions. The magnitude of the sidelobe response is dictated by the type of antenna employed (e.g., ordinary, low, or ultralow sidelobe antenna with and without sidelobe cancellation). The radar transmits a waveform that after reflection from targets is detected in a receiver noise background by a matched-filter receiver. The matched filter maximizes the received signal-to-noise ratio. The radar signal-to-noise ratio then depends on the energy received from the target and the receiver noise spectral density (proportional to the system noise temperature, which depends on the receiver noise figure and external noise intercepted by the radar antenna).

The second major type of radar that must be jammed by EA systems is the tracking radar. Tracking radars are usually given high priority in the hierarchy of EA threats because they are associated with the terminal phases of a weapon system. The fact that a tracking radar is locked onto a target implies that a weapon is directed at the target. The function of the EA system is to cause the tracking radar to break lock, which removes the guidance information being used by the weapon to converge on the target.

From the preceding models, there are two fundamental ways to introduce jamming into the radar. First, the receiver noise level can be raised by injecting external noise through the radar's antenna. This can be accomplished through either the radar antenna's one-way main-lobe or sidelobe responses. Noise jamming has the effect of obscuring the radar target by immersing it in noise. Second, spurious signals can be introduced into the radar's main-lobe or sidelobe response to confuse or deceive the radar with respect to the location of the real target. The radar's response to these spurious signals will be maximized if they closely replicate the actual radar transmitter waveform. Any deviation in either the envelope or phase (frequency) structure of the jamming signal with respect to the actual transmitted waveform will cause a mismatch loss, which must be compensated for by extra jamming power. For this reason, deception jammers usually repeat the radar's signal with both a coherently related time delay and frequency translation.

Noise jamming waveforms have the advantage against search radar that little need be known about the victim radar's parameters except its frequency range. The objective of the radar faced with noise jamming is to spread its transmitted energy over as wide a band as possible to force the jammer to dilute its effective radiated power spectral density (ERP/MHz) and to minimize its response to signals returned through the antenna sidelobes. Even with extremely wideband transmitted waveforms, the noise jammer holds the advantage with respect to main-beam jamming. The advent of ultralow sidelobe antennas and high-performance SLCs swings the advantage over to the radar when sidelobe noise jamming is employed. When the radar is successful in limiting the response to sidelobe noise jamming, a narrow strobe in the direction of the jammer is generated as the main antenna lobe of the radar scans over the jammer. This provides the capability of locating the jammer's position through the use of triangulation from multiple radar. When multiple jammers are employed, this capability is diminished due to the "ghosting" effect caused by the potentially ambiguous crossings of multiple strobes.

Repeater-type jamming waveforms have the potential to use jamming energy against search radar more effectively. Against search radar, the repeater or deception EA (DECM) system must intercept and store the essential spectral and temporal characteristics of the radar waveform. It must then generate synthetic targets that are synchronized with the scan and waveform pattern of the radar to create enough false targets to confuse the radar. As with noise jammers, this action is most easily accomplished when the jamming signals are injected into the main lobe of the radar's antenna pattern. However, sidelobe deception is highly desirable to prevent strobing in the jammer's direction.

The nature of tracking radar operation generally dictates a self-protection main-lobe DECM jamming approach. Also, since tracking radar performance generally improves in direct proportion to frequency throughout the microwave region, it is usual to find the tracking radar threat located in H, I, J, or K bands. Therefore, EA systems specifically directed to counter tracking radars are generally located in these bands. Coverage is also often extended to cover E, F, and G bands, where the acquisition radars that initially direct the tracking radars to the target are located. The rationale here is that if the acquisition radar can be jammed, then the tracking radar will not be able to locate the target.

Monopulse angle-tracking systems are inherently difficult to jam since angle error data are developed on each radar pulse making them insensitive to any form of amplitude modulation. Techniques used against monopulse angle tracking include cross-polarization, cross-eye, terrain-bounce, and towed decoy jamming. Of these techniques, the towed-decoy approach is currently favored.

3.3.1 Noise Jamming

The objective of noise jamming is to inject an interference signal into the enemy's electronic equipment such that the actual signal is completely submerged by the interference. This type of jamming is also called denial jamming or obscuration jamming. In principle, the optimal jamming signal has the characteristics of receiver noise, but in practice this may be difficult to achieve.

The primary advantage of noise jamming is that only minimal details about the enemy equipment need be known. A convenient classification of noise jamming is by the ratio of the jamming signal bandwidth to the acceptance bandwidth of the victim equipment. If the ratio is large, the signal is called barrage jamming; if the ratio is small, the signal is called spot jamming.

To analyze the effect of a noise jammer on a search radar, first consider the signal power received by the radar on a single pulse without jamming

$$S = \frac{P_T G_T G_R \sigma \lambda^2}{(4\pi)^3 R_t^4 L_R} \tag{3.1}$$

where

P_T = peak power transmitted, W

G_T, G_R = transmitter/receiver antenna gain

σ = RCS, m^2

λ = wavelength, m

R_t = target range, m

L_R = radar system losses

The internal noise power in the receiver is given by

$$N = kT_0 \cdot B \cdot F \tag{3.2}$$

where

k = Bolzmann's constant-1.38 $\cdot 10^{-33}$ (J/°K)

T_0 = noise temperature-290°K

B_r = noise bandwidth, Hz = $1/\tau$

F = noise factor

τ = radar pulsewidth, sec

The signal-to-noise power ratio from (3.2) is

$$S/N = \frac{P_T G_T G_R \sigma \lambda^2}{(4\pi)^3 R^4 L_R k T_0 \cdot B_r \cdot F} \tag{3.3}$$

The maximum radar range can be obtained by manipulating (3.3) as

$$R_{max} = \left[\frac{P_T G_T G_R \sigma \lambda^2}{(4\pi)^3 (S/N)_{min} L_R k T_0 F B_R} \right]^{1/4} \tag{3.4}$$

where $(S/N)_{min}$ is the single-pulse signal-to-noise ratio obtained from tables for the particular type of target (Swerling type 0, 1, 2, 3, 4) and desired probabilities of detection (P_d) and false alarm (P_{fa}).

Now consider a noise jammer with power

$$P_j = p_j B_j \tag{3.5}$$

where

p_j = jammer power density, W/Hz

B_j = jammer bandwidth, Hz

The jammer power density at the radar receiver is given by

$$p_{jr} = \frac{p_j G_j G_{sl} \lambda^2}{(4\pi)^2 R_j^2 L_j} \tag{3.6}$$

where

G_j = jammer antenna gain

G_{sl} = receiver antenna gain toward jammer

R_j = range between jammer and radar, m

L_j = jammer system losses

If we make the assumption that the jammer power density is much greater than the noise power density (i.e., $p_{jr} >> kT_0F$), then

$$R_{max} = \left[\frac{P_T G_T G_R \sigma \lambda^2}{(4\pi)^3 (S/J)_{min} L_R p_{jr} B_R}\right]^{1/4} \tag{3.7}$$

where $(S/J)_{min}$ is the signal-to-jam ratio for the appropriate target and detection statistics. Substituting (3.6) and (3.5) into (3.7) results in

$$R_{max} = \left[\frac{P_T G_T R_j^2 \sigma B_j G_R L_j}{4\pi (S/J)_{min} P_j G_j B_R G_{sl} L_R}\right]^{1/4} \tag{3.8}$$

Equation (3.8) is appropriate for stand-off (or stand-in) jamming calculations. R_{max} is often called the "burn through range" and indicates the range at which the radar overcomes the jamming effect. Note that the term (G_R/G_{sl}) represents the ratio between the peak antenna gain and the sidelobe level. For self-screening jamming, (3.8) can be simplified by letting $R_j = R_{max}$ and $G_{sl} = G_R$, resulting in

$$R_{max} = \left[\frac{P_T G_T}{4\pi (S/J)_{min}} \cdot \frac{\sigma}{P_j G_j} \cdot \frac{B_j}{B_r} \cdot \frac{L_j}{L_R}\right]^{1/2} \tag{3.9}$$

Equation 3.9 reveals several jamming principles. First, the term $(\sigma/P_j G_j)$ can be considered a figure of merit for the jammer. For the same jamming effect, if the target RCS (σ) is decreased (e.g., stealth), then the self-protection jammer ERP is decreased in the same proportion. Eventually no jammer is required if the radar cross section is low enough. The term (B_j/B_r) represents the ratio between the jamming and radar bandwidths. If this term approaches a value of 1 to 5, then the jamming is considered spot jamming. If $B_J >> B_R$, then barrage jamming is being used. It is to the advantage of the jammer (R_{max} is minimum) to be in a spot jamming mode.

The same comments apply to (3.8) for stand-off jamming with the additional observations that when R_j is increased (stand-off versus stand-in jamming), the burn-through range is increased, making the jamming less effective. Also, maximizing the ratio (G_R/G_{SL}) using ultralow sidelobes increases the burn-through range, thereby diluting the jamming effects.

▶ *Example 3.1*

Given an E/F-band (3-GHz) surveillance radar and stand-off jammer with parameters as given, use MATLAB to plot the target and jammer received signal strengths.

Radar	**Jammer**
$P_T = 150$ kW	$P_j = 2$ kW
$G_{T,R} = 40$ dB	$G_j = 13$ dB
$\sigma = 3$ m^2	$R_j = 300$ km
$\lambda = 0.1$ m	$B_j = 4$ MHz
$L_R = 10$ dB	$L_j = 7$ dB
$G_{SL} = 10$ dB	
$B_r = 80$ kHz	
$F = 5$ dB	

```
% Receiver Jamming Effects
%-------------------------
% rcvjam.m

clear,clc,clf;

Rmin=10; Rmax=300;    %  km

% Define Rtar(km) from Rmin to Rmax

Rtar=Rmin:Rmax;       % 1 km steps

% Radar and Target Parameters

Pt=1.5e5;             % w
Gr=1e4;
sigma=3;              % sq m
wl=.1;                % m
Lr=10;

Pr=(Pt*Gr^2*sigma*wl^2)./((4*pi)^3*Lr*(Rtar*10^3).^4);
Prdbm=30+10*log10(Pr);

% Jammer Related Parameters

Pj=1e4;               % w
Gj=10^(19/10);
Rj=3e5;               % m
Bj=4;                 % Mhz
```

```
Br=.08;                 % Mhz
Gsl=10;
Lj=10^(7/10);

Gsl=10^(Gsl/10);Lj=10^(Lj/10);
Prj=(Pj*Gj*(Br/Bj)*Gsl*w1^2)/((4*pi*Rj)^2*Lj);
Prjdbm=30+10*log10(Prj);

% Plotting Received Target and Jammer Signals

Prjdbm=Prjdbm*(Rtar./Rtar);
plot(Rtar,Prdbm,Rtar,Prjdbm);
ylabel('Received Signal (dbm)');
xlabel('Range (km)');grid;
title('SUPPORT JAMMING AND TARGET SIGNALS AT RADAR');
text(100,-72,'Standoff Jamming');
text(150,-101,'Main Beam Target Return');
end
```

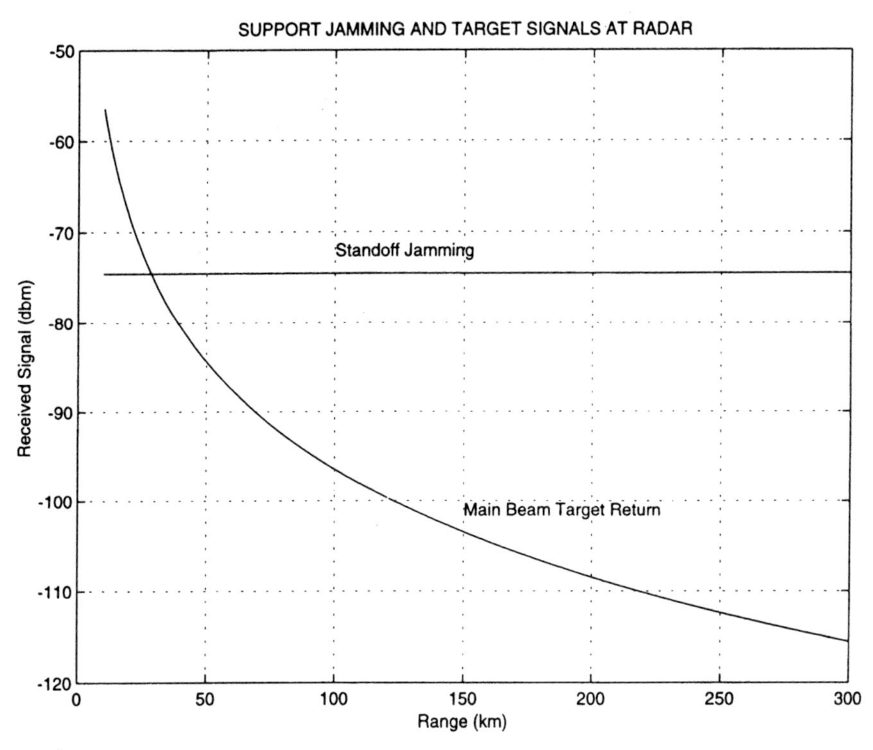

Noise jammers are relatively effective when applied through the radar's main antenna lobe. This is illustrated in Example 3.2, which plots the relative effects that various amounts of jamming ERP density (W/MHz) have on the radar's detection capability. In the self-screening case (main lobe jamming), the radar has little chance of penetrating through the jamming, so its main defense is to produce a strobe on the jammer indicating its angular position. This assumes that the radar is designed with adequately low sidelobes to reject jamming in other than its main beam response.

For stand-off jamming, Example 3.2 illustrates that a well-designed radar with ultralow sidelobes can make noise jamming relatively ineffective. Many radars employ sidelobe cancellation that further reduces the effects of sidelobe jamming. One way to improve the jamming response is to use "smart noise" jamming waveforms (see Figure 1.6) that are matched to the radar transmissions. This negates the processing gain achieved in most radars against noise that can range from 10 dB to 20 dB in conventional radar to 30 dB to 40 dB in pulse compression radars and up to 60 dB to 70 dB in PD radars.

▶ *Example 3.2*

Using the radar specified in Example 3.1, plot the reduction in radar range that results from using jammer ERP power densities (W/MHz) that vary from 0.001 to 10^6 W/MHz. Assume that a signal-to-interference ratio of 13 dB is required to achieve the specified probabilities of detection and false alarm. Plot the responses for self-screening jamming and stand-off jamming using regular and ultralow sidelobes of 50 dB below main-beam response. Remember to include the effects of receiver noise in addition to jamming noise since these will become important for small jamming signals.

```
% Effective Radiated Power Density
% --------------------------------
% erpdenb.m

clear;clc;clf;

% Generate ERP Density Vector

plmin=-4;plmax=6;
pl=plmin:.1:plmax;
pj=10.^pl; % w/Mhz

% Radar and Target Parameters

Pt=1.5e5;  % w
Gt=10^(40/10);
Gr=10^(40/10);
rcs=3;     % sq m
wl=.1;     % m
Lr=10^(10/10) ;
Br=.08;    % Mhz
Gsl=10^(10/10);
snr=10^(13/10);
nf=5;      % db

% Jammer Parameters

Pj=2e4;    % w
Gj=10^(13/10);
Rj=3e5;    % m
Bj=4;      % Mhz
Lj=10^(7/10);

% Calculate Normalized Range in Jamming

pn=-144+nf;
N=10^(pn/10);

% Jammer ERP density at Radar

prj=(pj*Gj*Gsl*(wl)^2)/((4*pi*Rj)^2*Lj);

Ro=(Pt*Gt*Gr*rcs*(wl)^2/((4*pi)^3*snr*N*Br*Lr))^.25;
R=((Pt*Gt*Gr*rcs*(wl)^2)./((4*pi)^3*snr*(N+prj)*Br*Lr)).^.25;

% Plot Range Reduction versus ERP Power Density

semilogx(pj,R/Ro);grid;
ylabel('Range Reduction - %');
xlabel('ERP Spectral Density - w/Mhz');
title('JAMMER ERP REQUIREMENTS FOR SOJ AND SSJ');
end
```

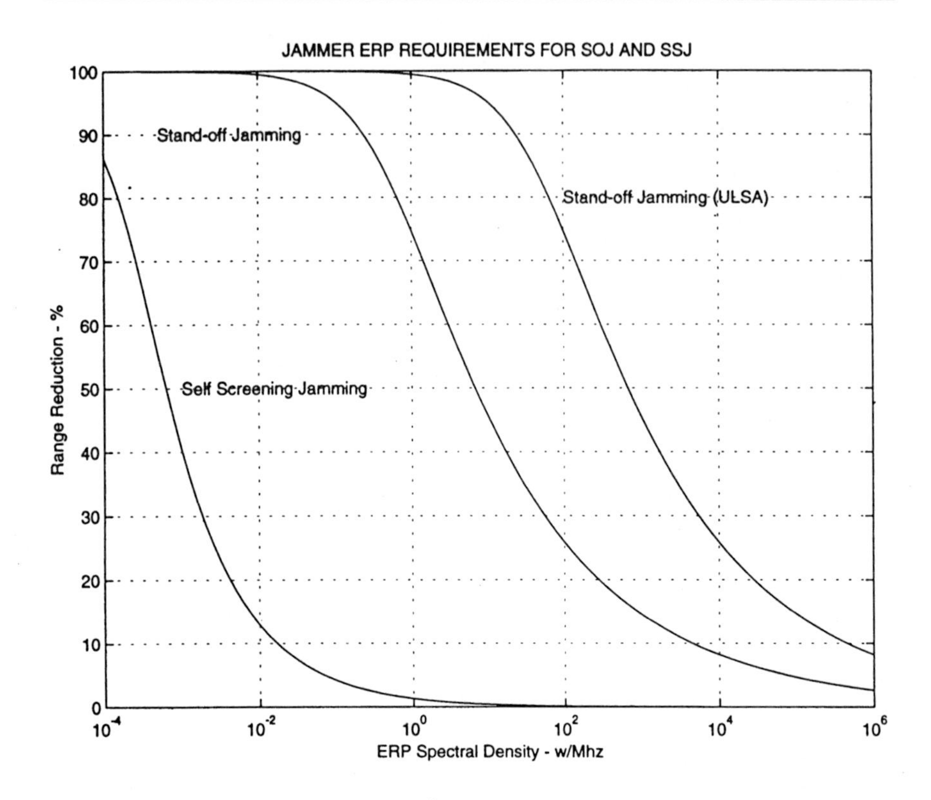

3.3.1.1 Noise Jammer Effectiveness

There are several factors that cause the effectiveness of noise jammers to be less than their theoretical capabilities. One significant factor is the need for a noise jammer to accommodate victim systems with various polarizations. This generally causes the jammer radiations to be either 45 degrees slant polarized or circularly polarized. Because most electronic systems are either horizontally or vertically polarized, this causes a typical 3-dB loss in jammer ERP. A more serious loss (e.g., 15 dB to 20 dB) would occur if the jammer antenna and the emitter (e.g., radar) antenna were orthogonally polarized, such as would occur if the emitter were right-hand circularly polarized and the jammer left-hand circularly polarized.

Another significant effectiveness factor relates to the quality of the jammer noise itself, which ideally should be as close to white Gaussian noise (e.g., receiver noise) as possible. Measurements on operational and developmental jammers show that the noise quality factor might cause losses (e.g., increase in required jammer ERP) as large as 17 dB when compared with the idealized noise waveform [1].

The most obvious method of obtaining a noise jammer signal is to pass band-limited Gaussian noise (e.g., receiver or thermal noise) through a high-power amplifier. This method of generating a jammer waveform is called *direct noise amplification* (DINA). The noise need not be generated at RF, but can be generated at base band (its spectral density appropriately shaped by band-limiting filters) and then heterodyned to RF where it is power amplified and radiated.

Based on information theory it is reasoned that white (uniform spectral density) Gaussian noise is the best noise-jamming waveform. This follows because white Gaussian noise has the maximum entropy, or uncertainty, of any random waveform for a specified average power. This conclusion is intuitively satisfying because it is statistically impossible to distinguish between receiver noise generated in the victim system and the externally injected jammer waveform.

However, there is a significant practical problem in generating high-power Gaussian noise since most high-power microwave amplifiers (e.g., traveling wave tubes) are limited in the peak power they can handle. In fact, many microwave power amplifiers are best operated in a saturated condition. If we adopt the information theory approach for this situation, we would generate noise whose amplitude probability distribution was uniformly distributed since this provides the maximum entropy under a peak power constraint. The utility of this "clipped" jamming waveform is highly questionable because it is the peaks, or extremes, of the noise that induce false alarms into the victim radar system.

The designer of a DINA jammer basically has two alternatives to consider. First, the high-power amplifier transmitter can be operated in a quasilinear mode that preserves the character of the Gaussian noise waveform. To accomplish this, the compression point of the power amplifier must be set in a region that is 7 dB to 10 dB above the *root mean square* (rms) value of the noise voltage. Alternatively, the power amplifier can be operated in a saturated condition, thereby maximizing the output power of the amplifier but causing the amplitude probability distribution of the noise to deviate substantially from a Gaussian distribution.

In the situation where the jammer noise bandwidth is much larger than the victim emitter bandwidth, the latter alternative is preferred. What happens in this case is that the amplitude probability distribution at the output of the victim receiver's IF amplifier tends to approach a Gaussian distribution, even though the noise input distribution is highly non-Gaussian due to the clipping of the noise peaks. This result follows from the central limit theorem, which states that the sum of a large number of independent random variables tends toward a Gaussian distribution under very general conditions.

The physical model that applies in this case consists of noise samples that excite the impulse response of the victim IF amplifier. The decorrelation time of the noise samples is inversely proportional to the jammer bandwidth. The assumed wide jammer bandwidth results in a short decorrelation time, which in turn results in a large number of noise samples within the duration of the IF amplifier's impulse response time. This provides the effect of adding many independent noise samples together and hence causes the output distribution to approach a Gaussian distribution. The rms value of this Gaussian distribution is proportional to the rms value of the clipped noise distribution present at the output of the jamming transmitter. This value depends on the clipping level of the transmitter. As an example, if the noise is clipped at the 1σ level, then the output power loss is approximately 3 dB. This loss is increased to about 7 dB if the noise is clipped at the $\sigma/2$ level and reduced to approximately 0.4 dB when clipped at the 2σ level. However, output power can be maximized by hard limiting. To summarize, the principal loss in this situation associated with power amplifier saturation is the reduction of average noise power due to the clipping of the noise peaks.

Another method of generating noise power is by frequency or phase modulation of the jammer signal with a random waveform. Power amplification saturation has little effect on an FM signal, so this method has the advantage that maximum output power can be extracted from the jamming transmitter's output stage. FM jammer noise can be generated using an output power amplifier, such as was used for the DINA technique, or using a rapidly tunable power oscillator, such as a *backward wave oscillator* (BWO) or carcinotron microwave source. Another method is to modulate the beam voltage of a TWT, producing a phase modulation on the carrier.

A block diagram of this technique is illustrated in Figure 3.7 as well as the mechanism that allows the buildup of a pseudo-Gaussian jamming waveform in the victim's IF receiver.

If the instantaneous frequency of the FM jammer signal looks approximately like a sine waveform for a number of cycles, it is natural to expect that the spectral density of the jammer signal will be approximately proportional to the probability density of the waveform that determines the instantaneous frequency. This can be demonstrated rigorously and applies under conditions where the frequency modulation index (rms frequency deviation divided by bandwidth of frequency deviation) is much greater than 1. The net result is that if a Gaussian-distributed process is used to frequency modulate a jammer transmitter with a wide phase deviation, then the output spectral density will have an approximately Gaussian shape.

Two questions arise with respect to the FM jammer-noise waveform described previously. First, is a Gaussian-shaped spectrum desirable? Second, how does the victim emitter receiver respond to the FM noise waveform?

The answer to the first question is that a uniform spectral density is desired to accommodate the uncertain location of the victim receiver's response

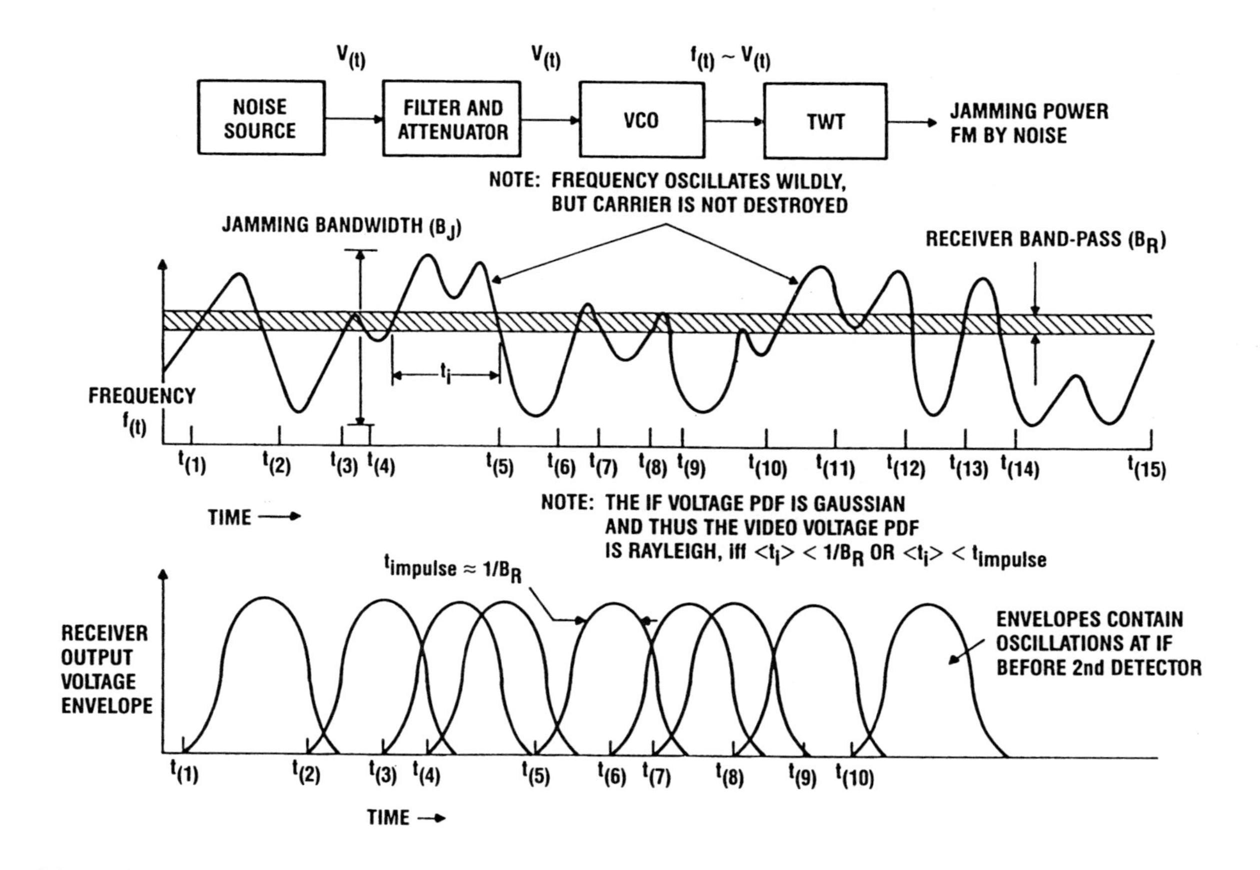

Figure 3.7 FM by noise-jamming technique.

band. One method used to achieve a uniform spectral density is to apply the Gaussian-distributed frequency control waveform through a nonlinear amplifier, which transforms the random Gaussian distributed waveform into a uniformly distributed random waveform. The proper nonlinear transfer function for achieving this response is the *error function* (erf). The nonlinear circuit that transforms Gaussian-distributed noise into uniformly distributed noise is referred to as an "erfer." The output noise is then called "erfed" noise. The erfer can be implemented using diode-resistor networks to accomplish a piecewise linear approximation to the erf(·) function.

Another means of achieving a uniform spectrum is to frequency modulate the jammer transmitter with a saw-toothed waveform that deviates the frequency over the band of interest. This will produce a relatively flat spectrum, but it is inappropriate as a jamming waveform. The output of the victim receiver to this waveform will be a regular sequence of pulses occurring at the repetition rate of the saw-toothed waveform. The shape of the pulse corresponds to the shape of the receiver's passband, while the pulse duration is a function of how long the instantaneous frequency of the jammer remains within the receiver passband. The action of the victim receiver to the linear swept-frequency waveform is analogous to that of a spectrum analyzer, where the roles of the swept local oscillator and input waveform are reversed. One of the problems with this waveform for jammer application is that deterministic (whereby the future can be predicted from the past) waveforms are generally highly susceptible to ECCM. In this case, one effective ECCM is a guard-band receiver whose passband is placed contiguous to the active victim receiver's passband. When the swept FM jammer waveform is picked up in the guard band, its output is used to blank the input to the active receiver for the duration of the time when the jammer's instantaneous frequency lies within the active receiver's passband. Another effective ECCM is the Dicke-fix receiver, which is described in the next section.

Randomness can be added to the jammer waveform by frequency modulation of the transmitter with a combined noise and saw-toothed waveform. If the frequency deviation of the jammer is much wider (i.e., wideband FM) than that of the victim receiver's passband, then the victim receiver's impulse response will be excited each time the jammer's instantaneous frequency sweeps through the receiver passband. If the receiver's impulse response duration (which is inversely proportional to the receiver bandwidth) is much greater than the jammer deviation rates, then a number of randomly spaced (and hence independent), overlapping impulse responses will be added together to form the victim receiver's output waveform. This output waveform meets the conditions of the central limit theorem and hence has an amplitude probability distribution that approaches a Gaussian distribution. Thus, a properly implemented wideband FM jammer waveform of this type will generally provide the same victim receiver response as an equivalent DINA jammer waveform, except that more average power can be provided when the output transmitter must be operated in a saturated mode. Because a continuous spectrum is desired for wideband FM noise jamming, the bandwidth of the noise component should be greater than half the repetition rate of the saw-toothed waveform.

Digital noise is sometimes used to control the instantaneous frequency of the FM noise jammer. The use of digital noise, which is generated using regenerative shift-register techniques, has the advantage that it can be synchronized to the victim radar's waveform. This provides the potential for generating "smart noise" (see Figure 1.6), which can give several advantages to the ECM designer. First, synchronizing the jamming waveform to the radar's transmission waveform causes the jamming signals to build up over the radar integration period, making the jamming signal more effective. Second, digital noise allows spot jamming to be performed against multiple threats by programming the transmission format, thereby avoiding intermodulation effects that might occur if two or more jamming waveforms in different parts of the band were radiated simultaneously. When two or more jamming signals are simultaneously applied to a TWT, an effect called "gain capture" occurs. This effect can suppress the power contained in the higher frequency signal from 8 dB to 10 dB when both the higher and lower frequency signals have equal amplitude at the input to the TWT. Digital noise is a form of power management that is key to the functioning of many modern EA systems.

▶ *Example 3.3*

The technique for producing FM noise described in Figure 3.7 is to be simulated using MATLAB. The first step is to generate a Gaussian noise voltage. Next, this signal is filtered using a six-pole elliptical filter to shape the noise bandwidth. The filter output is then applied to a voltage-controlled oscillator whose output represents the FM noise jammer output. The power spectrum of this signal is found using the MATLAB FFT function as is the power spectrum of the noise. A similar simulation is performed to determine the power spectrum of an "erfed" FM noise jammer. Note that the "erfed" FM jamming signal provides a more uniform spectrum than that of the FM jamming signal itself. Note also that random waveforms are being generated and spectrums will vary depending upon the particular random realization generated.

```
% FM Noise Spectrum
%----------------------
% fmnoise.m

clf;clc;clear;

% Define Time Vector -tmax to tmax

N=4096;tmax=1/2;
t=2*tmax*(-N/2:N/2-1)/N;

% Sample Frequency fs

fs=N/(2*tmax);

% Generate Gaussian Noise Vector - N(0,1)

n=randn(1,N);
subplot(2,2,1);plot(t,n);grid;
xlabel('Time'):ylabel('Noise Voltage');
title(['Gaussian Noise Voltage']);

% Generate 'Erfed' Noise

y=erf(n);

% Shape Noise With Elliptical Filter

 [b,a]=ellip(6,.5,.5,.005);
% [b,a]=butter(6,.005);
% [b,a]=cheby1(6,.5,.005);
v=filter(b,a,n);
verf=filter(b,a,y);

% Define VCO with k=50 Hz/v

f0=0;k=50;j=sqrt(-1);
f=f0+k*v;
f1=f0+k*verf;
x=cos(2*pi*f.*t)+j*sin(2*pi*f.*t);
xerf=cos(2*pi*f1.*t)+sin(2*pi*f1.*t);

% Find Power Spectrum for v(t)

V=fft(v);
Pv=V.*conj(V);
Pv=fftshift(Pv);

% Find Power Spectum for x(t) and xerf(t)

X=fft(x);XE=fft(xerf);
```

```
Px=X.*conj(X);Pxe=XE.*conj(XE);
Px=fftshift(Px);Pxe=fftshift(Pxe);

% Normalize Amplitude

Px=Px/max(Px);Pv=Pv/max(Pv);Pxe=Pxe/max(Pxe);
Pxdbm=30+10*log10(Px+1e-6);
Pvdbm=30+10*log10(Pv+1e-6);
Pxedbm=30+10*log10(Pxe+1e-6);

% Select Frequency Axis

fsam=fs*(-N/2:N/2-1)/N;

% Plot Noise Spectrum

subplot(2,2,2);
plot(fsam,Pvdbm);grid;
axis([-300,300,-30,30]);
xlabel('Hz');ylabel('DBM');
title(['Noise Spectrum']);

% Plot FM Noise Spectrum

subplot(2,2,3);
plot(fsam,Pxdbm);grid;
axis([-1000,1000,-30,30]);
xlabel('Hz');ylabel('DBM');
title(['FM Noise Spectrum']);

% Plot Erfed FM Noise Spectrum

subplot(2,2,4);
plot(fsam,Pxedbm);grid;
xlabel('Hz');ylabel('DBM');
axis([-1000,1000,-30,30]);
title(['Erfed FM Noise Spectrum']);
end
```

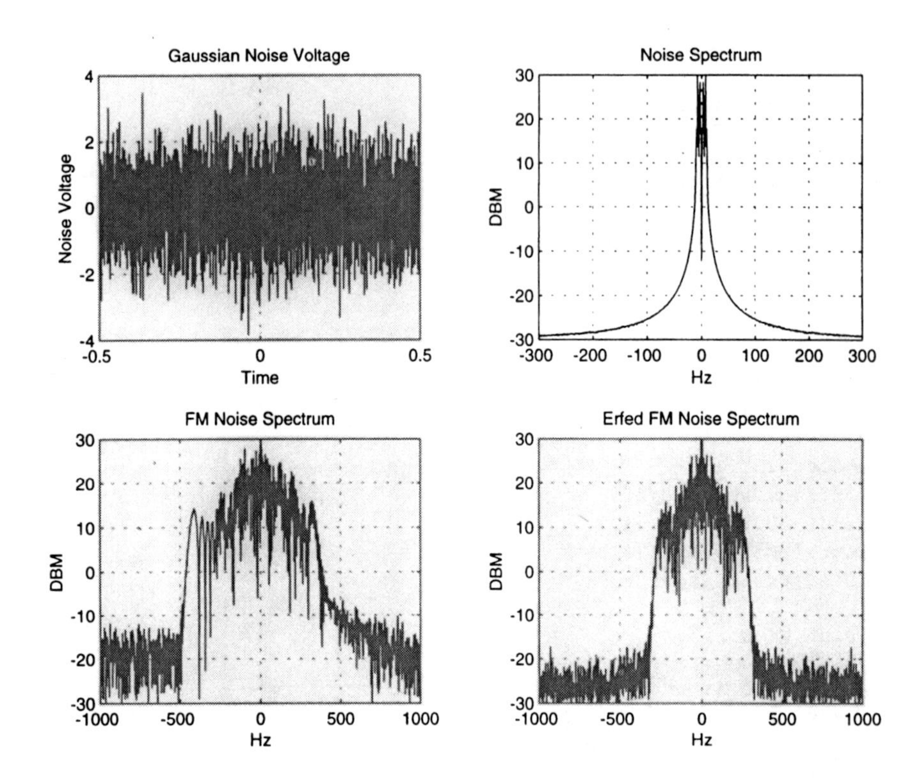

3.3.1.2 Dicke-Fix Receiver

A natural countermeasure to FM noise jamming is the Dicke-fix receiver. The Dicke-fix receiver consists of a wideband filter (generally set at 10 times the radar signal bandwidth) whose output is hard limited on receiver noise followed by a filter matched to the radar signal.

To understand the operation of a Dicke-fix receiver, consider FM noise with 1-GHz deviation being swept across a conventional receiver with a 1-MHz bandwidth in an average period of 50 μs. This will generate a 50-ns impulse every 50 μs that shock excites the receiver front end circuits to cause a noise output that blinds the radar receiver (see Figure 3.7). However, with a Dicke-fix receiver, the impulses are hard limited in the wideband portion of the receiver that inhibits their ability to shock excite the receiver. The only resulting action of the FM noise jamming is to suppress target (and also receiver noise) returns that are simultaneously present while the FM jamming signal sweeps through the receiver's wideband pass filter.

In the Dicke-fix receiver, the hard limiter maintains the rms value of the noise at a constant value in the wideband portion of the receiver. When this output is applied to the narrowband section of the receiver, its filtering action provides a dynamic range of about 10 dB while maintaining a constant false alarm rate.

A countermeasure against the Dicke-fix receiver is to introduce into the receiver a jamming signal that has modulation frequency components that lie within the wideband filter section of the receiver. This jamming signal has the effect of capturing the limiter that suppresses the true radar return signal. This effect is illustrated in Example 3.4, where the target return is suppressed by over 20 dB.

▶ *Example 3.4*

Using MATLAB, generate a square wave-modulated signal whose peak amplitudes are 2V and 4V. The carrier frequency is 256 Hz while the modulation frequency is 32 Hz. Limit the signal in a hard limiter at the 1-V level. Plot the power spectrum of the modulated and the limited signal. Note the suppression of the fundamental by the order of 20 dB, which illustrates the capture effect of the limiter.

```
% Dicke-Fix Countermeasure
% -------------------------
% dfcm.m

clf;clc;clear;

% Define Time Vector

tmax=.06;N=1024;
t=tmax*(1:N)/N;

% Sample Frequency

fs=N/tmax;

% Generate Modulated Signal

x=cos(2*pi*256*t)+3*square(2*pi*32*t).*cos(2*pi*256*t);

% Plot Waveform

subplot(2,2,1)
plot(t,x);grid;
xlabel('Time - sec');ylabel('Volts');
title(['Jam Waveform']);

% Limit Signal To Clip Level k
```

```
k=1;
for i=1:N;
   y(i)=x(i);
   if x(i)>k;y(i)=k;end;
   if x(i)<-k;y(i)=-k;end;
end;

% Plot Limited Waveform

subplot(2,2,3);
plot(t,y);grid;
xlabel('Time - sec');ylabel('Volts');
title(['Clipped Waveform']);

% Find x and y for M steps

M=64;A=[];B=[];
for i=1:M;
A=[A x];B=[B y];end;
x=A;y=B;

% Extend t vector

N=N*M;
t=M*tmax*(1:N)/N;

% Find Power Spectrums

Y=fft(y);X=fft(x);
Py=Y.*conj(Y);Px=X.*conj(X);
Py=fftshift(Py);Px=fftshift(Px);

% Select Central Frequency Axis

Px(3*N/4+1:N)=[];
Px(1:N/4)=[];
Py(3*N/4+1:N)=[];
Py(1:N/4)=[];
fax=fs*(-N/4:N/4-1)/N;

% Normalize Amplitude

Px=Px/max(Px);
Py=Py/max(Py);
Pxdbm=30+10*log10(Px+1e-6);
Pydbm=30+10*log10(Py+1e-6);

% Plot Power Spectrums

subplot(2,2,2);plot(fax,Pxdbm);grid;
xlabel('Frequency - Hz');
```

```
ylabel('Power - dbm');
title(['Power Spectrum']);
axis([0 ,600 ,-30,40]);
subplot(2,2,4);plot(fax,Pydbm);grid;
xlabel('Frequency - Hz');
ylabel('Power - dbm');
title(['Jammed Spectrum']);
axis([0,600,-30,40]);

end
```

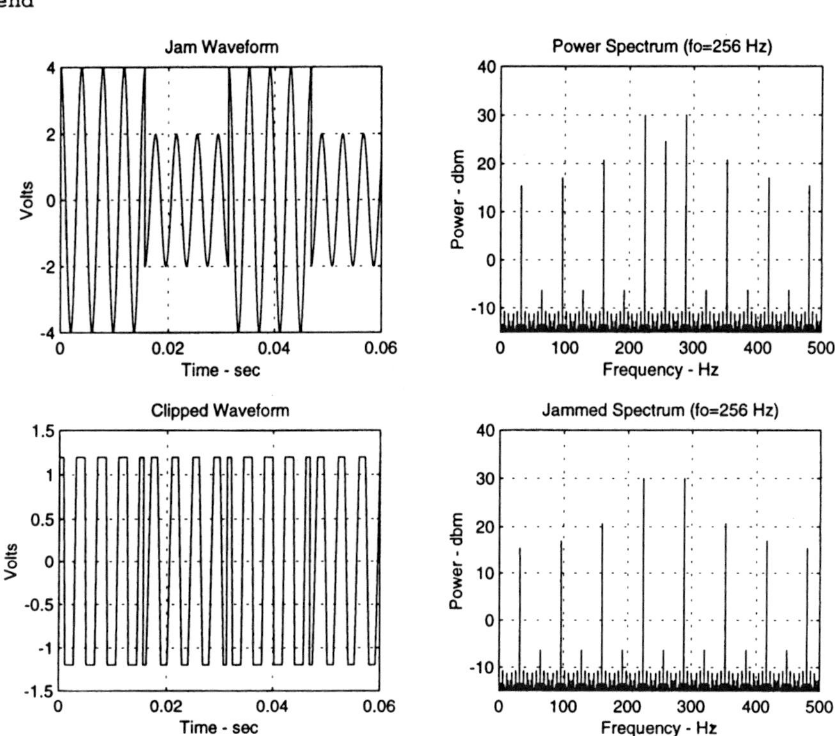

3.3.2 Deception Jamming

The objective of a deceptive EA (DECM) system against a radar is to mask the real target by injecting suitable modified replicas of the real signal into the victim system. DECM can be applied against both search and tracking radars. The primary advantage of this form of EA is that all the jamming power is absorbed by the victim radar and the radar's processing gain is either partially or entirely negated. Since the DECM signal penetrates into the radar's signal processor, it can be considered a direct attack on this radar function.

Deception-type jamming waveforms have the potential for more effective use of jamming energy against search radars. Against search radars, the repeater or deception EA (DECM) system must intercept and store the essential spectral and temporal characteristics of the radar waveform. It must then generate synthetic targets that are synchronized with the scan and waveform pattern of the radar to create enough false targets to confuse the radar. As with noise jammers, this action is most easily accomplished when the jamming signals are injected into the main lobe of the radar's antenna pattern. However, sidelobe deception is highly desirable to prevent strobing in the jammer's direction.

The deception-repeater jammer is obviously more complex than the noise jammer and its practicality resulted from the emergence of highly dense digital integrated circuits, which provided the vast amounts of computing power necessary to implement this type of jammer. The deception jammer is most effective (as compared to noise jammers) against modern radars that employ coherent integration techniques such as the PD and pulse-compression types. This occurs because radars employing coherent integration techniques have a large processing gain against noise (e.g., of the order of 20 dB to 60 dB) and hence attenuate the noise jammer signal by that amount while necessarily accepting any targetlike return (e.g., DECM jammer signal) unattenuated.

The ability of the DECM signal to penetrate the radar's sidelobe response is open to some question. This is a result of radar EP techniques, which are designed to prevent targetlike jammer signals from entering through the radar sidelobes. Most modern radars employ SLBs that use an auxiliary omnidirectional antenna whose response approximates that of the radar's antenna sidelobe response. If the signal in the radar does not provide a greater response than that in the auxiliary antenna, then the radar is blanked for the target pulse interval because a sidelobe response is indicated. The sidelobe blanker must respond only to targetlike signals because the extended temporal characteristics of noiselike signals could cause excessive blanking of the radar response. This is contrasted with SLCs that must respond only to high-duty factor signals such as noiselike signals entering the antenna sidelobes. In addition to the sidelobe blanker, there is always the possibility that the radar may use sidelobe cover transmissions whereby spurious transmissions are radiated by a separate omnidirectional antenna to prevent interception of the real radar signal by listening to the sidelobe transmissions.

A compromise between deception and noise jamming is the so-called "smart noise" jammer, whereby noise bursts about the radar's center frequency, which are timed to coincide with and cover the true target return, are transmitted by the jammer. This jamming waveform lacks the total effectiveness of the true repeater jammer and requires more knowledge of the victim radar than the true noise jammer. However, it makes better use of the available jamming energy than the true noise jammer and has a reasonable chance of being unaffected by either radar SLBs or SLCs.

Tracking radars that operate against maneuvering targets must generally provide a high data-sampling rate. This is reflected in the servo bandwidth (B) associated with the angular tracking loop, which might typically range from 1 Hz to 10 Hz depending upon the target characteristics. These bandwidths imply that the radar antenna must either dwell continuously on the target or scan over the target at a rapid rate (e.g., at a scan frame rate at least twice the value of B). In addition to angular tracking, it is necessary to discriminate against other targets within an angle resolution cell, which is generally accomplished using narrow range gates that straddle the target. A second servo bandwidth that is of the same order but generally higher than the angle tracking loop bandwidth is associated with the range tracker. Angle and range servo tracking loops are basic to all pulsed tracking radar.

Velocity or Doppler tracking loops are also often used in tracking radars that have a coherent mode of operation. In CW tracking radars, velocity discrimination provides the basic means of selecting the target while rejecting other targets and clutter also located within an angular resolution cell. In PD radars, velocity discrimination provides a means of rejecting clutter within an angle-range resolution cell and aids in the range-tracking function by allowing the accurate prediction of future positions of the range gates. Generally, target velocity information is required in fire-control systems to solve the weapon-guidance equations.

Other characteristics of tracking radars of importance to EA designers are the memory modes of operation generally incorporated into tracking radars. The angle and range outputs of the tracking radars are typically differentiated to provide angle and range rate data. These data are used to coast through fades and to aid in the rapid reacquisition of targets that are lost. In tracking radars with velocity loops, the differentiated range data are compared with the Doppler-generated range rate data to discriminate against deception-type jamming signals. Doppler-derived velocity data may be differentiated to derive acceleration data that can be used to discriminate against deception-type jamming signals with unrealistic accelerations.

The basic objective of a DECM system is to cause the tracking radars to break angular track. Because most tracking radars use narrow angle beams, this also results in a loss of all target range and velocity information. The typical reacquisition procedure is for the tracking radars to search those angle-range cells that might contain the target based upon the last measured angle and range position and rate data. Once this fails, a general reacquisition procedure is initiated that involves the use of an associated acquisition radar. All this can take appreciable time (e.g., about 10 sec or more), during which the weapon is without guidance data.

The classic procedure initially employed by a DECM against a tracking radar is to transmit back an amplified replica of the radar waveform that captures the radar tracking loops. The signal is then delayed in range at a rate that continuously increases, thus simulating a high-velocity accelerating target. At an appropriate time, the DECM signal is removed, causing the tracking radar to lose range track with the last measured target position and velocity grossly in error. This DECM mode is called *range gate pull-off* (RGPO).

RGPO by itself may not fool a PD radar, which checks the consistency of target velocity data obtained by differentiating range data and measuring target Doppler data. This can be corrected if the DECM induces a pseudo-Doppler frequency shift onto the repeated target signal, which simulates the Doppler signature of a realistic target. This mode, which captures the tracking radar's velocity loop, is called VGPO, and is the only effective mode in jamming a CW tracking radar that does not use range tracking loops.

RGPO and VGPO are only partially effective in causing a tracking radar to break angle lock. In fact, the tracking radar obtains accurate angle tracking data during all of the time the DECM is injecting false target data into the range and velocity tracking loops. The only time that angle errors result in the course of RGPO and VGPO jamming is during the reacquisition search time. Because the angle rates have been stored, this may result in only minimal angular errors. Therefore, for the DECM to be effective, generally some form of angular deception must be built into the system.

One form of angular tracking system employed in tracking radars induces an amplitude modulation onto the target signal by either conically scanning or sequentially lobing the radar antenna. A powerful EA technique against this form of tracking radar is for the DECM to induce an inverse modulation onto the synthetic target return. This causes the angle-tracking circuits in the radar to force the angle track away from the target's angular position. A secondary but important effect is that the angle rates obtained by differentiating the angle data cause the angular position of the antenna to be further displaced from the true angular position during the tracking radar's coast mode.

The effectiveness of EA against the sequential lobing type of tracking radar has resulted in *lobe-on-receive-only* (LORO) systems that prevent detection of their lobing rates. This type of tracking radar system is still susceptible to inverse modulation jamming, but the EA system must slowly scan its modulation through all possible lobing rates to find the actual rate used by the tracking radar. Many modern radars utilize monopulse angular tracking systems, which are not susceptible to amplitude modulation induced on the repeated signal.

Jamming of monopulse tracking radars involves inducing artificial glint onto the jamming return. This can be accomplished by transmitting the jamming signal through spatially dispersed antennas (e.g., antennas located on opposite wing tips of the aircraft) or by bouncing the jamming signal from a spatially dispersed reflector (e.g., terrain bounce) to create the effect of signals reaching the tracking radar antenna from different directions.

DECM systems are often implemented in the form of repeater jammers. As the name suggests, repeater jammers radiate replicas of the victim radar's signal, possibly delayed in time, modulated in amplitude and shifted in Doppler frequency as is appropriate to the jamming mode selected. The distinct characteristic of a repeater jammer is that the victim signal is coherently stored in the jammer. This is generally accomplished through some form of memory device, which either recirculates the signal in a broadband device (e.g., TWT memory loop) combined with a delay line device (e.g., SAW delay line) or in a digital memory (e.g., DRFM).

The block diagram of an advanced DECM type jammer that is directed against monopulse type tracking radars and seekers is depicted in Figure 3.8. Basically, this type of DECM jammer is equipped to provide range deception (RGPO), Doppler deception (VGPO), and angle deception (cross-polarization jamming).

Doppler deception requires that the jamming signal be cohered to the radar signal. This can be accomplished in a number of ways, but modern EA equipment uses either digital serrodyning or a DRFM.

Digital serrodyning employs a digital phase shifter as illustrated in Figure 3.9. The Doppler shift frequency (f_d) is related to the rate of change of phase with time in accordance with the relationship

$$f_d = \frac{1}{2\pi}\frac{d\varphi}{dt} \tag{3.10}$$

The Doppler offset frequency is determined by the reciprocal of the time (T) required ($f_d = 1/T$) to step the phase shifter through 2π radians. For example, if a five-bit digital phase shifter is employed that steps through 2π radians in 100 μs, then the phase will be advanced in 11.25-degree steps to produce a Doppler shift of 10 kHz.

A DRFM can be used to provide signal storage in a DECM system. The DRFM mitigates many of the deficiencies inherent in microwave memory loops [6]. DRFMs provide a memory mechanization that is independent of storage time and, hence, eliminate any deterioration of signal fidelity with delay time. Multiple simultaneous signals can be stored and replicated upon command. Pulse-compression and phase-coded signals with intrapulse modula-

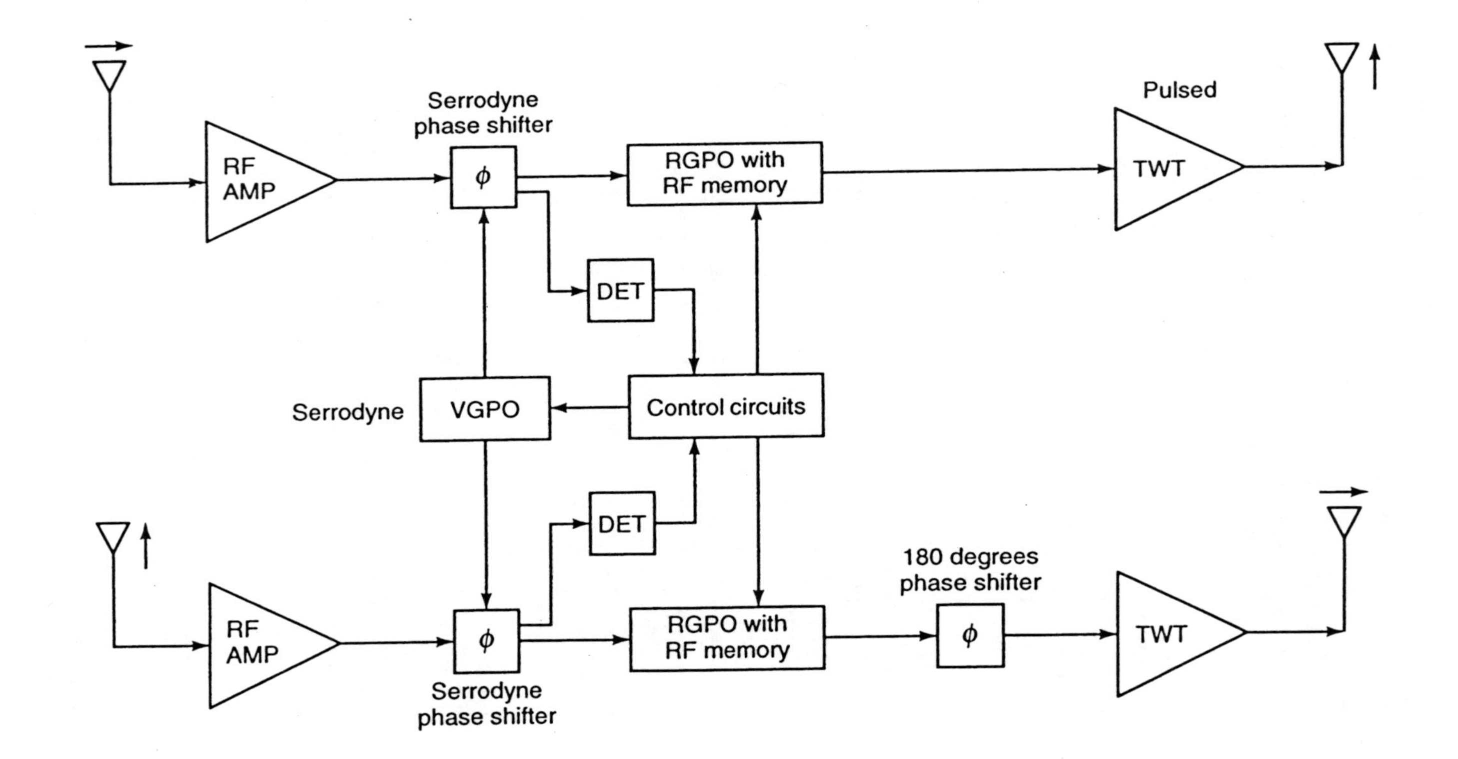

Figure 3.8 Advanced DECM block diagram.

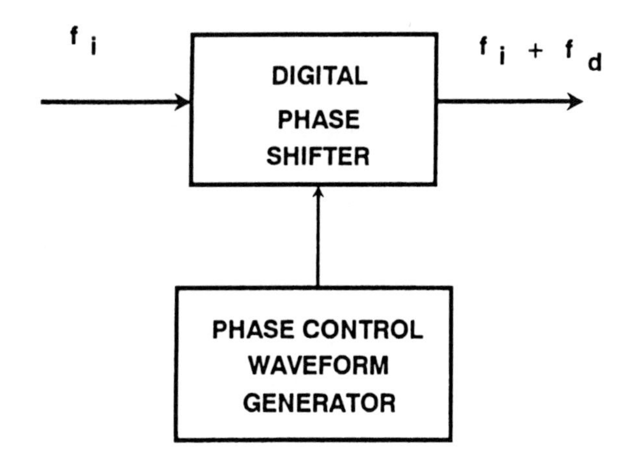

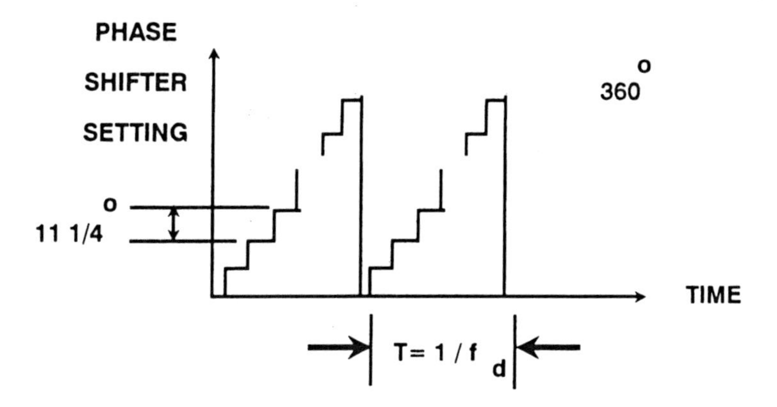

Figure 3.9 Serrodyning using a digital phase shifter.

tion can be replicated. Doppler shifts can be induced onto the replicated signal by simply offsetting the frequencies of the reference oscillator at the DRFM input, which translates the signal to baseband, and the reference oscillator at the output, which translates the signal to RF.

Range-gate deception or pull-off (RGPO) is a fundamental deception EA technique employed against automatic tracking radars. These radars, which are used for weapon-guidance and fire-control applications, generally employ early and late tracking gates that straddle the return echo. The gates are moved dynamically by a range servo, which follows the target by anticipating its future position using a range-rate or velocity estimate developed in the range tracker. The width of the early and late gates is generally of the order of a pulsewidth in the tracking mode and may be increased to several pulsewidths during acquisition and reacquisition modes. All other returns, except those returned within the tracking gates, are excluded by the range-tracking circuitry. This prevents spurious signals from entering the range-tracking circuitry and distorting the range estimate, but it also offers the opportunity for a deception jammer to operate in a range-gate stealing or track-breaking mode.

Using the RGPO technique, the deception jammer initially repeats the received radar pulse with minimum time delay. This allows the radar *automatic gain control* (AGC) circuitry to adjust to the stronger jammer signal, which has the effect of capturing the radar AGC that reduces the radar's sensitivity to the actual signal. Then, the deception jammer begins to introduce increasing amounts of time delay in the repeated signal. The range-gate circuitry in the radar begins to track the stronger jammer signal and gradually "walks-off" from the true target range.

This false target range information can result in significant aiming-guidance errors for anti-aircraft guns and for missiles that use command guidance because the ground-based computation of weapon lead-angle is strongly influenced by the target range as determined by the radar. However, the tracking radar angle circuitry still functions to point the radar antenna's boresight in the direction of the target. This angle information is sufficient for missile guidance systems using semiactive guidance, and hence RGPO by itself is not usually sufficient to cause the radar to break angle-lock.

In most DECM systems, RGPO is combined with some form of angle deception. The angle deception may be applied during the full RGPO cycle, or only after the delay has been increased. When the full delay cycle of the RGPO has been accomplished, the repeater is turned off. The radar then must accomplish a reacquisition cycle. If the target angle rates are high, or if angle deception has been accomplished, the radar must search in range and angle to reacquire the target. While the reacquisition process is being accomplished by the tracking radar, all tracking information is lost.

Most modern DECMs employ techniques designed to counter monopulse angle-tracking radars. These are required since many modern radars and missile seekers employ monopulse tracking due to their inherent resistance to amplitude modulated jamming and superior tracking performance. In the DECM block diagram illustrated in Figure 3.8, cross-polarization jamming is employed for this purpose. Cross-polarization jamming exploits the fact that some monopulse radars give erroneous angle-error information when the received signal (jammer) is orthogonally polarized with respect to the polarization of the radar receiving antenna. This technique is discussed in detail in Section 4.3.1.

The DECM configuration depicted in Figure 3.8 is designed to produce cross-polarized jamming independent of the attitude (e.g., angle) of the platform on which the jammer is located. To accomplish this, it uses orthogonal sets of input and output antennas that are mutually cross-polarized. A vector diagram of this configuration is illustrated in Figure 3.10. The input radar signal is decomposed into horizontal and vertical polarization components by the orthogonal input antennas. The horizontal component is processed, amplified in a TWT amplifier, and transmitted with vertical polarization. The vertical component is processed and amplified in a matched channel, phase shifted by 180 degrees, and transmitted with horizontal polarization. As indicated in the diagram, the resultant output signal is automatically cross-polarized with respect to the input signal.

3.3.2.1 Repeater Jamming Equations

The repeater jammer (see Figure 3.8) functions by first receiving the victim radar's signal and then reradiating this signal, which is amplified and possibly delayed in time and shifted in frequency. One of the most important factors to be considered in designing a repeater is the ratio of the deception repeater signal level to the radar target return as measured in the victim radar's receiver. The jam-to-signal ratio must be of the order of at least 7 dB to 10 dB to allow the jammer to capture the radar's tracking circuits. This requirement translates into a repeater gain and power output specification.

To calculate the jam-to-signal ratio at the radar receiver, we will first develop an expression for the repeated jammer signal. The transmitted radar signal received at the repeater jammer (P_{JR}) is given by

$$P_{JR} = \frac{P_T G_T \lambda^2 G_{JR}}{(4\pi R)^2 L_p} \tag{3.11}$$

where G_{JR} is the repeater jammer's receiving antenna gain, L_p is a loss term due to the possible use of different polarizations between the jammer and radar antennas and other losses, and $P_T G_T$ is the radar's ERP. The repeated jammer signal at the radar terminals (P_{RJ}) is then given by

$$P_{RJ} = \frac{P_T G_T^2 G_{JR} G_{JT} G_e \lambda^4}{(4\pi R)^4 L_p^2} \tag{3.12}$$

where G_{JT} is the repeater's transmitting antenna gain and G_e is the repeater's overall amplifier gain minus any losses other than polarization that are experienced by the repeated jamming signal and not by the radar return signal. The signal returned to the radar is given by

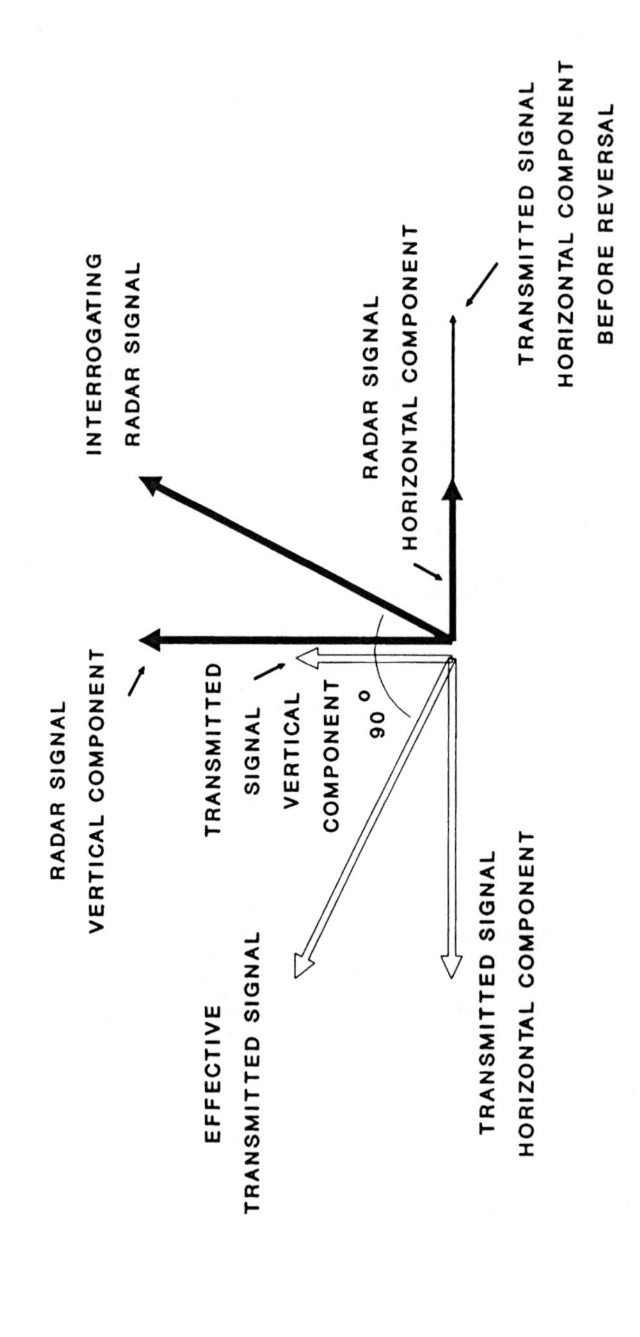

Figure 3.10 Cross-polarization jamming chart.

$$P_R = \frac{P_T G_T^2 \sigma \lambda^2}{(4\pi)^3 R^4} \tag{3.13}$$

where losses common to the repeated jammer signal are not explicitly shown. Then the jam-to-signal ratio is given by

$$\alpha = \frac{J}{S} = \frac{G_e G_{JR} G_{JT} \lambda^2}{4\pi\sigma L_p^2} \tag{3.14}$$

which, as expected, is independent of range. Next, define $\sigma_e = \alpha\sigma$, which expresses the magnified radar cross section desired from the repeater and compute the repeater gain as

$$G_{REP} = G_{JR} G_{JT} G_e = \frac{4\pi\sigma_e L_p^2}{\lambda^2} \tag{3.15}$$

which is similar to the expression for the capture area of an antenna. In addition, the repeater signal is sometimes commutated or gated at a high rate to prevent self-oscillation, which reduces the effective jammer signal average power. For a gated repeater with duty cycle $\beta \leq 1$, the repeater gain becomes

$$G_{REP} = \frac{4\pi\sigma_e L_p^2}{\lambda^2 \beta^2} \tag{3.16}$$

The power output of the repeater (P_J) can be found by multiplying P_{JR} by the repeater's electronic gain

$$P_J = \frac{P_T G_T \lambda^2 G_{JR} G_e}{(4\pi R)^2 L_p} \tag{3.17}$$

From (3.16) the electronic gain is given by

$$G_e = \frac{4\pi\sigma_e L_p^2}{\lambda^2 G_{JR} G_{JT} \beta^2} \tag{3.18}$$

Substituting this into (3.17) results in an alternative expression for the jammer's output power

$$P_J = \frac{P_T G_T L_p \sigma_e}{4\pi R^2 G_{JT} \beta^2} \tag{3.19}$$

▶ *Example 3.5*

It is desired to plot the signals received by a radar, which is subjected to repeater jamming, as a function of the range between the radar and the jammer. The characteristics of the radar and the repeater are given. Plot the repeater input signal, main and side lobe repeater jamming signals received at the radar receiver, and the main lobe skin return from a 5-m^2 target. Note the *jam-to-signal ratio* (JSR) in regions where the repeater is in a saturated and an unsaturated condition. What is the burn-through range (range where $J/S = 1$) for the repeater

Radar	**Repeater Jammer**
$P_T = 200$ kW	$P_j = 500$W (max)
$G_T = 37$ dB	$G_{JR} = 10$ dB
$\lambda = 0.1$ m^2	$G_{JT} = 10$ dB
$G_{SL} = 0$ dBi	$L_p = 3$ dB
	$\beta = 1$
	$\sigma_e = 50$ m^2

```
% Repeater Jamming Effects
%--------------------------
% repjam.m

clear,clc,clf;

Rmin=.1; Rmax=500;          %  km

% Define Rtar(km) from Rmin to Rmax

Rtar=Rmin:.1:Rmax;          % .1 km steps

% Radar and Target Parameters

Pt=1e6;                     % w
Gt=10^(37/10);
Gr=10^(37/10);
sigma=5;                    % sq m
wl=.1;                      % m
Lr=10^(10/10);
```

```
Gsl=10^(10/10);          % dbi (Near-In Sidelobes)

% Calculate Target Power Received at Radar

Pr=(Pt*Gr^2*sigma*wl^2)./((4*pi)^3*Lr*(Rtar*10^3).^4);
Prdbm=30+10*log10(Pr);

% Jammer Related Parameters

Pjsat=100;               % w
Gjr=10^(10/10);
Gjt=10^(10/10);
Lp=10^(3/10);            % Polarization Loss
Beta=1;                  % Choping Loss
Lj=10^(7/10);
effrcs=10*sigma;         % Effective RCS

% Calculate Power Received by Jammer

Pjr=Pt*Gt*wl^2*Gjr./((4*pi*Rtar*10^3).^2*Lp);
Pjrdbm=30+10*log10(Pjr);

% Calculate Repeater Gain

Grep=4*pi*effrcs*(Lp^2)/((wl^2)*Beta^2);

% Calculate Repeaters Electronic Gain

Ge=Grep/(Gjr*Gjt);

% Calculate Jammer Power

Pj=Pjr*Ge;
for i=1:length(Rtar);
   if Pj(i)>Pjsat;Pj(i)=Pjsat;end;
end;

% Calculate Jam Power at Radar Receiver

Prj=Pj*Gjt*Gr*wl^2./((4*pi*Rtar*10^3).^2*Lp);
Prjdbm=30+10*log10(Prj);

% Calculate Sidelobe Jam Power at Radar

Pjrsl=Pjr*Gsl/Gt;
Pjsl=Pjrsl*Ge;
for i=1:length(Rtar);
  if Pjsl(i)>Pjsat;Pjsl(i)=Pjsat;end;
end;
Prjsl=Pjsl*Gjt*Gsl*wl^2./((4*pi*Rtar*10^3).^2*Lp);
Prjsldbm=30+10*log10(Prjsl);
```

```
% Plotting Received Target and Jammer Signals

semilogx(Rtar,Prdbm,Rtar,Pjrdbm,Rtar,Prjdbm,Rtar,Prjsldbm);
ylabel('Received Signal (dbm)');
xlabel('Range (km)');grid;
title('REPEATER JAMMING AND TARGET SIGNALS AT RADAR');
axis([.1,1000,-100,50]);
text(2,35,'Repeater Input Signal');
text(20,-40,'Main Lobe Repeater Jammer');
text(.12,-75,'Side Lobe Repeater Jammer');
text(.4,10,'Main Lobe Skin Return');
text(6,-20,'Saturation');
text(.3,-7,'J/S=1');
```

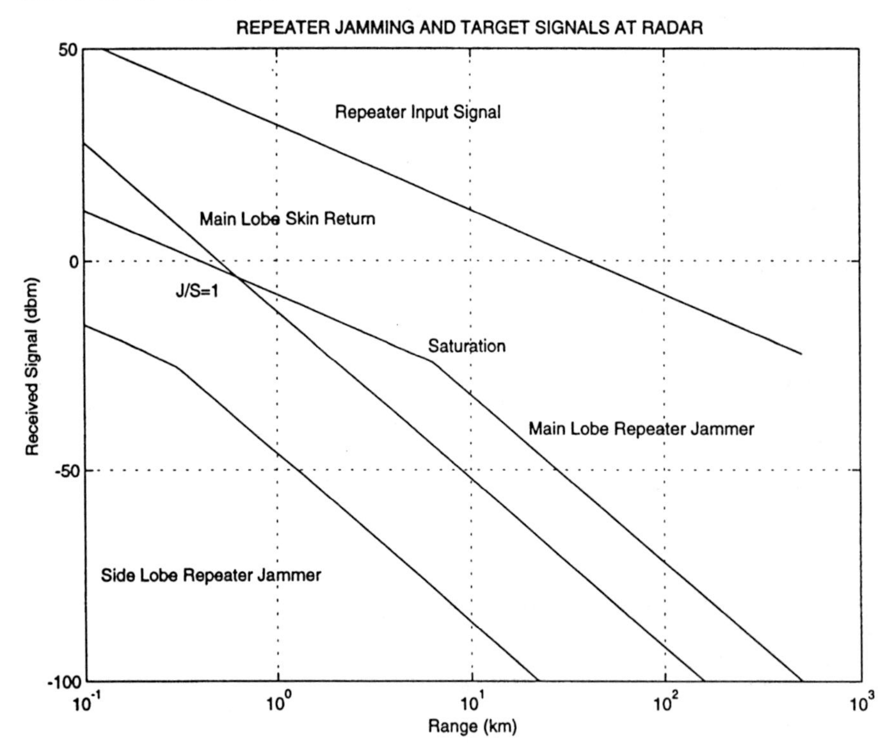

3.4 Transponder Jamming

A weakness of conventional RGPO repeater jammers is that an inherent delay exists due to transmission-line and TWT delays. This delay is typically of the order of 50 ns to 100 ns, which provides the possibility of tracking the real target echo instead of the repeater jamming signal using leading-edge tracking techniques.

Under certain circumstances, RGPO can be used to simulate an inbound target that is effective against leading-edge range trackers. This requires that the repeater anticipate the reception of the radar pulse and have sufficient storage capability to store the radar pulse for one PRI. The anticipation of the radar pulse generally requires that the radar use a PRF code that has a stable pattern so that the repeater timing can be initiated from the previous received radar pulse. Delay-line memory loops cannot provide the necessary storage time, so the use of a DRFM or a *voltage-controlled oscillator* (VCO) set-on oscillator is required. When a VCO set-on oscillator is employed, the DECM is usually called a transponder jammer rather than a repeater. Another form of range-denial jamming that can be used with unstable radar PRIs is cover jamming. With cover jamming, a relatively long noise-jamming pulse is radiated, which is initiated by the previous radar pulse, and has sufficient duration to cover any expected range of PRI jitter.

Transponder jamming is employed in many modern DECM systems since it generally provides a method to combat leading edge range tracking against unsophisticated radars. Frequency agility and PRF jitter are radar EP techniques to combat transponder jamming.

The key component associated with modern DECM transponder jamming systems is the fast-tuning VCO. VCOs allow a jammer to effectively function in a high-density environment and respond to a variety of sophisticated threats. The VCO receives a series of digital tuning signals that represent instructions to step tune to a specific frequency, dwell, and then tune to another threat frequency. To be effective, the VCO should be able to tune in 50 ns to 100 ns and have a set-on accuracy on the order of ±1 MHz [9].

Critical areas for VCOs are post-tuning drift, settling time, and repeatability. Post-tuning drift is the amount of frequency deviation incurred over some prescribed time (e.g., 150 ns to 50 μs) after the VCO has reached the prescribed frequency commanded by the digital input message. Settling time is the amount of time required for a VCO, after being switched from one frequency to another, to arrive within a specified error bandwidth around the new frequency. Repeatability is a measure of how accurately commanded frequencies are attained when attempting to repeat a given frequency at different times.

VCOs used in transponder modes of DECM systems are not required to be coherent with the victim radar's transmissions. Jamming of radars that use forms of coherent integration (e.g., PD and pulse-compression radars) require that coherent transmissions be radiated by the jammer. For this type of operation, modern DECM systems use DRFMs or DDS.

DDS technology is rapidly replacing VCOs in both EW and radar applications. In DDS systems, signals are generated digitally, providing for digital control of the signal phase, amplitude, and frequency. DDS sources generally consist of four functional components: a phase accumulator, *read-only-memory* (ROM), a *digital-to-analog converter* (DAC), and an output filter. The phase accumulator is a digital integrator that contains the signal phase. The signal phase is converted to digital amplitude information by the sine ROM that functions as a look-up table. The output of the ROM is converted to analog information in the DAC that is subsequently filtered to remove commutation components. The sample speed at which the digital circuitry commutates is determined by the bandwidth of the synthesized waveform (the Nyquist rate that is twice the highest frequency component to be synthesized). Using GaAs digital components, clock rates as high as 1.6 GHz are available [10].

The basic advantage of a DDS over a VCO is that complex intrapulse waveforms can be precisely generated, facilitating the synthesis of pulse compression and other coded waveforms. This allows jamming waveforms to build up in radar matched-filter networks, thereby negating the radar's processing gain while enhancing the JSR in the radar.

A simplified block diagram of a transponder jammer is depicted in Figure 3.11. An input receiver captures the signal environment. The environment is analyzed to collect those signals that are associated with a particular radar. These signals are then identified by comparison with a threat library and a jam decision is made. A VCO (or DDS) is tuned to the victim radar's frequency and temporally modulated to transmit the jamming signal at the appropriate time to cover the return from the shielded target.

If a DDS is employed, then coded waveforms can be generated that match pulse compression radar waveforms. Frequency diversity radars can be jammed by transmitting on all possible radar frequencies simultaneously. PRF agile radars can be jammed by extending jamming pulses to cover all possible temporal target positions. Transponder jamming is not appropriate against random frequency agile radars.

3.5 Support Jamming

The mission of escort and stand-off support jamming is to deny, delay, and degrade acquisition of strike aircraft while forcing early turn-on of terminal radars. This mission supports a broader strategy of general EW and defense suppression requirements that include:

- High-powered, wide-area jammers to degrade large numbers of EW/GCI/target acquisition radars and the C2W for their air defense network to function;
- DECMs to negate most SAM systems;

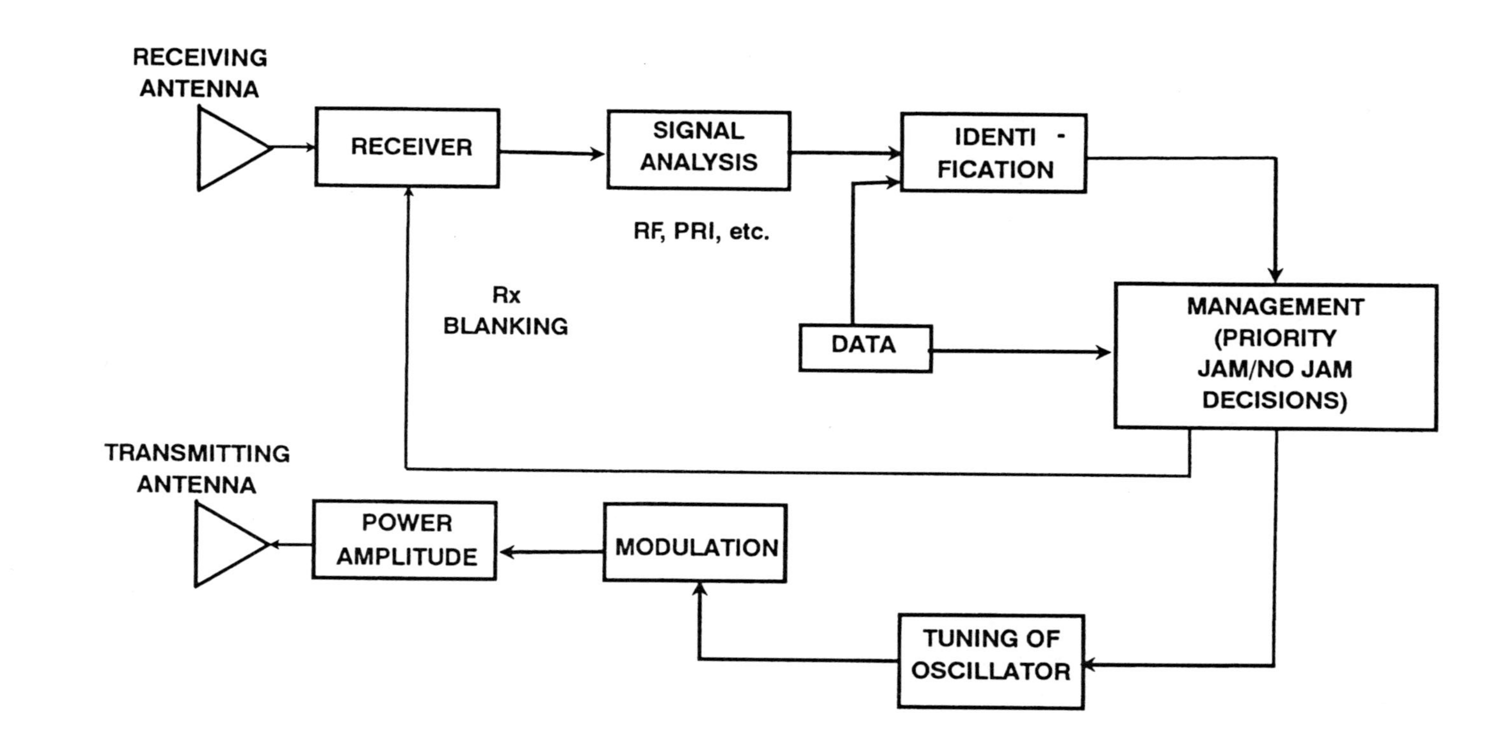

Figure 3.11 Simplified block diagram of transponder jammer.

- ARMs to destroy the smaller set of EW/GCI/target acquisition radar and SAMs that cannot be jammed with confidence (ARMs targeted with threat suppression aircraft).

Current support jamming systems have proven effective in recent conflicts against operational enemy air defense systems that used radar technology of modest capability. Support jammers promise to be more effective in suppressing air defenses against stealth targets. However, the rapid pace of modern radar developments and export of these technologies throughout the world has raised questions as to the effectiveness of support jamming in a high-technology radar environment. This concern has resulted in a number of support jammer upgrades that both increase the effectiveness of the jamming waveform and the complementary electronic support equipment that provides situational awareness to the jamming systems.

3.5.1 Issues in Support Jamming

The mission of support jammers dictates that jamming be accomplished in radar sidelobes; both to prevent strobing on the jamming aircraft and to shield spatially displaced strike aircraft. Modern radars employ *ultralow sidelobe antennas* (ULSAs), CSLCs, and SLBs to inhibit signals entering the radar through its sidelobes. ULSA antennas have rms responses less than –20 dBi while CSLCs can attenuate signals by 25 dB or more.

Figure 3.12 depicts the magnitude of this effect by comparing the range achieved in a stand-off noise jamming environment with the TPS-43E (i.e., a conventional sidelobe antenna system), the TPS-70 (i.e., an ULSA antenna system) and the TPS-70 with a CSLC. The illustration shows that the support jammer can significantly reduce the range of the TPS-43E. However, the introduction of a ULSA decreases the jamming effect to a small amount while the introduction of CSLC (in addition to the ULSA) suppresses the jamming effects.

A second area affecting jamming effectiveness, where modern radar technology has advanced, relates to coherency in the radar's transmitted waveform. Coherency allows the radar to provide large processing gains against interfering noiselike signals such as noncoherent jamming waveforms. Processing gain occurs both within a single radar pulse (intrapulse coherency) and between radar pulses (interpulse coherency). The former is associated with *pulse compression* (PC) type signals while the latter is concerned with processing Doppler type information in the radar.

The result of this increased processing gain is a dilution of the effective jamming power with respect to the available target power. PC processing gain

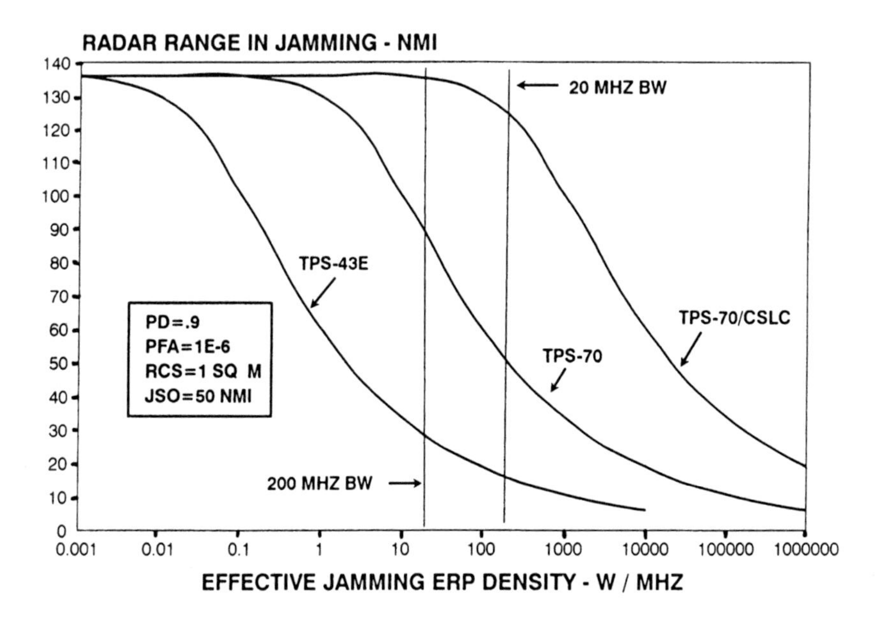

Figure 3.12 Radar range with support jamming.

is determined by the PC factor. Many current radars use 13-bit Barker code PC waveforms (i.e., the PC factor is 13). Radars using analog pulse compression networks generally have PC factors between 30 and 100. Newer PC radars using digital PC techniques have lager PC factors that can be of the order of 1000 or more.

Interpulse processing gain relates to the dwell time the antenna spends on the target. Doppler bandwidths can approach 50 Hz to 100 Hz based upon dwell time analyses. However, it must be considered that Doppler radars have to dwell at a single frequency for a CPI. The radar's dwell time can be filled with a number of CPIs, each associated with a different radar frequency that allows frequency agility to be applied to the radar's waveform. Thus, coherent processing gains for frequency agile Doppler radars are reduced from their maximum potential (CPI < dwell time) in order to allow frequency agile operation. The net result is that a larger Doppler bandwidth results from the radar's agile operation, but the jammer is forced to spread its energy over a wider frequency range to jam the radar.

A third issue concerning support jamming is the introduction of multiple simultaneous threats that result in the dilution of the available jamming energy available to address each threat. Processing of these advanced threats stresses the receiving and analysis portion of the support jammer's electronic support system (i.e., receiving system).

Continued effectiveness of support jammer operation against advanced threat technology requires:

- Effective use of available jamming power;
- Matched jamming waveforms;
- Advanced receiving techniques.

Effective use of jammer power is enhanced using "smart noise" (see Figure 1.6) that is matched to the radar's transmitted waveform. This essentially removes the processing gain advantage of modern radars against noiselike jamming interference. To achieve this objective, the support jammer must synthesize the victim radar's waveform. This involves either storage of the radar's signature in the jamming system or interception of the radar signal for subsequent analysis, storage, and retransmission.

Synthesis of the radar signature in the jammer can be accomplished by several methods. The first method involves storage of threat radar parameters based upon SIGINT information stored in the jammer's threat library. Interception of threat radars' signals in the jammer's analysis receiver allows the threat to be identified and the parameter set to be updated with the latest information relevant to the specific emitter to be jammed (e.g., frequency, pulse width, and PC factors). The jammer waveform is then synthesized from the signature data using a DDS and used to generate multiple synchronized random jamming pulses that are received through the radar's antenna sidelobes.

DDS jamming waveforms are not generally coherent with the radar pulse. However, they are quasi-coherent in that they are coherent over the intrapulse processing period of the radar. Hence, this type of jamming has the potential to mitigate the intrapulse processing gain of the radar. In addition, although not coherent with the radar from pulse-to-pulse, the overall interpulse processing gain advantage is reduced since the DDS jamming pulses are noncoherently integrated. The overall effectiveness of this approach is obviously dependent on how accurately the threat radar parameters can be determined in the jammer. We analyze this effectiveness as a function of the frequency and pulse width mismatch between the radar and the DDS synthesized signals.

The second method involves the actual storage of the threat radar signal in a DRFM. This approach has the obvious advantage that a coherent jamming waveform can be synthesized. However, the capacity of the DRFM to store a large number of waveforms is limited by the state-of-the-art in high-speed integrated circuitry. Also, the quantization ability of current DRFMs is limited, which results in intermodulation products between the various stored signals. In practical systems, this requires a frequency synthesis and thresholding of the stored wave forms to eliminate spurious signals and prevent dilution of available jamming power.

A third approach would be to use the DRFM to coherently store the threat radar signals. This signal could then be used to update the threat parameters and cohere a DDS that generates the actual jamming waveform. The advantage of this approach is that it allows the DDS to generate a coherent jamming signal that closely resembles the actual radar waveform. Since the signal is stored in digital form, it can be used to generate multiple randomly synchronized pulses that appear as coherent random targets when they enter the radar through its antenna sidelobes.

3.5.2 Direct Digital Synthesis Jamming

DDS architectures, as commercially implemented, generally utilize a phase accumulator, look-up table random access memory, and a digital-to-analog converter to generate the synthesized signals. In addition, for some applications, a frequency accumulator is placed in front of the phase accumulator to provide fast frequency changes and a phase added is placed at the output of the phase accumulator to generate phase modulation [11].

This configuration is ideally configured to provide linear frequency modulated (i.e., chirp waveforms) or phase-coded (i.e., Barker coded waveforms) pulse compression signals.

In operation, if a linear chirp signal is desired, the frequency accumulator receives a digital input for the start frequency and chirp rate and outputs the instantaneous frequency. The instantaneous frequency is applied to the phase accumulator that translates it to an instantaneous phase. Phase modulation can be simultaneously accomplished (if appropriate) using an added phase that modifies the phase signal from the phase accumulator. The phase accumulator drives a memory look-up table that converts phase to amplitude prior to analog conversion. DDSs can currently operate at clock rates exceeding 500 MHz, which provides an instantaneous bandwidth exceeding 250 MHz.

Use of the DDS in a jammer involves generating a digital message that is applied to the frequency and phase accumulators to synthesize the desired jamming waveform. The resulting digital synthesized waveform must then be translated to the threat radar's RF carrier frequency by mixing with an appropriate local oscillator signal. In general, by time multiplexing, a single DDS can be used to jam several threat radars simultaneously. Alternately, employing multiple accumulators, it is possible to simultaneously synthesize jamming waveforms for several threat radars at the same time.

The effectiveness of DDS-generated jamming waveforms depends upon the ability of the DDS to match the actual radar signal. Figure 3.13 shows the effect of a frequency mismatch against a 13-bit Barker-coded pulse compression waveform. A frequency mismatch of the order of 0.07 bandwidths (i.e.,

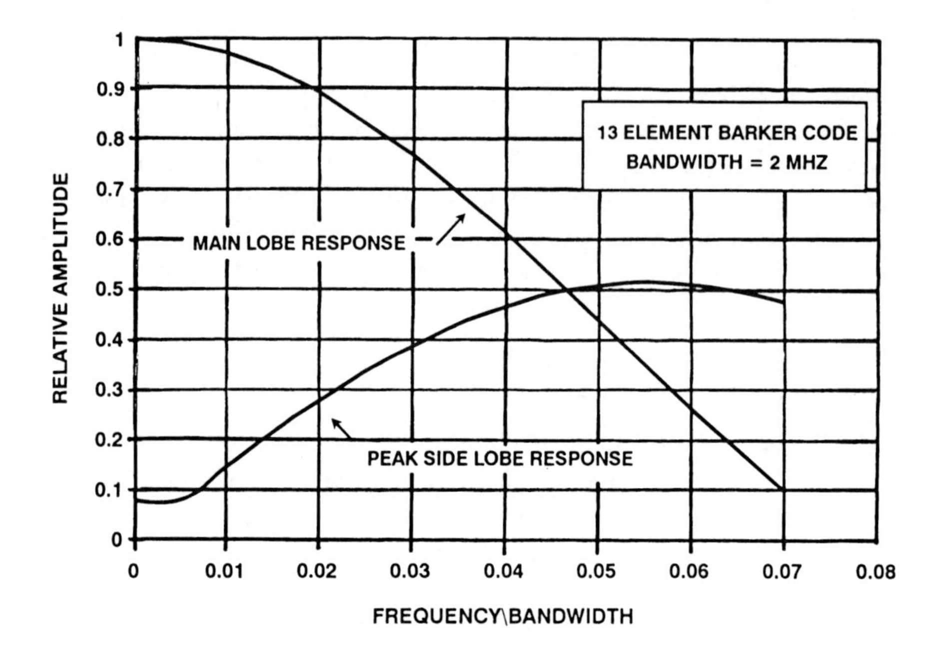

Figure 3.13 Barker-coded frequency mismatch.

140 kHz in the case of the TPS-43E/TPS-70) would negate the effect of the coherent jamming in this application. Linear FM chip signals are more tolerant of frequency mismatch than binary-coded waveforms. For example, a mismatch in frequency of 0.6 bandwidths results in an amplitude reduction of only 0.6 in the case of linear FM waveforms.

3.5.3 Digital Radio Frequency Memories

The basic function of the DRFM is to store and replicate RF signals. The DRFM process typically involves: (1) down converting the intercepted waveform to base band, producing in-phase and quadrature components of the signal waveform; (2) sampling and digitizing using an *analog-to-digital* (A/D) converter; (3) digitally storing the I and Q waveforms in memory; (4) reconstituting the I and Q waveforms, suitably modified for jamming, to analog form by *digital-to-analog* (D/A) conversion; and (5) reconstructing the RF jamming waveform using single sideband up conversion modulation of the I and Q samples using the original down conversion oscillator. In a commonly used variant, the I and Q samples are converted into phase samples that then require only a single A/D converter and memory for storage.

A number of issues exist when DRFMs are used in support jammers. Many of the available DRFMs use only 1-bit of quantization. This tends to generate spurious signals and causes intermodulation products when multiple signals are present. Instantaneous bandwidths generally range up to 500 MHz. When broadband DRFMs are used, current technology permits only on the order of 3 to 4 bits in tactical-sized units.

The advantage of DRFM-based support jamming is that both the phase and frequency of the threat radar can be stored. This allows coherent jamming of Doppler radars as well as jamming of frequency hopping and pulse compression radars. In some cases, radars that simultaneously transmit different frequencies on multiple antenna beams must also be jammed. This requires the storage of multiple signals within the DRFM at the same time.

Figure 3.14 illustrates the spectrum of a 1-bit DRFM with one signal, while Figure 3.15 depicts the same DRFM when 15 signals are stored. It is obvious that both intermodulation products and spurious signals are produced by the quantization of the stored signals. In addition, signal suppression occurs due to the hard-limiting operation. Figure 3.16 depicts the average loss incurred in the DRFM due to multiple signal storage and quantization.

3.5.4 Comparison of DDS and DRFM Support Jamming

Figure 3.17 illustrates a comparison of support jamming using DDS and DRFM techniques against a TPS-70 radar. One curve (TPS-70 DDS Jamming) assumes the DDS can tune accurately enough to negate the pulse compression processing gain and integration gain of the radar. To accomplish this, the support jammer would have to transmit 16 binary-coded pulses to ensure that the TPS-70 would be jammed on successive radar transmission. This requires the equivalent of a 32-MHz jamming spot.

A second curve assumes that the DDS is mismatched in frequency (see Figure 3.13) such that the jamming signal does not build up in the PC Network (i.e., a 140-kHz frequency offset). The 32-MHz jamming spot applies in this case. Also shown on the curves is the result of perfect DRFM jamming, which provides the effect of a 2-MHz bandwidth jamming spot. If multiple signals are stored in the DRFM and a one- to three-bit quantizer is used, the losses derived from Figure 3.16 would add to the effective jamming spot size. For example, if a one-bit quantizer and five signals were stored, then the effective jamming spot size would be about 12 MHz. The TPS-70 DDS jamming curve would apply in this case using the 12-MHz jamming bandwidth.

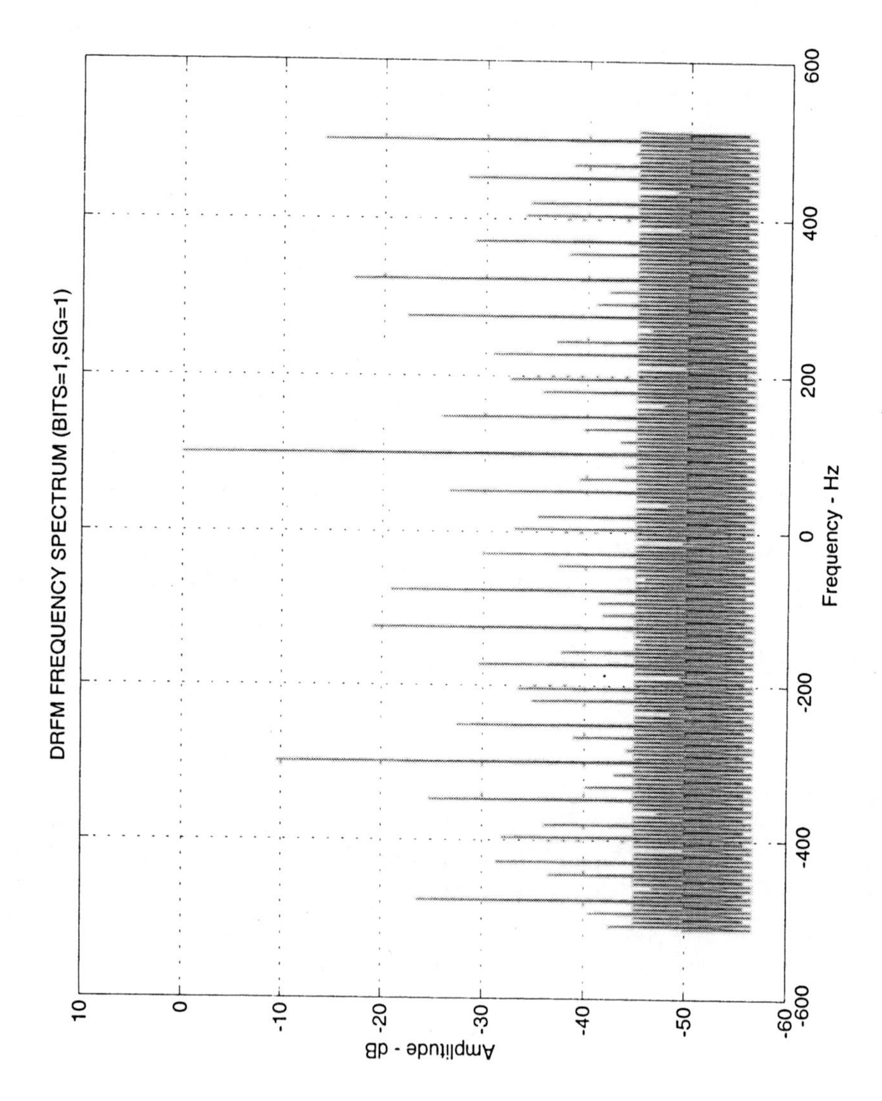

Figure 3.14 DRFM frequency spectrum (bits = 1, sig = 1).

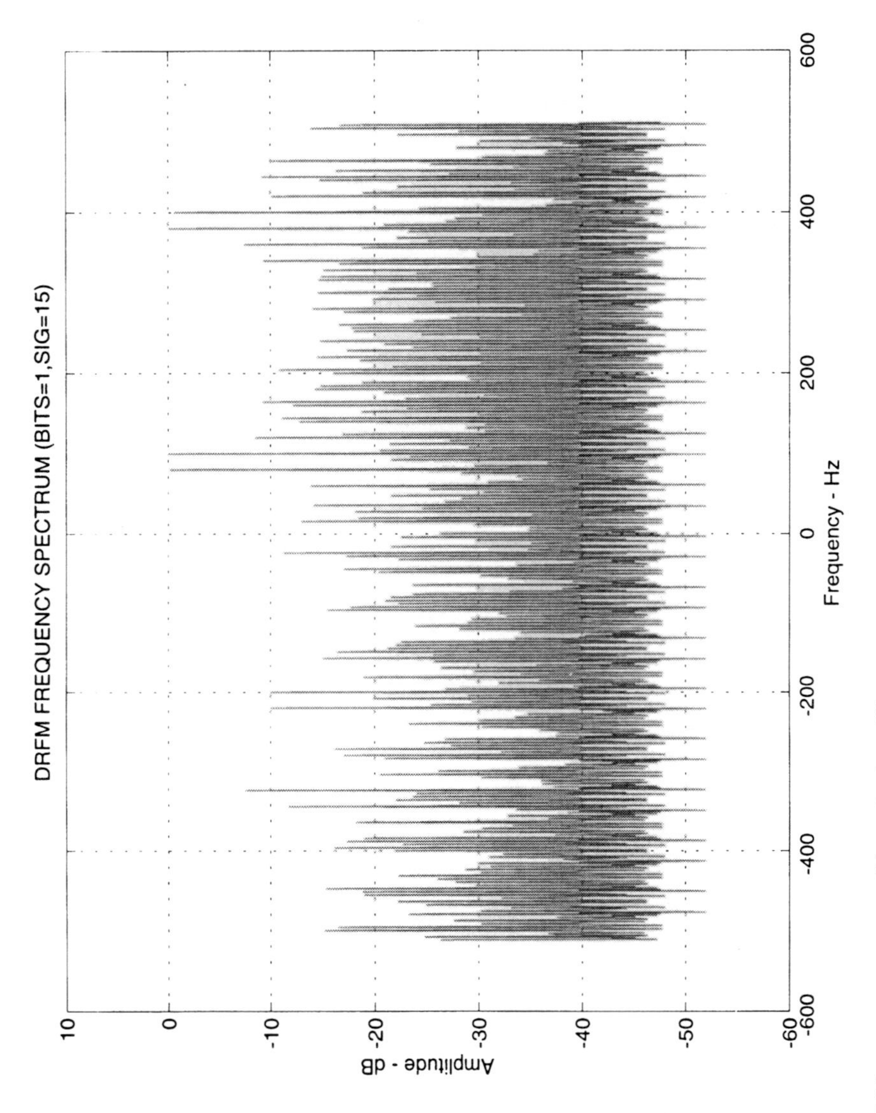

Figure 3.15 DRFM frequency spectrum (bits = 1, sig = 15).

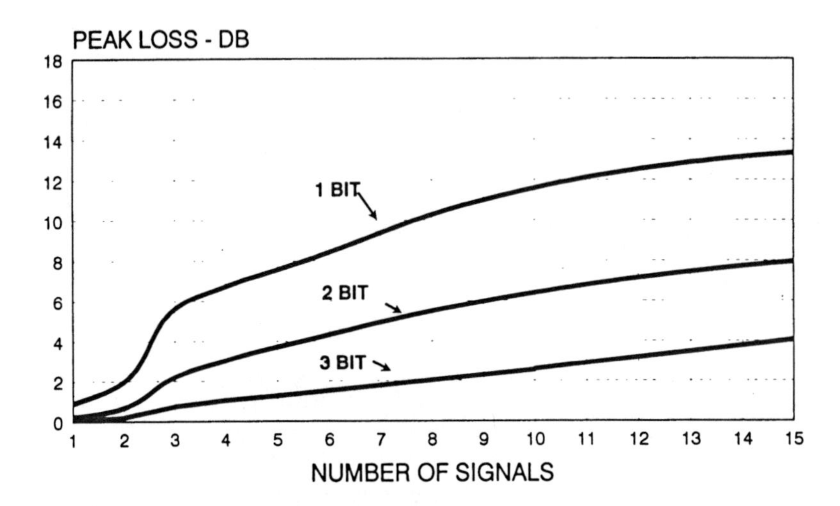

Figure 3.16 DRFM multisignal efficiency.

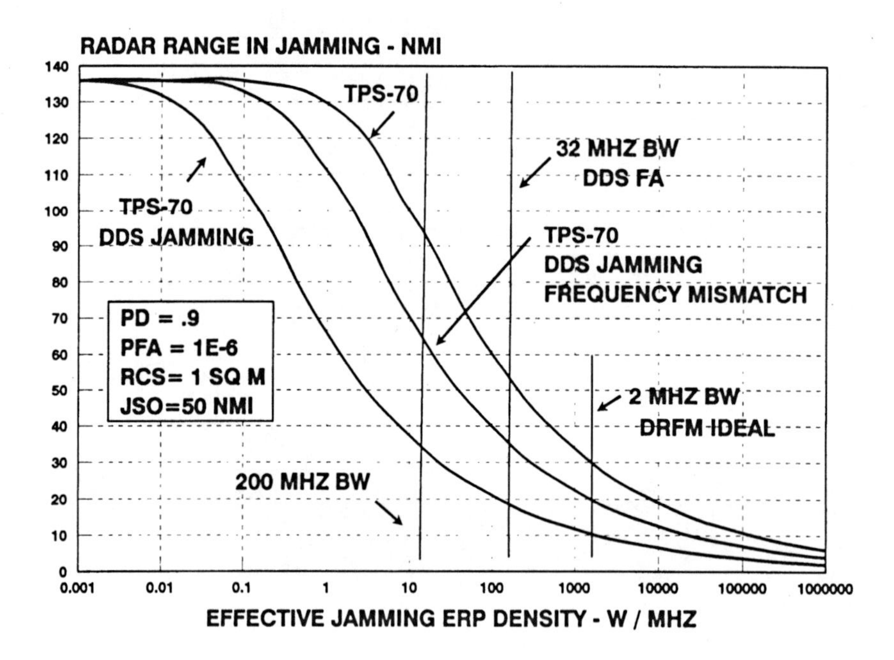

Figure 3.17 Radar range with DDS jamming.

3.6 Problems

1. Given the noncoherent search radar described in Table 3.4, compute the effective radiated power of a stand-off jammer with 30-MHz bandwidth to provide a 13-dB J/S ratio. The radar has an antenna sidelobe level toward the jammer of −20 dB relative to its peak gain. The jammer is at 200 km and is slant polarized (45 degrees from vertical).

Table 3.4
Radar Parameters

Quantity	Value
P_t	1 MW
$G_{t,r}$	35 dB
λ	0.5m
σ	10 m^2
F	3 dB
B_r	1 MHz
τ	1 μs
θ_{AZ}	6 degrees
PRF	360 pps
Scan rate	72 degrees/s
P_d	0.9
P_{fa}	10^{-6}
Target type	Swerling 1
L_r	12 dB
L_j	7 dB
Polarization	Horizontal

2. For the given jammer and noncoherent search radar parameters, determine the *jammer-to-noise* (J/N) ratio at a range of 100 km for a self-screening noise jammer.

Radar	Jammer
P_t = 1 MW	P_j = 100W (100% duty factor)
$G_{t,r}$ = 35 dB	G_j = 3 dB
f = 600 MHz	B_j = 10 MHz
PRF = 360 Hz	L_j = 5 dB
B_r = 1 MHz	polarization = slant 45 degrees
F = 5 dB	
n = 30 pulses	

3. For the noncoherent search radar described in Table 3.4 and jammer described in the previous problem, determine the jam-to-signal ratio for a 10-m^2 Swerling 1 target located at 40 km.

4. For the noncoherent search radar in Table 3.4 and jammer of Problem 2, what would the jam-to-signal ratio be for a 10-m^2 Swerling 1 target located at 40 km if the jammer were polarized at 30 degrees with respect to vertical and the radar were polarized horizontally?

5. Your target is protected by one airborne interceptor controlled by a GCI radar site that contains both a GCI and a height-finding radar. The radar parameters are as given. Each radar can quickly tune over its entire operating range and both are noncoherent. The height-finding radar must be cued by the GCI radar to locate a target in azimuth. The interceptor requires 5 min to reach the GCI site.

GCI Radar

P_t = 2.75 MW
$G_{t,r}$ = 30 dB
f = 1–1.1 GHz
PRF = 400 Hz
B_r = 250 kHz
θ_{AZ} = 3 degrees
$\dot{\theta}_{AZ}$ = 6 rpm
F = 5 dB
elevation = 200 ft
$(S/N)_{min}$ = 13 dB

Height-Finding Radar

P_t = 1/5 MW
$G_{t,r}$ = 33 dB
f = 2.2–2.4 GHz
PRF = 600 Hz
B_r = 1.2 MHz
ϕ_{el} = 1 degree
nod rate = 10 degrees/s
τ = 1 μs
elevation = 200 ft

There is a penetrator whose maximum speed is 500 knots with an RCS of 10 m^2 that is covered by a stand-off jammer capable of carrying two noise jammers. The characteristics of the noise jammers are given in Table 3.5. The penetrator cannot fly below 100 ft.

Table 3.5
Stand-Off Noise Jammers 1 and 2

Quantity	Jammer 1	Jammer 2
P_j	2 kW (100% duty factor)	1 kW (100% duty factor)
G_j	10 dB	13 dB
f	1–2 GHz	2–4 GHz
B_j	3–200 MHz	3–200 MHz

a. Without a jammer, how close can the penetrator fly before detection? What is its time to target after detection?
b. If the stand-off jammer is at 200 km with respect to the GCI site, above what altitude must it fly to be effective?
c. If a 3-dB J/S ratio is required to screen the penetrator against either radar; at what range will jammer 1 screen the target against the GCI radar? At what range will jammer 2 screen the target against the height-finder radar?
d. What penetration profile would be optimum?
e. Is it necessary (or desirable) to jam both radars? Explain your answer.

6. Given the tracking radar parameters, if transition to the tracking mode requires 20-dB S/N ratio on each pulse, at what range does the transition occur for a 5-m^2 target?

Tracking Radar

P_t = 50 kW	PRF = 1800 Hz
$G_{t,r}$ = 34 dB	polarization = vertical
f = 5.6 GHz	F = 5 dB
θ_{AZ}, ϕ_{el} = 2 degree (pencil beam)	scan axis gain = 31 dB
τ = 0.4 μs	

7. Given a 45-degree slant polarized 1-kW self-protection noise jammer with 20-MHz bandwidth, plot the resultant J/S ratio for a 5-m^2 target over interval of 1 to 20 nmi from the tracking radar described in the previous problem. Assume that the tracking radar angle servo's have a 10-Hz bandwidth. At what J/S ratio would the tracking radar start to break lock?

8. For the jammer described in Problem 7, assume that the RCS of the aircraft being tracked with the radar in Problem 6 varies with the angle as described below. Also assume that the jammer's antenna gain varies as $G(\theta) = G_{max}\cos^2\theta$, where θ is measured from the nose of the aircraft. For this situation, plot the range versus angle about the nose for achievement of a 10-dB J/S ratio.

θ (degrees)	σ_t (m^2)
0	10
15	8
30	6
45	5
60	30
75	40
90	60

9. Two aircraft, flying parallel, are separated by a distance L. A semiactive homing missile with a relative closing speed of 2000 m/s is fired at the aircraft. The missile can accelerate normally at 200 m/s^2 while its seeker has an antenna beamwidth of 10 degrees.

a. At what range will the missile resolve the individual aircraft? Assume the missile initially tracks a point midway between the aircraft.

b. Assuming the missile accelerates normally at its maximum rate toward one of the aircraft, as soon as the aircraft are resolved, determine the approximate miss distance.

c. What aircraft separation (L) maximizes the miss distance

10. A noncoherent search radar has a power output of 600 kW, pulsewidth of 0.4 μs, and operates at a frequency of 3 GHz. The antenna has an aperture of 15 ft^2 with 60% efficiency. The aircraft target has a 10-m^2 RCS, and the radar requires a single pulse SNR of 0 dB for detection.

a. If the aircraft carries a self-protection noise jammer that generates an ERP density of 10 W/MHz, find the burn-through range.

b. A stand-off noise jammer with an antenna gain of 10 dB and a jammer output of 4.5 kW/MHz is located 10 nmi from the radar. Assume that the jammer is in the 5-dB point in the radar antenna pattern and that it is 45-degrees slant polarized. At what range will the attack aircraft burn through the jamming?

c. What is the range of the radar on the attack aircraft in the absence of jamming?

d. The attack aircraft carries an ES receiver with an antenna with 0-dB gain. When it is at the maximum detection range of the radar, what is the received signal strength at the antenna terminals?

e. The attack aircraft uses its jammer against a pulse Doppler radar having an average power of 1 kW in the central line of its transmitted spectrum. The receiver noise bandwidth is 200 Hz. For a signal-to-noise ratio of 1, what is the burn-through range against the radar for a 10-W/MHz jamming power? Assume the same antenna gain as for the search radar described previously.

11. The burn-through range of a noise jamming attack aircraft against a particular fire-control radar using noncoherent integration is 5 nmi. Assume that the radar integrates 30 pulses. What would be the effect on burn-through range if the jammer uses coherent false targets instead of a noise waveform?

12. A UAV with a RCS of 0.5 m^2 is to carry a repeater to simulate a total RCS of 5 m^2 at a frequency of 3 GHz. Omnidirectional antennas are to be used for both transmitting and receiving. Find the required electronic gain for a duty cycle of 0.5 with a polarization loss of 3 dB.

13. A straight-through repeater has a gain of 75 dB. If it is to be redesigned as a gated repeater, since adequate antenna isolation is unavailable, what gain is required for a gated on time of 0.08 μs and off time of 0.05 μs.

14. What dynamic range is required for a ring-of-targets repeater to operate against a radar that has an antenna gain of 30 dB and an average sidelobe 10 dB below isotropic? What would be the effect on the dynamic range of employing a transponder rather than a repeater jammer?

15. A repeater jammer against a specific radar exhibits a burn-through range (J/S = 1) of 1 nmi. It is designed to provide a J/S = 10 at the radar. At what range is the full J/S ratio realized?

16. In a command-guided missile system, radars track the position of both the target and the missile. A guidance computer determines the missile trajectory necessary to cause the missile to collide with the target and transmits this information to the missile via a data link. Explain why a noise jammer on the target can provide an effective defense against this type of system.

17. Hard-kill and soft-kill mechanisms often interact within themselves and with each other. For example, chaff deployed to seduce an ASCM can also act to confuse a hard-kill radar-controlled gunfire system (i.e., CIWS). To detail these interactions, construct a threat interactive polygon. Associate each of the hard-kill/soft-kill mechanisms listed in Table 3.1 with each corner of the polygon. Then draw arrows from each mechanism that may affect the performance of another mechanism. Also include in this diagram the effects of the ship's satellite communication system that may interact with the other electronic-based mechanisms.

18. One of the most critical aspects of antiship missile defense is the reaction time to deploy countermeasures as the missile appears over the horizon. Using MATLAB, plot the time-to-impact (sec) versus the missile range-to-go (nmi) for missile velocities of Mach 0.8, 0.9, 1.2, 1.5, 2.0, 2.5, and 3.0. On these curves, plot lines representing the radar horizon for a missile height of 20 ft and an ES receiver height of 20 ft, and for a missile height of 100 ft and an ES receiver height of 100 ft.

19. Using the results of Problem 18, discuss the advantages and disadvantages of employing a Mach 3 missile versus the present Mach 1 versions.

20. An I-band (9375-MHz) active towed decoy is used to seduce antishipping missiles away from the ship. If the antiship missile seeker transmits a 0.5-μs, 10-kW pulse with an antenna gain of 23 dB, what decoy receiver sensitivity (dBm) is required for detection of the cruise missile signal at 20 nmi? Assume the decoy uses an omnidirectional antenna, has a 5-dB noise figure, 10-dB losses, and requires a signal-to-noise ratio of 15 dB for detection. What decoy transponder power is required to cover the ship signature (RCS = 2500 m^2) at 0.5 nmi? Assume a 10-times magnification factor is required.

21. Conical scan radars are vulnerable to inverse gain jamming. To reduce this vulnerability, an additional receive antenna beam is added that scans at the same rate as the main beam, but is displaced in angle by 180 degrees. The outputs of the two receive antenna beams are differenced to form the error detection signal. Explain how this scan-with-compensation technique works to reduce the vulnerability of conical scan radars to inverse gain jamming.

22. A high-power search radar generates a 4-MW peak power pulse of 6.5-μs duration. The antenna gain of its ultralow sidelobe antenna is 36 dB. The maximum sidelobe level is −5 dBi. It is desired to cover the sidelobe transmissions using a spurious cover pulse transmitted through an omnidirectional antenna. What power level is required for the cover pulse transmitter if it must produce a sidelobe cover level 10 dB higher than the actual sidelobe signals?

23. The ASR-9 radar generates 12 bits inphase and 12 bits quadrature data in each of its 700 radar resolution cells due to clutter and noise for each transmitted radar pulse. The PRF is 1200 Hz while the radar scans at 75 degrees/s. What is the data rate at the input to the radar's signal processor? What is the output data rate needed to provide a false alarm rate of 10^{-10}? How many false targets per second must be injected into the radar's sidelobes through jamming in order to overload the output data link of the radar whose capacity is 20,000 bits/s?

References

[1] Grimes, V., and J. Lok, "Shooting for New Targets," *Jane's Navy International,* July/August 1995.

[2] Lok, J., and S. Gourley, "The Harder Kill," *Jane's Navy International,* July/August 1995.

[3] Richardson, D., "Smarter Seekers," *International Defense Review,* Sept. 1995.

[4] Gill, T., "Self-Protection Calls for a Tow," *J. Electronic Defense,* Jan. 1997.

[5] Greenbaum, M., "The Complementary Roles of Onboard and Offboard EW," *J. Electronic Defense,* Nov. 1992.

[6] Annen, B., J. Herther, and R. Prevatt, "IDECM RFCM: A Common Solution to Multiservice Requirements," *J. Electronic Defense,* Nov. 1996.

[7] Neri, F., "New Technologies in Self-Protection Jammers," *J. Electronic Defense,* July, 1991.

[8] Seward, T., "Surface Ship EW," *J. Electronic Defense,* May 1991.

[9] Schleher, D. C., *Introduction to Electronic Warfare,* Norwood, MA: Artech House, 1986.

[10] Goldberg, B., "Reviewing Various Techniques for Synthesizing Signals," *Microwaves and RF,* May 1996.

[11] Adler, E., E. Viveiros, T. Ton, J. Kurtz, and M. Bartlett, "Direct Digital Synthesis Applications for Radar Development," IEEE International Radar Conf., Washington, DC, May 1995.

[12] Mansky, L., "Broadband Phased-Array Antennas," Norwood, MA: *Microwave J.,* September, 1984.

[13] *Aviation Week,* September 7, 1987, p. 88.

4

EA Against Modern Radar Systems

In this chapter, we examine EA against several of the most important advanced techniques employed in modern radar. These include *pulse compression* (PC), *pulse Doppler* (PD), monopulse, *ultralow sidelobe antennas* (ULSAs), and *coherent sidelobe cancellation* (CSLC). All these techniques make modern radars difficult to jam and require special EA techniques for effective jamming.

The use of PC and pulsed Doppler in radar systems represents the use of both intra- and interpulse coherence in the radar waveform. This produces not only improved radar performance but also a resistance to jamming waveforms that are mismatched to the radar transmitted waveform. This trend is driven by the rapid advancements in digital signal processing and the availability of stable microwave amplifiers. Both techniques provide a potentially large processing gain against noise-type signals. With PC, the processing gain is equal to the time-bandwidth product of the radar waveform, which might range from 30 (i.e., 15-dB gain) in older radars to over 1000 (i.e., 30-dB gain) in more modern radars. Pulse Doppler radars exhibit a similar processing gain equal to the product of the CPI and the radar's PRF. For example, if the radar's PRF were 50 kHz and the CPI were 20 ms, then the processing gain against noise would be 1000 or 30 dB. In principle, both techniques are compatible [1] and the processing gains in decibels are additive.

Monopulse is used primarily in target tracking radars and missile seekers. Its mechanization provides an inherent resistance to amplitude-modulated jamming waveforms that tends to limit jamming from point sources. Fundamental EA techniques against monopulse generally employ dual spatially displaced coherent sources [2] such as cross-eye, terrain bounce, and blinking [3]. Cross-polarization jamming can also be used to exploit weaknesses in certain forms of monopulse tracking systems susceptible to this technique. These include reflector-type antennas, small flat plate arrays susceptible to edge effects, and any antenna in an aircraft or missile radome.

ULSAs and CSLCs are forms of spatial filtering that eliminate jamming signals from the radar's sidelobe response. ULSA antennas are a proven operational technique that provide average sidelobes the order of −20 dBi or lower. ULSAs use a combination of careful aperture illumination design and control of mechanical tolerances to create a low sidelobe design. CSLCs are a form of adaptive antenna array processor that forms a null on jammers in the radar's sidelobes. Generally, one null can be formed due to the interferometric action between the main antenna and any auxiliary antennas employed. Thus, each auxiliary antenna deployed provides the potential for canceling one sidelobe jammer. CSLCs generally provide processing gains of the order of 20 dB to 30 dB, but their theoretical performance is potentially much higher [3]. A CSLC can be employed with an ULSA to provide a high degree of resistance to sidelobe jamming but tend to be generally applied with conventional sidelobe antennas.

4.1 PC

The basic concept of a PC waveform is to transmit a long pulse of duration τ_d with bandwidth B and process the target return in a matched filter that compresses the long pulse to a short pulse of duration $\tau_s = 1/B$. Since the matched filter is theoretically lossless, the energy at the input to the matched filter $(S \cdot T)$ must equal that at its output (P_{pc}/B). Thus, at the output of the matched filter, the pulsed compression power is given by

$$P_{pc} = (B \cdot T)S \tag{4.1}$$

where

S = received peak power

B = transmitted bandwidth

T = transmitted pulse width

The time-bandwidth product

$$\beta = BT \tag{4.2}$$

is a measure of the increase in peak power due to the PC. Note that for detection calculation purposes, either the uncompressed pulse factors (S, T) or the compressed pulse factors (P_{pc}, τ_s) give the same result since they contain the same energy. In fact, any system using a matched filter processor provides an output signal-to-noise ratio equal to $2E_R/\eta_o$, where E_R is the pulse energy $(S \cdot T)$ and η_o is the spectral density of the white receiver noise.

When noise jamming is employed, the PC network generally provides an advantage that is equal to the PC ratio $T \cdot B$, where the comparison is against a conventional radar with equal transmitter energy and matched jammer bandwidth. This can be rationalized by considering the matched filter signal-to-noise ratio $2E_R/\eta_o$, which is the same for both PC and conventional radar of equal transmitted energy. However, the jammer must support a bandwidth of B against the PC radar, resulting in a jammer power of $B \cdot \eta_o$ as compared to a bandwidth of $1/\tau$ for the conventional radar, resulting in a jammer power of η_o/τ. The ratio of the jammer powers for the PC radar as compared to the conventional radar is, therefore, the PC ratio $T \cdot B$, assuming that a narrow spot bandwidth $1/\tau$ would be usable for the conventional radar.

Common forms of PC waveforms are the linear FM and phase-coded types. With linear FM, the bandwidth of the waveform is closely approximated by the difference between the minimum and maximum frequencies of the chirp waveform. With phase-coded waveforms, which are similar to spread-spectrum communication signals, the PC bandwidth can be approximated by the chip rate.

A matched filter that processes signal $v(t)$ has a conjugate impulse response given by $v^*(-t)$. The output of the matched filter is then given by the convolution integral as

$$g(t) = \int_{-\infty}^{\infty} v(\tau)v^*(t - \tau)\, d\tau \tag{4.3}$$

which is also the *autocorrelation function* (ACF) of the signal. The Fourier transform of the matched filter output is given by the power spectral density $V(\omega) \cdot V^*(\omega) = |V(\omega)|^2$, where $V(\omega)$ is the Fourier transform of the signal. The output signal can alternatively be found as

$$g(t) = \frac{1}{2\pi}\int_{-\infty}^{\infty} |V(\omega)|^2 e^{j\omega t}\, d\omega \tag{4.4}$$

4.1.1 Linear FM Pulse Compression

A linear FM signal can be expressed in complex form as

$$\tilde{v}_t = \exp\left[-j\left(\omega_0 t + \frac{\alpha t^2}{2}\right)\right] \qquad -\frac{T}{2} < t < \frac{T}{2} \tag{4.5}$$

where

$$\alpha = 2\pi \frac{B}{T} \tag{4.6}$$

is the chirp slope. The instantaneous frequency of the signal is given by

$$f = \frac{1}{2\pi}\frac{d\phi}{dt} = f_0 + \frac{B}{T}t \tag{4.7}$$

The matched filter impulse response of the PC matched filter is given by

$$h(t) = v^*(-t) = \exp\left[-j\left(\omega_0 t - \frac{\alpha t^2}{2}\right)\right] \tag{4.8}$$

which indicates that the impulse response of the matched filter is the signal waveform with negative slope.

▶ *Example 4.1*

Using the complex representation for a linear FM signal given by

$$\tilde{s} = \cos\frac{\alpha t^2}{2} - j\sin\frac{\alpha t^2}{2} \tag{4.9}$$

1. Plot the PC power spectral density for a pulsewidth of 100 μs and a chirp band of 1 MHz (β = 100). What is the chirp bandwidth for this case?
2. Plot the matched filter output response on a logarithmic scale (dB). How far is the first-order sidelobe down from the main-lobe response? Why does this response resemble a sin(x)/x function?

```
% PULSE COMPRESSION SPECTRUM USING FFT
% -------------------------------------
% pcspec.m

clear;clf;clc;

% Input Chirp Bandwidth and Pulsewidth
```

```
bw=1e6;     % Bandwidth - Hz
T= 1e-4;    % Pulsewidth - sec

%Input FFT Points

lfft=1024;

% Set Time Base

t=-T/2:T/(lfft-1):T/2;

% Sample Complex Chirp

s= exp(-j*pi*bw*t.*t/T);

% Plot Chirp Waveform

subplot(221);
plot(t,real(s));grid;
title(['Chirp Waveform']);

% Find Fourier Spectrum of Chirp

ZY= fft(s,lfft);
Y=fftshift(ZY);

% Find Power Spectral Density

Psd=Y.*conj(Y)/lfft;
A=ceil(max(Psd));
l=length(Y);
f=(1/T)*(-(l-1)/2:(l-1)/2);

% Plot Power Spectral Density

subplot(222);
plot(f,Psd/A);grid
axis([-bw bw 0 1]);
title(['Pulse Compression Spectrum']);
xlabel('Frequency - Hz');
ylabel('Amplitude');

% Find Matched Filter Time Waveform

zh=ifft(Psd,lfft);
h=fftshift(zh);

%Plot Time Waveform

subplot(223);
plot(t,abs(h));grid;
```

```
axis([-T/16 T/16 0 1 ]);
title(['PC Matched Filter Output']);
xlabel('Time - Sec');ylabel('Amplitude');

%Plot Log of Time Waveform

zz=20*log10(abs(h));
subplot(224);
plot(t,zz);grid;
axis([ -T/16 T/16 -60 0]);
title(['PC Matched Filter Output']);
xlabel('Time - Sec');ylabel('Amplitude - DB');
```

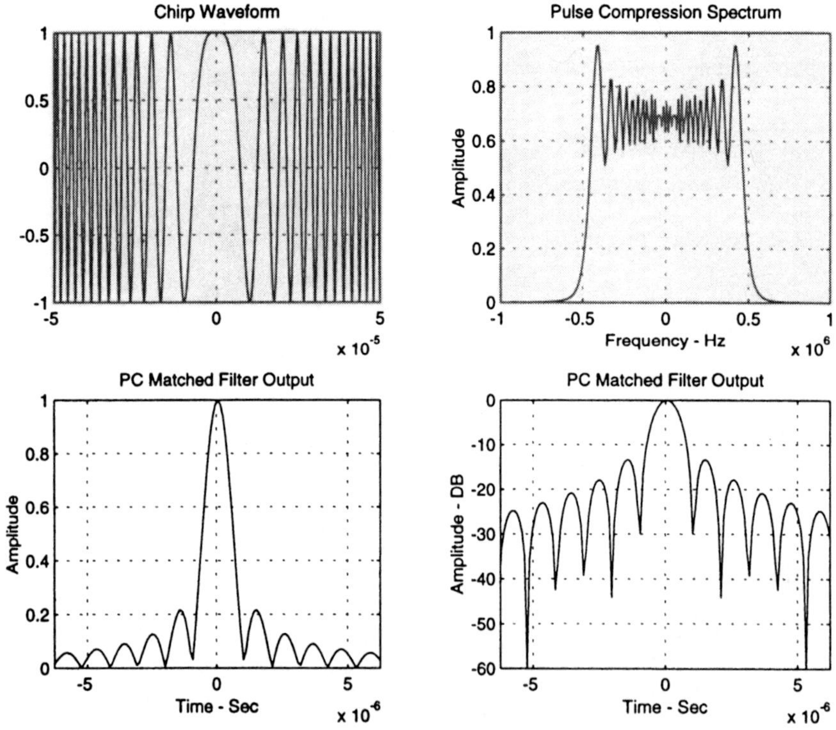

Example 4.1 shows that the output of the matched filter for a linear FM signal has relatively high time sidelobes. These time sidelobes are undesirable since small targets in the vicinity of large targets may be shadowed and go undetected. The solution to this problem is to weight the amplitude of the linear-FM signal with a window function [4]. However, as the sidelobes are reduced by windowing, the width of the main lobe of the PC signal is increased. Also, it is practically difficult to apply the window functions in the time

domain, so they are generally applied in the frequency domain by modifying the frequency response of the matched filter.

▶ *Example 4.2*

Extend Example 4.1 to apply Hamming, Hanning and Blackman window functions to reduce the time sidelobes at the output of a linear-FM matched filter. The weightings are given by

$$\begin{aligned}
&\text{Hamming:} && w(k) = 0.54 + 0.46 \cos\left(2\pi\frac{k}{N}\right)\\
&\text{Hanning:} && w(k) = 0.5\left(1 + \cos\left(2\pi\frac{k}{N}\right)\right)\\
&\text{Blackman:} && w(k) = 0.42 + 0.5 \cos\left(2\pi\frac{k}{N}\right) + 0.08 \cos\left(2\pi\frac{2k}{N}\right)\\
&\text{where} && k = -\frac{N}{2}, \ldots, +\frac{N}{2}
\end{aligned} \tag{4.10}$$

1. Plot the linear-FM matched filter response for Hamming, Hanning, and Blackman windows on a logarithmic scale. What are the relative time sidelobes and main-lobe widths for the various weighting?
2. Apply a Hamming window to the linear-FM chirp filter in the frequency domain. Plot the output power spectral density and matched filter time response. What time sidelobes and lobe spreading result from frequency windowing? Why are they different than time windowing?

```
% Pulse Compression Spectrum with Windows
% ---------------------------------------
% pcwina.m

clear;clf;clc;

% Input Chirp Bandwidth and Pulsewidth

bw=1e6;     % Bandwidth - Hz
T=1e-4;     % Pulse Width - sec

%Input FFT Points
```

```
lfft=1024;

% Set Time Base

t=-T/2:T/(lfft-1):T/2;

% Sample Complex Chirp

s= exp(-j*pi*bw*t.*t/T);

% Window Sample Chirp

w1=1;                                             % Uniform Weighting
w2=.54+.46*cos(2*pi*t/T);                         % Hamming Weighting
w3=.5*(1+cos(2*pi*t/T));                          % Hanning weighting
w4=.42+.5*cos(2*pi*t/T)+.08*cos(4*pi*t/T);       % Blackman Weighting

sw1=s.*w1;sw2=s.*w2;sw3=s.*w3;sw4=s.*w4;

% Find Fourier Spectrum of Chirp

ZY1=fft(sw1,lfft);ZY2=fft(sw2,lfft);
ZY3=fft(sw3,lfft);ZY4=fft(sw4,lfft);
Y1=fftshift(ZY1);Y2=fftshift(ZY2);
Y3=fftshift(ZY3);Y4=fftshift(ZY4);

% Find Power Spectral Density

Psd1=Y1.*conj(Y1)/lfft;
Psd2=Y2.*conj(Y2)/lfft;
Psd3=Y3.*conj(Y3)/lfft;
Psd4=Y4.*conj(Y4)/lfft;

% Find Matched Filter Time Waveform

zh1=ifft(Psd1,lfft);zh2=ifft(Psd2,lfft);
zh3=ifft(Psd3,lfft);zh4=ifft(Psd4,lfft);

h1=fftshift(zh1);h2=fftshift(zh2);
h3=fftshift(zh3);h4=fftshift(zh4);

% Plot Log of Time Waveform

zz1=20*log10(abs(h1));
zz2=20*log10(abs(h2));
zz3=20*log10(abs(h3));
zz4=20*log10(abs(h4));

C1=max(zz1);C2=max(zz2);
C3=max(zz3);C4=max(zz4);

subplot(221);
plot(t,zz1-C1);grid;
axis([ -T/16 T/16 -40 0 ]);
```

```
title(['PC Matched Filter - Uniform']);
xlabel('Time - Sec');ylabel('Amplitude - DB');

subplot(222);
plot(t,zz2-C2);grid;
axis([ -T/16 T/16 -80 0 ]);
title(['PC Matched Filter - Hamming']);
xlabel('Time - sec');ylabel('Amplitude - dB');

subplot(223);
plot(t,zz3-C3);grid;
axis([ -T/16 T/16 -80 0 ]);
title(['PC Matched Filter - Hanning']);
xlabel('Time - sec');ylabel('Amplitude - dB');

subplot(224);
plot(t,zz4-C4);grid;
axis([ -T/16 T/16 -120 0 ]);
title(['PC Matched Filter - Blackman']);
xlabel('Time - sec');ylabel('Amplitude - dB');
```


From Example 4.2, it is possible to deduce the resolution properties of the linear-FM waveform with different window functions at the output of the matched filter in the time domain. Applying the window function reduces the time sidelobes but increases the main lobe response width, thus making it more difficult to resolve two closely spaced targets. When a Doppler shift is present on the received signal, the filter is no longer matched and the output signal is distorted from its original form. The response of the matched filter to a range of frequencies can be determined by plotting the ambiguity function diagram for a particular waveform. In a three-dimensional plane, this diagram displays the response to time delay (range) in one dimension, Doppler shift (frequency shift) in an orthogonal direction, and the magnitude of the receiver response in the third orthogonal direction. The output of the matched filter can be expressed as

$$X(t_d, f_d) = \int_{-\infty}^{\infty} v^*(t) \cdot v(t - t_d) \cdot e^{j2\pi f_d t}\, dt \tag{4.11}$$

where t_d is the range delay and f_d is the Doppler frequency. Alternately, by applying Parseval's theorem, it can be expressed as

$$X(t_d, f_d) = \int_{-\infty}^{\infty} V^*(f) \cdot V(f - f_d) \cdot e^{j2\pi f t_d}\, df \tag{4.12}$$

The ambiguity function is then the squared magnitude $|X(t_d, f_d)|^2$.

▶ *Example 4.3*

Using MatLab, plot the ambiguity function for a linear-FM PC signal. Note that the matched filter response can be obtained using the cross-correlation function xcorr(x, y), where x is the chirp signal and $y = x \cdot \exp(-j2\pi \cdot f_d \cdot t)$. Also, use the Mesh command to generate a three-dimensional plot and the Contour command to observe the variation of time delay with frequency. Use a PC pulse width of 6 μs and a chirp frequency band of 0.3 MHz.

```
% Ambiguity Function for Linear FM Pulse
% -------------------------------------
% ambfuna.m

clc;clg;clear;

N=64;

% Pulse length and time vector

Tp=6;
t=-Tp/2:6/(N-1):Tp/2;

% Instanteous Frequency and deviation

df=.3;
f=t*(df/2/3); fs=N/6;  % Mhz

% Linear FM signal

j=sqrt(-1);
x=exp(j*2*pi*f.*t);

% Form Ambiguity Function

Bx=[];

for fd=-4*df:df/11:4*df;
   y=x.*exp(-j*2*pi*fd*t);
   chi=abs(xcorr(x,y));
   Bx=[chi;Bx];
end;

clg;mesh(Bx,[135 60]);
title(['Ambiguity Function for Linear FM']);
xlabel('time');ylabel('frequency');
zlabel('amplitude');

%subplot(212);contour(Bx);
%title(['Contour Plot']);
%xlabel('time');ylabel('frequency');
```

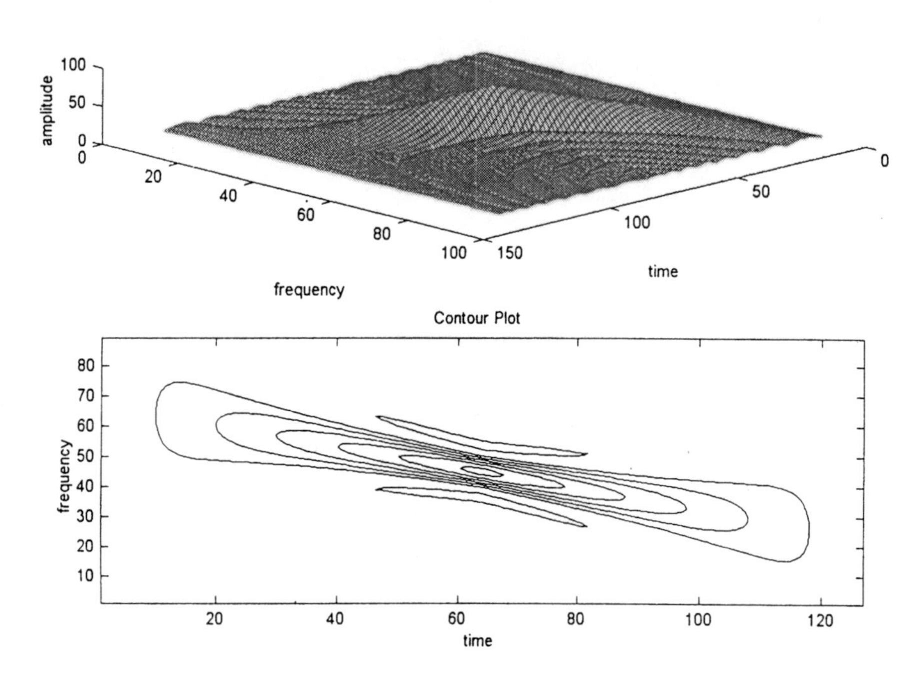

The ambiguity function for the linear-FM chirp shows that as the central frequency of the input signal is varied, the peak of the matched filter output varies in delay. Actually, the delay is a linear function of frequency. In the radar case, this range-Doppler coupling results in an apparent range shift with Doppler frequency that causes an erroneous range measurement. The actual target range may either lead or follow the apparent range depending upon whether the target Doppler is positive or negative.

The linear FM PC radar is more susceptible to repeater jamming than is a conventional radar. To see this, consider the ambiguity diagram depicted in Example 4.3. This diagram shows the output waveform of a PC radar for various mismatches in frequency. The output for the PC radar matched in frequency is shown along the central time axis of the diagram. By appropriately mismatching in frequency, the repeater jammer can generate a false target, which either leads or lags the occurrence of the matched target. Thus, this type of PC radar does not have the capability of a conventional radar to discriminate against a repeater jammer using leading-edge tracking. This makes it easier for a DECM to capture a PC radar than if a conventional waveform were employed.

Compressive ESM receivers also use linear FM PC techniques. The linear coupling of time delay with frequency allows the compressive receiver to separate multiple simultaneously occurring signals. This is equivalent to an instantaneous spectrum analysis of the signal set within a time window that corresponds to the shortest pulse width signal.

A block diagram of a digital system that accomplishes PC is depicted in Figure 4.1. Note that this system implements (4.4) using the FFT to find the Fourier transform of the signal and matched filter impulse response and the inverse FFT to convert the output signal back into the time domain.

In this method, a digitized version of the input signal at baseband (I and Q channels) is convolved with a digitized reference function that corresponds to the impulse response of the matched filter to the PC signal. This is accomplished in the frequency domain by first performing an FFT on the input signal, resulting in $V(\omega)$, which is then multiplied by the matched filter transfer function $V^*(\omega)$, thereby providing the PC output frequency spectrum $G(\omega) = V(\omega) \cdot V^*(\omega)$. This is then converted into the time domain by performing an inverse FFT on the output spectrum, resulting in $g(t) \leftrightarrow G(\omega)$. The advantage of this fully digital version is that large time-bandwidth product PC systems can be accomplished since the component limitations of analog implementations are bypassed.

4.1.2 Phase-Coded Pulse Compression

The most common form of this PC method uses binary phase coding. In this form of PC, a long pulse of duration T is divided into n segments each of width τ. The phase of each segment is chosen to be 0 or π in a pseudorandom manner. The time-bandwidth product of this PC signal is equal to the number of segments $n = T/\tau$. An advantage of this method is that a matched filter can be implemented using a simple tapped delay line (digital shift register) whose taps are spaced by the subsegment width τ and weighted by the phase (0 or π) corresponding to the appropriate sequence of random phases.

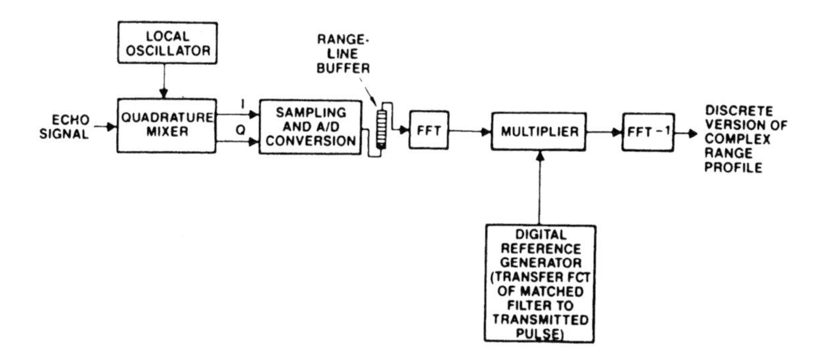

Figure 4.1 Digital PC. (*Source:* [15].)

Barker code binary sequences are often used in phase-coded PC systems. These sequences have the property that the voltage time sidelobes are equal and have magnitude $1/n$. Unfortunately, the longest known Barker code has length 13 (1, 1, 1, 1, 1, −1, −1, 1, 1, −1, 1, −1, 1), where a 1 corresponds to 0 radians phase shift and −1 to π radians phase shift.

When larger PC ratios are desired, linear shift register sequences are often used. These sequences are generated using shift registers with feedback. An N-stage register can generate a maximal length sequence of $n = 2^N - 1$. For large sequences, the peak voltage sidelobes are approximately $1/\sqrt{n}$ while the time-bandwidth factor is n. The sequences computed in Example 4.4 are not unique. For example, for $N = 5$, there are a total of six maximal length sequences [5].

▶ *Example 4.4*

Using MATLAB, construct a general program that generates maximum length sequence codes using feedback shift registers. Plot the auto correlation function for a five-stage 31-bit sequence. What is the maximum time sidelobe?

Number of Stages	*Feedback Stage*
2	1, 2
3	2, 3
4	3, 4
5	3, 5
6	5, 6
7	6, 7
9	5, 9
10	7, 10
11	9, 11

```
% Binary Maximal Length Sequence Code
% Program Generates a Random Sequence
% -----------------------------------
% maxlen.m

clear;clc;

% Code Length and Feedback (m=2-7;9-11)

m=input('length=');
a=m-1;
if m==5;a=3;end;
```

```
if m==9;a=5;end;
if m==10;a=7;end;
if m==11;a=9;end;
l=2^m-1;

% Start sequence

x= eye(1,m);
z=x;

% Generate sequence of zero-ones

for j=1:l;
    for i=2:m;x(i)=z(i-1);x(1)=z(a)+z(m);end;
    if x(1)==2;x(1)=0;end;
y(j)=z(m);z=x;end;

% Convert to sequence of 1,-1

for i=1:l;
    if y(i)==0;y(i)=-1;end;
end;

% Plot correlation

t=1:2*l-1;zz=xcorr(y);
plot(t,abs(zz));grid;
xlabel('Time');
ylabel('Voltage Correlation');
title(['Binary Code Autocorrelation']);

% Display Sequence

b=input('Display Sequence Yes=1;No=0;  ');
if b==1;disp(y);end;

end
```

```
» maxlen
length=5
Display Sequence Yes=1; No=0; 1
 Columns 1 through 12

-1 -1 -1 -1 1 -1 -1 1 -1 1 1 -1

Columns 13 through 24

-1 1 1 1 1 1 -1 - 1 -1 1 1 -1
```

```
Columns 25 through 31

1 1 1 -1 1 -1 1
```

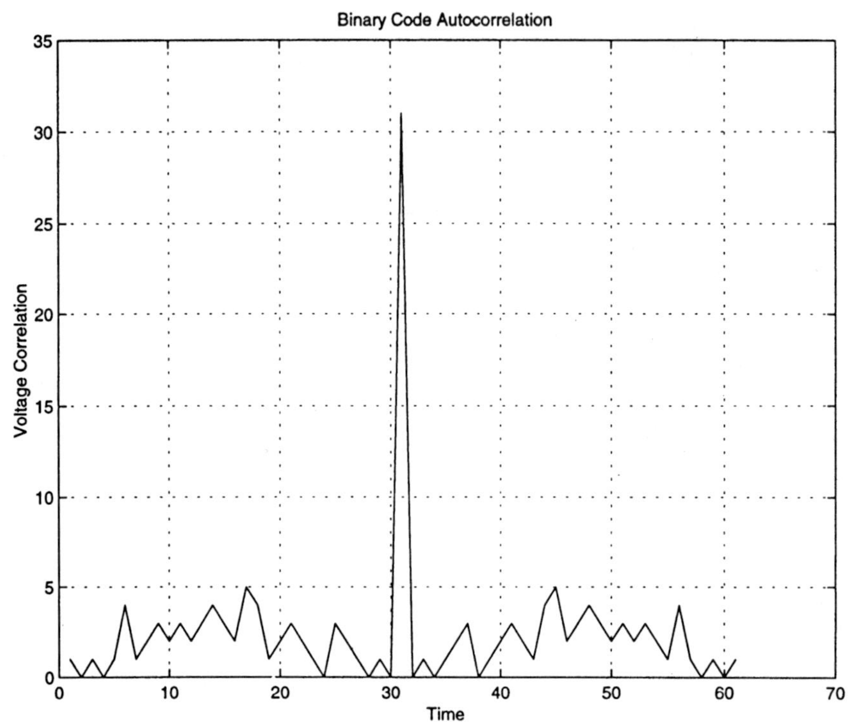

If Doppler processing is not required, the availability of different codes for phase-coded PC systems allows the coding to be changed on a pulse-to-pulse basis. This provides a counter to jammers that synthesize the PC waveform to generate false targets. However, DECM jammers that coherently repeat the intercepted waveform are still effective against this form of countermeasure.

Phase-coded PC waveforms are more sensitive than linear-FM waveforms with respect to Doppler frequency shift. Example 4.5 shows that for ratios of Doppler frequency to PC bandwidth as small as 0.05, significant distortion occurs in the compressed Barker-coded waveform and time sidelobes are much greater. Thus, phase-coded waveforms are not appropriate for radar that are designed to detect high-velocity targets. Also, EA jammers that synthesize phase-coded PC waveforms need accurate frequency set-on receivers to prevent dilution of the jamming signals and reduced effectiveness.

▶ *Example 4.5*

Using MATLAB, plot the matched filter output for a 13-bit Barker-coded PC waveform. Then replot the waveform for various ratios of Doppler frequency (f_d) shift to waveform bandwidth (B). Determine the f_d/B ratio where significant distortion occurs.

```
% Pulse Compression Mismatch
% --------------------------
% pcmissa.m

clear;clf;clc;

% Enter Barker code

xx=[1 1 1 1 1 -1 -1 1 1 -1 1 -1 1];

% Enter freqency offset fd/bw

r=[0 .02 .05 .07];
zz=[];

% Form input signal

for i=1:4;
rr=r(i);
dp=2*pi*rr;
  for n=1:13;
  y(n)=xx(n)*exp(j*n*dp);end;

% Form cross correlation

yy=xcorr(y,xx);
z=abs(yy);

if i==1;zz=z;
  elseif i>1;zz=[zz z];end;
end;

% Plot cross correlation

t=1:25;
subplot(221);
zx=zz(1:25);
plot(t,zx/max(zx)),grid;
title(['Barker Code Fequency Mismatch']);
axis([0 25 0 1]);
text(15,.9,['fd/bw=',num2str(r(1))]);

subplot(222);
zy=zz(26:50);
plot(t,zy/max(zy)),grid;
```

```
title(['Barker Code Frequency Mismatch']);
axis([0 25 0 1]);
text(15,.9,['fd/bw=',num2str(r(2))]);

subplot(223);
zw=zz(51:75);
plot(t,zw/max(zw)),grid;
title(['Barker Code Frequency Mismatch']);
axis([0 25 0 1]);
text(15,.9,['fd/bw=',num2str(r(3))]);

subplot(224);
zv=zz(76:100);
plot(t,zv/max(zv));grid;
title(['Barker Code Frequency Mismatch']);
axis([0 25 0 1]);
text(16,.9,['fd/bw=',num2str(r(4))]);
```

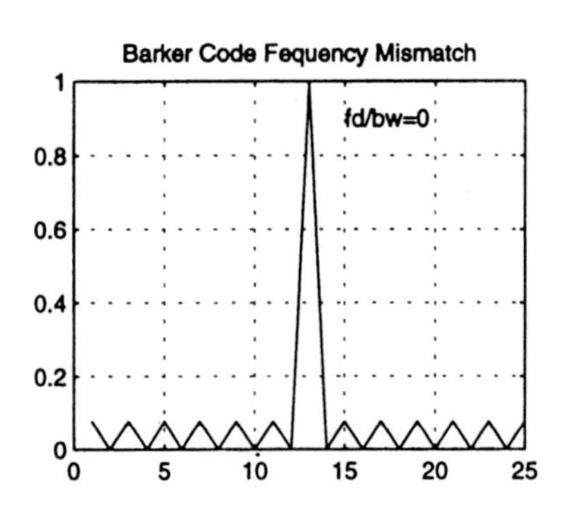

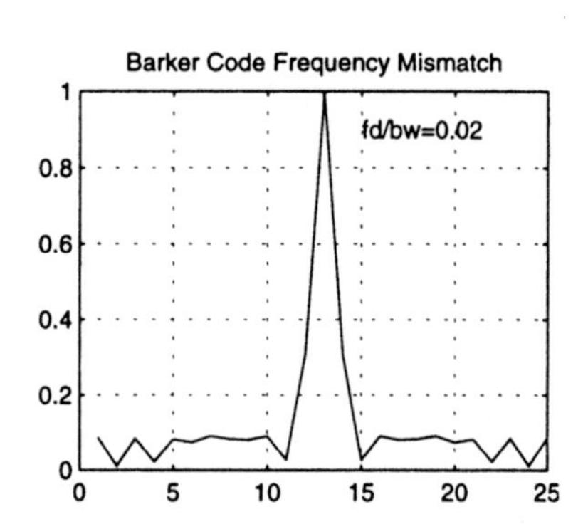

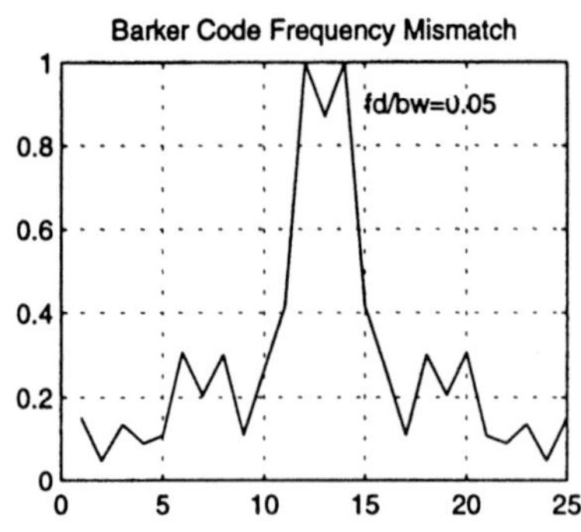

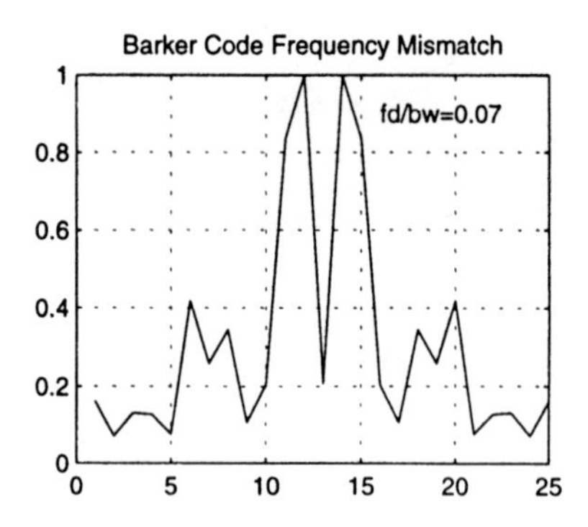

4.1.3 EA Against Pulse-Compression Radar

PC waveforms are commonly used in both surveillance and tracking radars. Jamming objectives depend on the type of radar to be jammed.

With surveillance radars, the object is generally to obscure the target by generating background interference or by synthesizing a false target that suppresses the detection of the true target. The latter procedure is sometimes called camouflage jamming.

For tracking radars, the objective is to generate a false target that captures the radar's tracking gates. The range-Doppler coupling properties of linear-FM PC radar makes these radars vulnerable to repeater jammers. In this mode of operation, the radar's signal is coherently shifted in frequency and repeated back to the radar. The frequency-shifted repeater signal then leads the true target return of the output of the matched filter, making leading-edge tracking EP techniques ineffective.

Noise jamming is generally the most inefficient jamming waveform against PC radar. This can be examined by considering the PC system's matched filter as a correlator. Noise that is completely decorrelated with respect to the signal does not achieve any correlation gain while a target that is fully correlated achieves full processing gain. Jamming waveforms that achieve some degree of correlation are more effective than random noise.

CW signals generally produce some degree of correlation in the PC matched filter. For example, a CW waveform at the carrier frequency is correlated with a phase-coded waveform approximately half the time, achieving a 6-dB power gain with respect to a noise waveform. Also effective is coherently repeating half the PC waveform, thereby generating two false targets out of the matched filter that straddle the true target return.

► *Example 4.6*

Determine the effect of CW and noise jamming on a 31-bit maximal length sequence phase-coded PC signal. Assume the CW signal is at the carrier frequency so that it produces components in phase with the central frequency of the phase-coded waveform. Assume the noise is also at the central frequency but that it produces uniform phase modulation $(0 - \pi)$ with respect to the phase-coded signal.

```
% Jamming of 31 Bit Shift Register Sequence
% With CW Noise
% ------------------------------------------
% cwjam.m

clear;clc;clg;

% Generate time vector

t=13/31*[1:1:61];

% Code order is 31-bit Maximal Length Sequence

x=[1 -1 1 1 -1 1 -1 1 -1 -1 1 1 1 1 -1 -1 -1 1 1 -1 ...
-1 1 -1 -1 -1 -1 -1 1 -1 1 1];

% CW-jamming in baseband with correct carrier
y=ones(1,31);

% Noise jamming in baseband with correct carrier
z=rand(1,31);

% Correlation of signal and jamming

acf1=xcorr(x);
acf2=xcorr(x,y);
acf3=xcorr(x,z);
effmv2=sum(acf2.*conj(acf2))/61;
effmv3=sum(acf3.*conj(acf3))/61;

% Plot jamming effects

subplot(211),plot(t,abs(acf1),':',t,abs(acf2));grid;
xlabel('Time in microsecs');
ylabel('Correlator Voltage Output');
title(['Pulse Compression Jamming With CW Jam Effect=', ...
num2str(effmv2)]);

subplot(212),plot(t,abs(acf1),':',t,abs(acf3));grid;
xlabel('Time in microsecs');
ylabel('Correlator Voltage Output');
title(['Pulse Compression Jamming with Noise Jam Effect=', ...
num2str(effmv3)]);
end
```

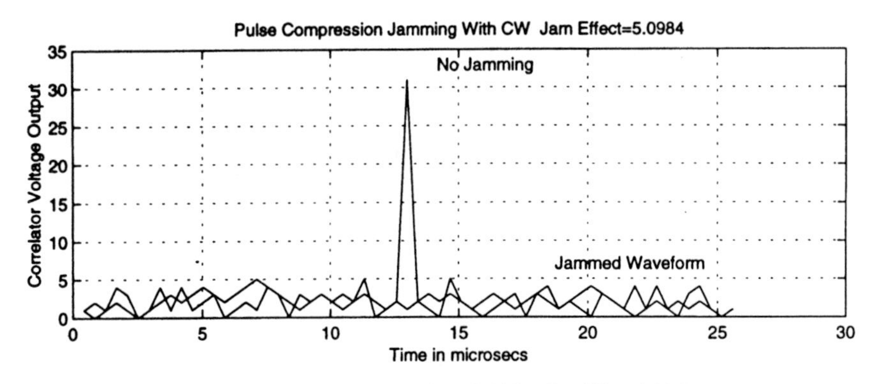

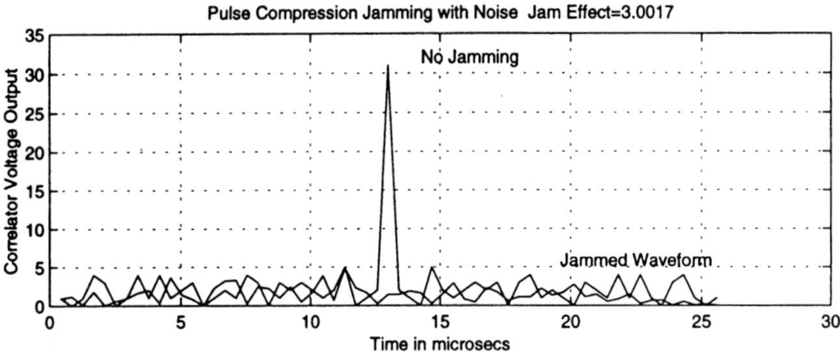

► *Example 4.7*

Determine the effect of half-code repeater jamming on a 31-bit maximal length sequence phase-coded PC signal. Use a combination of two 15-bit half codes plus one to fill the PC network.

```
% Pulse Compression Half Code Repeater Jamming
% -------------------------------------------
% pchalf.m

clear;clf;clc;

% Time Vector

t=13/31*[1:1:61];

% Maximal Length 31 Bit Sequential Code

x=[1 -1 1 1 -1 1 -1 1 -1 -1 1 1 1 1 -1 -1 -1 1 1 -1 ...
-1 1 -1 -1 -1 -1 -1 1 -1 1 1];

% Half Code Length

y=[1 -1 -1 1 1 1 1 -1 -1 -1 1 1 -1 -1 1];
```

```
% Jamming uses two halfs plus 1 to fill

z=[y y 1];

akf1=xcorr(x);
akf2=xcorr(x,z);

% Plot Jamming Effects

plot(t,abs(akf1),':',t,abs(akf2));grid;
xlabel('Time');ylabel('Correlator Voltage Output');
title(['Repeater Jamming with Half Pulses']);
text(14,30,'No Jamming');
text(18,14,'Repeater Jamming');
```

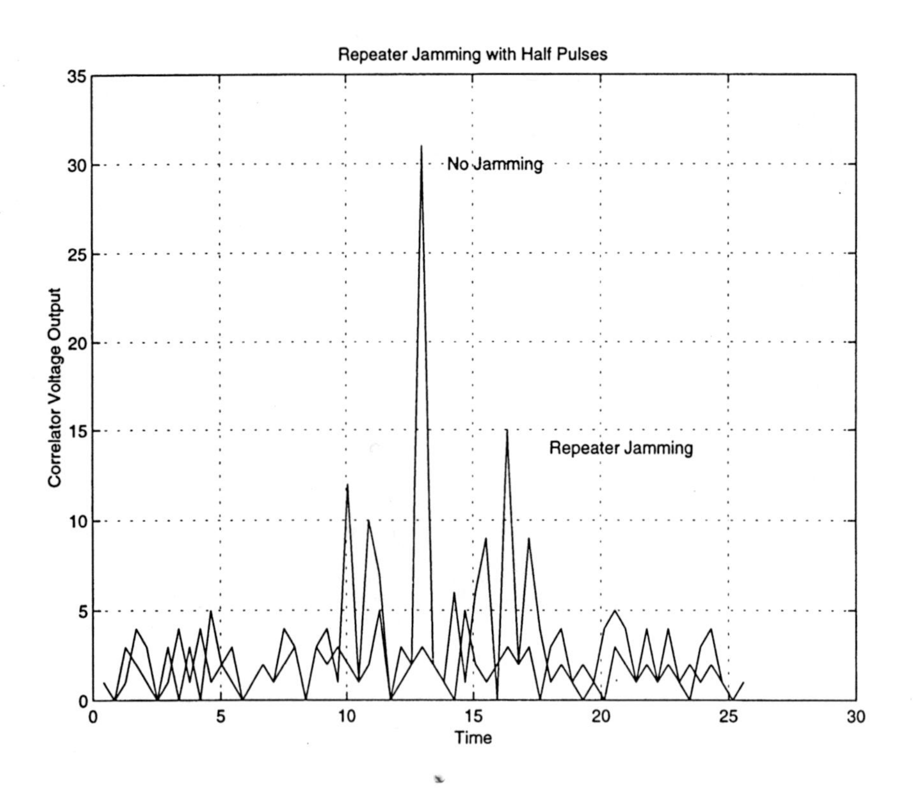

The first step in attacking a PC radar is to ascertain its parameters. There are two methods employed for this purpose. Older ES systems that employ envelope detectors destroy the phase (or frequency) information within the signal and hence rely on the fact that PC radars have long pulse widths for identification. Newer systems employ circuitry that measures the slope of the linear-FM chirp or the chip rate of phase-coded waveforms. These systems generally correlate the signal with a delayed replica in a phase detector to determine its phase properties similar to the *instantaneous frequency measurement* (IFM) receiver described in Section 6.3.

A block diagram of a general system to jam PC surveillance radars is depicted in Figure 4.2. This jamming system stores the input PC signal received from the radar in a *digital RF memory* (DRFM), where it is continually repeated and amplified throughout the radar's interpulse interval to create many false targets. Since the false jamming targets are coherent with the radar, they negate the processing gain of the PC network with respect to the radar target. Also, since they are synchronized with the radar's PRF, they receive the benefit of any post-detection integration performed by the radar.

In Figure 4.2, the DRFM can be replaced with a DDS to produce a similar effect. This is described in Section 5.4 and involves storage of the

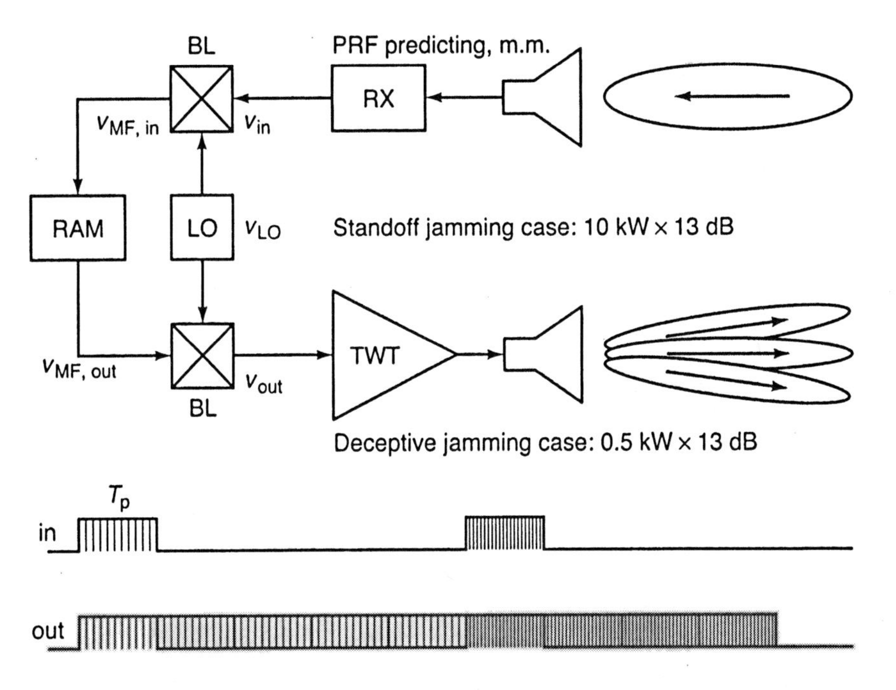

Figure 4.2 General PC jammer. (Skolnik, M., *Radar Handbook,* 2nd Ed., © 1990 McGraw Hill. Reprinted with permission.)

parameters associated with each threat radar in the jammer's library. The jammer receiver then identifies the particular PC radar to be countered, measures its critical parameters (i.e., such as the radar's frequency), draws its remaining parameters from the threat library, synthesizes a pseudocoherent replica of the threat radar's waveform in the DDS, and transmits this replica throughout the radar's interpulse interval. The effectiveness of the DDS approach depends upon how accurately the radar parameters are known or measured. As described in Section 3.5, frequency set-on is important for DDS-based jammers that counter phase-coded PC radar. This is a consequence of the rapid deterioration of the matched filter output of a phase-coded PC waveform with frequency off-set.

Repeater jamming against tracking radars using PC waveforms requires capture of the radar's tracking gates. To accomplish this, the DECM must repeat coherent replicas of the radar's PC waveform, possibly shifted in frequency, which can be varied with time. The procedure is to initially repeat the received radar pulse with minimum time delay. This ensures that the jammer pulse enters the radar's tracking gates along with the true signal and allows capture of the radar's *automatic gain control* (AGC) circuits. Then the DECM begins to introduce increasing amounts of time delay in the repeated signal. The range-gate circuitry in the radar begins to track the stronger jammer signal and gradually "walks-off" from the true target range.

Most tracking radars include two *electronic protection* (EP) devices against repeater jammers. They generally employ leading-edge range tracking that functions against just the leading edge of the received signal. Hence, DECM signals that initially include significant delay (i.e., greater than 20 ns to 50 ns) lag with respect to the true signal are rejected in favor of the true signal. In addition, tracking radars perform radial velocity checks by comparing the radial velocity derived from the received Doppler with that obtained by differentiating tracking range data. Hence, repeaters must induce realistic frequency shifts onto the repeated signal that are consistent with the range "walk-off" program. Also, some tracking radars differentiate Doppler to obtain a measure of the acceleration characteristics of the target. If this does not check with that of realistic targets, then the signal is rejected.

A block diagram of a repeater jammer using fiber optic delay lines is depicted in Figure 4.3. In this system, the RF input signal modulates a laser. The laser signals excite various lengths of monomode fiber optic cable that are interconnected using optical switches. The laser signal is demodulated using a photo optical diode to recover the RF signal. The delays available from the fiber optic coils are generally related in a binary manner. For example, if the shortest coil is 0.1 μs. in length, and eight coils are used, the delay available from the longest coil is 12.8 μs. Using this technique, the delay can be switched

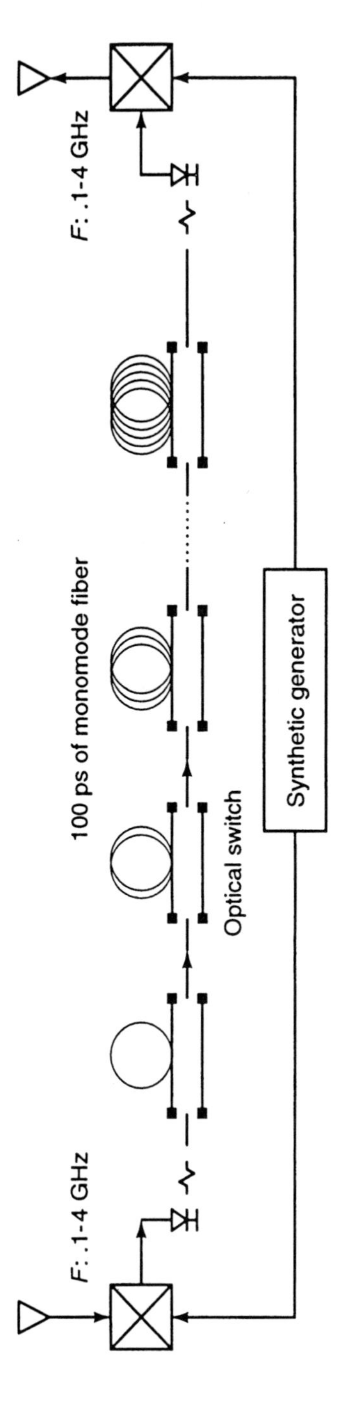

Figure 4.3 Repeater jammer using fiber optic delay lines.

in 0.1-μs steps to a maximum delay of 25.5 μs. With this method, a dynamic range of approximately 35 dB with 2-GHz instantaneous bandwidth can be achieved.

The simulation of Doppler shift in the previously described repeater jammer can be accomplished using mixers to down-convert and up-convert signals before and after propagation through the optical delay lines. The frequency offset between the down- and up-converted signals then represents the differential Doppler shift imparted onto the signal. A Doppler shift can also be imparted onto the jamming signal using a digital phase shifter as described in Section 3.3.2.

Alternately, a DRFM can be used to store PC signals in a DECM repeater jammer. With this method, input signals are converted to digital form by sampling at or above the Nyquist rate and stored in a digital memory. They are then clocked out when needed (minimum delay is the order of 10 ns to 20 ns) and converted to analog form for transmission. Practical implementation of DRFMs involves down-converting signals to baseband where they can be efficiently stored digitally. Doppler shifts can be imparted onto the signal by an appropriate frequency difference between down-conversion and up-conversion oscillator frequencies.

Storage of PC waveforms imposes some restrictions on DRFMs. This occurs because digital memory capacity for high clock speeds is limited. To accommodate long PC signals, the waveform is strobed and short durations of the complete signal are stored. When strobed, the short durations of stored signal are read out on a head-to-tails basis, thus creating a step-approximation to the whole waveform. For a linear-FM PC signal, the step approximations are smoothed in the radar's PC network. With phase-coded waveforms, the approximation is performed at the chip rate providing good resemblance to the real waveform. However, in this case, some phase discontinuities may exist at code transitions.

▶ *Example 4.8*

Using MatLab, plot the output of a linear-FM PC matched filter to signals with frequency offset. What offset frequencies produce output signals that lead or lag the signal with no frequency offset? Why is this effect useful in negating leading edge range trackers?

```
% PULSE COMPRESSION WAVEFORM WITH FREQUENCY OFFSET
% -----------------------------------------------
% pcoff.m

clear;clf;clc;
```

```
% Input Chirp Bandwidth,Pulsewidth and Offset Frequency

bw=1e6;     % Bandwidth - Hz
T= 1e-4;    % Pulsewidth - sec
fo=2e5;     % Frequency Offset - Hz

%Input FFT Points

lfft=1024;

% Set Time Base

t=-T/2:T/(lfft-1):T/2;

% Sample Complex Chirp

s= exp(-j*pi*bw*t.*t/T);

% Find Fourier Spectrum of Chirp

ZY= fft(s,lfft);
Y=fftshift(ZY);

% Find Chirp Power Spectral Density

Psd=Y.*conj(Y)/lfft;
A=ceil(max(Psd));
l=length(Y);
f=(1/T)*(-(l-1)/2:(l-1)/2);

% Find Matched Filter Chirp Time Waveform

zh=ifft(Psd,lfft);
h=fftshift(zh);

% Input Frequency Offset Signals

s1=exp(-j*2*pi*fo*t)+exp(j*2*pi*fo*t);
so=s.*s1;

% Find Fourier Spectrum of Offset

ZZY=fft(so,lfft);
YY=fftshift(ZZY);

% Find Power Spectral Density of Offset

Psd1=YY.*conj(Y)/lfft;

% Find Offset Matched Filter Time Waveform

zzh=ifft(Psd1,lfft);
hh=fftshift(zzh);

% Plot Time Waveform
```

```
plot(t,abs(h),t,abs(hh));grid;
axis([-T/4 T/4 0 1 ]);
title(['PC Matched Filter Output']);
xlabel('Time - Sec');ylabel('Amplitude');
text(.2e-5,.95,'No Offset');
text(-2.4e-5,.85,'Repeater With Offset Frequency');
text(.4e-5,.85,'Repeater With Offset Frequency');
```

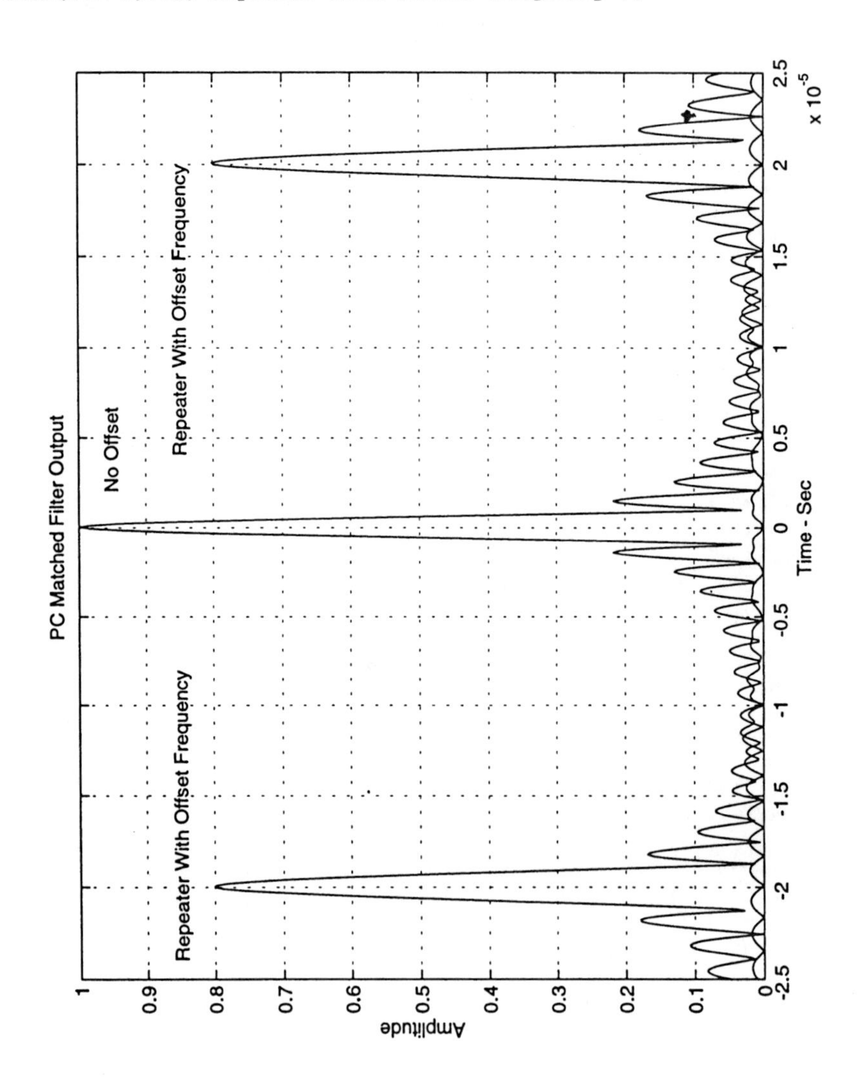

4.2 Pulsed Doppler Radar

With the rapid advance of digital signal processing and stable microwave power sources, the use of PD radars has dominated the modern radar field, particularly in airborne radar applications. These radars use the Doppler effect to extract targets from background interference on the basis of their velocity signature, providing modern airborne radars with "look-down–shoot-down" capability.

The magnitude of the Doppler shift frequency can be computed by noting that the phase of the return from a scatterer, when received at the radar, is given by $\phi = 4\pi r/\lambda$ radians, where r is the range to the scatterer and λ is the radar's apparent wavelength. The Doppler shift can then be found as

$$f_d = \frac{1}{2\pi}\frac{d\phi}{dt} = \frac{2v_r}{\lambda} \tag{4.13}$$

where v_r is the target's radial velocity. This result is accurate except for high-velocity targets where a correction factor $2 \cdot (v_r/c)^2 \cdot f_t$ applies.

▶ *Example 4.9*

For an airborne radar, plot the constant Doppler contours on the ground, called "isodops," as a function of the ratio v_g/v_{ac}, which varies from 0 to 0.9 in 0.1 increments, where v_g is the Doppler velocity on the ground and v_{ac} is the aircraft's velocity. How would you use this plot to determine the frequency spectrum of ground clutter?

```
% PLOT OF CONSTANT DOPPLER CONTOURS - ISODOPS
% ------------------------------------------
% isodop.m

clear;clc;clf;

% Set Aircraft Parameters

ro=.1;                      % vg/vac ratio
h=10000;                    % feet

% Calculate Doppler Contour

for i=1:9;
rr=ro+.1*(i-1);
phi=acos(rr);               % depression angle
a=tan(phi);
xmin=h/a+1e-3;
x=xmin:30000;
y1=sqrt(a^2*x.^2-h^2);
q=1:xmin;
Ay=zeros(size(q));
Ay1=[Ay y1];
```

```
% Plot Doppler Contour

Ax=xmin*ones(size(q));
x1=[Ax x];
plot(x1,Ay1,'k',x1,-Ay1,'k',-x1,Ay1,'k',-x1,-Ay1,'k');
grid;axis([-3e4 3e4 -5e4 5e4]);hold on;
end;hold off;
title('Constant Doppler Contours - Isodops');
xlabel('Ground Track Range - feet');
ylabel('Cross Track Range - feet');
text(2.5e4,.2e4,'Ground Track');
```

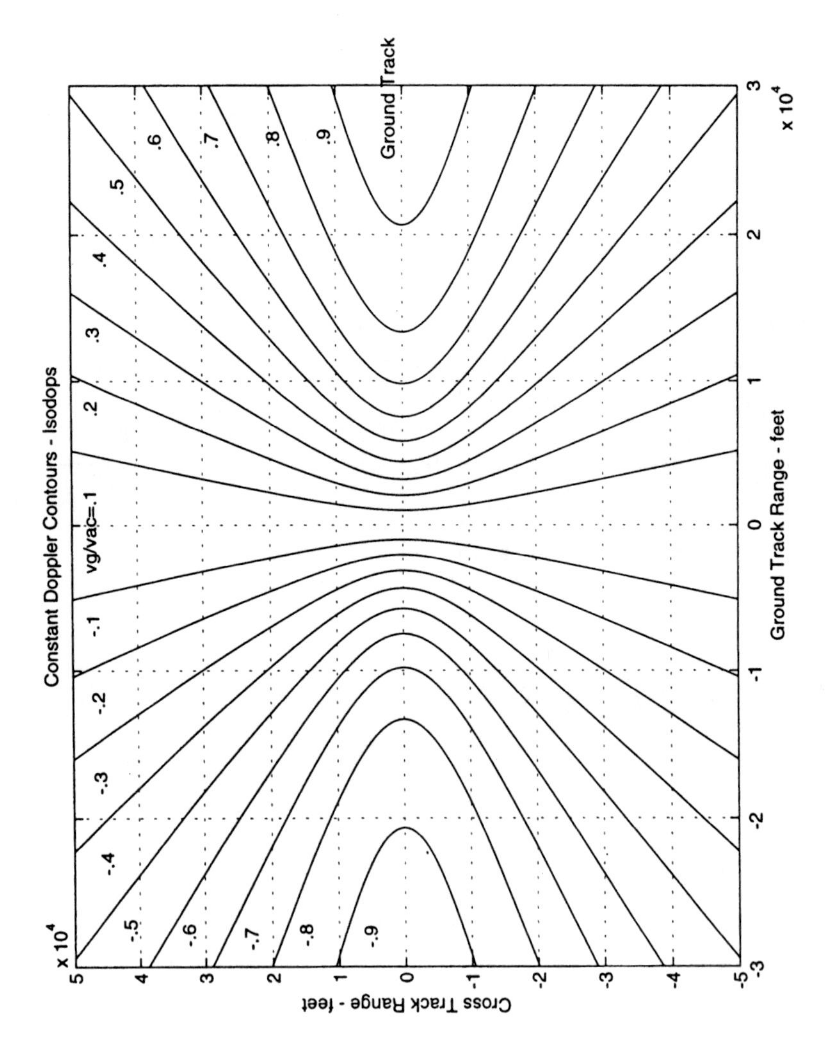

A block diagram of a simple pulse Doppler radar is depicted in Figure 4.4 [13]. The transmitter generates a train of phase stable pulses at a particular PRF. The signals returned from the target illuminated by a PD radar are generally processed in a set of narrowband filters that coherently integrate the signals to which they are tuned. The filters both resolve and enhance signals within a particular velocity band. In search radar applications, the bank of contiguous Doppler filters is used to cover the velocity band of interest to extract specific targets and reject clutter and targets outside that velocity band. Once the target is acquired, only one filter may be needed for tracking. In tracking radar applications, such as missile seekers, a speed gate may be used that basically consists of a narrowband Doppler filter that is caused to follow the targets Doppler using a servo technique.

The bandwidth (B_d) of a Doppler filter is inversely proportional to the time in which a signal is processed within the filter which is called the *coherent processing interval* (CPI). For search radar, the CPI must be shorter than the time (T_d) the antenna dwells on the target. The number of pulses coherently integrated is given by CPI · PRF, resulting in a processing gain with respect to noise of

$$PG = 10 \log(CPI \cdot PRF) \tag{4.14}$$

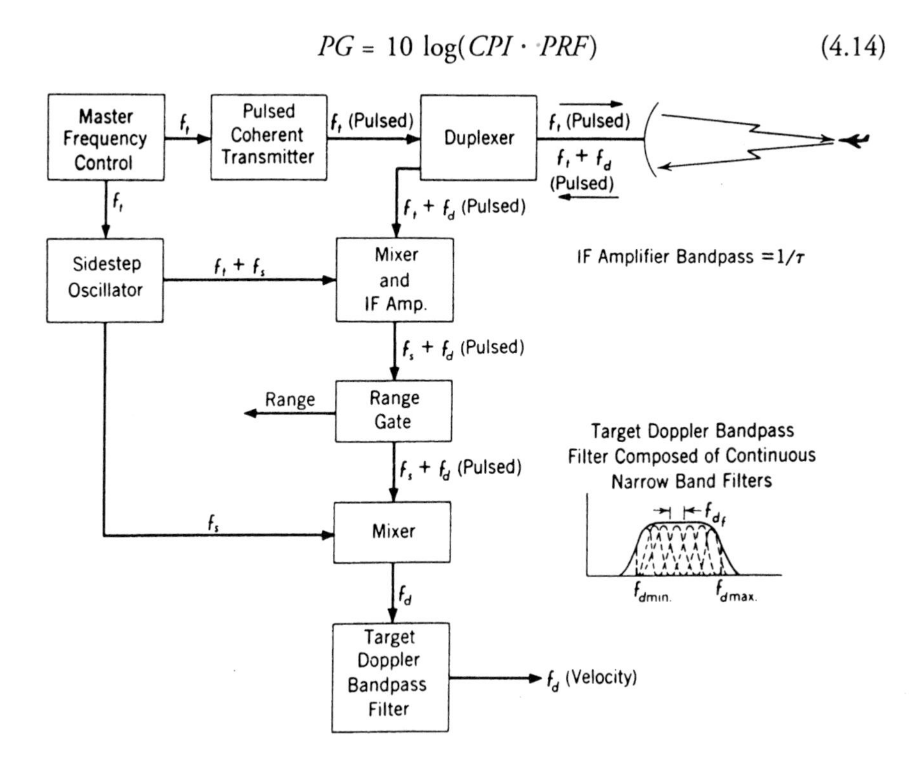

Figure 4.4 Block diagram of pulsed Doppler radar. (*Source:* [13].)

Using mixed predetection (coherent) and postdetection integration (matched to T_d), it is possible to provide good overall detection performance for the PD radar while changing the transmitted frequency or *pulse repetition frequency* (PRF) on a block-to-block (i.e., CPI) basis. The block-to-block frequency changes are useful in applying frequency agility to the PD radar and PRF changes are useful in resolving range ambiguities.

As an example, consider a ground-surveillance PD radar that rotates at 36 degrees/s, has an azimuth beam width of 3.6 degrees, and a PRF of 50 kHz. The dwell time of this radar is 3.6 degrees/(36 degrees/s) or 100 ms. If the CPI for this radar is set equal to the dwell time, then the Doppler filter bandwidth is 10 Hz while the processing gain is 5000 or 37 dB. From an EA standpoint, when noise jamming is used, this processing gain must be compensated by employing additional jamming ERP. When deceptive jamming is used, the deceptive jamming signal must be highly accurate in frequency so that it can compete with the true signal within the Doppler filter's 10-Hz bandwidth.

The preceding example illustrates how employing interpulse coherence makes the PD radar very difficult to jam unless the jammer system employs equivalent coherence. This has resulted in the emergence of the DRFM as a primary component in EA systems against PD radar.

PD radars can be categorized as low, medium, or high PRF, according to whether Doppler (low PRF), range (high PRF), or range and Doppler (medium PRF) are ambiguous. From an EA viewpoint, it is important to understand these PD modes since each presents different jamming opportunities.

Unambiguous range (nmi) in a radar is given by $R_u = 80.91/\text{PRF (kHz)}$, while unambiguous velocity (knots) is given by $v_u = 971 \cdot \lambda(\text{m}) \cdot \text{PRF (kHz)}$, resulting in

$$R_u v_u = \frac{2.3565 \cdot 10^4}{f_{\text{GHz}}} \tag{4.15}$$

Equation (4.15) illustrates that it becomes more difficult to achieve unambiguous range and velocity as the frequency of the radar is increased. For example, if the radar must detect targets with unambiguous velocities as high as 600 knots, then at a frequency of 1 GHz the unambiguous range is 40 nmi. If the frequency is raised to 10 GHz, then the unambiguous range is reduced to 4 nmi. Further, for an airborne radar, the appropriate velocity to consider is the closing velocity between the radar and the target (i.e., 1200 knots for a 600-knot platform velocity), which compounds the problem.

▶ *Example 4.10*

Using the expression for range-velocity ambiguities given in (4.15), plot a curve of unambiguous velocity versus unambiguous range using frequency as a parameter.

```
% Unambiguous Range-Velocity Chart
% --------------------------------
% unamb.m

clf;clc;clear;

% Define Range Vector

Rmin=3;     % Nmi
Rmax=1000; % Nmi
R=Rmin:Rmax;

% Define Frequency Values

f=[.1 .3 .6 1 3 6 10 16 35 60]; % Ghz

% Compute Unambiguous Velocity

for k=1:10;
v=2.3565e4./(R*f(k));
vr(:,k)=v';
end;

% Plot Unambiguous Range/Velocity

loglog(R,vr,'k');grid;
xlabel('Unambiguous Range - nmi');
ylabel('Unambiguous Velocity - knots');
title(['Unambiguous Range - Velocity']);
axis([ 5 1000 10 3000 ]);
```

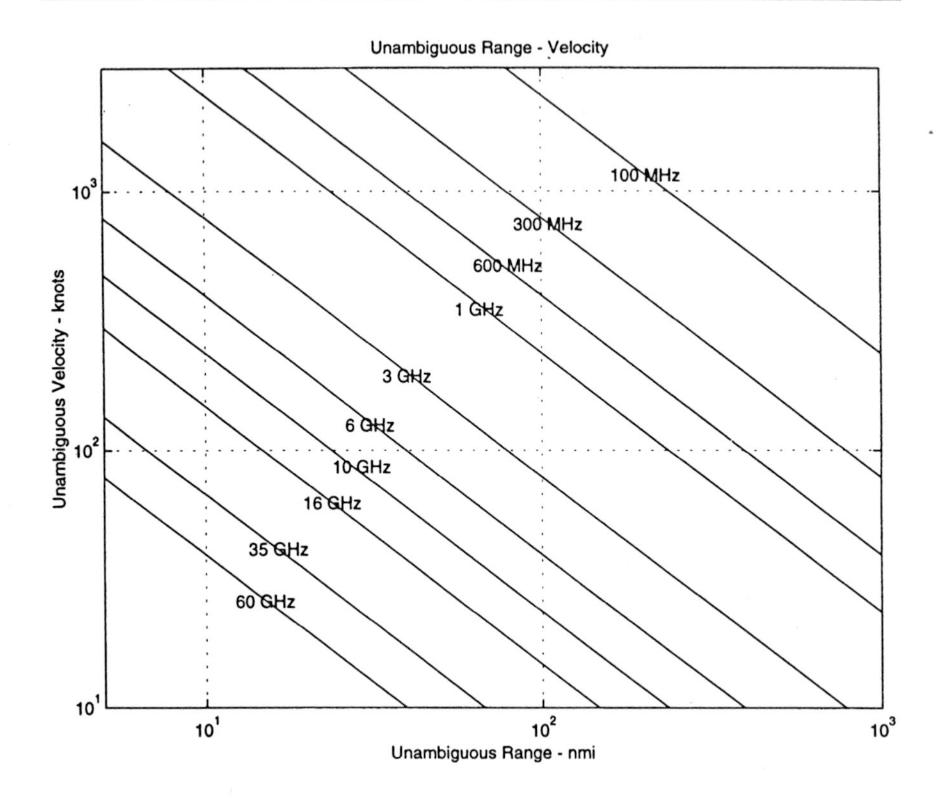

Figure 4.5 depicts the three PRF types in range-Doppler space for an *airborne intercept* (AI) application. In general, the interpulse period is divided into range cells (range gates) that correspond to the radar's transmitted bandwidth ($\tau_e = 1/BW$), and the Doppler space is divided into Doppler cells (Doppler filters), the bandwidth of which is inversely proportional to the dwell time of the radar antenna's mainbeam on the target. The number of range gates is greatest for a low-PRF radar, whereas only a small number is used for a high-PRF radar. Conversely, only a small number of Doppler filters is used in a low-PRF design, whereas a large number are used in a high-PRF design. Although the interpulse period is generally filled with range gates for all types, the low-PRF type is the only design that fills the frequency region between PRF lines with Doppler filters. In medium-PRF designs, those Doppler frequencies occupied by ground vehicular traffic are generally excluded from the Doppler frequency coverage region. Generally, the high-PRF design fills only the Doppler-clear region ($2v_r/\lambda < f_d < PRF - 2v_r/\lambda$) with Doppler filters, whereas the medium-PRF design excludes the region occupied by moving vehicular targets from the Doppler coverage.

The principal advantage of a low-PRF design is its ability to restrict clutter intake to just that which is in the same range cell as the target. For

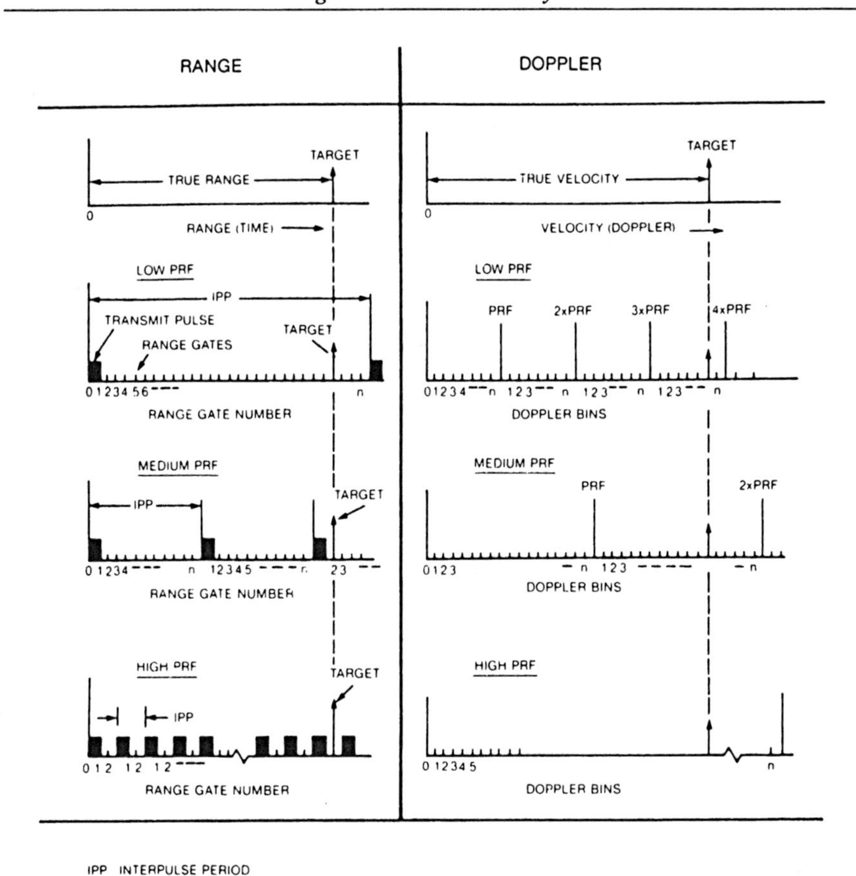

Figure 4.5 Pulsed Doppler radar in range-Doppler space. (*Source:* [16].)

example, a rain or chaff clutter cell several miles in extent would not interfere with targets not directly in the cell, whereas, in a medium- or high-PRF design, this same rain cell could clutter the entire unambiguous range interval. Also, sidelobe and main-lobe clutter can be restricted by increasing the range resolution of the radar and are not magnified by folding ambiguous clutter returns from short range into the same range cell as the target.

Other features of the low-PRF design include the ability to use a *sensitivity-time-control* (STC) function that allows normalization of the clutter return over the full range of the radar and the fact that range can be directly measured without generating any spurious targets. STC cannot be used in medium- or high-PRF designs because it would attenuate long-range targets of interest as well as near-range clutter. This results in a low-PRF system processor requiring less dynamic range, which allows better control of false alarms. In medium-

and high-PRF designs, a range correlation system is required to resolve the resulting ambiguous ranges. When multiple targets are present, they can confuse the range correlator and generate spurious targets called *ghosts.*

On the negative side, in a low-PRF design, unless main-lobe clutter is separated in range from the targets the radar encounters, it can be rejected only on the basis of differences in Doppler frequency. The highly ambiguous Doppler response causes a lack of Doppler visibility, requiring the use of staggered or randomly jittered PRF to eliminate blind velocity zones. In airborne radars, the frequency of the mainbeam ground clutter is generally translated to zero frequency. The spread of the main-lobe clutter spectrum is then determined primarily by the aircraft's velocity, azimuth antenna beamwidth, and azimuth look angle. At azimuth angles significantly off the aircraft's ground track (i.e., 30 degrees to 90 degrees), the main-lobe spectrum generally occupies a large fraction of the frequency region between PRF lines. The wide Doppler notch blinds the low-PRF airborne PD radar to many targets of interest.

The use of high-PRF radar is most prevalent in airborne applications where targets are closing on the radar. The high-PRF radar allows the average transmitter power to be maximized by increasing the PRF. At I-band, typical PRFs for airborne radar are the order of 100 kHz to 300 kHz with duty factors as high as 50%. To achieve the same amount of average power with a low-PRF radar, subject to a transmitter peak power limitation, requires increasing the radar's pulse width and employing large amounts of PC to provide the degree of range resolution essential to low-PRF radar operation.

A feature of the high-PRF waveform in an airborne application is that slow-moving ground-based targets can be eliminated because such a wide Doppler spectrum is available for target processing. For example, in an X-band high-speed interceptor (i.e., ground speed equal to 1500 knots), the sidelobe clutter spectrum will extend out to 50 kHz. If the expected targets close on the interceptor at a maximum of 600 knots, then an additional 20 kHz of clear Doppler must be provided. If a PRF of 150 kHz, for example, is selected, the lower zone of 80 to 100 knots can be discarded to eliminate ground targets. The result would have only minimal effect in the detection of high-speed closing aircraft targets.

High-PRF airborne radar designs generally operate in a velocity-only search detection mode for initial detection. In this mode, range ambiguities are not resolved and a maximum detection range equivalent to that of a CW radar (i.e., with an average power of $P_A = P_p d_u$) is achieved. When range information is desired, the high-PRF design must employ frequency modulation of the RF carrier (e.g., linear or sinusoidal frequency modulation) or multiple PRFs during the radar's dwell time on target to resolve the range. This requirement may cause an estimated 20% to 25% degradation in detection range compared to that achieved when range information is not required. A significant limitation of the high-PRF waveform for airborne applications is its limited ability to detect targets moving at velocities that cause their Doppler frequencies to appear in the sidelobe clutter regions. The clutter-free detection regions occur in the forward or closing aspects of the target; whereas for rear or chasing aspects, the target must compete with large sidelobe clutter returns, particularly at low altitudes.

With the high-PRF waveform, range folding causes short-range clutter to compete with long-range targets. In particular, all the sidelobe clutter folds into a small ambiguous range interval due to the many range ambiguities. For example, with a PRF of 150 kHz, the unambiguous range is only 0.5 nmi and, because sidelobe clutter can be well above receiver noise out to beyond a range of 10-nmi, more than 19 ambiguities of clutter will be folded into the PRF interval. Hence, at low altitudes, sidelobe clutter generally will obscure targets located within its velocity domain (i.e., $f_d = \pm 2v_r/\lambda$).

In summary, a high-PRF waveform provides ghost-free target Doppler, the capability to reject slow-moving targets, and a clutter-free target detection region that generally facilitates detection of closing targets. However, the limited all-target aspect detection, especially at low altitudes in tail chases, generally restricts the use of this waveform to closing targets. In addition, high-PRF designs require complicated methods for resolving range ambiguities and may also suffer from eclipsing problems, where target returns are lost, either totally or partially, because the receiver is blanked when the target is received due to the high duty factor of the waveform.

Medium-PRF airborne radar was conceived as a solution to the problems of detecting tail-aspect targets in the presence of both main-lobe and strong sidelobe clutter, thereby providing good overall coverage from all aspects. For modest operating ranges, the PRF can be set high enough to provide adequate separation between the periodic repetitions of the main-lobe clutter spectrum without incurring particularly severe range ambiguities. Main-lobe clutter can then be rejected through Doppler filtering, and individual targets can be extracted from sidelobe clutter using a combination of Doppler filtering and range resolution discrimination. Also, ground moving targets, being similar in characteristic to mainlobe clutter, can be rejected using a medium-PRF design.

The key feature of medium-PRF airborne radar operation is its ability to detect tail-aspect moving targets in a sidelobe clutter background. Increasing the PRF over low-PRF operation opens the clear region between mainlobe clutter responses that occur at multiples of the radar's PRF. This makes it easier to construct practical Doppler filters to both reject main-lobe clutter and bracket the target's response, thereby rejecting that sidelobe clutter not in the target's Doppler resolution cell (e.g., $B_D = 1/T_d$, where T_d is the time on target).

However, in the process of increasing the PRF, range ambiguities are created that allow close-in main-lobe and sidelobe clutter to enter the range resolution cell containing the target. Main-lobe clutter can be rejected through the use of a comb filter (e.g., MTI canceller) just as for a low-PRF design except that the rejection capability must be increased due to the ambiguous clutter. Sidelobe clutter must be rejected through increased Doppler resolution (e.g., bandpass filtering), which serves the dual function of attenuating the clutter outside the target's Doppler resolution cell while minimizing the unavoidable sidelobe clutter intake within the Doppler filter containing the target. A prime consideration in the design of medium-PRF radar involves the minimization of the unavoidable sidelobe clutter intake that ultimately limits the radar's performance.

The clutter intake into a medium-PRF airborne radar is regulated through sorting by both range and Doppler frequency. Sorting by range is accomplished using range gates that isolate the returns received from relatively narrow strips of ground at constant range. Because of the range ambiguities, the return passed by each gate will come from several strips of grounds.

Sorting sidelobe clutter by Doppler frequency is accomplished by applying the output of each range gate to a bank of Doppler filters. They will isolate the returns received from strips of ground lying between lines of constant angle relative to the radar's velocity, which are called *isodops* (see Example 4.9). Because of Doppler ambiguities, any one filter will pass the returns from several strips.

The medium-PRF waveform combines some of the attributes of both high- and low-PRF waveforms, which for airborne applications can provide slow-moving target rejection, all-target-aspect coverage, sidelobe clutterfree regions, and accurate range information. The range and Doppler ambiguities of a medium-PRF design cause the returns from short-range sidelobe clutter to fold into the same Doppler regions occupied by target returns. However, the essence of medium-PRF waveform design is to establish clear regions within the range-Doppler space by the judicious placement of multiple medium PRFs.

To resolve both range and Doppler blind zones occurring in medium-PRF airborne radar, the radar's PRF is cycled through a fixed number of fairly widely spaced PRFs while the radar's main lobe dwells on the target. For example, the radar might be cycled through eight different PRFs. If a target is in the clear on any three PRFs and its echoes exceed the detection threshold on all three, the target will be deemed detected. The range ambiguities and all spurious target ranges will then be resolved. The optimum PRF code is a function of the operational situation such as radar altitude, clutter levels, and speeds that must be determined for each specific radar.

For a target to be detected during a particular transmitted PRF of a medium-PRF code, it must be simultaneously in both a Doppler-clear region and in a range-clear zone. If the target's Doppler frequency falls in a Doppler blind zone, its return will not pass through a Doppler filter and be detected, despite the fact that it is in a range-clear zone. If the target is in a range blind zone, although its return may pass through a Doppler filter, it still will not be detected because the accompanying sidelobe clutter through the filter will cause the detection threshold to be greater than the target response. Figure 4.6 depicts a range-Doppler matrix of a representative medium-PRF radar that shows the zones for which the radar is both range-clear and Doppler-clear for at least three of eight widely spaced PRFs. Note that for a tracking radar, the PRF can be adaptively controlled to place the target in clear range and velocity zones so that only one burst at this PRF need be transmitted rather than the long, diverse bursts required by a search radar.

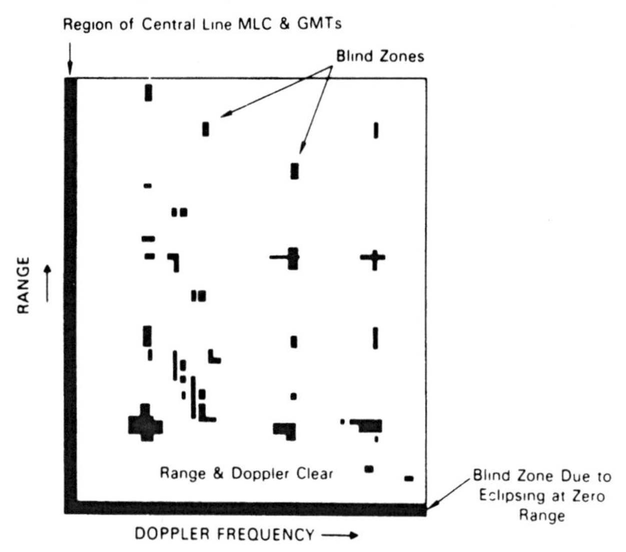

Figure 4.6 Range-Doppler regions in a medium PRF radar that are clear on at least three of eight PRFs. (*Source:* [17].)

► *Example 4.11*

Using MATLAB, plot the ambiguity diagram for a pulsed Doppler radar. Consider an eight-pulse burst and plot time for ±4 PRI intervals and frequency for ±2 PRFs.

```
% Ambiguity Function for Pulsed Doppler Waveform
% ----------------------------------------------
% ambfunb.m

clc;clf;clear;

N=2048;j=sqrt(-1);

% Waveform Parameters

Tp=1e-6;        % pulse width - sec
T=10e-6;        % period - sec
f=2e7;          % carrier frequency - Hz
Tx=8*T;         % burst length - sec

% Set Time Sampling

t=-Tx/2:Tx/(N-1):Tx/2;

% Sample Coherent Burst Signal

df=100*Tp/T;
zz=(1+square(2*pi*t/T,df))/2;
x=zz.*exp(j*2*pi*f*t);

% Form Ambiguity Function

Bx=[];

fr=1/T;
for fd=-2*fr:fr/11:2*fr;
  y=x.*exp(-j*2*pi*fd*t);
  chi=abs(xcorr(x,y));
  Bx=[chi;Bx];
end;

%subplot(211),
mesh(Bx,[-37.5 30]);
title(['Ambiguity Function for Pulse Doppler Waveform']);
xlabel('time');ylabel('frequency');
zlabel('amplitude');

%subplot(212);contour(Bx);
%title(['Contour Plot']);
%xlabel('time');ylabel('frequency');
```

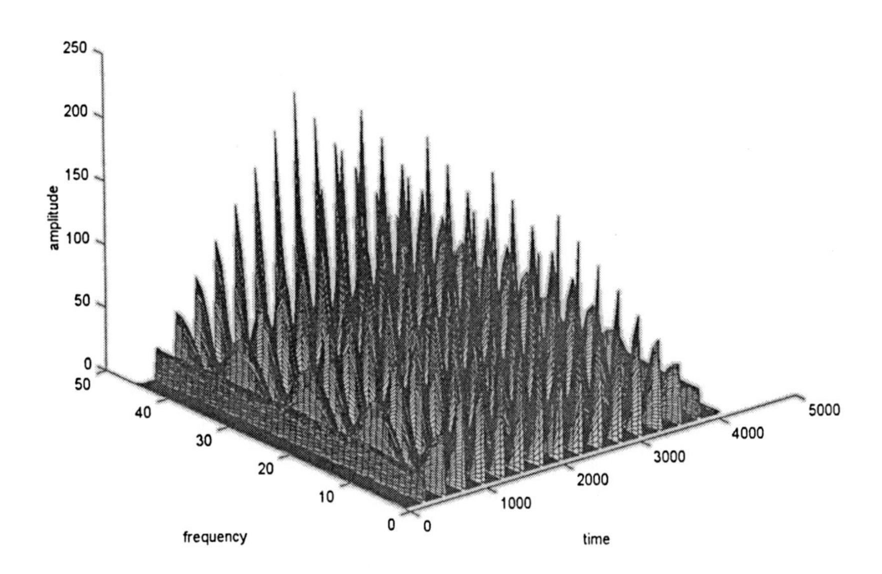

In a medium-PRF design, the number of PRFs processed during an antenna dwell time (the time the antenna takes to scan over the target) is constrained on the high side by the available processing time. If the time becomes too long, the antenna scan rate drops below an acceptable value. On the low side, the number of PRFs per dwell must be sufficient not only to resolve range but also to ensure good Doppler visibility.

An action that can reduce sidelobe clutter is to increase the radar's range resolution by narrowing the radar's effective pulsewidth and correspondingly narrowing the radar's range gates. Narrowing the radar's effective pulse width reduces the sidelobe clutter level in the same proportion by which the pulsewidth is narrowed. Additional range gates are required to provide full range coverage and more Doppler filters must be formed because a separate bank of Doppler filters is required for each range gate.

One way to narrow the effective radar pulse width without reducing the average transmitted power is PC. A common practice is to maximize the average transmitter power without incurring an unacceptable loss due to eclipsing and then use enough PC to achieve the desired range resolution. An alternative approach that minimizes eclipsing is to transmit very narrow pulses of higher peak power.

The sidelobe clutter with which a target must compete can be further reduced by narrowing the passbands of the Doppler filters. To retain all the power in the target return, the filters must be wide enough to pass the target's spectrum. The target spectrum generally is inversely proportional to the radar's time on target (T_d) plus an allowance for internal target motion and acceleration effects.

An important form of sidelobe clutter in medium-PRF airborne radar results from large RCS targets on the ground (e.g., buildings, water towers, and other specular reflectors), which are particularly prevalent in urban or mountainous areas. Also included in this class of clutter is unwanted moving ground vehicular traffic that, because of the range and Doppler ambiguities, conflicts with the detection of airborne targets.

Although an airborne radar is vulnerable to such targets when operating at low PRF, it is more vulnerable at medium PRFs due to the severe range ambiguities in this type of operation. One way to reduce these unwanted targets is to provide the radar with a guard channel that consists of a separate receiver processor whose input is supplied by a small horn antenna mounted on the main radar antenna. The width of the horn's main lobe is sufficient to encompass the entire region illuminated by the radar antenna's principal sidelobes, and the gain of the horn's main lobe is greater than any of the main antenna's sidelobes. Any detectable target in the radar antenna's sidelobes will produce a stronger output from the guard receiver than from the main receiver. Conversely, any detectable target in the main lobe will produce a stronger signal than in the guard receiver. Consequently, by comparing the output of the two receivers and inhibiting the output of the main receiver when that of the guard receiver is stronger, we can eliminate the unwanted targets in the radar's sidelobe region.

4.2.1 EA Against PD Radar

The PD radar generally has good resistance against EA. The ability to resolve targets in Doppler frequency allows rejection of chaff and other correlated interference, which are separated from the target return by more than a Doppler filter bandwidth. The matched-filter aspect of PD operation provides the effect of coherent integration, which optimizes the target-to-jamming ratio in the presence of noise jamming. The ability of a PD radar to extract radial target velocity through direct Doppler measurement and by differentiating range measurements makes it necessary for a deception jammer to induce a realistic Doppler signature on the simulated target return. The large number of pulses generated by a PD radar provides problems for many types of EW intercept receivers that rely on deinterleaving pulse trains to identify threat radar.

The only significant disadvantage against EA is the necessity of maintaining a stable transmitter frequency and PRF for a time equal to the inverse of the Doppler filter bandwidth. This allows a jammer the time to tune to the first transmitted pulse and then jam subsequent pulses in a spot jamming or deception mode. The vulnerability in this mode is generally to main-beam jammers, since appropriate sidelobe cancellation and blanking can be applied to reduce the effect of sidelobe jammers. Also, a sidelobe cover mode can be used whereby multiple spurious frequencies are radiated in the sidelobe direction through an auxiliary antenna at the same time that the main radar pulse is transmitted.

Efficient use of jamming ERP against a PD radar requires that the jamming waveform have sufficient correlation over the processing interval (i.e., CPI) of the PD radar to inject sufficient energy into a Doppler filter to cause it to report a target or to suppress the true target. This indicates that, from an energy viewpoint, the PD radar jammer must maintain a frequency stability greater than the Doppler filter bandwidth (i.e., 50 Hz to 500 Hz) over a time interval corresponding to the CPI time (i.e., 2 ms to 20 ms). While this might be adequate to deceive older type analog PD radars, more precise frequency control is indicated for modern PD radars using FFT digital signal processing techniques. Modern radars form multiple Doppler filters covering the frequency range of interest. Signatures of most military type targets within a particular range gate fall within a single or adjacent Doppler filter or, if due to *jet engine modulation* (JEM), have a distinctive pattern. Hence, responses in multiple Doppler filters are indications of the presence of interference or jamming signals. This forms a criterion for the purity of Doppler radar jamming signals in that they must not be detectable in multiple Doppler filters. Since Doppler filters formed digitally exhibit a sidelobe structure of the order of 20 dB to 30 dB below the main lobe, this indicates that PD jammers must not generate spurious signals above this level (i.e., 20 dB to 30 dB below the main filter response).

PD radars are used in both search and track applications. EA against PD search radars generally requires the simultaneous storage of radar signatures for many radars. This can be accomplished in several ways. The most universal is to employ a DRFM. In support jamming applications where several radars must be jammed simultaneously, this can lead to intermodulation problems and possible detection of the jamming signals (see Section 3.5). An alternate method is to employ a DDS using multiple accumulators, each tuned to a particular identified threat. This requires an extensive threat library and an accurate frequency set-on receiver (see Section 3.5). Another way is to employ multiple voltage-controlled oscillators that are locked to each threat using phased-locked loop techniques. This technique is used in several operational EA systems but is not as accurate or efficient as jammers using DDS or DRFM digital techniques.

Doppler filters in many modern radars are synthesized using the FFT. The normalized transfer function of the FFT is given [1]

$$|H(\omega)| = \left|\frac{\sin(N\omega_d T/2)}{\sin(\omega_d T/2 - k\pi/N)}\right| \tag{4.16}$$

where N is the number of pulses or FFT points, ω_d is the Doppler radian frequency, T is the interpulse interval (PRI), and k is the Doppler filter number. The FFT transfer function is illustrated in Example 4.12, where the cross-coupling between the Doppler filter channels is evident due to the high sidelobe levels of the filters (i.e., −13.3 dB for first-order sidelobe). The input to the FFT can be windowed (see Example 4.12) to produce lower sidelobes at the expense of a broader filter and less Doppler discrimination between channels. One way to implement the windowing is to post-filter-combine adjacent channels given by

$$H_\omega(\omega) = H_k(\omega) + \alpha[H_{k+1}(\omega) + H_{k-1}(\omega)] \tag{4.17}$$

where $\alpha = 0.5$ for Hanning weighting and $\alpha = 0.426$ for Hamming weighting [1]. The reduced sidelobe Doppler filters are also shown in Example 4.12 and indicate a discrimination capability of the order of 30 dB as a reasonable estimate for modern radars.

▶ *Example 4.12*

Using MATLAB, plot the transfer function for a bank of Doppler filters using a 32-point FFT and uniform weighting. Next, plot the transfer function for the same situation as before, but use postcombining with Hamming weighting. What is an estimate of the level of spurious content of a jamming signal that could be detected using the reduced sidelobe filters?

```
% FFT Doppler Filter Response
% ---------------------------
% fftfila.m

clf;clc;clear;

% Define Frequency Vector

N=32;        % FFT Points
PRF=2000;
```

```
T=1/PRF;
fd=(1/(2*T))*(0:1024-1)/1024;

% Compute Filter Response - Uniform Weights

for k=1:16;
x=2*pi*fd*T/2+1e-10;
H=sin(N*x)./sin(x-((k-1)*pi/N));
Hd=H.*H;
Pdb(:,k)=10*log10(Hd'/(N^2)+1e-2);
end;

% Plot Filter Response

subplot(211);
plot(fd,Pdb,'k');grid;
xlabel('Frequency - Hz');
ylabel('Amplitude - dB');
title(['FFT Filter Response - Uniform Window']);
axis([ 0 1000 -20 0]);

% Compute Filter Response - Hanning Weights

a=.5;

for k=1:16;
x=2*pi*fd*T/2+1e-10;
H(k,:)=sin(N*x)./sin(x-(k-1)*pi/N);
end;

for l=2:15;
Hw(l,:)=H(l,:)-a*(H(l+1,:)+H(l-1,:));
Hd(l,:)=Hw(l,:).*Hw(l,:);
Pldb(:,l)=10*log10(Hd(l,:)'/(N^2)+1e-6);
end;

% Plot Filter Response

subplot(212);
plot(fd,Pldb,'k');grid;
xlabel('Frequency - Hz');
ylabel('Amplitude - dB');
title(['FFT Filter Response - Hanning Weighting']);
axis([ 0 1000 -50 0 ]);
```

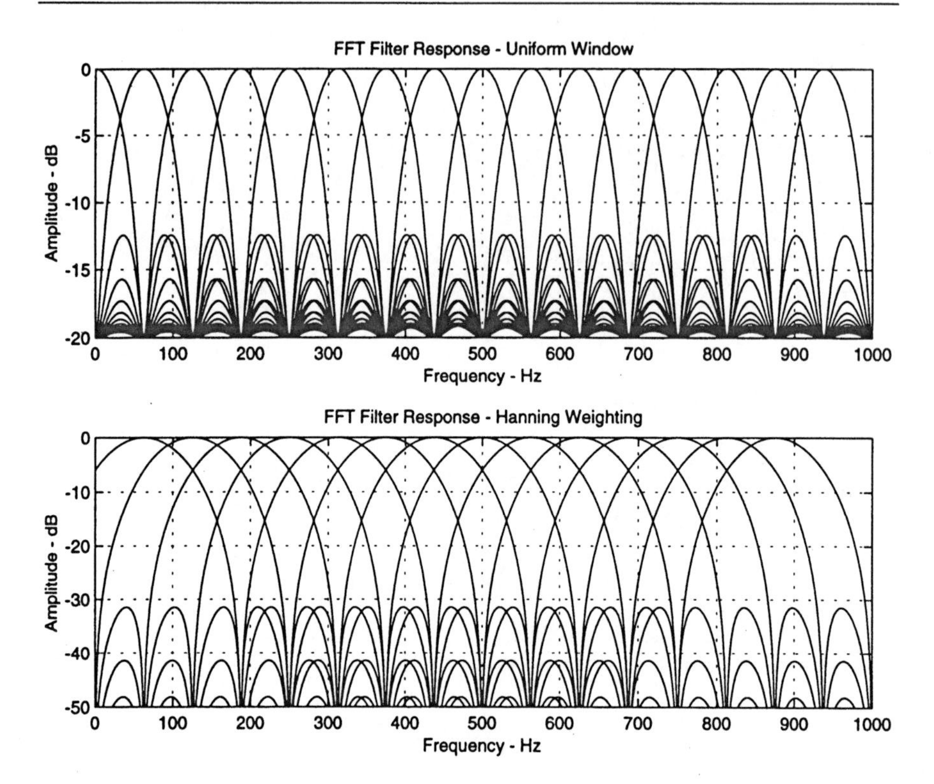

PD radars are extensively used in fire control systems to supply targeting data for missile systems. In some systems, the PD radar is located in the missile (e.g., AMRAAM) and acts to provide guidance data directly to the missile control system. Other systems using semiactive PD radars have the active PD radar that is tracking the target act as an illuminator for a bistatic receiver located in the missile. In either case, narrow "speed gates" (Doppler tracking filters) are employed to track the target. The object of the deceptive EA system is then to capture the speed gates, causing the missile to lose guidance information. This VGPO procedure is a classic technique employed against PD tracking radar and missile seekers.

VGPO is a technique used in DECM systems for use against automatic velocity tracking radars that capture the victim radar's velocity gate (speed gate), walks it off in velocity, and then turns off, leaving the radar's velocity gate with no signal. The PD radar must then go through its reacquisition mode to attempt to reacquire the target. The effectiveness of the DECM system is proportional to the time that the missile can be kept in its reacquisition mode. In modern PD radars, the use of multiple Doppler filters, formed digitally, speeds this reacquisition process and alerts the radar to the presence of jamming signals.

The bandwidth (B_d) of the speed gate must be sufficient to encompass Doppler frequency deviations caused by acceleration between the target and the missile and analog components in the servo system. This frequency deviation is given by

$$\Delta f_d = \frac{19.63 \cdot g}{\lambda B_d} \tag{4.18}$$

where g is the acceleration in g's, λ is the radar's wavelength, and B_d is the Doppler filter's bandwidth. If we balance (4.18) such that the Doppler filter bandwidth and the Doppler frequency deviation are equal for an I-band radar with a 10-g missile acceleration, then the resulting Doppler bandwidth of the "speed gate" is 80 Hz. The preceding discussion indicates that Doppler frequency bandwidths of the order of 100 Hz are reasonable for aircraft targets.

The prior discussion points to the need in DECMs that operate against PD radar to have low spurious spectral components (greater than 30 dB down from the carrier) in order to prevent detection. An examination of the DECM block diagram shown in Figure 3.5 indicates that three major components have the potential to generate spurious phase components. These are (1) the TWT microwave amplifier where incidental modulation (e.g., power supply ripple) on the helix causes phase modulation, (2) the DRFM that stretches signals through head-to-tail repetition, and (3) the serrodyne frequency offset device (see Chapter 3) that generates the deceptive Doppler signature.

Example 4.13 illustrates the effect of phase modulation on the helix of the TWT where ±4-degree peak phase modulation causes spurious components of the order of 30 dB down from the carrier at the frequency of the disturbance and its harmonics. These point to the need for good filtering of power supplies whose ripple frequencies range from 50 Hz to 400 Hz for conventional supplies to up to 100 kHz for switching power supplies.

▶ *Example 4.13*

Using MATLAB, plot the spectrum of a phase-modulated sine wave given by

$$V(t) = \cos\left(\omega_t + 2\pi \cdot \frac{\Delta\phi}{360} \cdot \sin(\omega_m t)\right) \tag{4.19}$$

where the maximum phase deviation $\Delta\phi$ is selected to give spectrum levels 30 dB below the carrier at the modulating frequency ω_m. Extend this example to multiple phase modulation components. How do these multiple components interact?

```
% Spectrum for Phase Modulated Sine Wave
% -------------------------------------
% phspec.m

clf;clc;clear;

j=sqrt(-1);

% Sample with fs

fs=2048;      % Hz
t=(1:fs)/fs; % sec

% Baseband and 400 Hz noise

f0=0;fm=400; % Hz
dfi=4;        % Degrees

x=exp(j*(2*f0*t+dfi/360*2*pi*sin(2*pi*fm*t)));

% Find spectrum

X=fft(x,fs);
X=fftshift(X);
Pxx=X.*conj(X);

Pxx=Pxx/max(Pxx);
Pxxdb=10*log10(Pxx+1e-10);

% Plot over frequency axis +/- fs/2

f=(-fs/2:fs/2-1);

plot(f,Pxxdb);grid;
xlabel('Frequency - Hz');
ylabel('Power - dB');
title(['Spectrum for 400Hz Phase Modulation']);
axis([-1500, 1500, -80, 0 ]);
```

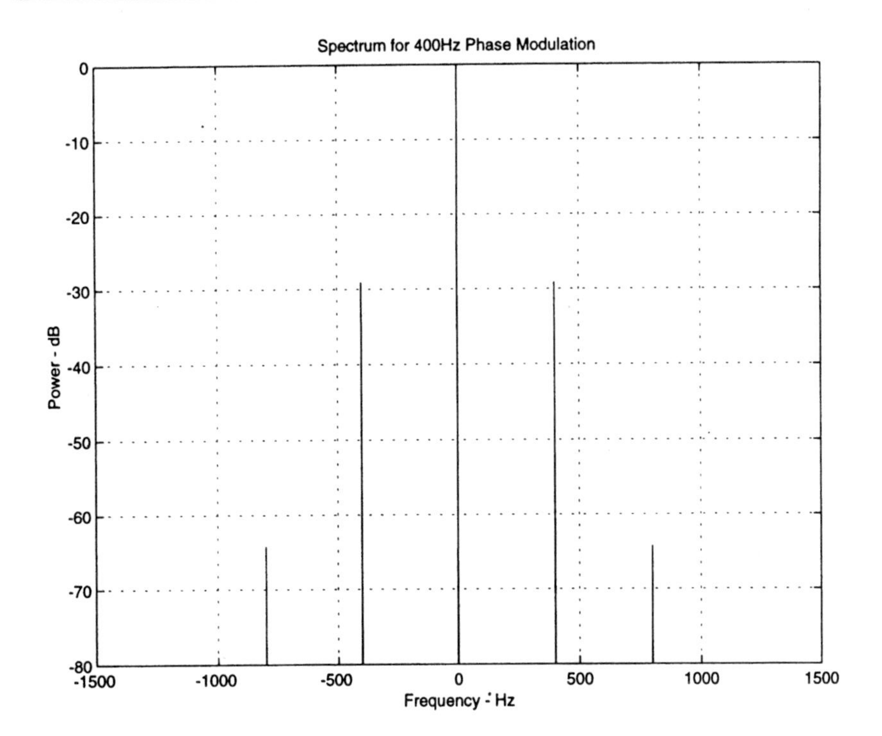

Example 4.14 indicates the spurious frequency components caused by head-to-tail stretching of stored radar signatures in a DRFM. This example illustrates that the random phase modulation using this procedure must be held to under 30 degrees to provide spurious components down 30 dB from the carrier using a 1-bit DRFM. This points to the need for development of multiple bit DRFMs and tight phase control in this type of operation.

► *Example 4.14*

Using MATLAB, find the spectrum of a CW signal that is reconstructed by storing a small portion of the signal and then repeating this portion a number of times to form the full signal. The phase discontinuity caused by the difference in storage time and the number of phase cycles of the signal can be modeled by an initial random phase of each segment. Assume the signal is extended,eight times and find the initial random phase to reduce spurious spectral components to 30 dB below the carrier. Assume a 1-bit DRFM.

```
% Spectrum of CW Signal with Random Phase
% ----------------------------------------
% cwspec.m
```

```
clc;clf;clear;

% Time vector with N elements in 1 second

N=2048;t=(1:N)/N;fs=N;

% Base band

 w=0;

% Maximum phase deviation in degrees

d=30;
fi=d/360*2*pi*(rand(1,N)-.5);

% Extension to M periods

M=8;
for i=1:N;fi1(i)=fi(1+fix(i/M));end;

% Form signal

j=sqrt(-1);
x=cos(w*t+fi1)+j*sin(w*t+fi1);

% Find spectrum

X=fft(x,N);
X=fftshift(X);
Pxx=X.*conj(X);
Pxx=Pxx/max(Pxx);
Pxx=10*log10(Pxx+1e-6);
f=fs*(-N/2:N/2-1)/N;

% Plot spectrum

plot(f,Pxx);axis([-1500 1500 -60 0]);
title(['CW Spectrum with Random Phase '...
 '  Dev.=',num2str(d),'degrees']);
xlabel('Frequency - Hz');
ylabel('Power - dB');grid;
```

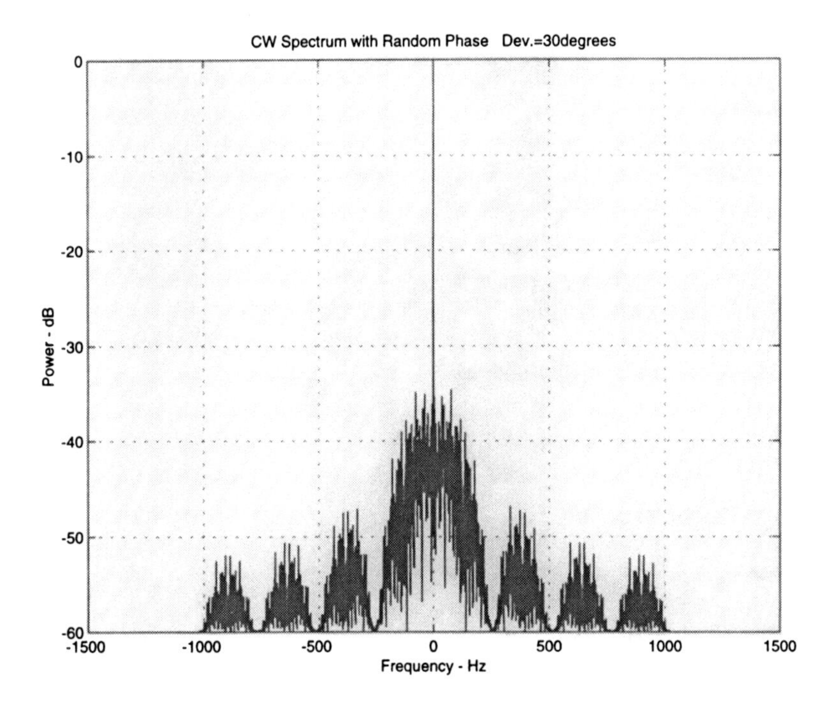

Serrodyning is a process used in DECM repeater jammers to translate the frequency of the repeated signal, thereby generating a false Doppler signal. This is accomplished by inducing a linear phase change with time, $\phi(t) = kt$ onto the signal

$$v(t) = \sin(\omega t + \phi t) = \sin(\omega + k)t \tag{4.20}$$

There are several ways that serrodyning is accomplished.

The digital phase shifter as depicted in Figure 3.8 can be stepped in phase at a linear rate, thereby causing a frequency shift in the output signal. Since the phase is stepped in discrete increments ($\Delta\phi = \pi/2^{b-1}$, where b is the number of phase bits), spurious phase components are generated in the process determined by the number of bits utilized.

▶ *Example 4.15*

Using a b-bit digital phase shifter that provides a $360/2^b$-degree phase increment per step, plot the resultant frequency spectrum for this serrodyne method. How many bits are required to ensure that the spurious spectral components are 30 dB down from the main signal?

```
% Digital Phase Shifter Serrodyning
% --------------------------------
% digserro.m

clf;clf;clear;

% Set Input Parameters

b=4;      % Quantizing Bits
N=128;    % Sample Points

% Define time vector 0 to tmax

tmax=1;
t=tmax*(0:N-1)/N;

% Sample frequency fs

fs=N/tmax;

% Define step ramp z which runs
% from 0 to tmax with amplitude 1

B=2^b;
y=B*t/tmax;
z=floor(y)/B;

% Define Phase Modulated Signal

j=sqrt(-1);
fo=0;
x=exp(j*2*pi*(fo*t+z));

% Find x for M steps

M=64;A=[];
for m=1:M;
A=[A x];end;
x=A;

% Extend for t vector

N=N*M;
t=tmax*(0:N-1)/N;

% Spectrum for x(t)

X=fft(x);
Px=X.*conj(X);
Px=fftshift(Px);
```

```
% Select central frequency axis

Px(3*N/4+1:N)=[];
Px(1:N/4)=[];
fax=fs*(-N/4:N/4-1)/N;

% Normalize amplitude

Px=Px/max(Px);
Pxdbm=10*log10(Px+1e-6);

plot(fax,Pxdbm);grid;
xlabel('Frequency - Hz');ylabel('Amplitude - dbm');
title(['Digital Phase Shifter Serrodyning' ...
'   ( Bits =',num2str(b),')']);
axis([-40 40 -60 0]);
```

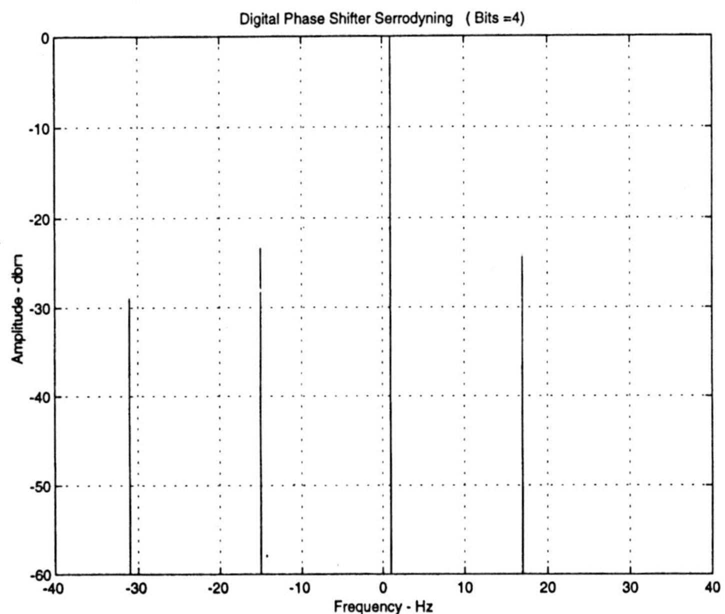

Another method for serrodyning is to modulate the helix of a TWT with a sawtooth waveform. The sawtooth amplitude must be adjusted to provide 2π-radian phase shift while its repetition rate is set at the desired Doppler frequency offset. Spurious components are caused by the finite flyback time associated with the sawtooth waveform and the deviation from 2π-radian phase shift achieved by the actual system. Example 4.16 illustrates that the flyback time of the sawtooth wave must be less than 2% of the total period for the spurious components to be 30 dB down from the output.

▶ *Example 4.16*

Plot the spectrum that results from serrodyne modulation of a TWT using a sawtooth waveform with finite flyback time so that the period is $T_1 + T_2$, where T_2 is the flyback time. What is the ratio of $T_2/(T_1 + T_2)$ that suppresses spurious components by more than 30 dB?

```
% TWT Serrodyning With Offset
% --------------------------------
% twtserro.m

clf;clc;clear;

% Set Input Parameters

r=.02;  % Ratio T2/T1+T2
N=128;  % Sample Points

% Define time vector 0 to tmax

T2=r/(1-r);
tmax=1+T2;
t=tmax*(0:N-1)/N;

% Sample frequency fs

fs=N/tmax;

% Define ramp z which runs from
% 0 to T1=1 with amplitude 1 and
% flyback from t=1 to t=1+T2

for k=1:N;
if t(k)<=1;z(k)=t(k);
else z(k)=1-(t(k)-1)/T2;end;
end;

% Plot Serrodyne Waveform

t1=tmax*(0:N)/N;
zz=z;
zz(N+1)=0;

subplot(211);
plot(t1,zz);grid;
xlabel('Time - sec');
ylabel('Amplitude');
title(['Serrodyne Waveform']);
```

```
% Define Phase Modulated Signal

j=sqrt(-1);
fo=0;
x=exp(j*2*pi*(fo*t+z));

% Find x for M steps

M=64;A=[];
for m=1:M;
A=[A x];end;
x=A;

% Extend for t vector

N=N*M;
t=tmax*(0:N-1)/N;

% Spectrum for x(t)

X=fft(x);
Px=X.*conj(X);
Px=fftshift(Px);

% Select central frequency axis

Px(3*N/4+1:N)=[];
Px(1:N/4)=[];
fax=fs*(-N/4:N/4-1)/N;

% Normalize amplitude

Px=Px/max(Px);
Pxdbm=10*log10(Px+1e-6);

subplot(212);
plot(fax,Pxdbm);grid;
xlabel('Frequency - Hz');ylabel('Amplitude - db');
title(['TWT Serrodyne Spectrum' ...
'    (Flyback Ratio =',num2str(r),')']);
axis([-30 30 -40 0]);
```

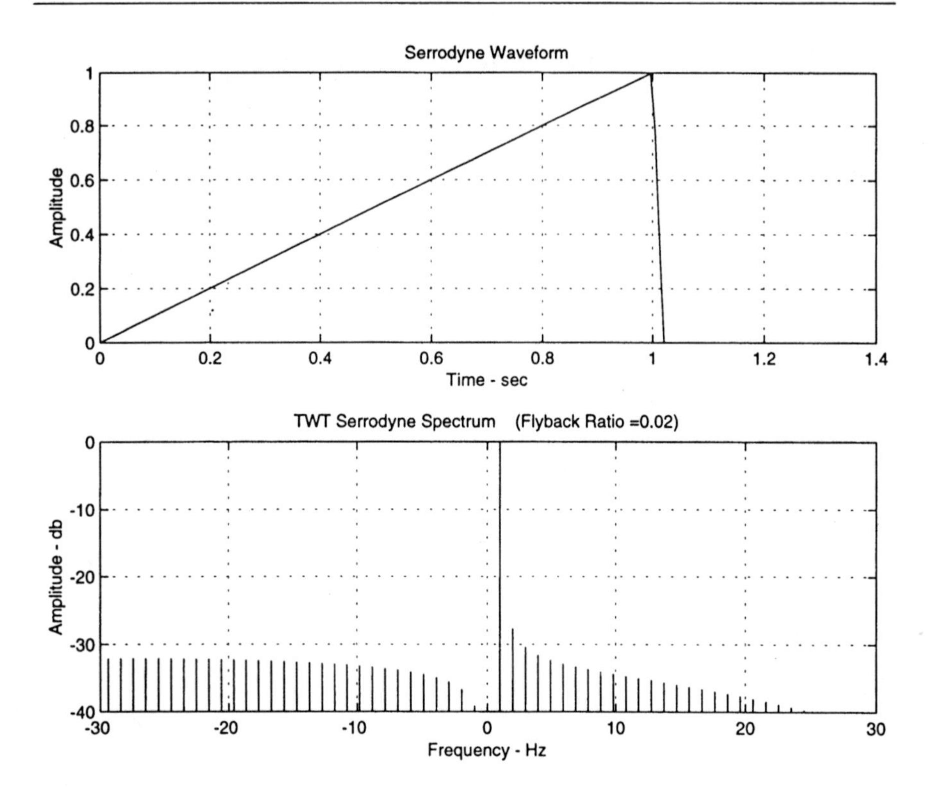

Another method of generating deceptive Doppler signals is employed in DRFMs. In DRFMs, digital storage is generally accomplished at baseband since this results in the least amount of memory for a given signal. Signals are translated to baseband by down-converting using a stable local oscillator. They are then up-converted for transmission after readout of the digital memory by up-conversion using a stable sidestep oscillator. The frequency offset between the down-converting and up-converting local oscillators corresponds to the deceptive Doppler shift induced on the signal. The generation of spurious components depends upon the stability of the oscillators and the circuitry employed in the single sideband conversion processes. These can be held to the order of 30 dB through careful design.

The tactics involved in applying deceptive Doppler signals into PD radar are dependent upon the type of PD radar involved. Low-PRF PD search radars have good range resolution capability and reject jamming signals whose Doppler is in a frequency band around the radar's PRF. Also, they usually employ PC waveforms. When the jamming technique uses false targets, both factors must be taken into account. Another technique used against low-PRF PD search radars is range-bin masking. This technique exploits the radar signal processing circuitry to suppress radar targets. As the name implies, an extended cover pulse with a Doppler signature within the Doppler filter passband is placed over the radar target. This signal passes through the clutter rejection circuitry into the radar CFAR circuitry, which raises the detection threshold to suppress the target.

Against high- and medium-PRF PD search radars, an effective technique is to excite multiple range gates with multiple deceptive Doppler returns. This overloads the radar signal processor, which must sort out all range ambiguities to determine the true target range. One way to generate multiple Doppler targets is to divide down the radar's PRF. For example, if the radar's PRF is divided down by four (i.e., every fourth pulse repeated), then three additional deceptive targets are generated within the radar's range-Doppler matrix.

4.3 Monopulse

Generally, an effective DECM employs both range and angle deception techniques. Range deception techniques are somewhat independent of radar implementation as compared to angle deception techniques, which must be tailored to a specific radar implementation. For this reason, range deception is almost universally employed in deception jammers, but its effect is primarily limited to introducing false range information into the victim radar. While the false range information is being absorbed by the radar, it still provides accurate angle information. It is only when the radar's range gate has been captured and the DECM is turned off that angle information is denied to the victim radar. Reacquisition in the range dimension may be rapid (i.e., of the order of milliseconds) if the radar is pointed in the direction of the target. For this reason, it is appropriate to introduce false angle information into the radar at the same time RGPO is being attempted. If the radar is forced to search in both angle and range during reacquisition, then this cycle is appreciably lengthened and the radar is rendered ineffective during this period.

Modern radars and missile seekers generally use monopulse tracking systems. Monopulse tracking systems (sometimes called simultaneous lobing) form an angular-error estimate on each return pulse, thereby rendering the system insensitive to amplitude fluctuations on the data. This improves the radar performance and eliminates the possibility of amplitude modulation jamming (i.e., inverse gain) so effective against conical scanning type radar. Further, this property makes monopulse radars effective in tracking noise jamming signals that are employed against missile seekers having "home-on-jam" modes.

Monopulse tracking radars fall into three general types: amplitude-comparison or simultaneous-lobing, phase-comparison or interferometer, or a combination of both types. The amplitude-comparison type involves the formation of multiple antenna beams simultaneously in space where each beam is squinted or displaced in angle from an adjacent beam. The angular measurement is made by comparing the power received in overlapping adjacent beams. An angle error is indicated by the difference in amplitude of the signals returned in the adjacent beams. The amplitude comparison can be made at any stage in the radar (RF, IF, or video), but RF comparison is typical because it provides the highest tolerance against component instability.

In phase-comparison monopulse, the antenna elements are displaced by a number of wavelengths from each other along a common baseline. If the target is located off a perpendicular bisecting the baseline, then the return signal will reach one element of the antenna before it reaches any other element. This results in a time difference between signals as they are intercepted by different elements of the antenna, which is functionally related to the angle of the target from the antenna's boresight. This time difference translates into a phase difference, which depends on the antenna element spacing and the signal's RF wavelength.

For either phase or amplitude monopulse, four antenna elements are required to provide tracking in both azimuth and elevation angles. By offsetting the antenna elements about the boresight and using a combination of phase and amplitude monopulse, it is possible to accomplish azimuth and elevation tracking using only two antenna elements. Of the three types of monopulse techniques described, amplitude monopulse is by far the most common.

Jamming techniques against monopulse radars can generally be divided into two categories. The first relies on imperfections in the monopulse design or implementation that are exploited by the jammer. These involve such techniques as image jamming, skirt frequency jamming, and cross-polarization jamming, which reverse the sense of the angle-error signal developed by the monopulse angle discriminator [1]. The second category involves multiple source techniques whose objective is to distort the EM wave's angle-of-arrival at the monopulse antenna such that either the monopulse tracker is caused to point away from the target's angular direction or spurious modulation is introduced into the tracker's servo system that causes a break-lock situation. These techniques against monopulse radars include blinking, formation jamming, cross-eye, and terrain bounce. Needless to say, it is more difficult to jam monopulse radars than sequential-lobing types (including conical scan and TWS variations). This supports the current trend in tracking radar, which is toward monopulse implementation.

In amplitude comparison monopulse the radar antenna generally forms four squinted offset antenna beams, two in azimuth and two in elevation coordinates. The monopulse pattern forms a sum pattern by combining all beams and two difference patterns, one in azimuth and one in elevation. The radar transmitter radiates through the sum pattern while the tracking receiver uses signals that are formed through both the sum and difference antenna patterns.

Considering only the azimuth dimension (the elevation is similar), we can determine the sum and difference pattern by considering the rectangular offset apertures depicted in Figure 4.7 [6]. The sum and difference patterns are given by taking the Fourier transform of the illumination functions

$$\Sigma = (v_a(t) + v_b(t))/\sqrt{2} \qquad (4.21)$$
$$\Delta = (v_a(t) - v_b(t))/\sqrt{2}$$

The resulting sum and difference antenna patterns are given in Example 4.17. Note that the sidelobes of the difference pattern are relatively high due to the use of rectangular illumination functions.

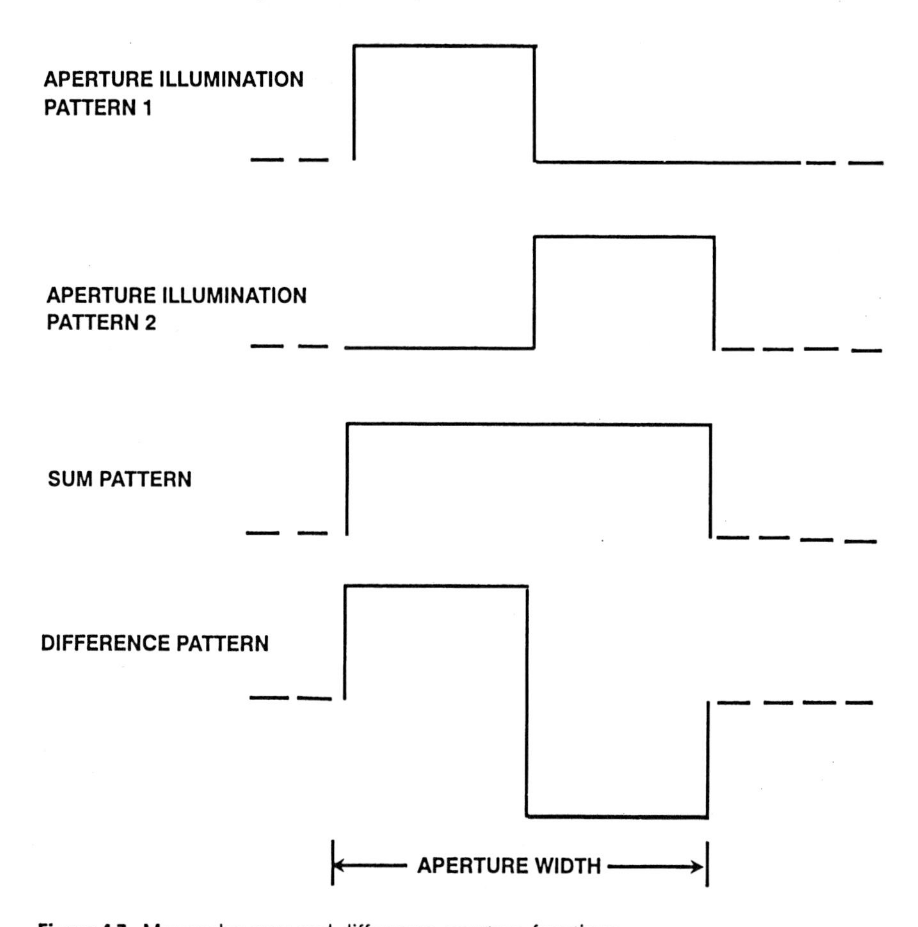

Figure 4.7 Monopulse sum and difference aperture functions.

► *Example 4.17*

Using MATLAB, plot the monopulse sum and difference patterns for the illumination function shown in Figure 4.7. Over what angular region of the sum pattern can the monopulse tracker lock onto the target? Is the monopulse tracking system stable when tracking in the sidelobes on a strong target or signal?

```
% Monopulse Antenna Pattern
% --------------------------
% monpat.m

clear,clc,clf;

% Normalized Aperture Width
na=4;

% Sampling Frequeny=Number elements per norm aperture
fs=8;

% Norm aperture with N elements
N=fs*na;
xna=na*(-1/2:1/(N-1):1/2);

% Illumination Function

wxna(1:N/2)=ones(1,N/2);
wxna(N/2+1:N)=-ones(1,N/2);
wxnb(1:N/2)=ones(1,N/2);
wxnb(N/2+1:N)=ones(1,N/2);

% Fill with M/2 zeros front and back

M=1024;
xna=na*(-1/2:1/N+M-1:1/2);
wxna=[zeros(1,M/2) wxna zeros(1,M/2)];
wxnb=[zeros(1,M/2) wxnb zeros(1,M/2)];

% Beam Functions from -fs/2 to fs/2 in sine space

Nfft=length(wxna);
Esine=fft(wxna,Nfft);
Esine=fftshift(Esine);

Esum=fft(wxnb);
Esum=fftshift(Esum);

% Azimuth vector

sinfi=fs/4*(-Nfft/2:Nfft/2-1)/Nfft;
```

```
% Azimuth vector in radians

fi=asin(sinfi);

% Beam gain functions

Gfi= Esine.*conj(Esine)/Nfft;
Gfs=Esum.*conj(Esum)/Nfft;

Gfi(1:Nfft/2)=sqrt(Gfi(1:Nfft/2));
Gfi(Nfft/2+1:Nfft)=-sqrt(Gfi(Nfft/2+1:Nfft));
Gfs=sqrt(Gfs);

% Plot Monopulse Antenna Pattern

plot(fi,Gfi,fi,Gfs);grid;
axis([ -.25 .25 -.8 1]);
ylabel('Amplitude');
xlabel('Angle - radians');
title(['Monopulse Antenna Patterns']);
text(.04,.8,'Sum Pattern');
text(-.22,.6,'Difference Pattern');
```

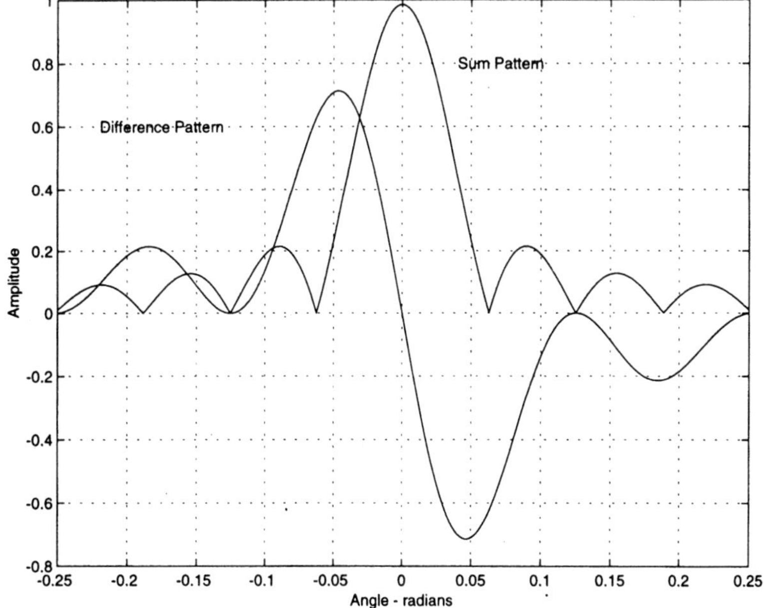

The monopulse ratio (r) is given by

$$r = \frac{|\Delta|}{|\Sigma|} e^{j\phi} \tag{4.22}$$

where Σ and Δ are the complex sum and difference voltages and ϕ is the phase angle between the sum and difference channels. Since both Σ and Δ are proportional to the target or signal strength, the monopulse ratio is independent of the amplitude of the target. In monopulse tracking systems, the real part of the monopulse ratio given by

$$\mathrm{Re}(r) = \frac{|\Delta|}{|\Sigma|}\cos\phi \tag{4.23}$$

is used to sense the magnitude and direction of the error between the antenna boresight and the target position. This error is then applied to a servo system that drives the antenna to null the output of the monopulse ratio detector to zero for accurate tracking. Note that when tracking a target, the monopulse antenna's boresight is driven orthogonally to the incident wave front.

From an EA viewpoint, a fundamental method for deceiving a monopulse radar in angle is through phase front distortion. Since a point source can only produce a spherical wave front, it generally is necessary to employ multiple sources to produce wave front distortion. This can be accomplished using geometrically dispersed multiple antennas or multipath signals (i.e., terrain bounce and chaff reflections) to simulate the effect of multiple antennas.

4.3.1 EA Against Monopulse Radars

The nature of monopulse tracking radars whereby angle error data are developed on each radar pulse makes them inherently difficult to jam. Some monopulse angular jamming techniques, such as skirt and image jamming and cross-polarization jamming, are designed to exploit weaknesses in the implementation of the monopulse radar. Other jamming techniques, such as cross-eye, terrain or ground bounce, and blinking or formation jamming, are designed to attack weaknesses fundamental to all monopulse tracking systems. In general, it is better to attack fundamental weaknesses rather than to rely on design weaknesses. Exploitation of design weaknesses implies a detailed knowledge of the design of the victim radar and is always susceptible to modification of the design to correct those weaknesses.

For example, image jamming, or jamming at the image frequency of the monopulse radar, depends on the fact that the phase angle at IF between two signals of the image frequency is the reverse of that which would appear at the IF if the two signals were at the normal response frequency of the receiver. In a phase-comparison monopulse system, this reverses the polarity of the error and causes the antenna to be driven away from the target if the jamming power exceeds the signal power. If the monopulse radar is equipped with an image rejection filter or mixer, this form of jamming is rendered ineffective and hence cannot be considered as a dependable jamming technique.

Cross-polarization jamming exploits the fact that some monopulse radars give erroneous angle-error information when the received signal (jammer) is orthogonally polarized with respect to the polarization of the radar-receiving antenna. This technique exploits a design weakness generally associated with reflector-type antennas whose response to cross-polarized signals (called Condon lobes) is significantly different than that of the normal polarization response. This situation is depicted in Example 4.18, where the resulting angle-error discriminator response to the cross-polarized signal is actually the reverse of the response to the normal polarized signal. This causes the antenna to be driven away from the cross-polarized jamming signal rather than toward a signal with normal polarization.

As shown in Figure 4.8, the magnitude of the cross-polarization response is significantly reduced from that of the normal polarization response. Data on parabolic antennas indicate that the cross-polarized response is reduced by 15 dB to 30 dB with respect to the normal polarized response, while comparable data for a hyperboloid lens shows a reduction of 30 dB to 45 dB. This results in a requirement that the ratio of cross-polarized jamming power to target signal power be a minimum of 20 dB, and preferably of the order of 30 dB to 40 dB for effective cross-polarized jamming.

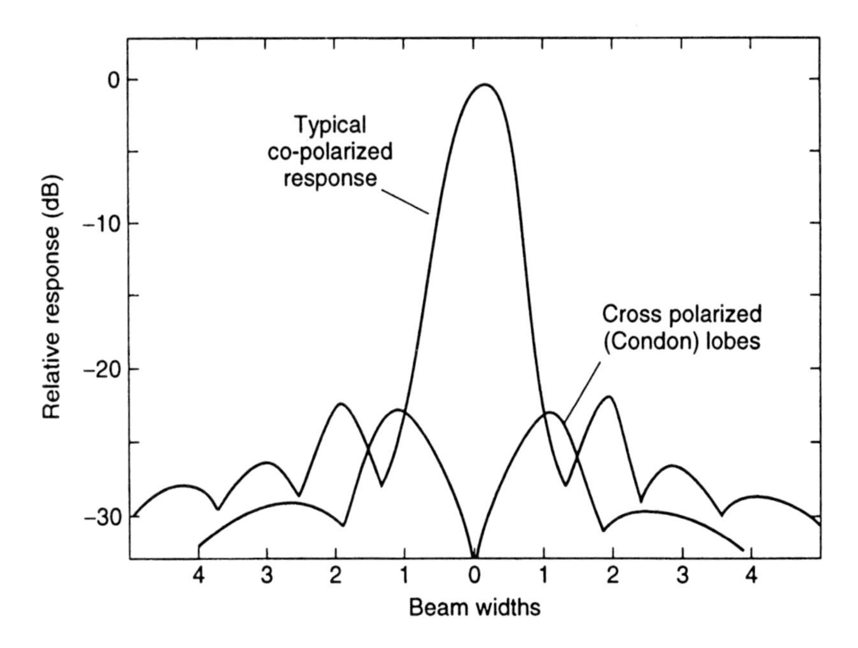

Figure 4.8 Cross-polarization response of a parabolic antenna.

In addition, any deviation in the polarization of the jamming signal from that of an exactly cross-polarized signal results in a component with normal polarization. This normal-polarized component is of the order of −41.2 dB for 0.5-degree polarization rotation, −35.2 dB for 1-degree rotation, and −29.1 dB for 2-degree deviation from orthogonality. If the normal-polarized component (which receives full response) is greater than the cross-polarized response (which is highly attenuated), then the jamming signal acts as a beacon rather than a jammer. Thus, orthogonality requirements are quite severe for cross-polarization jamming signals with requirements generally stated as ±5 degrees, which correspond to a −21.2-dB normally polarized response.

▶ *Example 4.18*

Use the fact that the cross-polarized antenna pattern is approximately 45 degrees displaced from the copolarized pattern to simulate the antenna patterns depicted in Figure 4.8 for copolarized and cross-polarized signals. Plot the difference pattern for an amplitude monopulse tracking system. Will the monopulse tracking system be stable for cross-polarized signals?

```
% Cross Polarizes Monopulse Antenna Pattern
% ----------------------------------------
% xpol.m

clear,clc,clf;

% Normalized Aperture Width

na=4;

% Sampling Freqeuny=Number elements per norm aperture

fs=8;

% Norm aperture with N elements

N=fs*na;
xna=na*(-1/2:1/(N-1):1/2);

% Illumination Function

wxna(1:N/2)=ones(1,N/2);
wxna(N/2+1:N)=-ones(1,N/2);
wxnb(1:N/2)=ones(1,N/2);
wxnb(N/2+1:N)=ones(1,N/2);
```

```
% Fill with M/2 zeros front and back

M=1024;
xna=na*(-1/2:1/N+M-1:1/2);
wxna=[zeros(1,M/2) wxna zeros(1,M/2)];
wxnb=[zeros(1,M/2) wxnb zeros(1,M/2)];

% Beam Functions from -fs/2 to fs/2 in sine space

Nfft=length(wxna);
Esine=fft(wxna,Nfft);
Esine=fftshift(Esine);

Esum=fft(wxnb);
Esum=fftshift(Esum);

% Azimuth vector

sinfi=fs/4*(-Nfft/2:Nfft/2-1)/Nfft;

% Azimuth vector in radians

fi=asin(sinfi);

% Beam gain functions

Gfi= Esine.*conj(Esine)/Nfft;
Gfs=Esum.*conj(Esum)/Nfft;

Gfi(1:Nfft/2)=sqrt(Gfi(1:Nfft/2));
Gfi(Nfft/2+1:Nfft)=-sqrt(Gfi(Nfft/2+1:Nfft));
Gfs=sqrt(Gfs);

% Approximate Cross Pol Pattern

Gpc=[Gfi(1:528) -Gfi(529:1056)];

% Form Cross Pol Sum and Difference Patterns

AA=[ zeros(1,15) Gpc(1:1041)];
BB=[ Gpc(16:1056) zeros(1,15) ];

Gdif=BB-AA;
Gsum=BB+AA;

% Plot Co- and Cross Pol Paterns

subplot(2,1,1);
plot(fi,Gfs,fi,Gpc);grid;
axis([ -.25 .25 0 1 ]);
title('Co-Polarized and Cross-Polarized Patterns');
```

```
ylabel('Amplitude');
xlabel('Angle - radians');
text(.025,.9,'Co-Polarized');
text(.085,.5,'Cross-Polarized');

% Plot Co- and Cross Pol Difference Patterns

subplot(2,1,2);
plot(fi,Gfi,fi,Gdif);grid;
axis([ -.25 .25 -1 1]);
title('Co- and Cross-Polarized Monopulse Difference Patterns');
ylabel('Amplitude');
xlabel('Angle - radians');
text(.04,.6,'Cross-Polarized');
text(-.075,.8,'Co-Polarized');
```

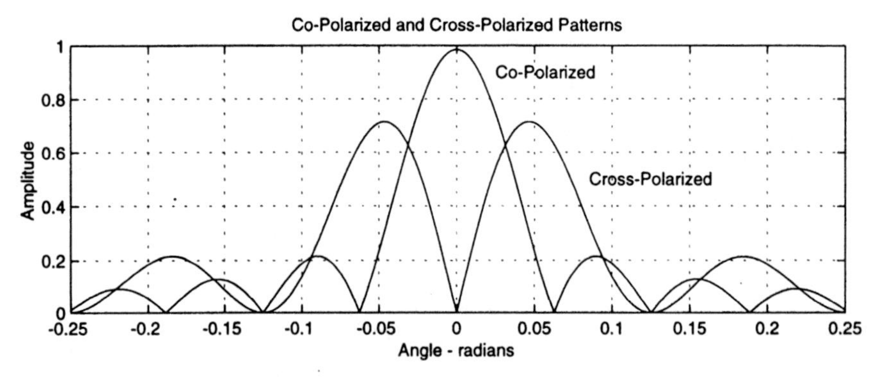

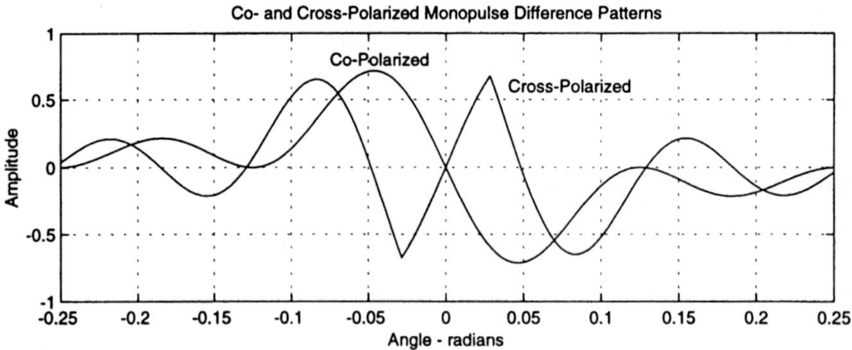

Cross-polarization jamming can be defeated by employing a radar antenna that is not susceptible to cross-polarized jamming signals. Examples of this type of antenna are flat-plate planar arrays and conventional antennas that use a polarization screen to prevent entry of the cross-polarization component. On the other hand, many radars use radomes, which tend to generate high cross-polarization components, making them susceptible to cross-polarization jamming.

A more fundamental approach to jamming a monopulse radar, or in general, any angle-tracking radar including sequential-lobing types, is to use spatially dispersed jamming sources. The phase characteristics of the spatially dispersed jammers can be either coherently or incoherently related. Coherent jamming sources imply a deterministic or synchronized relationship between the phases of the multiple jammers, while incoherent jamming sources imply a random relationship. Coherent jammers have the unusual property that the jammer's apparent power centroid can lie in a direction that is outside the solid angle containing the jammers, while the power centroid for incoherent jammers must always lie within this solid angle.

When the jamming sources are constrained to a single platform (e.g., aircraft or ship), the solid angle subtended by the jammers is necessarily small, and coherent jamming is necessary to generate a large angular error. On the other hand, if the jamming sources are located on multiple spatially dispersed platforms, then the subtended solid angle containing the jammers is large and incoherent jamming can be used to generate large angular errors. When spatially dispersed jammers are used, only those that intercept the main beam of the tracking radar are effective.

Consider the jamming of a tracking radar by two spatially separated incoherent jammers located within the radar's main-lobe beamwidth. Under the assumption that the tracking radar's angle-error discriminator has a linear characteristic, it can be shown that the angular tracking error (θ_e) from the midpoint of the sources is given by $\theta_e = \Delta\theta \cdot (\alpha^2 - 1)/2(\alpha^2 + 1)$, where α is the relative voltage amplitude ratio of the two jammers and $\Delta\theta = L \cos \psi/R$ with L the jammer separation, R the range between the tracking radar and the midpoint of the line between the jammers, and ψ the angle between the radar pointing axis and a perpendicular to the midpoint of the line between the two jammers. This relationship indicates that if the incoherent jammer sources have equal value, the tracking radar will point at the midpoint between the sources. If one jamming source is stronger than the other, the tracking radar antenna will tend to point in a direction that is closer to the stronger source.

One form of multiple-source incoherent jamming is called formation jamming. With this type of jamming, two or more aircraft or other jamming sources (e.g., decoys or multiple reflectors) are located within the beam of the monopulse tracking radar in the same range cell. As the aspect angles to these jammers change and the various jammer signal strengths change, then the apparent direction of arrival of the composite signal will wander back and forth between the jammers. If the composite jamming signals are synchronized to arrive simultaneously (within a range resolution cell) at the tracking radar and are sufficiently strong with respect to the skin returns from the targets carrying the jammers, then the real targets will be obscured from the tracking radar. Once separation between targets exceeds the radar beamwidth, the radar can track either target and the jammer becomes a beacon. Thus, careful coordination between the multiple jammers is required for successful formation jamming.

Another form of multiple-source incoherent jamming is called blinking jamming. This method attempts to attack the tracking dynamics of the angle-tracking radar and hence may be effective against some types of monopulse radar as well as other tracking radars. Basically, this method turns on multiple spatially dispersed jammers one at a time at a rate that is within the passband of the angle-tracking servo (e.g., 0.1 Hz to 10 Hz). As the tracking radar transfers from one jamming source to another, the step response of its angle-tracking servo is excited. If the angle-tracking servo is designed to be critically damped, then the radar antenna will smoothly move between the various jamming sources with a settling time proportional to the reciprocal of the servo bandwidth. However, if the angle tracker exhibits high overshoot (underdamped design), then exciting its servo at a rate close to its bandwidth will cause an ever-increasing perturbation, which will eventually result in the radar's breaking lock. Conversely, if the blinking rate is much higher than the angle tracker's servo bandwidth, then the angle servo will tend to average the signals from the various jamming sources, resulting in the radar antenna taking a position that corresponds to the angular center of mass of the jammers within its main-lobe antenna response.

The critical parameter with blinking jamming is the rate at which the jamming sources are commutated. Too high a rate causes the tracking radar to average the data and tends to minimize the tracking error. Too low a rate allows the tracker to accurately determine the angular position of each of the jammers. The best situation is a rate that is of the order of the tracking servo's bandwidth (e.g., 0.1 Hz to 10 Hz). However, it is not a simple problem for the jammer to attain the optimal blinking rate. One possibility is to vary the blinking rate over the range of possible values until maximum tracking error is observed at the jammer. This type of jamming is similar to that for swept audio jamming against conically scanning radar.

Coherent jamming is best performed from two synchronized sources separated by a baseline L in an interferometric configuration as depicted in Figure 4.9. From Example 4.17, it is recognized that the difference pattern Δ is approximately linear to about the 3 dB points in the sum pattern. Using this approximation, the angles measured by the monopulse tracker (θ_1, θ_2) to the two coherent jamming sources are given by

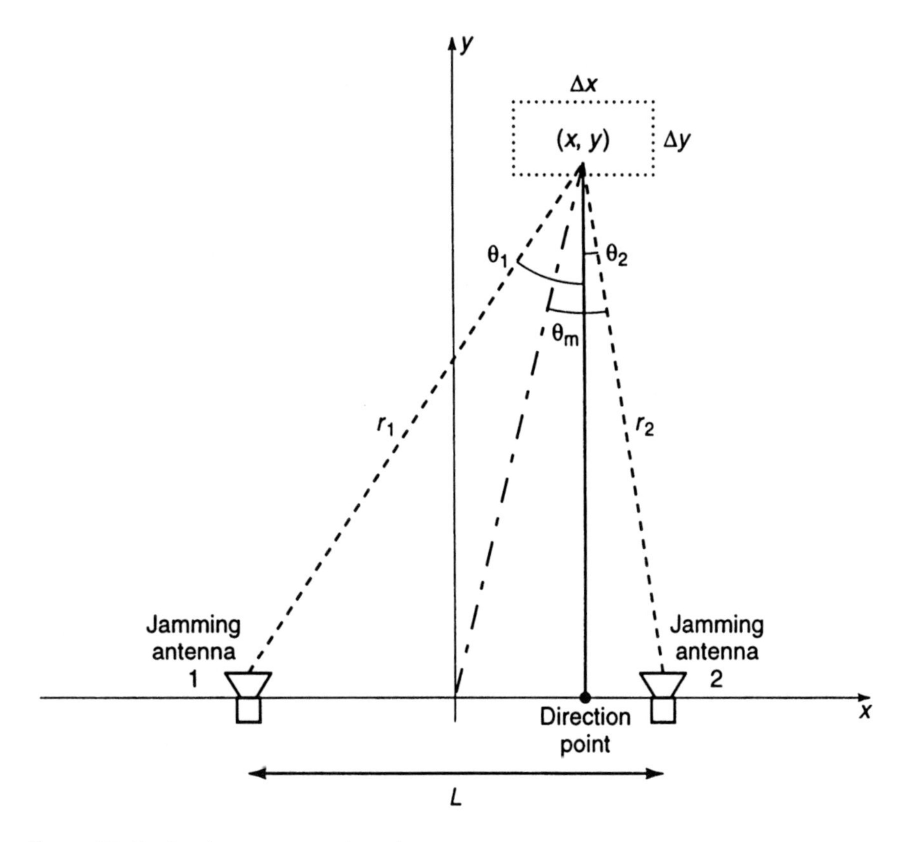

Figure 4.9 Dual-coherent source jamming.

$$\begin{aligned}\Delta_1 &= km\ \theta_1\Sigma_1 \\ \Delta_2 &= km\ \theta_2\Sigma_2\end{aligned} \tag{4.24}$$

where *km* is a scale factor. The indicated tracking angle (θ_i) of the monopulse tracker is then equal to

$$\theta_i = \frac{1}{km}\frac{\Delta}{\Sigma} = \frac{\Delta_1 + \Delta_2}{\Sigma_1 + \Sigma_2} = \frac{\theta_1\Sigma_1 + \theta_2\Sigma_2}{\Sigma_1 + \Sigma_2} \tag{4.25}$$

The ratio Σ_2/Σ_1 can be written in complex form as

$$\frac{\Sigma_2}{\Sigma_1} = ae^{j\phi} \tag{4.26}$$

where a is the amplitude ratio and ϕ is the phase between the jamming signals received from the two coherent jamming sources. The indicated tracking angle can then be written as

$$\theta_i = \frac{\theta_1 + ae^{j\phi}\theta_2}{1 + a} \tag{4.27}$$

Note that at $a = 0$ (jamming source two shut-off) $\theta_i = \theta_1$ as expected. Multiplying the second term by $(1 - ae^{j\phi})/(1 - ae^{j\phi})$ and applying Euler's formulas results in

$$\theta_i = \theta_m - \frac{\Delta\theta}{2}\frac{1 - 2ja \sin \phi - a^2}{1 + 2a \cos \phi + a^2} \tag{4.28}$$

for the indicated angle. The real part of (4.28) indicates the indicated tracking angle of a monopulse radar subjected to dual coherent source jamming as

$$R_e(\theta_i) = \theta_m - \frac{\Delta\theta}{2}\frac{1 - a^2}{1 + 2a \cos \phi + a^2} \tag{4.29}$$

The second part of (4.94) represents the miss angle (θ_{miss}), so the miss distance is given by

$$r_{\text{miss}} = R \cdot \tan\left[\frac{\Delta\theta}{2}\frac{1 - a^2}{1 + 2a \cos \phi + a^2}\right] \tag{4.30}$$

Using the small angle approximation to tan(·) and recognizing that $\Delta\theta = L \cos \psi / R$ results in a miss distance

$$r_{\text{miss}} = \frac{L \cos \psi}{2}\frac{1 - a^2}{1 + 2a \cos \phi + a^2} \tag{4.31}$$

where L is the separation between dual coherent jamming sources, a is the ratio of the amplitude of the jamming sources, ϕ is the phase difference between the jamming sources, and ψ is the angle between the radar pointing axis and a perpendicular to the midpoint of the line between the jammers.

▶ *Example 4.19*

Using MATLAB, plot (4.31) with phase (ϕ) as the abscissa and miss distance normalized to the source separation (L) as the ordinate for values of jamming amplitude ratios $a = 0.4$ to 2.5. Assume $\psi = 0$. What values of a and ϕ result in the maximum jamming effect?

```
% Normalized Miss Distance for Crosseye
% -------------------------------------
% missdis.m

clear;clc;clf;

% Input Phase Error

phi=0:.1:360; % degrees

% Input Dual Source Amplitude Ratio

for i=1:13;
a=.4+.1*(i-1);

% Compute Miss Distance

l=cos(phi*pi/180+1e-3);
r=(1-a^2)./(2*(1+2*l*a+a^2));
if a==1;r(1800)=15;end

% Plot Miss Distance

rr(:,i)=r';end;
plot(phi,rr,'k');grid;
A=max(max(rr))+1;
axis([0 360 -A A]);
title('Normalized Miss Distance for Crosseye');
ylabel('Miss Distance - r / L');
xlabel('Dual Source Phase Mismatch - degrees');
```

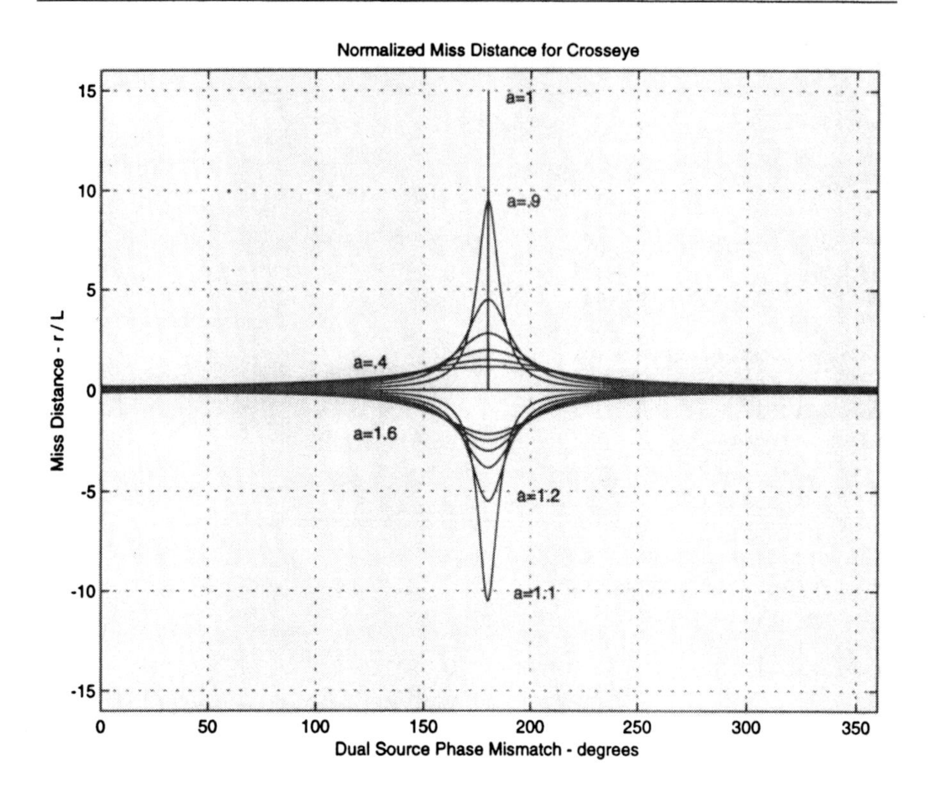

Considerable insight into the conditions for maximum jamming effectiveness can be gained by an examination of the relationship given by (4.31). First, the numerator contains a zero singularity at $a = 1$, which can only be removed by an identical singularity in the denominator. This is accomplished by setting $\phi = 180$ degrees, resulting in θ_e being proportional to $(1 + a)/(1 - a)$, which in turn is maximized for $a = 1$. Thus, the conditions that result in maximum jamming effectiveness are that the jamming sources have the same amplitude (power) and a phase separation of 180 degrees. The relationship indicates that these conditions will produce an infinite error. However, the equation applies only over the linear range of the angle-error discriminator, which generally extends only to within the radar's 3-dB beamwidth. Beyond the linear range, the angle-error discriminator saturates, which limits the angular error produced by this technique to within the radar's 3-dB beamwidth.

Now let us examine the first term in the expression for the angle error that provides insight into the sensitivity of coherent source angle jamming. The numerator is directly proportional to the interferometer source space (L) and the cosine of the angle (ψ) between a perpendicular to the interferometer baseline and the radar tracking axis. The numerator can be increased by moving the jamming sources further apart (hence, L is increased), which is the reason that jamming sources for coherent angle jamming are located on the wingtips for aircraft installations, or on the fore and aft and port and starboard portions of naval ships. The numerator reaches its maximum value when $\psi = 0$, indicating that the jamming is most effective when the tracking radar is located broadside to the interferometric baseline. Alternatively, the jamming is ineffective when the jamming sources are in a parallel line with the tracking radar axis ($\psi = 90$ degrees). The denominator is proportional to the range (R) between the jammer and the tracking radar. This indicates that the jamming becomes increasingly effective as the range decreases, which contradicts the general expectation for most jamming situations.

Although the angular error expression gives insight into the conditions and sensitivities for maximum jamming effectiveness, it gives little insight into the particular physical phenomenon that allows large angular errors to be generated by coherent source jamming. The physical reasons are apparently related to a sharp wavefront distortion that occurs in the vicinity of interferometric nulls. Photographs of this type of distortion have been synthesized on a computer and obtained experimentally [7]. The large angular tracking error is caused as the tracking radar attempts to align itself on a perpendicular to the highly distorted wavefront.

Coherent source jamming imposes severe requirements both on the amplitude match between the sources and maintaining a phase differential of 180 degrees. To examine these sensitivities, we can separate the fundamental angle-error relationship into two terms. The first term is given by $\Delta\theta = L \cos \psi / R$, which represents the projected angle (in radians) subtended by the jamming sources as seen by the radar. The second term is a magnification factor given by $\Delta/(\Delta^2 + \delta^2)$, where $\Delta = 1 - a$ and $\delta = 180$ degrees $- \phi$ represent the deviations from the desired design values. This expression was obtained by neglecting higher order terms in both the numerator and denominator. An examination of this factor indicates that the maximum magnification is given by $1/(1 - a)$.

Consider an example of an airborne coherent source jammer whose wingtip installation is separated by 20 ft while the range is 10 nm. The jamming sources then subtend an angle of 0.33 mrad. If an angular tracking error of one degree is desired, then a magnification factor of 53 is required. This can be obtained with various combinations of a and ϕ, but assume that $a = 0.99$ is a practical value for the amplitude match. A maximum magnification of 100 is obtained when $\delta = 0$. Then it is found that a phase match of 0.54 degrees is required to obtain the desired magnification of 53. This example illustrates the high phase and amplitude stabilities required to obtain large magnification factors.

A major impediment to preserving phase coherency in dual coherent source jamming systems is the final output, high-powered TWT. The TWT *amplitude-to-phase modulation* (AM/PM) characteristic is typically 1 to 3 degrees per dB. Therefore, as much as 45 degrees of phase shift may be obtained if the input signal varies over a 15-dB dynamic range. This inherent TWT limitation potentially can be reduced by preceding the TWT with a limiter that compresses the 15 dB to a few tenths of a decibel. Limiters in I and J bands that operate over octave bandwidths and have the necessary AM/PM characteristics are not currently available. As an alternative, limiters can be implemented at lower frequencies over minimum bandwidths of several gigahertz and translated to the higher operating frequency using mixing techniques.

The practical implementation of coherent source angle jamming uses a technique called cross-eye. A block diagram of this technique is depicted in Figure 4.10. Also shown is the geometry for an off-center-line jamming situation that motivates this particular implementation. The cross-eye technique uses two separate repeater paths. Each path contains receiving and transmitting antennas, a transmission line to connect the antennas, and an amplifier to generate the jamming signal. In addition, one path contains a 180-degree phase shifter that generates an interferometer null in the direction of arrival of the victim radar's signal. Also, phase and amplitude controls are contained in one path so that the two repeater paths can be adjusted for phase and amplitude match.

The advantage of the cross-eye implementation is that this configuration ensures that the signals radiated by the two coherent jamming sources arrive

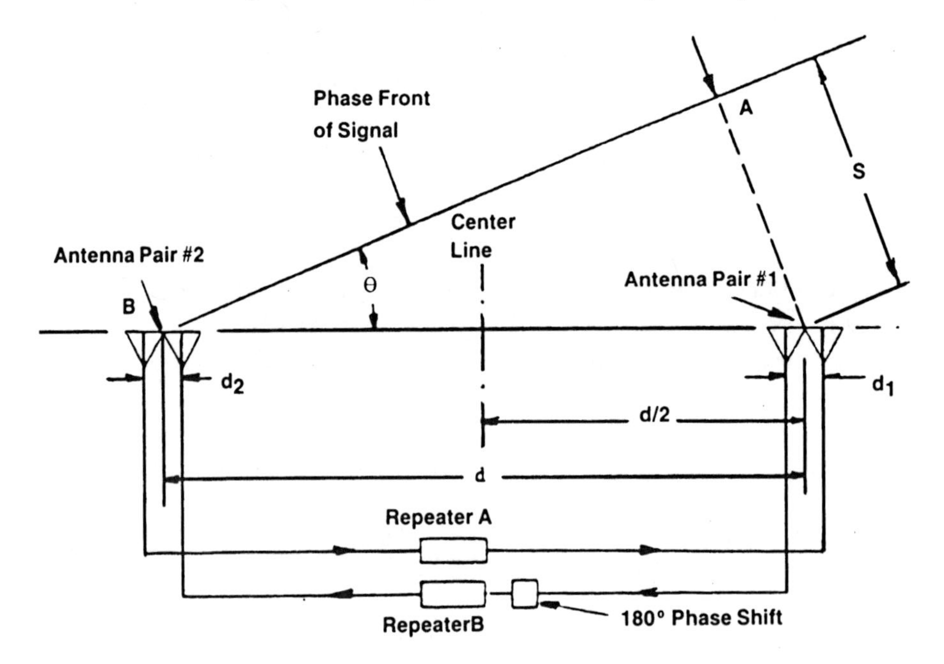

Figure 4.10 Cross-eye repeater jammer. (*Source:* [3].)

at the victim radar matched in amplitude and 180 degrees out of phase, independent of the angle of arrival of the victim radar's signal at the jammer. If the crossed configuration were not used, then a differential phase shift equal to $L \sin \psi / \lambda$ would exist between the jamming signals received at the victim radar, even though the jamming signals were matched in amplitude and 180 degrees out of phase at the jammer. This differential phase shift is caused by the additional path length (S) experienced by one jamming signal with respect to the other for off-axis signals. In the previous example, if the victim radar were at a frequency of 10 GHz, an aircraft yaw of 0.1 degree would cause a differential phase shift of about 20 degrees. If this were not compensated, the magnification factor would be reduced to a negligible amount. With the cross-eye configuration as depicted in Figure 4.10, it can be seen that both receiving and transmitting phase fronts at the jammer are parallel, thereby providing compensation for the differential phase shift caused by off-axis operation. The price paid for this compensation is an additional time delay (1 ns per foot of baseline) in the repeater path that enhances the possibility that the radar may be able to track the leading edge of the real target signal before the jammer signal arrives.

Another factor that complicates cross-eye jamming is that successful operation creates an interferometric null between the jamming signals in the direction of the victim radar. The jamming signals must compete with the real target return to capture the radar's angle tracker. To accomplish this, the angle noise caused by the real radar target must perturb the radar's antenna off the jamming signal null by an amount sufficient for a positive jamming-to-signal ratio to be generated. This results in a jamming-to-signal ratio requirement of at least 20 dB for successful cross-eye operation.

► *Example 4.20*

Given dual coherent jamming sources separated by 20m, plot the field points and phase front distortion for cross-eye jamming. Use a frequency of 10 GHz and a relative source amplitude ratio of 0.8. Note that the monopulse tracking antenna attempts to align itself perpendicularly to the phase front. What tracking error is indicated by the phase front plots?

```
% Field Pattern for Cross-Eye Jamming
% ------------------------------------
% crosseye.m

clf;clc;clear;

% System parameters

k=.8;

f0=9e+9;c=3e+8;t=0;

P1=400;G1=100;
P2=P1/k;G2=100;

% Antenna separation

a=20;

x1=-a/2;y1=0;
x2=a/2;y2=0;

% Field extent (dx,dy) about (x,y)

x=0;y=2000;
dx=8;dy=.1;

% Number grid points along axis

n=50;

Ex=[];E=[];

% Define loops

for yf=y-dy/2:dy/(n-1):y+dy/2+dy/(n-1)/2;
   for xf=x-dx/2:dx/(n-1):x+dx/2+dx/(n-1)/2;

% Compute magnitudes of sinal vectors

r1=sqrt((x1-xf)^2+(y1-yf)^2);
r2=sqrt((x2-xf)^2+(y2-yf)^2);

% Compute field density at (xf,yf);

p1=P1*G1/4/pi/(r1^2);
p2=P2*G2/4/pi/(r2^2);

% Compute field strengths at (xf,yf);

E1=sqrt(277*p1);
```

```
E2=sqrt(277*p2);

% Compute field strength with 180 deg phase

e1=E1*sin(2*pi*f0*(t-r1/c)+pi);

% Compute field strength at (xf,yf)

e2=E2*sin(2*pi*f0*(t-r2/c));

% Combine vector fields at (xf,yf)

e=e1+e2;

% Matrix in x-direction

Ex=[Ex;e];

end;

Ext=Ex';
Ex=[];

% Matrix in y-direction

E=[Ext;E];

end;

% Plot sum fields

kx=10*log10(k);
subplot(211),mesh(E,[-45 45],[1 1 1/2]);
title(['Jamming sep=20 m , rel eff=',num2str(kx),...
'dB ,wl=3 cm,  phase front at 2 km']);
 xlabel('x');ylabel('y');zlabel('z');

subplot(212),contour(E,1);
title(['Jamming sep=20 m , rel eff=',num2str(kx),...
'db ,wl=3 cm, phase front at 2 km']);
xlabel(['Total ',num2str(dx),' m']);
ylabel(['Total  ',num2str(dy*100),' cm']);
```

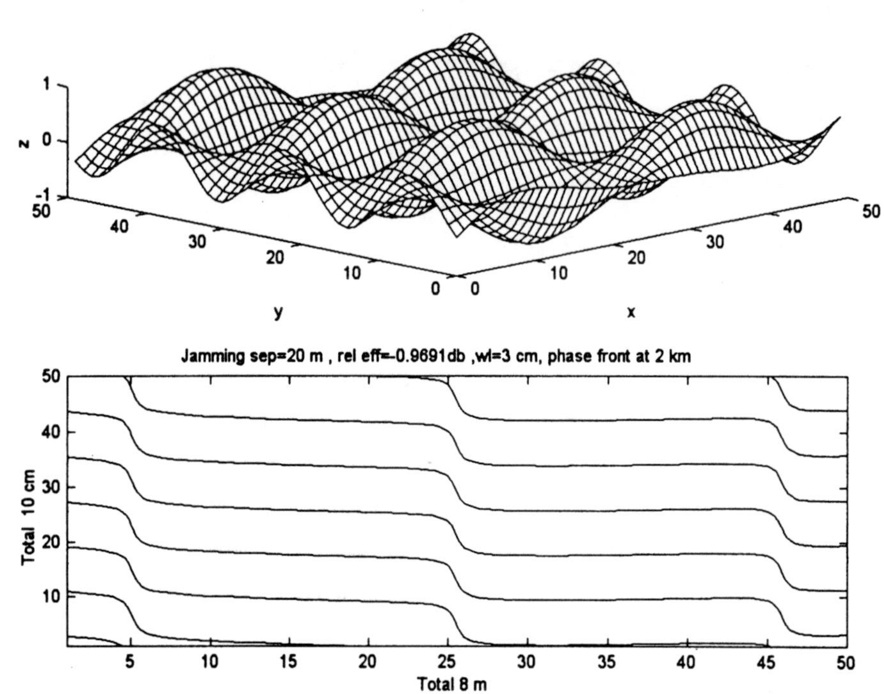

Another monopulse jamming technique is called terrain or ground bounce. The basic concept of this technique is depicted in Figure 4.11 for an engagement involving a jammer aircraft and a semiactive missile. Terrain bounce is used in the elevation plane to illuminate the Earth's surface in front of and below the jammer aircraft so that the semiactive missile angle tracker homes on the illuminated ground spot and not the jammer aircraft. The requirements

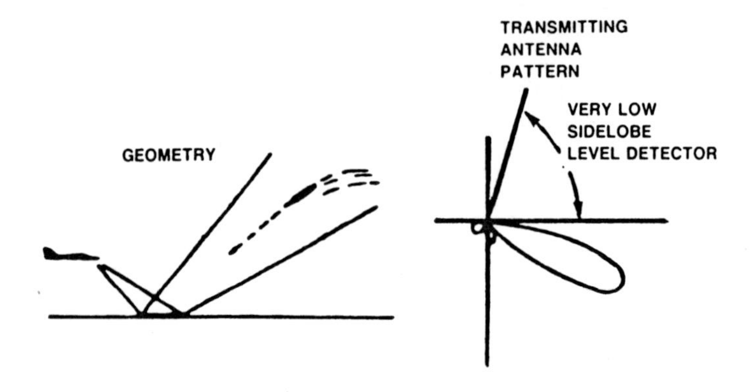

Figure 4.11 Terrain bounce jamming. (*Source:* [3].)

for terrain bounce jamming include a narrow elevation plane beamwidth, a broad azimuth beamwidth to extend the jamming coverage sector, high RF power to overcome the losses associated with the terrain propagation path, and very low sidelobes at the horizon and above to prevent the missile from beaconing on the jammer.

In addition to the terrain bounce operation depicted in Figure 4.11, it is also possible for low-flying aircraft to jam ground radar using multipath reflections from the ground. One method of accomplishing this type of operation is to form an interferometric null on the ground radar in the elevation plane in a similar manner to that used with the cross-eye technique. This prevents the ground radar from beaconing on the jammer and provides a ground bounce signal for the radar to track.

4.4 Coherent Sidelobe Cancelers

The control of sidelobe noise jamming using ultralow sidelobe antennas is not usually sufficient to completely protect a surveillance radar against sidelobe jamming. In addition, most operational radar do not use ultralow (less than –40 dB) or low (–30 dB to –40 dB) sidelobe antennas and have antenna sidelobes in the −13- to −30-dB region with average sidelobes of 0 dBi to –5 dBi. SLC is a coherent processing technique that has the potential of reducing noise jamming through the antenna sidelobes and is employed in a number of operational radars for this purpose. Present SLCs have the capability of reducing sidelobe noise jamming by 20 dB to 30 dB, but their theoretical performance is potentially much higher.

The operation of a sidelobe canceller is best understood by relating its operation to that of an adaptive antenna array processor. In an adaptive array, a weighting factor (phase and amplitude) is applied to each of the antenna elements, which are then combined in a summing network. By controlling the weighting factors through an adaptive-loop antenna pattern, nulls can be generated in any direction due to the interferometric action between any two sets of elements. Thus, the number of nulls that can be formed is equal to the number of elements minus one, which is sometimes referred to as the number of degrees of freedom of the array. The same adaptive array action can be applied to groups of elements combined into subarrays, which reduces the processing requirements for a phased array to a reasonable size. Thus, an adaptive antenna array has the potential of placing antenna pattern nulls in the direction of sidelobe jammers while still maintaining the main-lobe pattern, thereby reducing the effects of the jammer at the output port of the antenna.

While adaptive arrays are applicable to phased arrays, they are not appropriate for conventional single-element antennas. However, by adding auxiliary antennas to a conventional radar, an adaptive array type of action can be formed between the main antenna and the added antennas. This configuration is called a SLC, and the only requirement is that the auxiliary antennas must have a greater response in the direction of the jammer than the sidelobe response of the main antenna in that direction.

One of the limitations of the SLC is that the number of degrees of freedom is usually low, since only a small number of auxiliary antennas can be practically added to the main antenna (e.g., Patriot, five SLC arrays; AEGIS, six SLC arrays). Because the maximum number of sidelobe jammers that can be handled is equal to the number of auxiliary antennas, the cancellation system is easily saturated. This problem is compounded by any jammer multipath from objects in the proximity of the radar that add an additional degree of freedom for each multipath signal that has an angular direction significantly different than that of the main jammer. Another complication occurs for antennas whose cross-polarization response is significantly different than its main polarization response. This causes two orthogonally polarized auxiliary antennas to be added for each degree of freedom of the SLC system.

Sidelobe cancellation loops are best suited to cancel narrowband jamming signals. When wideband jamming signals (with respect to the radar's receiving channels) impinge on the antenna array (main and auxiliary antennas) at angles other than normal to the array, the signal appears to span an arc in angle. Thus, a single null is not adequate to cancel a wideband jammer and closely spaced multiple nulls are required in this case. Typically then, the number of sidelobe cancellation loops might be specified at up to three times the number of expected jamming sources [9]. The implementation of more than three or four cancellation loops presents a difficult operational design problem [10].

To understand the detailed operation of a CSLC, consider the block diagram depicted in Figure 4.12 [11]. Targets are received in the main antenna lobe, while jamming interference is received in the antenna's sidelobe response. Jamming signals are also intercepted in a number of auxiliary antennas whose gain in the direction of the jammers exceeds that of the sidelobe gain of the main antenna. It is further assumed that the target signals intercepted in the auxiliary antennas are insignificant compared to the jammer's strength. This is generally a good assumption, but in some implementations a directive auxiliary antenna is employed to enhance the jamming signals.

The jamming signals intercepted in each auxiliary antenna are weighted in amplitude and phase (i.e., complex weights) to form a sum that is then subtracted from the signal in the main antenna. The weights are controlled by an adaptive processor to minimize the jammer signal power at the output

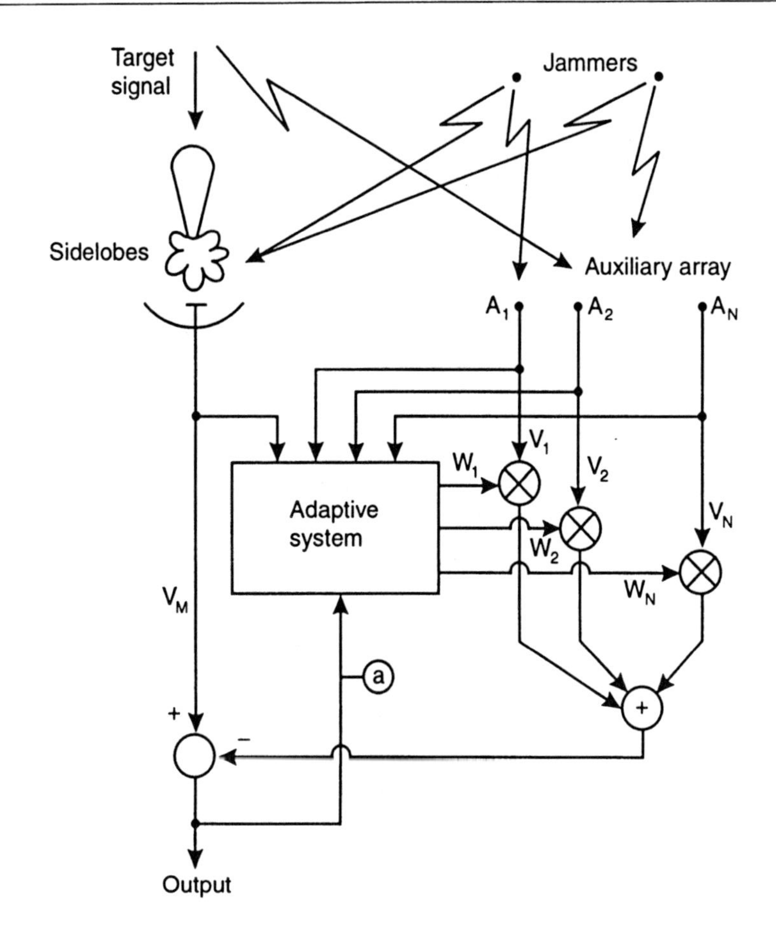

Figure 4.12 Coherent sidelobe canceller block diagram. (Farina, A., *Electronic Counter-Countermeasures,* © 1990 McGraw-Hill.)

of the system. In effect, if the CSLC performs optimally, it will adjust its weights so that the jammer signals in the main antenna are completely canceled by the jamming signals received in the auxiliary antenna.

In the absence of receiver noise, the signal at the output of the CSLC is composed of three components

$$\mathbf{Z} = \mathbf{S} + \mathbf{J} - \mathbf{J}_a^T \mathbf{W} \tag{4.32}$$

where **S** is the signal vector, **J**` is the jamming signal vector, $\mathbf{J}_a^T = (J_{a1}, J_{a2}, \ldots, J_{ak})$ is the jamming signal in the auxiliary antennas, and $\mathbf{W} = (w_1, w_2, \ldots, w_k)$ are the complex weights. The CSLC power output is given by

$$P_z = E(Z^2) = P_S + P_j - 2\mathbf{P}^T\mathbf{W} + \mathbf{W}^T\mathbf{R}\,\mathbf{W} \tag{4.33}$$

where P_S is the signal power, P_j is the jamming power in the main lobe, $\mathbf{P}^T = E(\mathbf{J}\,\mathbf{J}_a^T)$, and $\mathbf{R} = E(\mathbf{J}_a\,\mathbf{J}_a^T)$ is the jammer's covariance matrix.

Using the orthogonality principle, the optimal weights result when the data is orthogonal with the error

$$E(\mathbf{Z} \cdot \mathbf{J}_a) = E((\mathbf{S} + \mathbf{J} - \mathbf{J}_a^T\,\mathbf{W}) \cdot \mathbf{J}_a) = 0 \tag{4.34}$$

Performing the indicated operations and noting that the signal and jammer signals are uncorrelated results in

$$\mathbf{P} - \mathbf{R}\,\mathbf{W} = 0 \tag{4.35}$$

The optimal weights for best cancellation are then

$$\mathbf{W}_0 = \mathbf{R}^{-1}\mathbf{P} \tag{4.36}$$

Substituting (4.36) into (4.33) results in

$$P_Z = P_S + P_j - \mathbf{P}^T\mathbf{W} \tag{4.37}$$

The jammer residual power at the CSLC is given by $\Delta J = P_j - \mathbf{J}_A^T\,\mathbf{W}$, so the *cancellation ratio* (CR) equals

$$CR = \frac{P_j}{P_j - \mathbf{P}^T\mathbf{W}} \tag{4.38}$$

If $P_j = \mathbf{P}^T\mathbf{W}$, the cancellation is perfect.

When noise is present, the CSLC output signal becomes

$$\mathbf{Z} = \mathbf{S} + (\mathbf{J} + \mathbf{N}) - (\mathbf{J}_a + \mathbf{N}_a)^T\mathbf{W} \tag{4.39}$$

where $\mathbf{N}$ is the main and $\mathbf{N}_a$ the auxiliary channel noise vectors. Applying the orthogonality theorem to determine the optimal weights results in

$$\mathbf{W}_0 = (\mathbf{R} + \sigma_n^2\,\mathbf{I})^{-1}\mathbf{P} \tag{4.40}$$

and the cancellation ratio becomes

$$CR_n = \frac{P_j}{P_j - \mathbf{P}^T(\mathbf{R} + \sigma_n^2\mathbf{I})^{-1}\mathbf{P}} \tag{4.41}$$

where σ_n^2 is the receiver noise power in the main channel and **I** is the identity matrix.

For a single auxiliary antenna CSLC, the covariance matrix for no noise degenerates to the scalar

$$R = E(J_1^2) = \sigma_j^2 \tag{4.42}$$

where σ_j^2 is the jammer's power. We assume a Gaussian correlation function given by

$$\rho = \exp(-(\Delta t/\tau_c)^2) \tag{4.43}$$

where Δt is the differential time between the jamming signal in the main and auxiliary channel and $\tau_c = 1/B_c$ where B_c is the channel's bandwidth. The differential time of arrival Δt is given by

$$\Delta t = \frac{d \sin \theta}{c} \tag{4.44}$$

where d is the distance between the main and auxiliary antennas and θ is the angle the jammer makes with respect to a perpendicular from a line connecting the two antennas. The cross-correlation $\mathbf{P}^T$ between the main and auxiliary jamming signals is then found as

$$P = E[\mathbf{J}\,\mathbf{J}_1] = \rho\, \sigma_j^2 \tag{4.45}$$

The optimal weight is given by

$$\mathbf{W}_0 = \mathbf{R}^{-1}\mathbf{P} = \rho/\sigma_j^2 \tag{4.46}$$

resulting in a cancellation ratio for a single CSLC with no noise of

$$CR_1 = \frac{\sigma_j^2}{\sigma_j^2 - \sigma_j^2 \rho^2} = \frac{1}{1 - \rho^2} \tag{4.47}$$

Table 4.1 tabulates the single CSLC cancellation ratio without receiver noise for various values of the correlation coefficient ρ.

When receiver noise (σ_n^2) is present in the single CSLC, then $R = \sigma_j^2 + \sigma_n^2$ and $P = \rho\sigma_j^2$, resulting in an optimal weight of

Table 4.1
Single CSLC Cancellation Ratios for Various Correlation Coefficient Values (No Noise)

ρ	CR_1 (dB)
0.9	7.2
0.99	17
0.999	27
0.9999	37

$$W_0 = \frac{\rho \sigma_j^2}{\sigma_j^2 + \sigma_n^2} = \alpha \rho \tag{4.48}$$

where

$$\alpha = \frac{\sigma_j^2}{\sigma_j^2 + \sigma_n^2} = \frac{JNR}{1 + JNR} \tag{4.49}$$

The cancellation ratio then becomes

$$CR_1 = \frac{\sigma_j^2}{\sigma_j^2 - \sigma_j^2 \alpha \rho^2} = \frac{1}{1 - \alpha \rho^2} \tag{4.50}$$

Table 4.2 tabulates the single CSLC cancellation ratios with receiver noise for various values of the correlation coefficient ρ.

An examination of Table 4.2 indicates that jamming signals can only be canceled using CSLCs to the noise level when a perfect correlation between

Table 4.2
Single CSLC Cancellation Ratios for Various Correlation Coefficient Values
($JNR = \sigma_j^2 / \sigma_n^2$)

	CR (dB) for JNR =				
ρ	10 dB	20 dB	30 dB	40 dB	∞
0.9	5.8	7.0	7.2	7.2	7.2
0.99	9.6	15.3	16.8	17	17
0.999	10.3	19.3	25.2	26.8	27
1	10.4	20	30	40	∞

the jamming signals in the main and auxiliary channels exists. With imperfect correlation and high JNRs, the jamming signals will exceed the receiver noise level by a considerable amount.

Analysis of the performance for a double CSLC with receiver noise can be performed using the general formulations given in (4.39) to (4.41). The covariance matrix becomes

$$\mathbf{R} = (\sigma_j^2 + \sigma_n^2)^2 \begin{bmatrix} 1 & \alpha\rho^4 \\ \alpha\rho^4 & 1 \end{bmatrix} \tag{4.51}$$

while the cross-correlation becomes

$$\mathbf{P}^T = \sigma_j^2 [\rho \ \rho] \tag{4.52}$$

The resulting optimal weights are

$$\mathbf{W}_0^T = \begin{bmatrix} \dfrac{\alpha\rho}{1 + \alpha\rho^4} & \dfrac{\alpha\rho}{1 + \alpha\rho^4} \end{bmatrix} \tag{4.53}$$

The cancellation ratio can be found as

$$CR_2 = \frac{1 + \alpha\rho^4}{1 + \alpha\rho^4 - 2\alpha\rho^2} \tag{4.54}$$

Table 4.3 tabulates the double CSLC cancellation ratios with receiver noise for various values of the correlation coefficient ρ. It indicates that increased cancellation ratios are available with a double CSLC as compared to a single CSLC.

Table 4.3
Double CSLC Cancellation Ratios for Various Correlation Coefficient Values ($JNR = \sigma_j^2 / \sigma_n^2$)

	CR_2 (dB) for JNR =				
ρ	10 dB	20 dB	30 dB	40 dB	∞
0.9	11.1	15.6	16.5	16.6	16.6
0.99	13.1	22.7	31.5	36	36.9
0.999	13.2	23.0	33	42.8	51
1	13.2	23.0	33	43	∞

Note that the preceding analysis assumes that optimum weights are achieved by the CSLC. When closed-loop CSLCs are employed, high loop gain is required to allow optimal weights to be achieved. Also, the settling time to achieve optimal weight settings are a function of the strength of the jamming waveforms.

▶ *Example 4.21*

Using MATLAB, plot the cancellation ratios achieved with single and double CSLC for various correlation coefficients (ρ = 0.9 to 0.9999) with no noise and a *JNR* = 10 dB to 50 dB.

```
% Cancellation Ratio for CSLC
% ---------------------------
% cslcpera.m

clear;clc:clf;

% Input Correlation Coefficients

e=1:5;
rhox=1-1./10.^e;

% Input Jam-to-Noise Ratio in DB

jnrdb=10:50;
jnr=10.^(jnrdb/10);

% Calculate Single Loop CR

for i=1:5;
rho=rhox(i);
a=jnr./(jnr+1);
cr1=1./(1-a*rho^2);
cr1db=10*log10(cr1);
CR1(:,i)=cr1db';
end;

% Plot Single Loop CR

subplot(1,2,1),
plot(jnrdb,CR1);grid;
title('CSLC Cancellation Ratio - Single Loop');
xlabel('Jam-to-Noise Ratio - dB');
ylabel('Cancellation Ratio - dB');

% Calculate Double Loop CR

for i=1:5;
rho=rhox(i);
```

```
a=jnr./(jnr+1);
cr2=(1+a*rho^4)./(1+a*rho^4-2*a*rho^2);
cr2db=10*log10(cr2); CR2(:,i)=cr2db';
end;

% Plot Double Loop CR

subplot(1,2,2),
plot(jnrdb,CR2);grid;
title('CSLC Cancellation Ratio - Double Loop');
xlabel('Jam-to-Noise Ratio - dB');
ylabel('Cancellation Ratio - dB');
```

4.4.1 EA Against Coherent Sidelobe Cancelers

The CSLC has a number of inherent defects that can be exploited by support jammers. First, as previously discussed, the number of degrees of freedom in a CSLC is generally limited. Thus, if multiple jamming signals can be induced into the jammer at various angles through, for example, multipath reflections, then the CSLC becomes overloaded and its performance becomes severely degraded.

Another factor that affects CSLC performance is any decorrelation that exists between the jamming signals observed in the main antenna channel and those in the auxiliary antenna channels. This decorrelation becomes prevalent when the bandwidths of the two channels are mismatched as generally occurs since the main channel is matched to receive the radar targets while the auxiliary channel is matched to receive the jamming waveform. An examination of Tables 4.1 through 4.3 shows that performance degrades rapidly when decorrelation occurs.

Perhaps the most vulnerable characteristic of CSLCs is their transient response. Most CSLCs are designed to counter CW jamming signals. As such, they employ servo loops with narrow bandwidths that take a long time to settle. Generally, CSLCs are supplemented with sidelobe blankers that function to blank pulse signals. However, the sidelobe blanker must be judiciously employed because if it functions against a CW signal, it would enhance the jamming by blanking all the targets. A jamming technique that is often effective against radar is the transmission of multiple synchronized false targets in the sidelobe region. The CSLC cannot respond to these targets because of its transient response; but if a sidelobe blanker is employed, it will blank any main-lobe target that corresponds to a false sidelobe target.

Another support jamming tactic that attempts to exploit the transient behavior of the CSLC is to employ blinking jamming. Blinking refers to the synchronized commutation of jamming transmissions among a group of spatially displaced jammers. When properly timed, the various closed loops associated with each auxiliary antenna in a CSLC will never settle to a steady-state condition, thus severely degrading its performance.

A more general vulnerability of the CSLC is the cross-polarization response of the main and auxiliary antennas. If these are not matched, as is generally the case, then the CSLC will not cancel the copolarized interference. This is of particular concern if the main antenna has a large copolarized response since most jammers employ either circular or slant polarized jamming radiations with large cross-polarized components. To protect against this situation, many radars employing CSLCs have both vertical and horizontal auxiliary antennas to allow for cancellation of both the vertical and horizontal components of the jamming signal.

Another EA tactic that has been suggested using cross-polarization jamming attacks the sidelobe blanking system that is generally associated with CSLC systems [12]. This main-lobe jamming technique generates alternating polarization jamming cover pulses that are switched within a radar pulse width. The cross-polarized response of the main antenna is generally nulled for a jamming signal in the direction of the antenna's boresight. The higher cross-polarization response in the auxiliary antenna will activate the sidelobe blanker, which will partially blank the real target return, thereby limiting detection.

4.5 Problems

1. A PC radar transmits a 40-μs linear FM pulse with 25-MHz bandwidth. The radar has a 5-dB noise figure and produces a 20-dB signal-to-noise ratio at the radar receiver output on a 10-m^2 target. What jammer power at the radar receiver output is required to produce a 0 dB jam-to-signal ratio? How does this compare to the jamming power that would be required if the linear FM chirp waveform is replaced with a conventional pulse of the same duration?

2. Assume that both the PC and conventional radar in Problem 1 employ pulse-to-pulse frequency agility with 200-MHz bandwidth. How do the relative jamming powers required for 0 dB jam-to-signal ratio now compare?

3. Consider a binary phase code modulated radar with 50-kW power output at 5.5 GHz, 36-dB antenna gain, 10-MHz chip bandwidth, and a PC ratio of 128. Further assume that a 200-MHz main beam broadband noise jammer provides 10-dB jam-to-signal ratio before the matched filter. What is the burn-through-range of the radar?

4. For the radar in Problem 3, assume that main beam support jamming with 200-W power and 10-dB antenna gain is located 60 km from the radar. What is the burnthrough range if CW jamming tuned to the carrier frequency of the radar is employed?

5. Using MATLAB, plot the ambiguity function for a 13-bit Barker-coded PC waveform.

6. A 10-GHz pulse Doppler radar transmits 3 kW of average power at a PRF of 300,000 pulses per second. Its antenna gain is 35 dB and it processes targets using a 2048-point FFT that employs Hamming weighting (processing loss is 3.1 dB, 3-dB bandwidth is 1.3 with respect to uniform weighting). The noise figure is 4 dB. Find the maximum range at which a 2-m^2 target produces a 15-dB signal-to-noise ratio assuming the only loss is the windowing loss.

7. A noise jammer with 10-MHz bandwidth is located on the target detected by the radar in Problem 6. What jammer ERP is required to reduce the detection range to 2 km (signal-to-jam ratio of 15 dB)? Assume that radar and jammer losses are equal except for the windowing loss and that the jammer has a 3-dB polarization loss.

8. A DRFM is used on the jammer of Problem 7 to concentrate jamming energy into the Doppler filter containing the target. What ERP is now required to reduce the detection range to 2 km? How does this compare with the ERP required by the noise jammer?

9. A 10-GHz aircraft radar illuminates a target for a semiactive radar homing missile. The radar illuminator transmits 4-kW CW and has 33-dB gain. The missile seeker uses a 50% efficient 6-in-diameter circular array, has 500-Hz effective bandwidth, and 9-dB receiver noise figure. If a 4-m^2 target is 20 nmi from the illuminator, find the range at which the missile can effectively track the target assuming a 20-dB signal-to-noise ratio is required. Assuming a noise jammer with 1-MHz bandwidth is used to protect the target aircraft, what jammer ERP is required to reduce the target lock-on range to 1 nmi assuming the only loss is a 3-dB jammer polarization loss?

10. To prevent the seeker in Problem 9 from tracking the noise jammer, assume that a swept velocity noise waveform is used to capture the seeker's speed gate. This sweeps bandpass noise through the expected Doppler band covered by the seeker. What is the maximum Doppler frequency seen by the seeker assuming that the target and illuminating aircraft travel at 500 knots and the missile at 1000 knots? What frequency range should be swept by the jammer? What bandpass and sweep rate should be used? How much jammer ERP is required if a jam-to-signal ratio of 20 dB is necessary to capture the seeker's speed gate?

11. Develop a MATLAB program for a step approximation of a linear FM chirp waveform that might be used in a DRFM to conserve memory space. Sample the chirp given in Example 4.1 at eight equally spaced intervals. Then reconstruct the chirp by filling in the missing portions of the waveform in a head-to-tails manner as in Example 4.14. Use a random phase discontinuity in the reconstructed waveform of 30 degrees for each segment. Then replot and compare the spectrum and matched filter output as shown in Example 4.1.

12. Using the MATLAB program for digital phase shifter serrodyning given in Example 4.15, plot the peak spurious level versus the number of phase shifter bits from 1 to 8.

13. Generate an amplitude monopulse difference pattern by subtracting two sin x/x patterns separated by their 3-dB beamwidth. Show that the sign of the error signal is always correct in the sidelobes. Will this be true for other beam spacings? For what other antenna patterns will this not be true?

14. Assume it is desired to defend a naval ship against an ASCM using a monopulse seeker. A cross-eye system is to be used. The cross-eye antennas have a baseline of 30m, while the ship length is 600m. The phase mismatch is 1/2 degree. What amplitude mismatch ratio is required to cause the monopulse seeker to point off the ship at a range of 1 km?

15. The Condon lobes of a monopulse parabolic reflector antenna have a magnitude 20 dB down from the peak of the main lobe. A cross polarization jammer is used against the monopulse tracking system. What is the maximum polarization mismatch that can be tolerated to prevent the jammer from acting like a beacon instead of a jammer?

16. Cassegrain twist reflector antennas are used on several antiship missile seekers to protect against cross-polarization jamming. A fixed vertically polarized monopulse feed structure illuminates a fixed parabolic reflector formed using a vertical wire grid. The reflected wave is then directed against a flat plate twist reflector that changes the antenna's polarization from vertical to horizontal. The horizontally polarized wave is then radiated through the transparent parabolic reflector. Discuss how this configuration is resistant to cross-polarization jamming.

17. The reduction of radar detection range due to noise jamming can be deduced as $[(J + N)/N]^{1/4}$, where J is the jamming power and N is the noise power at the receiver terminals. Plot the range reduction factor as a function of the ratio of jammer-plus-noise to noise.

18. Assume that a radar requires a 20-dB signal-to-noise ratio for reliable target detection at maximum radar range. A noise jammer operates in a radar sidelobe that is 20 dB down from the main lobe response. What jam-to-signal power is required to reduce the radar detection range to 30% of its full value (see Problem 17)? What sidelobe cancellation ratio is required to restore the detection range to 70% of its full value?

References

[1] Schleher, D. C., *MTI and Pulsed Doppler Radar,* Norwood, MA: Artech House, 1986.

[2] Neri, F., "New Technologies in Self-Protect Jammer," *J. Electronic Defense,* July 1991.

[3] Schleher, D. C., *Introduction to Electronic Warfare,* Norwood, MA: Artech House, 1986.

[4] Harris, F., "On the Use of Windows for Harmonic Analysis with the Discrete Fourier Transform," *IEEE Proc.,* Vol. 66, No. 1, Jan. 1978.

[5] Berkowitz, R., *Modern Radar,* New York: John Wiley, 1965.

[6] Sherman, S., *Monopulse Principles and Techniques,* Norwood, MA: Artech House, 1984.

[7] Dunn, J., and D. Howard, "Target Noise," in M. I. Skolnik (ed.), *Radar Handbook,* New York: McGraw-Hill, 1970.

[8] Hyberg, P., "Radar Countermeasure Technology," A Compendium for the Swedish Military Academy, FOA Report A 10052-1.1, Dec. 1993.

[9] Chapman, D., "Adaptive Arrays and Sidelobe Cancelers: A Perspective," *Microwave J.,* Aug. 1977.

[10] Prior, J., and N. Woodward, "Performance of Current Radar Systems in an EW Environment," *Military Micro Conf.,* 1976.

[11] Farina, A., and F. Studer, "Evaluation of Sidelobe Canceller Performance," *IEEE Proc.,* Vol. 129, Pt. F, No. 1, Feb. 1982.

[12] Van Brunt, L., *Applied ECM,* Volume 2, Dunn Loring, VA: EW Engineering, 1982.

[13] Povejsil, D., R. Raven, and P. Waterman, *Airborne Radar,* Princeton, NJ: Van Nostrand, 1961.

[14] Skolnik, M., *Radar Handbook,* 2nd Ed., Ch. 9, New York: McGraw Hill, 1990.

[15] Weher, D. R., *High Resolution Radar,* Norwood, MA: Artech House, 1987, p. 139.

[16] Long, M. W., "Medium PRF for the AN/APG-66 Radar," *Proc. of the IEEE.*

[17] Stimson, G. W., *Introduction to Airborne Radar,* El Segundo, CA: Hughes Aircraft Company, 1983, p. 448.

5

Digital Radio Frequency Memory

The increasing use of coherent radars employing both Doppler and PC techniques has driven the need for coherent jamming sources that can replicate the signals of the radar threats against which they are directed. In this way, jamming waveforms can overcome the substantial processing gain (30 dB to 60 dB) achieved by coherent radars. *Digital RF memory* (DRFM) allows for the storage of intercepted radar signatures in a digital memory and has been the subject of intensive research in the EW field. The concept of a DRFM is simple, but its implementation provides a number of technical challenges.

In this chapter, we discuss the operation of *direct digital synthesizers* (DDS), which provide the capability to digitally synthesize any radar waveform contained in their libraries. Jamming waveforms synthesized by DDS are not fully coherent with a particular radar's waveform, but rather are coherent within themselves. This generates a requirement for accurate set-on in frequency to allow these jamming signals to penetrate the radar's signal processor.

The general objective of an EA system is to inject sufficient energy within the victim radar's acceptance bandwidth to prevent it from accomplishing its intended function. For a conventional pulsed radar, the acceptance bandwidth is inversely proportional to the radar's pulsewidth. This establishes jammer tuning requirements that are generally not too stringent (e.g., on the order of 1 MHz) and can be satisfied by VCO type exciters.

Advanced type radars generally utilize the coherence within the radar waveform to perform coherent integration. When radar Doppler processing is involved, the radar Doppler bandwidth is inversely proportional to the time the radar dwells on the target (e.g., on the order of 50 Hz to 500 Hz). Therefore, to jam a Doppler radar the EA system must be able to tune such that appreciable jamming energy is concentrated within this narrow acceptance bandwidth. A somewhat analogous situation occurs with PC radar that transmits a coherent coded waveform, which is subsequently match filtered in the radar. In this case, unless the jammer can closely emulate the radar's coded waveform, the jam-to-signal energy ratio will be reduced by the radar's PC factor.

This discussion points to the need, when jamming coherent radar, for an RF memory, which allows precise matching of the EA waveform with respect to the victim radar's waveform. A device that satisfies this need is the DRFM. In theory, the DRFM stores in-phase and quadrature samples of the intercepted radar waveform. If the samples are taken and stored at a rate that is at least twice that of the information bandwidth of the radar signal, then in accordance with the sampling theorem the signal can be reconstructed at some later time without any information loss. It should be noted that the uses of DRFMs are not restricted to coherent jammers: they are equally effective against conventional pulse radar and, therefore, are an acceptable substitute for VCOs in this application.

5.1 DRFM Architectures

There are two basic forms of DRFM architectures. The wideband structure depicted in Figure 5.1 covers the full band as determined by the DRFM bandwidth and stores signals occurring anywhere in the band. The multiple narrowband structure depicted in Figure 5.2 contains many narrowband DRFMs, each of which stores individual threat radar signatures. The structures

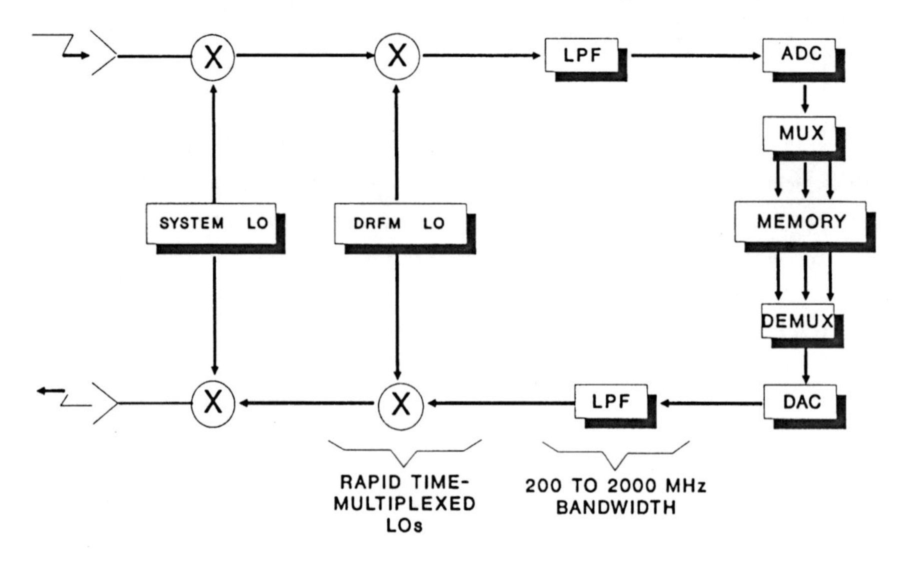

Figure 5.1 Wideband DRFM structure.

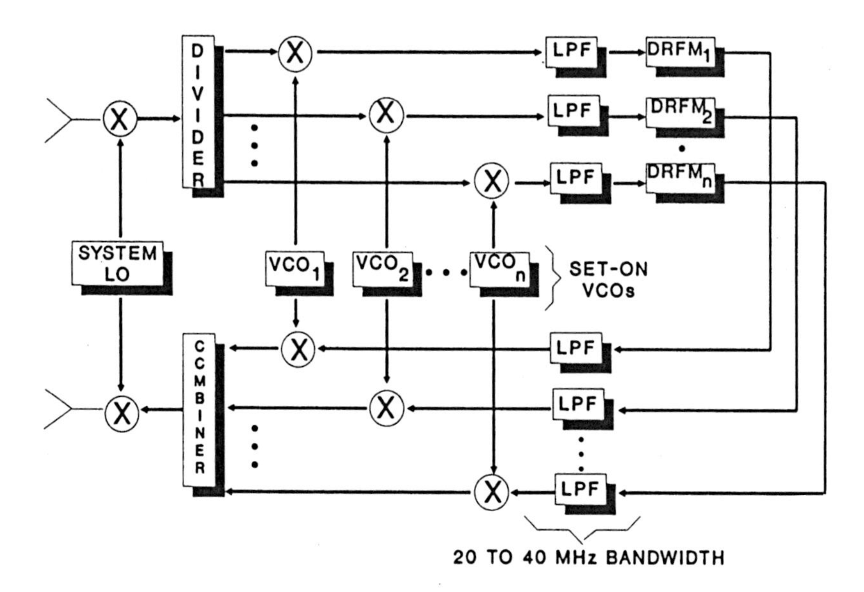

Figure 5.2 Multiple narrowband DRFM structure.

are compared in Figure 5.3. The difference between the structures relates to the technology needed to implement each type. The wideband DRFM requires a state-of-the-art A/D converter that operates at high speed while the narrowband DRFM must only cover the widest instantaneous radar bandwidth involved (i.e., order of 100 MHz) and uses multiple, relatively low-speed, A/D converters and memory devices.

The advantage of the narrowband DRFM structure is that it is readily implementable using existing digital componentry. This includes multibit A/D converters (see Figure 6.11) that allow high fidelity to prevent detection of the replicated signal. Only one threat signal is generally stored in each DRFM preventing intermodulation between signals that would generate spurious components (see Figure 3.14). Also, controller logic is simplified by allocating independent DRFMs for different threats.

Disadvantages of the narrowband DRFM are its limited flexibility in handling wideband frequency agile and PC signals that exceed its instantaneous bandwidth. Also, its ability to handle multiple threats is determined by the set-on speed of the *voltage controlled oscillators* (VCOs) that direct the signals into a particular narrowband DRFM. The cost of the VCOs is also an issue since these components are not needed in the wideband implementation.

The advantage of the wideband DRFM is its flexibility in replicating any threat radar signal within its frequency coverage. These include wideband frequency agile and PC threats not covered by the narrowband DRFM approach. Also, it is generally not susceptible to overloading in the presence

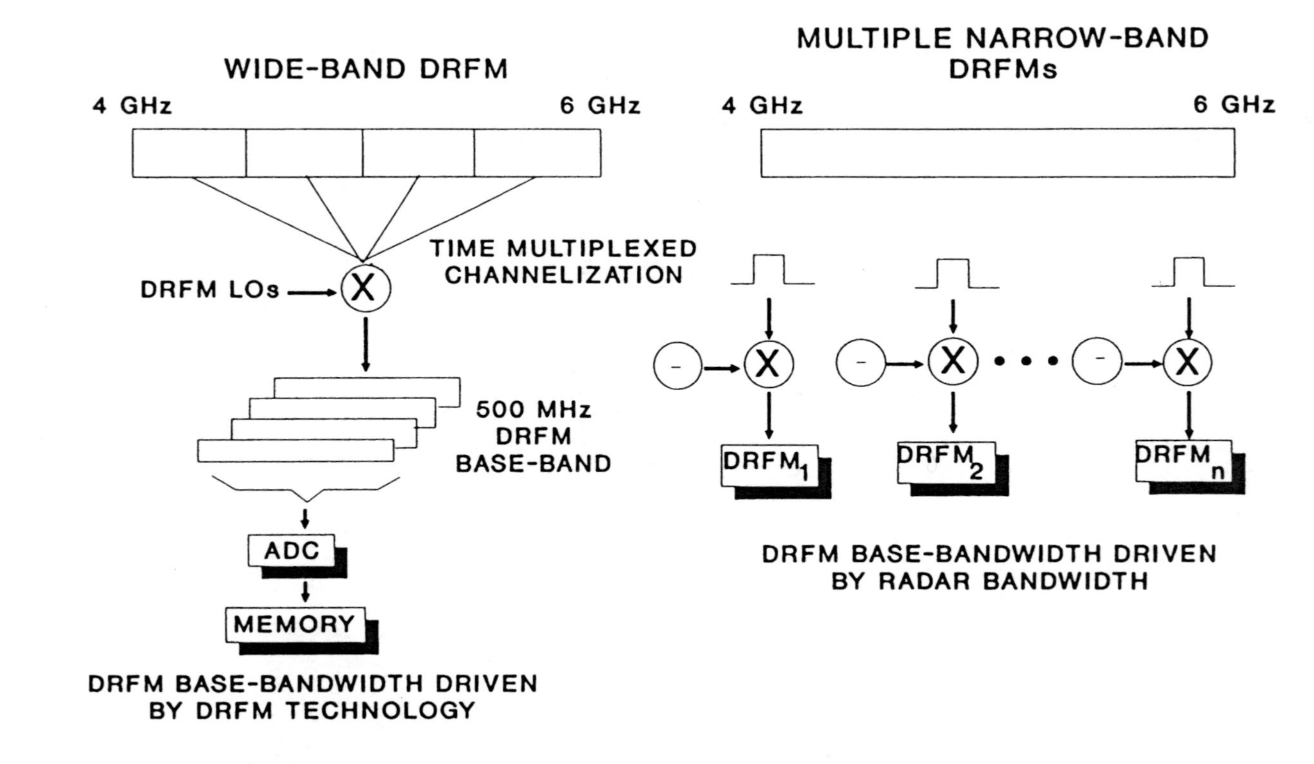

Figure 5.3 DRFM structure comparison.

of multiple threats. Because of its potential to provide a cost-effective solution, most current DRFM technology research is directed toward the wideband approach.

On the other hand, a number of technology issues are involved with the wideband approach. First is the development of high-speed multibit A/D converters. Multiple bits are required, both to reduce spurious responses due to the quantizing action of the A/D and also to prevent small-signal suppression. Complex high-speed controllers are required to route signals into and out of the *random access memory* (RAM). This is particularly difficult in a dense threat environment where multiple stored signal conflicts exist. High-speed multiplexing and demultiplexing is required to switch the DRFM local oscillator over the coverage band.

In summary, emphasis in DRFM research continues to focus on wideband technology improvements. Recent A/D converter technology has advanced to a point where wideband, single-channel, multiple-bit DRFM operation in a 500-MHz to 1-GHz instantaneous bandwidth with four to six bits is available. Advances in *gallium arsenide* (GaAs) digital circuit devices that offer high-speed, wide-bandwidth operation while providing radiation hardness continue to improve DRFM technology. Multiple DRFM units with moderate instantaneous bandwidth provide an alternative for near-term applications.

A sampling of some currently available DRFM devices is given in Table 5.1. Weight is important for tactical applications and tends to increase with bandwidth and memory length. Memory length determines how many threat signals can be stored. Special methods are used to store CW and PC signals because of their length. A small portion of the CW signal is stored and then replicated by recirculating the stored signal from head to tail (see Example 4.14 for spurious output due to this type of operation). Linear FM PC signals are strobed and samples stored throughout the waveform. The PC signal is then replicated in a similar manner to the CW signal, which results in a step approximation to the waveform. For phase-coded PC signals, samples are stored for each chip of the code. The signal is then replicated in each chip in the same manner as is used for CW signals. The spurious level of a DRFM is important in many EA operations since high spurious levels allow the radar to detect that it is being jammed. Example 4.12 indicates that a PD radar using a FFT processor can determine spurious levels that are on the order of 30 dB below the level of the main signal.

Many currently operational DRFMs avoid using an A/D converter by hard-limiting the signal before storage. This is equivalent to using a one-bit A/D converter or, equivalently, representing the frequency of the signal by the zero crossings of the signal.

A block diagram of a basic 1-bit DRFM covering the 8- to 16-GHz band is depicted in Figure 5.4. The incoming signal is first down-converted to a

Table 5.1
DRFM Characteristics by Type

Type	Bandwidth	Bits	Memory Length	Spurious Length	Delay Resolution	Weight
Anaren 450130	500 MHz	3	250 μs	−17 dBc	50 ns	6 lbs
Tecmus TCM-875	1000 MHz	4	340 μs	−15 dBc	8 ns	300 lbs
Kor 1027	800 MHz	4	145 μs	−24 dBc	4 ns	25 lbs
Whittaker MIP-130	460 MHz	2	64 μs	−71 dBc	64 ns	7 lbs
Whittaker MIP-810	110 MHz	8	1048 μs	—	4 ns	2 lbs

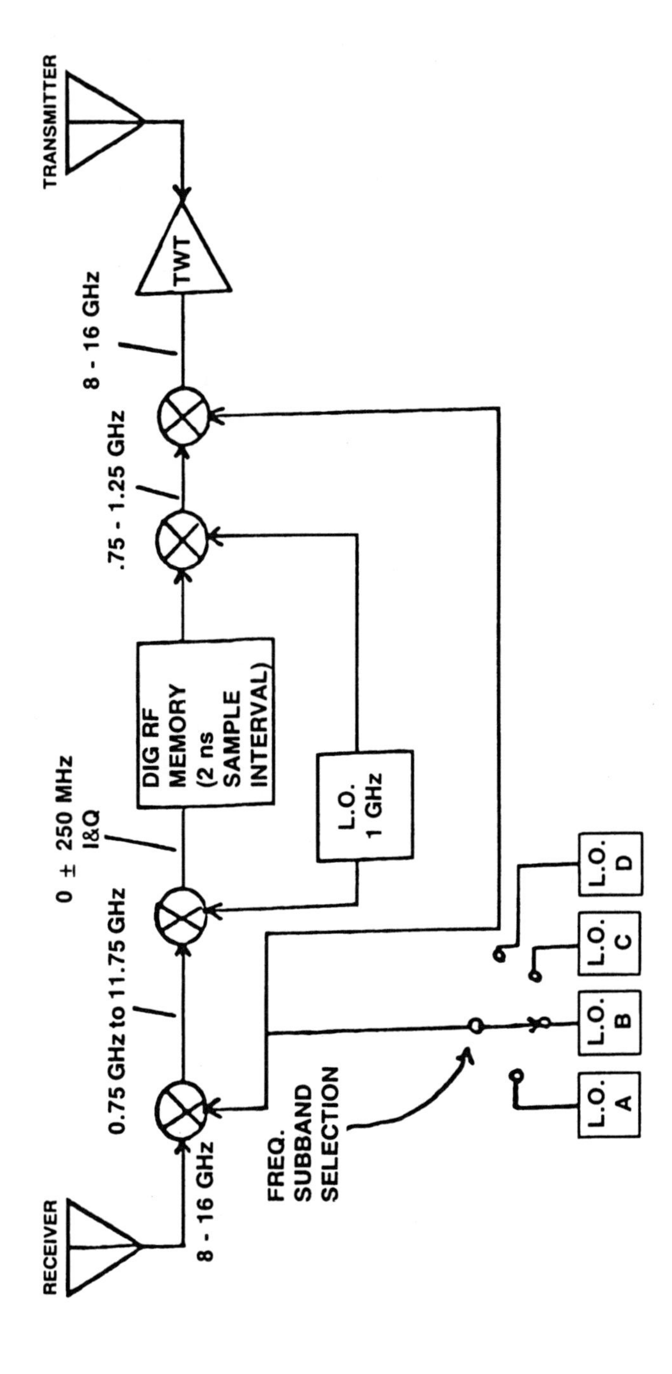

Figure 5.4 Block diagram of a basic one-bit DRFM. (*Source:* [1].)

500-MHz band centered about 1 GHz by selecting an appropriate local oscillator. The resulting down-converted signal is quadrature mixed against a 1-GHz local oscillator to provide 250-MHz baseband *in-phase* (I) and *quadrature* (Q) signals. The I and Q baseband signals are one-bit quantized and sampled with a 500-MHz clock (2-ns sampling interval). The samples are then stored in a digital RAM for future use. To replicate the signal, the inverse process is implemented, whereby the stored digital samples are clocked out at a 500-MHz rate and up-converted with the same oscillators that were used in the down-conversion storage process. Note that the DRFM output signal replicates the instantaneous frequency (embodied in the zero crossings) of the incoming signal, but that its power spectrum is distorted by the one-bit quantization process. Use of multibit quantization would allow a closer representation of the signal's power spectral density, but the greater digital storage requirements would result in reduced bandwidth.

In a practical system, the leading edge of the RF pulse is used to start the memory loading cycle. The trailing edge of the pulse is used to determine pulsewidth and to stop the memory read-in. The start and stop addresses are stored to control the read-out. The memory size determines the maximum number of emitters which can be stored. For example, if a 256 by 64 (16,384) bit *emitter-coupled-logic* (ECL) RAM were used with the prior 500-MHz clock rate, then 16.384 μs of emitter pulsewidth with 500-MHz instantaneous bandwidth could be stored. Because the memory is nondestructive, the DRFM is capable of reading out a stored signal within the order of 10 ns to 20 ns, which establishes the minimum delay time of the DRFM. Additionally, the stored signal can be stored for an indefinite period of time and read out as many times as desired or until the threat signal is unloaded from the memory.

The DRFM is thus a precise threat waveform storage device capable of being used as an exciter in jamming transmitters, which function against both coherent and conventional radar. Its frequency precision is determined by the stability of the local oscillators used to down-convert the signal into the digital baseband and then subsequently up-convert the signal for retransmission. A small offset in the retransmission local oscillator can be used to simulate a Doppler shift or to apply frequency modulation to the jamming signal. The one-bit mode of digital storage causes spurious spectral components in the output, which are 10 dB to 20 dB below the carrier. The nondestructive read-out allows retransmission with delays as short as 10 ns to 20 ns, which is an important factor in RGPO jamming. The coherent stretching feature allows anticipatory jamming, the simulation of radar clutter, and minimizes the amount of digital memory required to store long-duration radar waveforms.

5.2 DRFM Fundamentals

As the name implies, the function of a DRFM is to store an RF signal in a digital memory. Basically, this implies representing the signal by a set of numbers that can be stored and manipulated to produce jamming signals that resemble the victim radar's signature, and then reconstituted for transmission at RF.

The first part of the process employed by the DRFM is to convert the input RF signal into equivalent low-pass I and Q components. This is accomplished using synchronous detectors (i.e., phase detectors) that multiply the input RF signal with quadrature reference signals whose frequency is centered in the middle of the band to be sorted. As is known from information theory, all the information within the signal is contained within the resulting low-pass I and Q components, provided the signal is band limited at baseband to a frequency band less than half the sampling frequency.

The next step in the process is to convert the I and Q components into digital numbers using an A/D converter. In order to accomplish this digital conversion, the low-pass signal must be sampled at a frequency that is at least twice its highest frequency component (i.e., Nyquist rate). This indicates the advantage of converting to baseband signals since this results in the lowest possible sampling rate consistent with the bandwidth of the signal. Also, note that separate I and Q converters are necessary to digitize the signal.

The block diagram of a basic DRFM quadrature channel system is depicted in Figure 5.5. Assume that a 0.25-μs pulse is to be stored and that the frequency band coverage of the DRFM is 500 MHz. As shown in Figure 5.5, the 500-MHz band is converted into I and Q channels with 250-MHz bandwidth. The sampling frequency of the A/D converter is then 500 MHz to preserve all the information in the pulse. The A/D converters would generally quantize the I and Q signals using four to six bits, which would reduce the spurious signal content and prevent small-signal suppression if multiple signals were stored simultaneously. The signal is then represented by 125 (500 MHz by 0.25 μs) I and 125 Q, four- to six-bit samples, or a total of 250 samples, that are stored in digital memory.

When the signal is needed, the I- and Q-channel memories are clocked at the 500-MHz rate and the resulting signals passed through low-pass filters, D/A converted, up-converted to the original RF, and transmitted.

Analog down-converters are generally used in the basic Q-channel DRFM due to their ability to process wideband signals. The major deficiency of analog down-converting is that it is difficult to achieve good phase and amplitude balance between the channels.

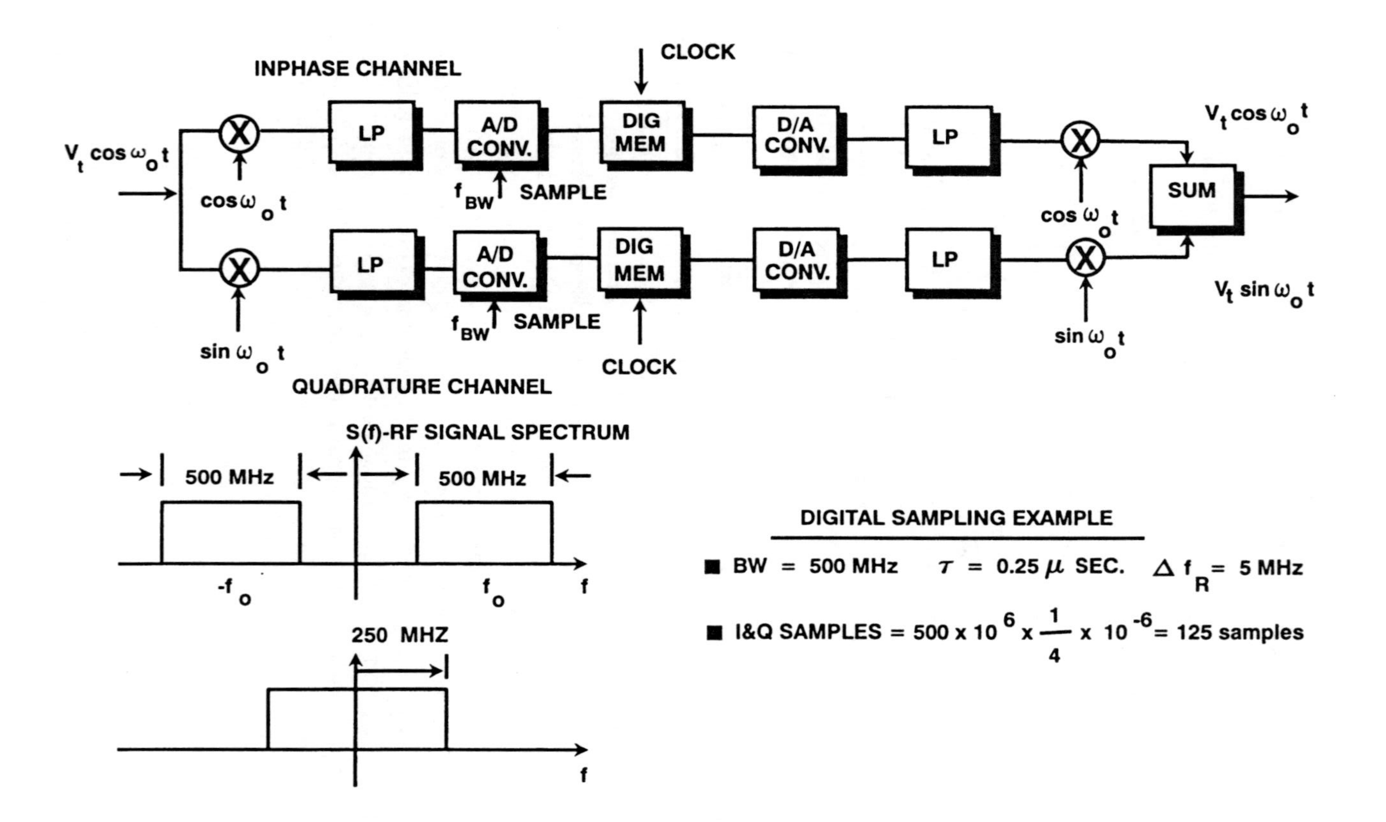

Figure 5.5 DRFM quadrature channel system.

There are two basic approaches to building analog I/Q down-converters. In both cases, two mixers are used in combination with a single local oscillator. In the most common solution (see Figure 5.5), a 90-degree phase shifter is inserted at the output of the local oscillator to provide quadrature local oscillator signals (i.e., $\cos(\omega_0 t)$ and $\sin(\omega_0 t)$). The alternate solution divides the input signal into two paths and places a 90-degree phase shifter in one path, while using the same local oscillator signal in both channels. The difficulty of this approach in wideband applications is the problem of maintaining a 90-degree phase shift over a wide frequency range.

Another approach that provides excellent phase and amplitude balance is the Hilbert transform algorithm [6]. In this approach, the signal is down-converted into the appropriate frequency range using a single analog mixer. The signal is then digitized in an A/D converter, and the resulting digital signal is processed using the Hilbert transform, which provides an analytic signal output containing a digital I-channel signal and a digital Q-channel signal. The balance of the quadrature channels is then perfect due to the stable nature of the digital design. The only problem with this approach is that Hilbert transformers are generally limited in operating speed and may not be able to support wideband operation. Future developments may correct this deficiency.

A further complication of the design depicted in Figure 5.5 is the translation of the input bandpass signal to baseband by centering the local oscillator signal in the middle of the DRFM passband. With this approach, signals in both the upper and lower portions of the band fold into the common baseband. However, any other oscillator frequency placement results in a larger video bandwidth and, consequently, a more complex A/D converter design.

The ambiguity can be resolved by measuring the phase differential between the I- and Q-channels. If the Q-channel phase leads the I-channel phase, then the signal frequency is higher than the local oscillator; while if it lags, the frequency is lower than the local oscillator signal.

To consider the quadrature channel process, the RF signal can be represented as

$$v(t) = x(t) \cdot \cos \omega_0 t - y(t) \cdot \sin \omega_0 t \tag{5.1}$$

where $x(t)$ and $y(t)$ are slowly varying signals compared to the carrier radian frequency ω_0. The signal spectrum is derived from (5.1) as

$$\begin{aligned} V(\omega) = & \frac{1}{2}[X(\omega - \omega_0) + jY(\omega - \omega_0)] \\ & + \frac{1}{2}[X(\omega + \omega_0) - jY(\omega + \omega_0)] \end{aligned} \tag{5.2}$$

where $X(\omega)$ and $Y(\omega)$ are the Fourier transforms of $x(t)$ and $y(t)$. Equation (5.2) can be rewritten in complex notation as

$$V(\omega) = \frac{1}{2}[\tilde{V}^*(\omega - \omega_0) + \tilde{V}(\omega + \omega_0)] \tag{5.3}$$

where $\tilde{V}(\omega) = X(\omega) + jY(\omega)$. The respective bandpass and lowpass spectrums are depicted in Figure 5.5.

After mixing with the quadrature local oscillators ω_{LO} and lowpass filtering, the signal in the I-channel is

$$v_I(t) = \frac{1}{2}[x(t) \cdot \cos(\omega_0 - \omega_{LO})t - y(t) \cdot \sin(\omega_0 - \omega_{LO})t] \tag{5.4}$$

while the Q-channel signal is

$$v_Q(t) = \frac{1}{2}[x(t) \cdot \sin(\omega_0 - \omega_{LO})t + y(t) \cdot \cos(\omega_0 - \omega_{LO})t] \tag{5.5}$$

When the local oscillator frequency matches the signal frequency $\omega_0 = \omega_{LO}$ then

$$\begin{aligned} v_I(t) &= x(t) \\ v_Q(t) &= y(t) \end{aligned} \tag{5.6}$$

The signals are up-converted by multiplying with the quadrature reference local oscillators as

$$\hat{v}(t) = v_I(t)\cos(\omega_{LO}t) + v_Q(t)\sin(\omega_{LO}t) \tag{5.7}$$

The I component is given by

$$\begin{aligned} \hat{v}_I(t) = &\frac{1}{2}\left[x(t) \cdot \frac{1}{2}(\cos(\omega_0 t) + \cos(\omega_0 - 2\omega_{LO})t)\right] \\ &- \frac{1}{2}\left[y(t) \cdot \frac{1}{2}(\sin(\omega_0 t) + \sin(\omega_0 - 2\omega_{LO})t)\right] \end{aligned} \tag{5.8}$$

while the Q component is

$$\hat{v}_Q(t) = -\frac{1}{2}\left[x(t) \cdot \left(-\frac{1}{2}\right) \cdot (\cos\omega_0 t - \cos(\omega_0 - 2\omega_{LO})t)\right. \tag{5.9}$$
$$\left. + \frac{1}{2} \cdot y_t(\sin(\omega_0 t) - \sin(\omega_0 - 2\omega_{LO})t)\right]$$

When $\omega_{LO} = \omega_0$, the output replicates the input signal given by (5.1).

To consider the effect of phase and amplitude distortion between channels, assume that the input signal is given by $x(t)\cos(\omega_0 t)$. The output I-channel signal is given by

$$\hat{v}_I(t) = x_I(t) \cdot \cos[(\omega_0 - \omega_{LO})t + \Delta\phi] \cdot \cos(\omega_{LO} t) \tag{5.10}$$

while the Q-channel signal is given by

$$\hat{v}_Q(t) = -x_Q(t) \cdot \sin((\omega_0 - \omega_{LO})t) \cdot \sin(\omega_{LO} t) \tag{5.11}$$

where $\Delta\phi$ is the phase difference between channels and $x_I(t) \neq x_Q(t)$ is the amplitude imbalance. The resulting distorted output signal can be written as

$$\hat{v}(t) = [x_I(t) \cdot \cos(\Delta\phi) + x_Q(t)]\cos(\omega_0 t) - x_I(t) \cdot \sin(\Delta\phi) \cdot \sin(\omega_0 t) \tag{5.12}$$

If the frequency of the local oscillator (ω_{LO}) and the input signal (ω_0) are mismatched, then

$$\hat{v}(t) = [x_I(t)\cos(\Delta\phi) - x_Q(t)] \cdot \cos(2\omega_{LO} - \omega_0)t \tag{5.13}$$
$$- x_I(t) \cdot \sin(\Delta\phi) \cdot \sin(2\omega_{LO} - \omega_0)t$$

The ratio (α) of the sideband power to the signal power can be computed as

$$\alpha = \frac{[x_I(t) \cdot \cos(\Delta\phi) - x_Q(t)]^2 + x_I^2(t) \cdot \sin^2(\Delta\phi)}{[x_I(t) \cdot \cos(\Delta\phi) + x_Q(t)]^2 + x_I^2(t) \cdot \sin^2(\Delta\phi)} \tag{5.14}$$

which can be expressed in terms of the amplitude imbalance $-x_Q(t)/x_I(t)$ as

$$\alpha = \frac{[\cos(\Delta\phi) - x_Q(t)/x_I(t)]^2 + \sin^2(\Delta\phi)}{[\cos(\Delta\phi) + x_Q(t)/x_I(t)]^2 + \sin^2(\Delta\phi)} \tag{5.15}$$

As discussed in Section 5.1, the spurious sideband power should be held to the order of 30 dB to prevent detection of the jamming signal. Example 5.1 illustrates that the phase difference must be held to under 2 degrees and the amplitude balance to better than 20% to achieve 30-dB spurious rejection.

▶ *Example 5.1*

Using MATLAB, plot contours for a quadrature-channel DRFM with −10-, −15-, −20-, −30-dB spurious response as a function of the phase difference and amplitude imbalance in accordance with (5.15).

```
% DRFM Phase-Amplitude Imbalance
% ------------------------------
% drfmbalc.m

clear;clf;clc;

% Input Spurious Levels

spur=-[10 15 20 30];  % db
spur=10.^(spur/10);

% Define Phase Imbalance Vector

phi=-40:.1:40;          % deg
phir=phi*pi/180;        % radians

% Calculate x=xq/xi Amplitude Imbalance

for k=1:4;
b=(1+spur(k))*cos(phir)/(1-spur(k));
xx=b+sqrt(b.^2-1);
cc=real(20*log10(xx));
aa=find(cc);
l=length(aa);
bb=cc(aa(1)-1:aa(l)+1);
dd=phi(aa(1)-1:aa(l)+1);

% Plot Phase-Amplitude Contour

plot(bb,dd,'k',-bb,dd,'k');
axis([-6 6 -40 40]);
xlabel('Amplitude Ratio xq/xi - db')
ylabel('Phase Difference - degrees');
title(['Phase-Amplitude Balance for Sideband Suppression']);
text(5,20,'-10 dB');
text(2.85,12,'-15 dB');
text(1.6,7,'-20 dB');
```

```
text(.4,4,'-30 dB');
grid on;hold on;
end;
```

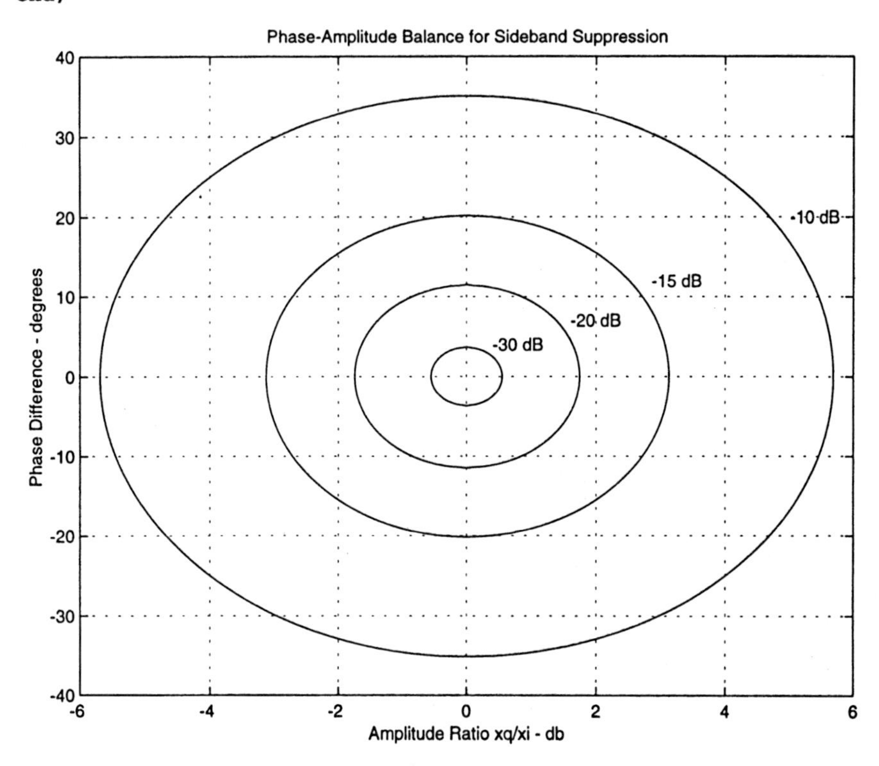

While the spurious components created by imbalances in the phase-and-amplitude match between channels in a quadrature-type DRFM can be eliminated by judicious design, another phenomena called quantization noise is an inevitable consequence of signal digitization. The level of quantization noise depends upon the number of bits used in the A/D converter. Its rms value expressed as a signal-to-noise ratio in decibels is given by

$$SNR_{\text{dB}} = 6b + 1.76 \tag{5.16}$$

which indicates that at least five bits must be used in an A/D converter to prevent spurious components greater than −30 dB. A two-tone test illustrated by Example 5.2 indicates that the order of four bits is required to reduce spurious levels to less than −30 dB.

▶ *Example 5.2*

Using MATLAB, plot the output waveform of a sine wave after 4- and 6-bit A/D conversion. Perform a two-tone test by calculating the spectrum after A/D conversion. How many bits are required for the spurious intermodulation products to be below −30 dB?

```
% A/D Converter Spectrum
% -----------------------
% adconb.m

clear;clf;clc;

% Set Input Parameters

N=2048;         % Sample Points
f=80;           % Hz
b=[3 5];        % Quantizing Bits

% Define Time Vector

t=(10/f)*(1:N)/N;

for k=1:2;

% Define Input Signals

Em=2^b(k);
xn=(Em/2)*sin(2*pi*f*t);

% Quantize Signal - Mid-point Quantizer

qn=(2*floor(xn-1e-6)+1)/Em;
  if xn<0;qn=(2*floor(xn+1e-6)+1)/Em;end;

qn=qn/max(qn);

% Store Quantized Signal

xx(:,k)=qn';

% Find Quantized Signal Spectrum

X=fft(qn,N);
X=fftshift(X);
Pxx=X.*conj(X);
Pxx=Pxx/max(Pxx);
Pz=10*log10(Pxx+1e-6);

% Store Quantized Signal Spectrum
```

```
yy(:,k)=Pz';

end;

% Plot Quantized Signal

subplot(221);
plot(t,xx(:,1));grid;
axis([0 .04 -1.25 1.25]);
xlabel('Time - sec');
ylabel('Voltage');
title(['Quantized Sine Wave (3 bits)']);

% Plot Quantized Signal Spectrums

fx=(f/10)*(-N/2:N/2-1);
subplot(223);
plot(fx,yy(:,1));grid;
axis([0 2500 -60 0]);
xlabel('Frequency - Hz');
ylabel('Power - db');
title(['Quantize Signal Spectrum (3 bits)']);

subplot(222);
plot(fx,yy(:,2));grid;
axis([0 2500 -60 0]);
xlabel('Frequency - Hz');
ylabel('Power - db');
title(['Quantized Signal Spectrum (5 bits)']);

% Find Two Tone Signal Spectrum

Em=2^b(2);
f1=80;f2=200;     % Hz
xn=(Em/4)*(sin(2*pi*f1*t)+sin(2*pi*f2*t));
qn=(2*floor(xn-1e-6)+1)/Em;
if xn<0;qn=(2*floor(xn-1e-6)+1)/Em;end;
qn=qn/max(qn);
X=fft(qn,N);X=fftshift(X);
Pxx=X.*conj(X);
Pxx=Pxx/max(Pxx);
Pz=10*log10(Pxx+1e-6);

subplot(224);
plot(fx,Pz);grid;
axis([0 2500 -60 0]);
xlabel('Frequency - Hz');
ylabel('Power - db');
title(['Two Tone Power Spectrum (5 bits)']);
```

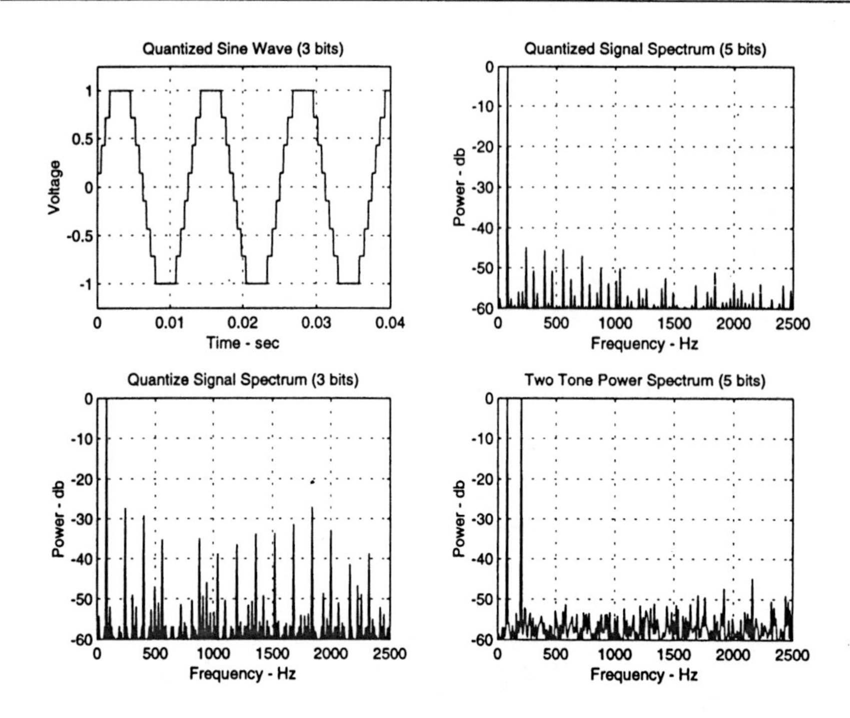

▶ *Example 5.3*

Using MATLAB, plot the spectrum from the output of a quadrature-channel DRFM. For a single signal, what is the spurious level using a one-bit A/D converter? What is the spurious level when five signals are present in the DRFM? Are any of the signals suppressed? What results when a four-bit A/D converter is used?

```
% Program Plots Output Spectrum of DRFM
% Using I and Q Quadrature Channels
% --------------------------------------
% drfmfreq.m

clear;clf;clc;

% Sample Input Signal

N=2048;     % Samples
t=(1:N)/N;
fs=100;     % Hz
j=sqrt(-1);

% Generate input signals

B=[1 5];    % Quantizing Bits
num=[1 5];  % Number Input Signals
```

```
for bb=1:2;
for nn=1:2;
z=exp(j*2*pi*fs*t);

for k=1:num(nn)-1;
zz=exp(j*2*pi*fs*(1+.2*k)*t);
z=z+zz;end;

% Quantize Input Signals

M=2^(B(bb)-1);
for i=1:N;
xx(i)=(M*real(z(i)))/num(nn);
yy(i)=(M*imag(z(i)))/num(nn);
if xx(i)>0;x(i)=ceil(xx(i))/M;
      else x(i)=floor(xx(i))/M;end;
if yy(i)>0;y(i)=ceil(yy(i))/M;
      else y(i)=floor(yy(i))/M;end;end;
m=nn+2*(bb-1);
w(m,:)=x+j*y;

% Find Spectrum

X=fft(w(m,:),N);
X=fftshift(X);
Pxx=X.*conj(X);
Pxx=Pxx/max(Pxx);
Pz=10*log10(Pxx+1e-6);
P(:,m)=Pz';
end;
end;

% Plot DRFM Spectrum

f=(-N/2:N/2-1);

subplot(221);
plot(f,P(:,1));grid;
title(['DRFM Spectrum
(Bits=',num2str(B(1)),',Sig=',num2str(num(1)), ')']);
axis([-500 500 -40 0]);
xlabel('Frequency - Hz');
ylabel('Amplitude - dB');

subplot(223);
plot(f,P(:,2));grid;
axis([-500 500 -40 0]);
xlabel('Frequency - Hz');
ylabel('Amplitude - dB');
title(['DRFM Spectrum
(Bits=',num2str(B(1)),',Sig=',num2str(num(2)),')']);
```

```
subplot(222);
plot(f,P(:,3));grid;
axis([-500 500 -40 0]);
xlabel('Frequency - Hz');
ylabel('Amplitude - dB');
title(['DRFM Spectrum
(Bits=',num2str(B(2)),',Sig=',num2str(num(1)),')']);

subplot(224);
plot(f,P(:,4));grid;
axis([-500 500 -40 0]);
xlabel('Frequency - Hz');
ylabel('Amplitude - dB');
title(['DRFM Spectrum
(Bits=',num2str(B(2)),',Sig=',num2str(num(2)),')']);
```

DRFM Spectrum (Bits=1,Sig=1)
DRFM Spectrum (Bits=5,Sig=1)
DRFM Spectrum (Bits=1,Sig=5)
DRFM Spectrum (Bits=5,Sig=5)
Amplitude - dB
Frequency - Hz
0
-10
-20
-30
-40
-500
0
500

A limitation in the quadrature-channel DRFM is that it requires two A/D converters. In wideband applications, the multibit A/D converter is the most costly component. One solution is to use hard limiting (one-bit A/D conversion), but this results in excessive spurious responses and small signal suppression when multiple signals are present. In the quadrature-channel DRFM, both phase and amplitude information are preserved; while in the one-bit DRFM, only phase information is used. Extending this concept, it is possible to encode the phase using multiple bits. This is referred to as a phase-sampling DRFM and has the advantage that it only requires one A/D converter. A further advantage is that, since amplitude information is not used, it has the capability of handling a wide dynamic range of signals. The most serious disadvantage is that of small-signal suppression when multiple signals are present.

To consider the phase-sampling DRFM process, we can write (5.1) as

$$v(t) = A(t)\cos(\omega t + \phi(t)) \tag{5.17}$$

where the amplitude process

$$A(t) = \sqrt{x^2(t) + y^2(t)} \tag{5.18}$$

and the phase process

$$\phi_t = a\tan\frac{y(t)}{x(t)} \tag{5.19}$$

Equation (5.17) indicates that a band-limited signal has two degrees of freedom, amplitude and phase. For most radar signals $A(t)$ is a fixed magnitude that only becomes important in a DRFM when multiple signals are present. Hence, adequate operation can be accomplished for single signals by storing only the phase of the signal.

A block diagram of a phase sampling DRFM is depicted in Figure 5.6. As with the quadrature-channel DRFM, the input signal is down-converted to baseband using quadrature reference oscillators. The down-converted signal is then processed in a phase detector that produces the operation arctan $(y(t)/x(t))$. The phase is digitized in an A/D converter and stored in a digital memory. The jamming signal is produced by reading the phase information into an D/A converter that provides the instantaneous phase between the $x(t)$ and $y(t)$ quadrature components. The $x(t)$ and $y(t)$ components are synthesized from this information, up-converted by mixing with the reference local oscillators, and combined into the final signal.

In the phase-sampling DRFM, the I signal component is given by

$$v_I(t) = A(t) \cdot \cos(\omega t + \phi(t)) \cdot \cos\omega_{LO}t \tag{5.20}$$

while the Q signal is

$$v_Q(t) = A(t) \cdot \cos(\omega t + \phi(t)) \cdot \sin\omega_{LO}t \tag{5.21}$$

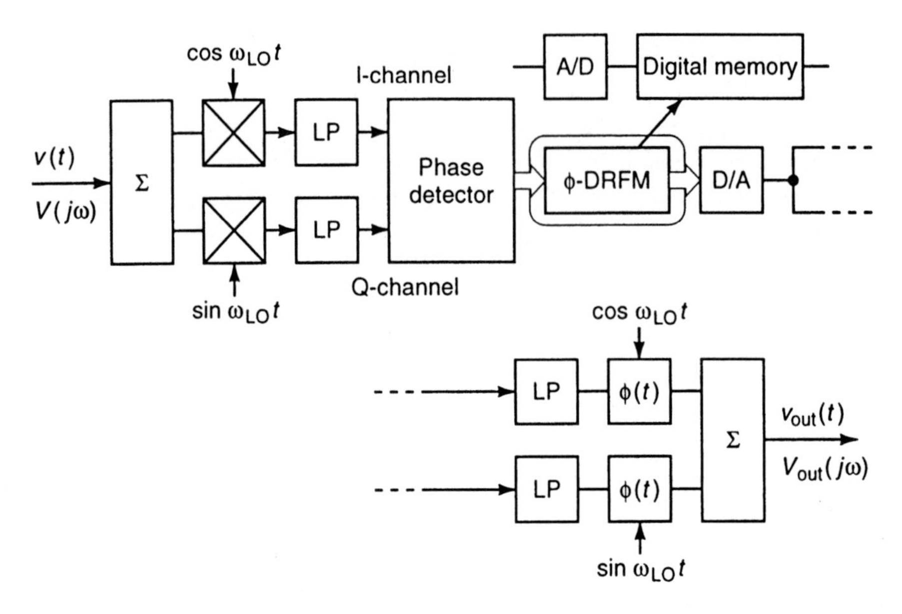

Figure 5.6 Phase-sampling DRFM block diagram.

After low pass filter, these signals become

$$v_{\mathrm{I}}(t) = \frac{A(t)}{2} \cdot \cos[(\omega - \omega_{\mathrm{LO}})t + \phi(t)] \tag{5.22}$$

and

$$v_{\mathrm{Q}}(t) = -\frac{A(t)}{2} \cdot \sin[(\omega - \omega_{\mathrm{LO}})t + \phi(t)] \tag{5.23}$$

The instantaneous phase is then found for $\omega = \omega_{\mathrm{LO}}$ as

$$\phi(t) = \arctan\frac{v_{\mathrm{Q}}(t)}{v_{\mathrm{I}}(t)} \tag{5.24}$$

and is independent of the amplitude $A(t)$, which is the same in both quadrature-channels.

The process of reconstructing the signal is depicted in Figure 5.7 for a three-bit A/D converter. The phase for this case is stored in 45-degree increments. In summary, the phase-sampling DRFM, in comparison with the quadrature-channel amplitude and phase-sampling system using an equal number of bits, provides the same instantaneous frequency coverage, but with only one

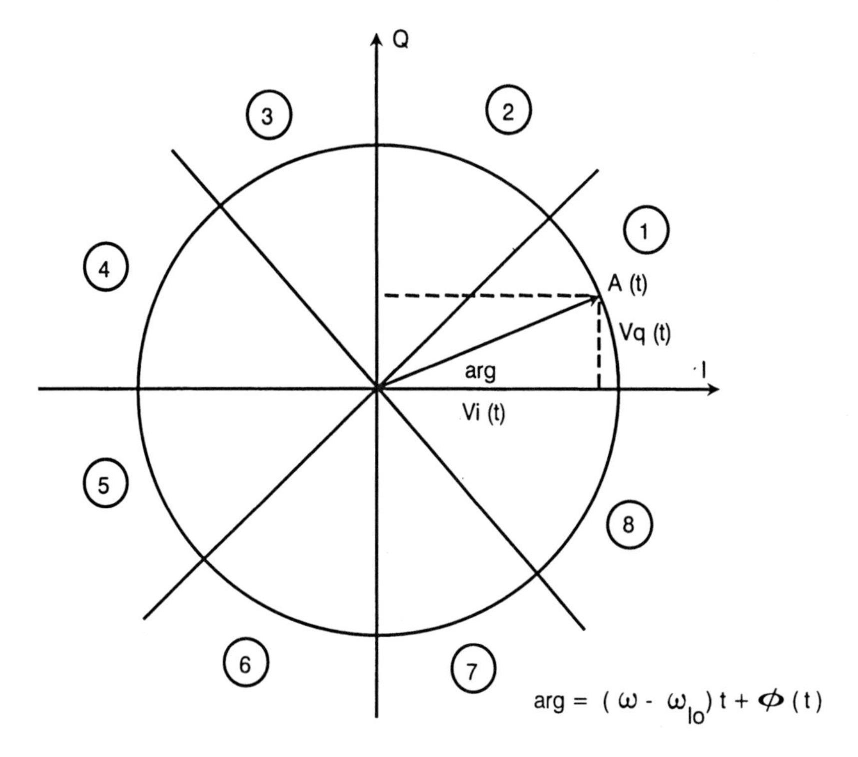

Figure 5.7 Reconstruction of quadrature signal.

A/D converter. As shown in Example 5.4, the spurious level is somewhat greater than a quadrature-channel DRFM and signal suppression occurs when multiple signals are present.

► *Example 5.4*

Using MATLAB, plot the spectrum from the output of a phase-sampling DRFM. For a single signal, how many bits are required to provide similar spurious response to the four-bit quadrature-channel DRFM? What happens to the spurious response and signals when five signals are simultaneously present?

```
% Program Plots Output Spectrum of DRFM
% Using Phase Quantization
% ------------------------------------
% drfmphsb.m

clear;clf;clc;

% Sample Input Signal
```

```
N=2048;    % Samples
t=(1:N)/N;
fs=100;    % Hz
j=sqrt(-1);

% Generate input signals

B=[1 5]; % Quantizing Bits
num=[1 5]; % Number Input Signals

for bb=1:2;
for nn=1:2;
z=exp(j*2*pi*fs*t);

for k=1:num(nn)-1;
zz=exp(j*2*pi*fs*(1+.2*k)*t);
z=z+zz;end;

% Quantize Input Signals Using Mid-Point Quantizer

M=2^(B(bb)-1);
for i=1:N;
ph(i)=atan2(imag(z(i)),real(z(i)));
qph(i)=(floor(M*((ph(i)/pi)-1e-6))+.5)/M;
if ph(i)<0;qph(i)=(floor(M*((ph(i)/pi)+1e-6))+.5)/M;end;
end;
m=nn+2*(bb-1);
w(m,:)=cos(pi*qph)+j*sin(pi*qph);

% Find Spectrum

X=fft(w(m,:),N);
X=fftshift(X);
Pxx=X.*conj(X);
Pxx=Pxx/max(Pxx);
Pz=10*log10(Pxx+1e-6);
P(:,m)=Pz';
end;
end;

% Plot DRFM Spectrum

f=(-N/2:N/2-1);

subplot(221);
plot(f,P(:,1));grid;
title(['DRFM Spectrum
(Bits=',num2str(B(1)),',Sig=',num2str(num(1)), ')']);
axis([-500 500 -40 0]);
xlabel('Frequency - Hz');
ylabel('Amplitude - dB');
```

```
subplot(223);
plot(f,P(:,2));grid;
axis([-500 500 -40 0]);
xlabel('Frequency - Hz');
ylabel('Amplitude - dB');
title(['DRFM Spectrum
(Bits=',num2str(B(1)),',Sig=',num2str(num(2)),')']);

subplot(222);
plot(f,P(:,3));grid;
axis([-500 500 -40 0]);
xlabel('Frequency - Hz');
ylabel('Amplitude - dB');
title(['DRFM Spectrum
(Bits=',num2str(B(2)),',Sig=',num2str(num(1)),')']);

subplot(224);
plot(f,P(:,4));grid;
axis([-500 500 -40 0]);
xlabel('Frequency - Hz');
ylabel('Amplitude - dB');
title(['DRFM Spectrum
(Bits=',num2str(B(2)),',Sig=',num2str(num(2)),')']);
```

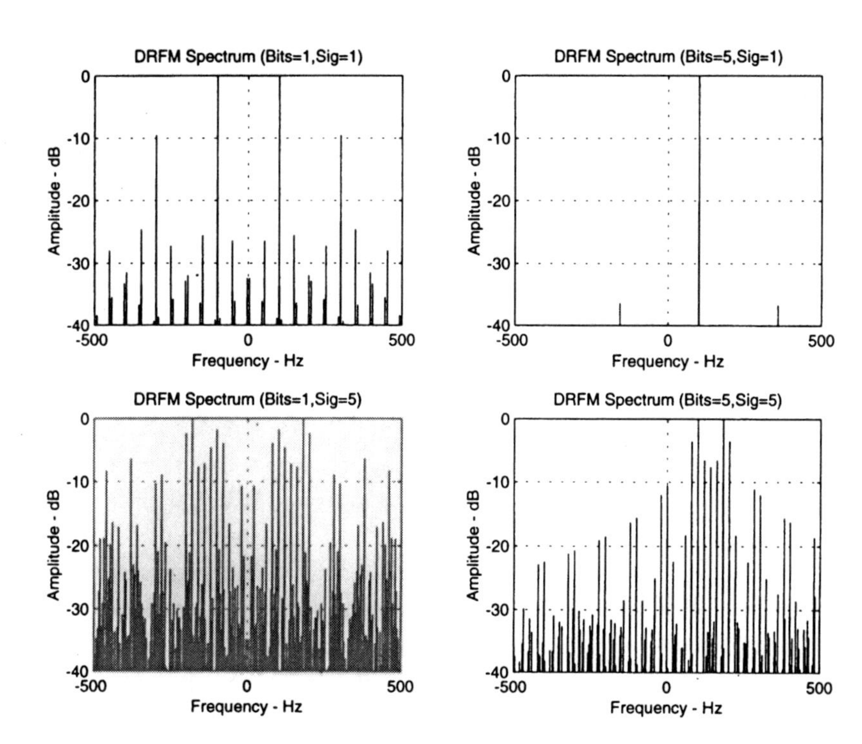

5.3 DRFM Sampling Techniques

In general, wideband DRFMs are preferred since they allow signals to be stored from radar using wideband modulation techniques such as frequency agility or wideband PC. However, an examination of Figure 6.11 indicates that wideband A/D converters generally imply a low-resolution capability (i.e., the A/D converter has a low number of bits). This in turn raises the spurious level out of the DRFM that allows detection of the replicated signal. Thus, techniques that allow wideband operation with reduced bandwidth components are often employed in DRFMs.

One technique that allows lower component bandwidths is the series-parallel architecture. The fundamental action of this technique is depicted in Figure 5.8. It basically employs a tapped delay line, with taps at $\Delta t = 1/f_s$, such that multiple sample points can be taken simultaneously. If five taps are employed, as illustrated in Figure 5.8, then the switch has to close at only one-fifth the rate to store all the samples. This means that a 500-MHz DRFM needs to have circuitry that operates at only 100 MHz. Actually, a factor of 12 is possible using this technique. The penalty paid for the reduced bandwidth is a magnification of the required hardware. However, the reduced bandwidth allows the use of A/D converters with a greater number of bits, which reduces spurious components and the speed at which the memory must clock the signals. The reduced memory speed often allows the use of conventional RAM integrated circuits rather than high-speed GaAs components.

Figure 5.9 illustrates the technique of using a shift register instead of the analog tapped delay line. This configuration uses a single A/D converter to digitize the signal but reduces the memory speed by a factor of eight. Note that for multiple-bit A/D converters, parallel channels are employed (one channel per bit), maintaining the reduced memory bandwidth.

Employing the configuration shown in Figure 5.9, a one-bit configuration can result in a dramatic increase in DRFM bandwidth using low-speed memory components. For example, using a hard limiter followed by a tapped delay line in place of the A/D converter in Figure 5.9 can result in a bandwidth magnification of $12 \times 8 = 96$ for the DRFM. This means that memory components clocked at 12.5 MHz can be used in a one-bit DRFM with 500-MHz bandwidth.

An implementation of a four-bit 500-MHz DRFM using series-parallel architecture is depicted in Figure 5.10. Twelve parallel channels each with a 100-MHz four-bit A/D converter are used. The sampling rate for the MOS type memory is reduced to 12.5 MHz using an eight-bit shift register. The signal is recovered using a four-bit D/A converter operating at 100 MHz.

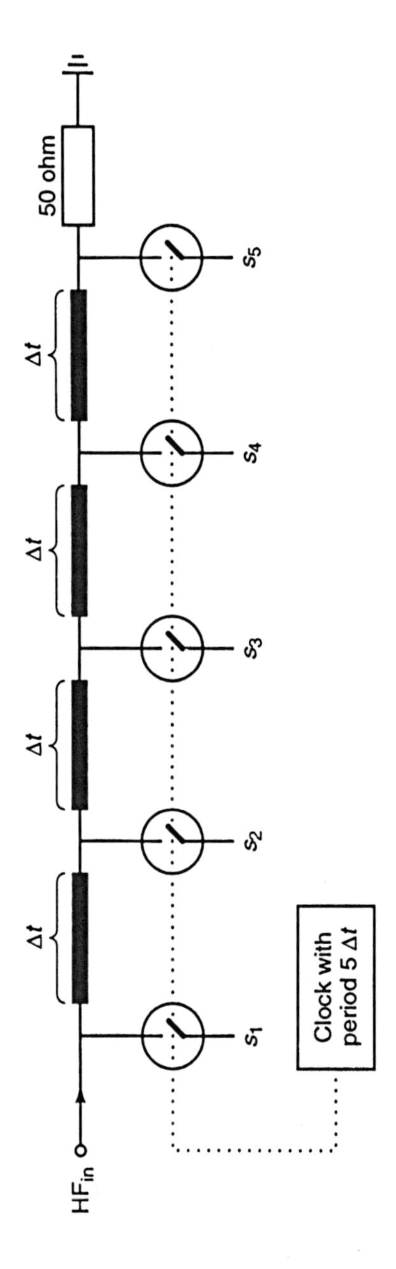

Figure 5.8 Series-parallel sampling technique.

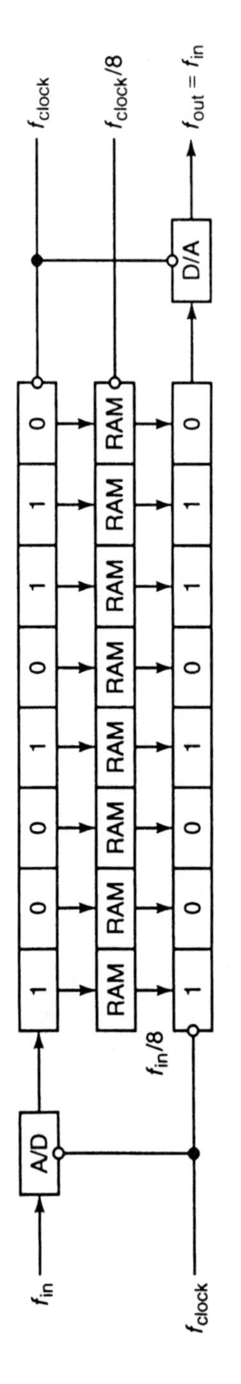

Figure 5.9 Shift register technique for series-parallel conversion.

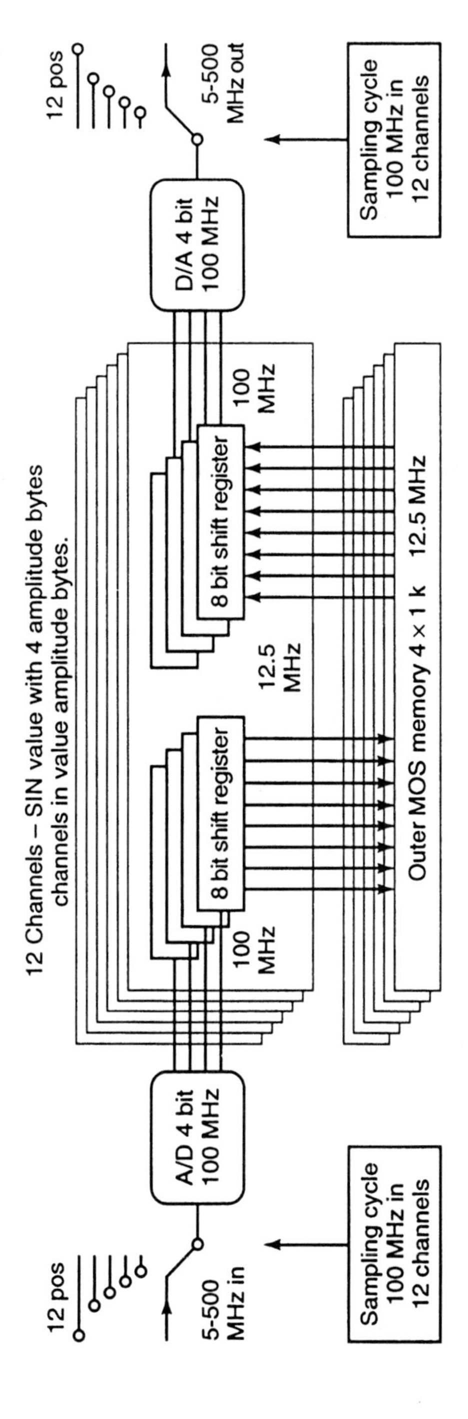

Figure 5.10 Architecture for multibit 500-MHz DRFM.

5.4 Direct Digital Synthesizer

With a DDS, it is possible to digitally synthesize an arbitrary waveform. For EW purposes, this allows the synthesis of radar or communications waveforms from a set of stored parameters. The fidelity of the replication is then a function of the accuracy of the parameter description. Since the DDS waveform is generated from a stable clock frequency, it is quasi-coherent or coherent within itself. Thus, it will build up in radar-matched filters provided its central frequency closely matches the central frequency of the radar or communications system against which it is directed. This allows the jamming signal to neutralize the processing gain of the radar or communications systems signal processor.

DDS techniques are ideally suited to generate PC waveforms. This was discussed in Section 3.5 on support jamming. However, as discussed in Section 3.5, PC systems are generally sensitive to Doppler frequency shift. From an EW viewpoint, this means that if the jamming signal is generated with a frequency offset (i.e., apparent Doppler shift), then it may not build up in the pulse compression (PCI) network. Phase-coded PC waveforms are most susceptible to this effect (see Figure 3.13). Ambiguity diagrams can be generated to examine the sensitivity of a particular waveform to a frequency offset.

The concept of a DDS can be understood through an examination of (5.17), which defines an arbitrary signal as composed of independent amplitude (A_t) and phase (ϕ_t) elements. The phase element couples into the signal through a sinusoidal transformation, while the amplitude element simply multiplies the transformed phase element. For most radar signals, the amplitude is a constant and only the phase, which is given by

$$\theta_t = \omega_0 t + \phi_t \tag{5.25}$$

is pertinent. Hence, most signals of interest can be formed via the use of a phase accumulator, whose output defines the instantaneous phase of the signal, followed by a look-up table mapping function (i.e., sin θ_t) that determines the value of the signal for each particular value of phase. Multiple simultaneous signals can be formed using a phase accumulator and look-up table for each signal and then summing these signals.

A block diagram of a DDS configured for chirp waveforms generation is depicted in Figure 5.11. The standard DDS consists of a phase accumulator, look-up table memory, and a D/A memory. The chirp generator adds a frequency accumulator, prior to the phase accumulator, which provides fast frequency changes. The frequency accumulator receives a digital input for the start frequency and chirp rate and provides the signal's instantaneous frequency. This instantaneous frequency is applied to the phase accumulator, which trans-

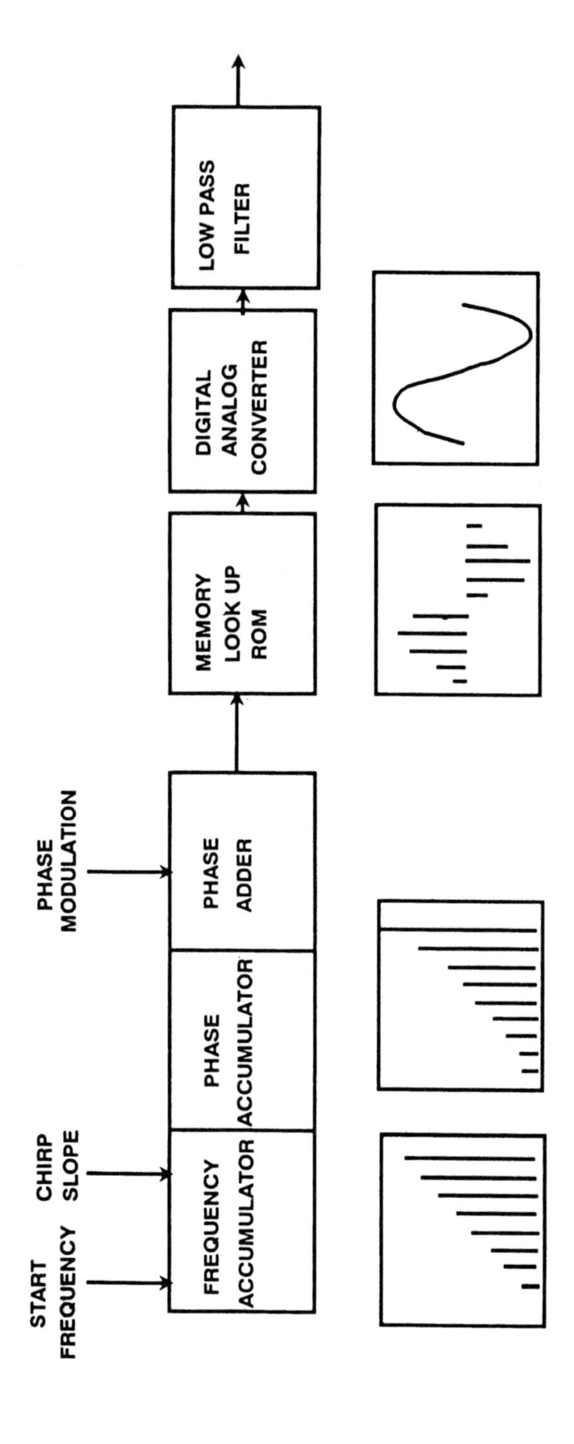

Figure 5.11 Block diagram of a direct digital synthesizer.

lates it into an instantaneous phase. A phase adder is placed at the output of the phase accumulator to generate phase modulation. The phase accumulator drives a memory look-up table that converts digital phase to digital amplitude prior to analog conversion. The final D/A converter converts the output to analog form, which is then filtered to remove any commutation effects caused by the digital process. All the digital devices within the chirp synthesizer are clocked synchronously with a master clock typically operating in the 500- to 600-MHz frequency range.

The frequency resolution achievable by a DDS depends upon the clock frequency and size of the accumulator in bits. For a 500-MHz sampling rate with a 24-bit accumulator, the minimum frequency setup size is $f_s/2^b$ = 29.8 Hz for a typical DDS. Since a new frequency is generated every 2 ns, this results in a chirp rate resolution of 14.9 kHz/μs. Phase noise of the DDS is the same as the phase noise associated with the clock generator. Spurious levels generally depend on the size of the D/A converter producing spurious signals of level $-6 \cdot N$ dBc where N is the number of bits used in the D/A converter.

5.5 Advanced DRFM Architecture

The architecture of an advanced DRFM configured as an *application-specific integrated circuit* (ASIC) is depicted in Figure 5.12. In this design, the complete DRFM (excluding A/D and D/A converters and RAM) is contained on a single GaAs ASIC. The GaAs ASIC controls the A/D and D/A converters and RAM and modulates both the I- and Q-channels by multiplying the delayed RF signals with independently synthesized modulation waveforms. The GaAs ASIC has a 100-ns maximum signal delay and operates using a 500-MHz clock frequency, providing 200-MHz bandwidth [7].

The characteristics of the DRFM are given in Table 5.2. It accepts up to a 200-MHz bandwidth input signal generating both I/Q digitized signals that are sampled at a 500-MHz rate. The signal can be delayed, Doppler shifted, phase and amplitude modulated. The core of the system is a DDS modulator capable of processing 12-bit I/Q signals at a 500-MHz clock rate. The modulator provides fine signal delay in 2-ns steps. Up to 134-ms signal delay is provided via an external memory. The DDS provides Doppler shift modulation as great as ±250 MHz with 0.1-Hz resolution. Note that the preceding characteristics do not include the effects of the A/D converter, which may limit the overall performance depending upon the type and frequency selected (see Figure 6.11). The ASIC contains about 500,000 transistors and dissipates about 65W.

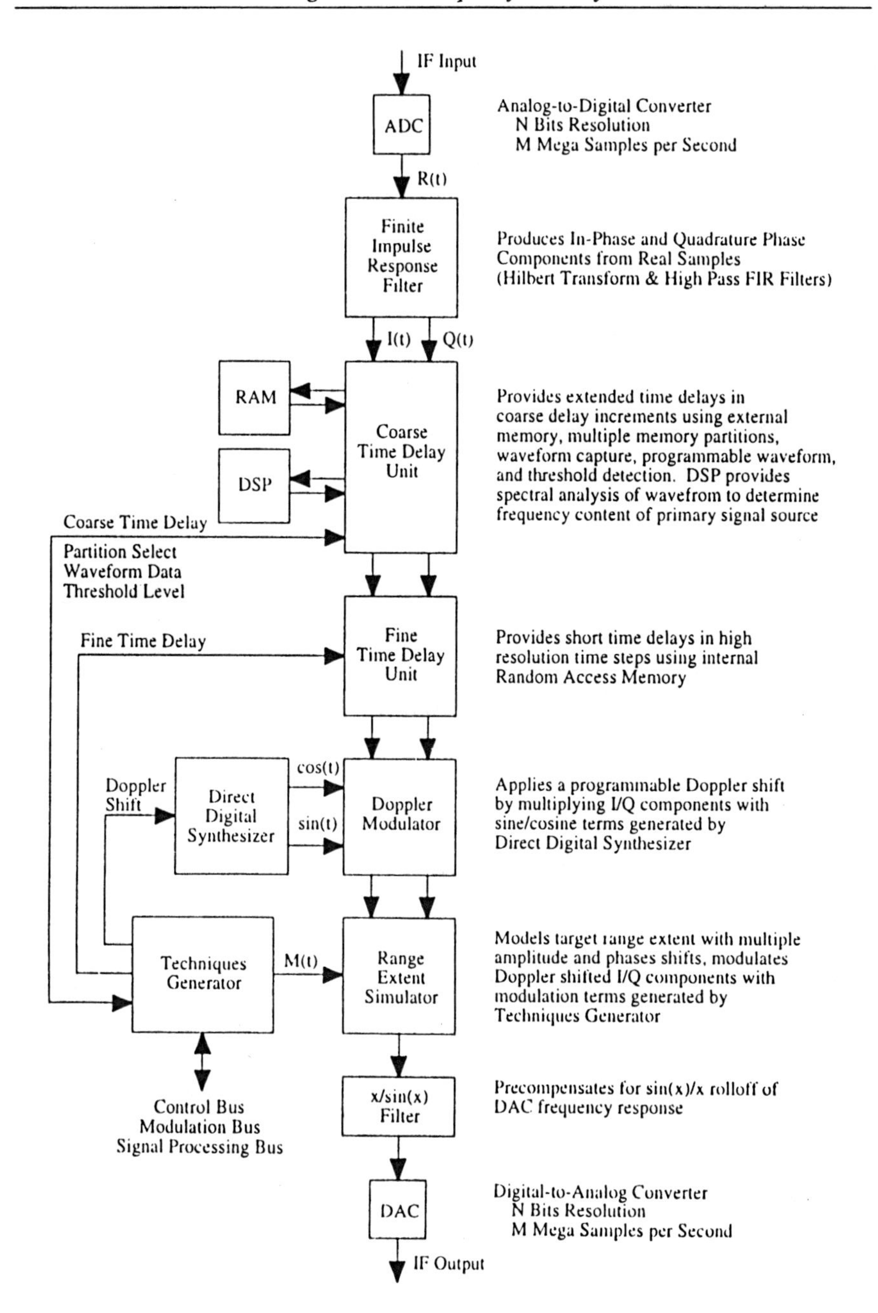

Figure 5.12 Advanced DRFM architecture. (*Source:* [7].)

Table 5.2
Advanced DRFM ASIC Specification

Bandwidth	200 MHz maximum
Center frequency	User-specified
I/Q sample rate	500 MHz
I/Q sample resolution	1–12 bit (depends on input frequency)
Spurious response	< –80 dBc
Delay	4 ns to 134 ms
Delay resolution	2 ns
Doppler shift	±250 MHz
Doppler resolution	0.1 Hz
Amplitude modulation	14 bits
Phase modulation	14 bits (0.022 degrees)
Modulation update rate	250 MHz

The complete DRFM architecture depicted in Figure 5.12 contains a number of interesting features. A Hilbert transform filter is used to form the I- and Q-channels and provides excellent balance control, allowing low spurious output. An external DSP complementing the time delay unit provides spectral analysis of the signal. Coarse time delay is provided externally, while fine time delay is provided using internal RAM. A DDS is used to provide precise Doppler modulation on the signal. A *techniques generator* (TG) provides the capability for RGPO complemented by matched VGPO. The D/A converter's $\sin(x)/x$ roll-off is compensated via a precompensating filter.

5.6 Voltage Controlled Oscillators

Another key microwave component associated with many ECM systems is the fast-tuning VCO which allows a jammer to effectively function in a high-density environment and respond to a variety of sophisticated threats. The VCO receives a series of digital tuning signals that represent instructions to step tune to a specific frequency, dwell, and then tune to another threat frequency. To be effective, the VCO should be able to tune in 50 ns to 100 ns and have a set-on accuracy of the order of ±1 MHz.

Critical areas for VCOs are post-tuning drift, settling time, and repeatability. Post-tuning drift is the amount of frequency deviation incurred over some prescribed time (e.g., 150 ns to 50 μs) after the VCO has reached the prescribed frequency commanded by the digital input message. Settling time is the amount of time required for a VCO, after being switched from one frequency to another, to arrive within a specified error bandwidth around the new frequency.

Repeatability is a measure of how accurately commanded frequencies are attained when attempting to repeat a given frequency at different times.

VCOs generally are used in transponder modes of EA systems that are not required to be coherent with the victim radar's transmissions. Jamming of radars that use forms of coherent integration (e.g., pulsed Doppler and PC radars) require that coherent transmissions be radiated by the jammer. For this type of operation, modern EW systems use DRFMs or DDS waveform generators.

The high-threat density confronting modern EA systems requires almost instantaneous jamming response against multiple threats. This is accomplished in modern DECM systems using a transponder mode of operation that allows jammer tuning to the threat frequency in very short times (e.g., on the order of 100 ns). A key component used in this type of EA system is the VCO.

In a typical mode of operation, an associated ES system identifies the threat parameters, which include the radar's frequency, power, PRI, pulsewidth, direction, and any special modulation characteristics. This information is used in the EA computer to recognize the threat, establish priorities, and determine the jamming response. The response is in the form of a time-phased command to the EA transmitter's exciter (e.g., VCO) which establishes the EA system's jamming frequency and associated amplitude, frequency, and temporal modulation.

For example, if ten radars were to be jammed, then an ideal VCO for EW exciter use would tune to each upon command and a single VCO could then be used to jam ten radars. If certain radars operated simultaneously from different directions, the VCOs could be increased in number and used simultaneously in different directions as required to satisfy EA design requirements.

The ES receiver used to determine the frequency of the victim radar can also be used to establish the tuning of the VCO. This provides a method where only the relative frequencies between the victim radar and the jammer are involved and avoids the measurement of absolute frequency. To understand this mode of operation, assume that a *digital IFM* (DIFM) ES receiver measures the threat radar frequency with 2-MHz accuracy. The reference is then a digital word that is stored, representing the desired output of the DIFM receiver when jamming this particular threat radar.

Upon command, the EA exciter VCO tunes until the DIFM output digital word is matched to the reference word. Because the DIFM measurement accuracy is 2 MHz, the absolute error between the victim radar's frequency and the jammer frequency will be a maximum of 4 MHz. The most probable error will generally be much less than the maximum because any bias error in the 2-MHz measurement accuracy will be eliminated in this method. After the VCO is set on frequency, any post-tuning drift results in further error. Also, at the time tuning is attempted, the VCO must be relatively stable for meaningful final set-on and accurate jamming.

The previous system requirements can be translated into VCO requirements generally requiring fast settling times, low post-tuning frequency drift, and accurate frequency repeatability. Typically, varactor-tuned oscillators are used that can tune rapidly over octave bands to some set-on frequency. In some cases, varactors with hyperabrupt junctions are used that provide a linear voltage *versus* frequency tuning curve. Basic oscillators used in VCOs are transistor-multiplier oscillators, Gunn oscillators, and FET oscillators. An important aspect of circuit design for a fast-set-on-time VCO is the thermal design of the semiconductor devices. One of the reasons for the success of Gunn diode oscillators in VCO applications is that the distance between the active region of the Gunn device and the heat sink is very low, resulting in fast response to changes in power dissipation.

In a VCO, after a change in tuning voltage, the frequency does not immediately reach its steady-state final value, but there is some frequency drift, called *post-tuning drift* (PTD) that may last for a considerable period of time. This is bothersome in DECM applications requiring RGPI or anticipatory jamming. There are many factors affecting PTD arising from the driver amplifier, varactor diode, the active oscillator device, and circuit characteristics. Thermal effects in the varactor are apparently a major factor in causing post-tuning drift.

The varactor in the VCO may be required to dissipate a significant amount of RF power. When tuning from one end of the band to another, this dissipated power will change considerably. The change in power causes a temperature change at the varactor junction, which reflects in a capacitance change due to the temperature coefficient of the varactor capacitance. This effect can produce changes in the output frequency of several megahertz. However, another effect that is associated with reverse current flow in the varactor can produce much larger frequency changes than thermal drift. The cause of this phenomenon is not well understood.

Several concerns associated with PTD are generally reflected in the specification for VCOs used for EA systems. First, the magnitude of PTD is a function of the dwell time at its prior frequency setting. Second, the direction of PTD is a function of the direction in which the new frequency setting is approached. These factors lead to a conclusion that repeatability can be made independent of prior dwell time and direction of approach by always specifying the maximum prior dwell time for PTD and always approaching the new frequency from the same direction or not specifying repeatability until after settling time is complete.

A typical specification for a VCO used in a self-protection jammer is given in Table 5.3. This VCO is designed to cover the 2- to 8-GHz band through appropriate wideband mixing and ultimately will cover the 2- to 18-GHz band. Another approach not covered here is to use an extremely fast VCO, which can settle in the order of 25 ns to the incoming frequency. This allows the following of frequency agile radars.

Table 5.3
VCO Specification for Self-Protection Jammer

Frequency	2.5 to 5.3 GHz
Settling time	Within 3 MHz in 100 ns
Post-tuning drift	Maximum ±1 MHz from 160 ns to 50 μs
Repeatability	Within ±1 MHz of PTD curve from 160 ns to 50 μs for a 125-ms period
Harmonic output	Lower than –20 dBc
Spurious outputs	AM and FM noise lower than –20 dBc; other –60 dBc
Linearity	2%

5.7 Problems

1. Using the MATLAB program adconb.m, plot the peak spurious level as a function of the number of bits (1 to 8) used in the A/D converter.

2. Given a 10-GHz Doppler radar operating with a 25-kHz PRF and having a CPI of 20 ms, what additional J/S ratio is required if a noncoherent jamming pulse is used relative to using a coherent DRFM jammer signal? Assume a noncoherent integration gain exponent of 0.65.

3. In an I-Q type DRFM with 1-dB amplitude match, what phase imbalance can be tolerated to restrict the spurious level to less than 30 dB down from the carrier?

4. Given a Doppler radar operating at 10 GHz with a CPI of 1 ms, and employing 10% frequency agility from one CPI to the next, consider a deceptive repeater jammer that induces a fixed frequency offset to simulate a Doppler shift on the false target. At what simulated target velocity will the radar be able to detect that the target is not real?

5. Using MATLAB, find the spurious spectrum of a linear FM PC signal that is approximated using a stepped-frequency approximation. Plot the spurious sidelobe level that results from the step approximation versus the number of approximation steps. How many steps are required to reduce the spectral sidelobe level to less than –30 dB? Note that this method is used in DRFMs to allow the chirp PC waveform to be stored with a minimum number of memory bits.

6. In the basic quadrature channel DRFM (see Figure 5.5), show that if the input signal is lower in frequency than the down-converting local oscillators, then the Q-channel signal phase will lag the I-channel signal phase.

7. Modify the block diagram of Figure 5.5 using a Hilbert transform approach instead of the quadrature down-converters. Use the y = hilbert(x) function in MATLAB to demonstrate the formation of the I- and Q-channel signals.

8. A DRFM is used in a repeater jammer that employs both RGPO and VGPO. The RGPO uses a 12-μs parabolically shaped walk period that increases the apparent target acceleration from 0 g's to 10 g's. What Doppler profile is required for VGPO to match the RGPO characteristic?

9. In support jamming, multiple signals may be simultaneously stored in a DRFM. Use the MATLAB program drfmfreq.m to study the efficiency of the DRFM (i.e., signal power/signal power + spurious power) for one- to five-bit A/D converters and 1, 5, 10, and 15 simultaneous signals.

10. A 13-bit Barker-coded PC signal is stored in a DRFM by sampling a small portion of each chip and then reconstituting the complete waveform through head-to-tails recirculation. What spurious levels result if a 30-degree random phase discontinuity exists at each segment of the Barker code? (*Hint*: See Example 4.14, cwspec.m MATLAB program.)

11. Construct a matched filter for a 13-bit Barker code using MATLAB. What distortion of the output signal occurs when the DRFM waveform of Problem 10 is applied to the matched filter?

12. Construct a matched filter for a linear FM chirp signal with a time bandwidth product of 100. What is the response of the matched filter to an eight-step DRFM approximation of the chirp signal, assuming perfect phase match? Repeat for a 30-degree random phase discontinuity between steps.

References

[1] Schleher, D. C., *Introduction to Electronic Warfare,* Norwood, MA: Artech House, 1986.

[2] Schleher, D. C., *MTI and Pulsed Doppler Radar,* Norwood, MA: Artech House, 1991.

[3] Hyberg, P., "Radar Countermeasure Technology," A Compendium for the Swedish Military Academy, FOA Report A 10052-1.1, Dec. 1993.

[4] Adler, E., E. Viveiros, T. Ton, J, Kurtz, and M. Bartlett, "Direct Digital Synthesis Applications for Radar Development," *IEEE International Radar Conf.,* Washington, DC, May 1995.

[5] Goldberg, B., "Enhancing the Performance of DDS Signal Sources," *Microwaves and RF,* June 1996.

[6] Tsui, J., *Digital Techniques for Wideband Receivers,* Norwood, MA: Artech House, 1995.

[7] McMillian, G., *Digital RF Memory (DRFM),* Austin, TX: Systems & Processes Engineering Corp., 1996.

6

Electronic Warfare Support

The proliferation, lethality, and pace of modern missile systems have made them a major threat to all platforms whether they are on land, sea, or air. They represent the most sophisticated and deadliest form of weapon and, without countermeasures against them, are capable of overwhelming an opposing force. To counter these threats, most platforms require some form of warning device to (1) alert the crew of imminent danger and (2) provide information necessary to launch an effective countermeasure.

Against modern weapons, the warning and countermeasure's launch time is measured in seconds for aircraft and tens of seconds for ships and other surface vehicles. This requires an automatic response against most missiles directed against a platform since there generally is not enough time for human intervention. To respond with a countermeasure, the type of weapon and its location and velocity must be ascertained. For RF guided weapons, this information can be determined by ES systems. IR/EO guided missiles require a MAW system (see Section 7.3.3).

The general procedure for ES against RF guided weapons is to (1) intercept the radiations associated with the weapon, (2) separate these signals from other intercepted signals, (3) measure the parameters of the selected signals, (4) compare the parameters against a stored set of threat parameters to identify the type of sensors associated with the intercepted signals, and (5) identify the weapon that is defined by the identified sensors. Sometimes the complete procedure can be simplified if distinctive signals (i.e., such as missile data link signals) are intercepted.

In addition to identifying the signal, ES systems also provide information on the angular direction of the intercepted signal and, in some situations, its location. Signal location generally requires accurate *direction finding* (DF) capability from multiple platforms or multiple measurements from a single platform

6.1 Signal and Threat Environment

The signals of most interest to ES systems are those associated with weapon systems. These include primarily radar and data link signals, but also include communications signals. The expected frequency range of these signals is shown in Figure 1.4. In general, radar and data link signals are more structured than communications signals and hence become the prime target for ES systems.

The frequency range of most radar weapon-related signals is in the microwave region of 2 GHz to 18 GHz. There is also minor activity in the millimeter-wave window bands around 35 GHz, 94 GHz, 140 GHz, and 220 GHz. These frequency bands are associated with short-range weapons (i.e., antiaircraft guns and beam rider missiles), and hence most millimeter-wave extensions of ES systems extend to only 40 GHz. There is also activity in the EO/IR and laser radar bands that are covered in Section 7.3. In the lower end of the frequency spectrum, there are radars in the VHF and UHF bands. These search radars generally have low resolution and hence do not in themselves constitute a threat to conventional military platforms. However, the ability of these radars to detect stealth targets has increased the importance of radars in the 80-MHz to 1-GHz band because they can designate targets to fire control systems. Those search radars in the lower microwave region (1 GHz to 2 GHz) are important since they provide initial acquisition of targets for weapon systems. In some ES systems associated with EA systems, these signals are needed to set on jammers and the ES frequency coverage range is lowered to 500 MHz to intercept these radars.

Most radars are of the conventional pulsed type that generate RF pulses that vary from 50 ns to several hundred microseconds. The radars with longer pulsewidths are generally of the *pulse compression* (PC) type that use some form of intrapulse coding. These generally fall into three waveform classes: (1) the FM chirp, (2) the binary phase-coded pulse, and (3) the polyphase-coded pulse. ES systems have difficulty in decoding these types of signals since they must be processed by a matched filter whose exact characteristic is unknown by the ES receiver.

Radars of the *pulsed Doppler* (PD) type have high-duty cycles producing between 10,000 and 500,000 pulses per second. The large number of pulses produced by PD radars tends to overload the processing capability of the ES signal processor. To combat this, special receivers are employed that search for and lock onto high-duty factor signals. These signals are then processed separately and removed from the signal set (i.e., by trapping). The remaining signals are then of the conventional pulsed type that can be processed by the usual ES signal processor. The preceding procedure also applies to CW radar signals and missile data links.

The *pulse repetition frequency* (PRF), or its reciprocal PRI, is another important parameter that often is used in the identification of a radar type. The PRF can range from 100 Hz to over 500 kHz (see Example 4.10 for the ambiguous range of the radar that is associated with the PRI). Many older radars use stable PRFs, but it has been recognized that this is a disadvantage against EA. Therefore, modern radars generally randomly jitter their PRFs. However, PD radars must maintain a stable PRF over their processing interval (i.e., CPI). MTI radars can use a randomly jittered PRF that varies from pulse-to-pulse, but some older types with analog processors use a staggered PRF (i.e., a PRF that varies with a regular pattern).

Low probability of intercept (LPI) radars use a waveform that is intended to be severely mismatched to that of the intercept receiver. One currently operational type (i.e., the Pilot radar) uses an FM-CW waveform [1]. This waveform sweeps over 40 MHz in 1 ms and then repeats itself providing a 46-dB processing gain to the radar as compared to an intercept receiver with a 40-MHz bandwidth. A sensitive ES receiver (i.e., –80 dBm) located on a platform can detect this signal, but at a considerably smaller range than the platform that can be detected by the radar. This is the essence of LPI operation.

A general solution to the LPI radar intercept problem would be for the ES receiver to employ a matched filter that is tuned to the LPI radar's waveform. This is generally difficult to implement due to the presence of other signals (i.e., conventional pulsed radars, pulsed Doppler, and CW radars) in the signal set that occur simultaneously with the LPI signal. Also, LPI radar signals generally employ high duty cycle waveforms that in themselves are difficult to process in ES systems.

Frequency is another radar parameter generally used in the process of emitter identification. Many of the older radars use relatively stable frequencies. However, these tend to be concentrated in narrow radar bands sanctioned by ITU treaty restrictions. Modern military radars generally use some form of frequency agility (i.e., random frequency selection) or diversity (i.e., random selection of a number of fixed frequencies). Intercept receivers have little difficulty in detecting frequency agile radars. However, there is some difficulty in associating which pulses belong to a particular emitter in a frequency agile pulse train. PD and *moving target indication* (MTI) radars must maintain a constant frequency over their CPI that enhances their intercept susceptibility.

Jammers employed in EW systems work almost continuously. The sensitive ES system that supplies information to the jammer must generally be blanked when the high-powered jammer is operating and vice versa. This dilemma is solved by providing a look-through period in which the jammer is deactivated and the ES system is free to collect information such as whether the signals being jammed are still in operation. The randomly selected look-through period is generally less than 5% or less duty cycle with a period of several to 10 ms.

6.2 Parameters Measured by the ES System

Each instantaneous signal intercepted by the ES system must be characterized by a set of parameters. This provides the information required to associate a set of signals belonging to a particular emitter and to identify that emitter among other emitters whose signals have been intercepted. The parameters generally measured by the ES system for a pulsed signal are depicted in Figure 6.1. These include carrier (RF) frequency, *pulse amplitude* (PA), *pulsewidth* (PW), *time of arrival* (TOA), and *angle of arrival* (AOA). Also, in some systems, polarization of the input signal is measured. *Frequency modulation on-the-pulse* (FMOP) is another parameter that can be used to identify a particular emitter and also can be used to determine the chirp rate or phase coding of a PC signal.

CW signals are generally identified as those signals whose pulse lengths exceed several hundred microseconds. TOA measures are made with respect to an internal clock on the leading edge of the pulse. The parameters measured

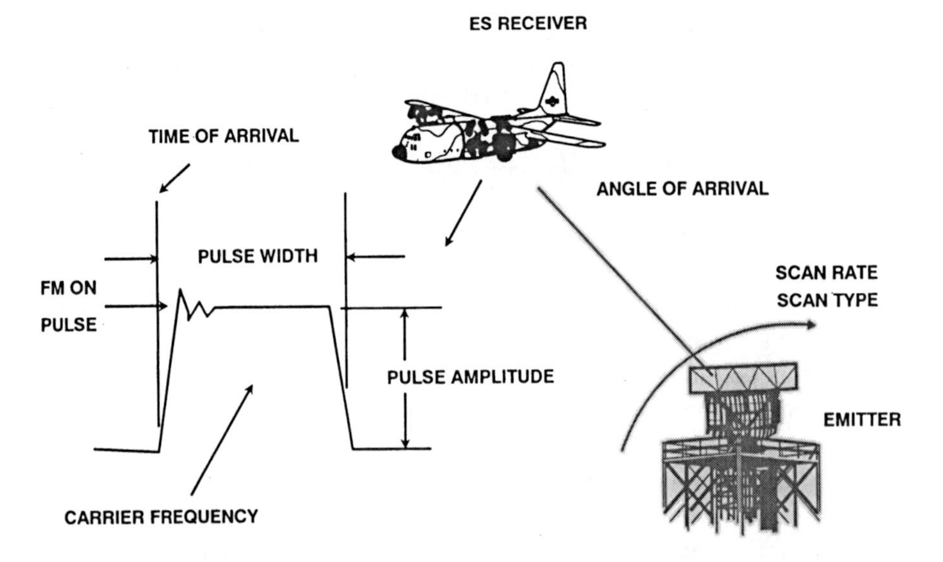

Figure 6.1 Parameters measured by ES receiver.

on a single intercepted pulse are called *pulse descriptor words* (PDW). The PDW form a set of vectors in parameter space. By matching vectors from multiple pulses, it is possible to isolate those signals associated with a particular emitter. This process is called deinterleaving.

Once a signal is isolated, another set of signal parameters associated with the emitter can be determined. These include the PRF or its pattern (from multiple TOAs), antenna beamwidth (from multiple PAs), antenna scan rate or type (from multiple PAs), mode switching (from multiple PWs and TOAs), and range (from multiple AOAs). These derived parameters (sometimes called emitter beam descriptors) are useful in identifying the particular emitters intercepted by the ES system and the threat they represent. Observation in the ES system of mode switching is particularly useful since it often conveys the intent of a particular weapon (e.g., an increase in PRF from a particular emitter generally indicates a transition from search to track).

6.2.1 Pulse Deinterleaving

Pulse deinterleaving is the process of sorting the pulses (or signals) intercepted by the ES system's receiver into individual sets associated with particular emitters. This process is illustrated in Figure 6.2. To accomplish this sorting, each intercepted pulse must be compared with all the other intercepted pulses to determine whether they originated from the same radar. The common parameters used to accomplish this task are the center frequency (or RF) and the AOA. Other parameters used to a lesser extent are PW and difference in TOA, which leads to an estimate of the PRI.

For a conventional pulsed radar, the objective is to measure the carrier frequency of the pulse. If the input is a chirp signal, then the information of interest is the starting and ending frequencies and the PW that allow the chirp rate to be calculated. If the signal is phase coded, both the carrier frequency and the chirp rate are of interest. With frequency agility signals, the average or center frequency of the signal and its excursion band are of interest.

The measurement of frequency is important information used both in sorting and jamming. The accuracy of the frequency measurement determines the resolution that can be achieved in matching various intercepted signals to determine whether they belong to the same emitter. Also, accurately knowing the frequency of the victim radar or communications system allows the jammer to concentrate its energy in the desired frequency range, making the jammer more effective.

The theoretical limit to the accuracy of frequency measurement for a rectangular RF pulse is given by

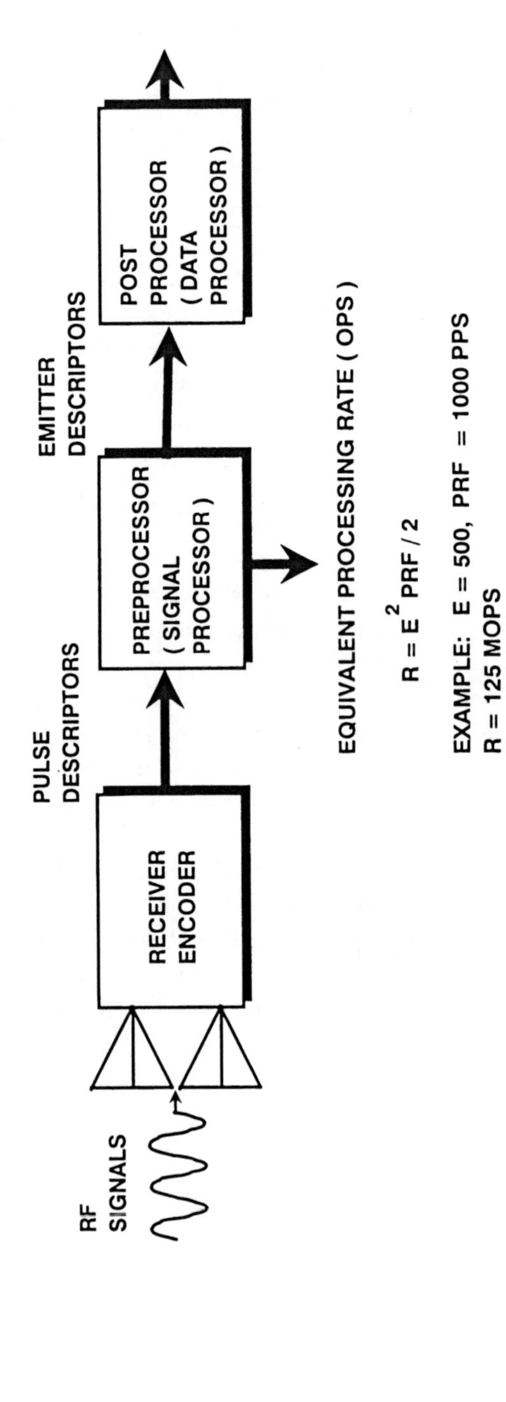

Figure 6.2 EW preprocessing.

$$\Delta f_{\text{rms}} = \frac{\sqrt{3}}{\pi \cdot \tau (S/N)^{1/2}} \tag{6.1}$$

where τ is the PW and S/N is the signal-to-noise ratio. In ES receivers, the measurement of frequency is generally made within the shortest PW expected, which is the order of 50 ns. This requires a 20-MHz receiver bandwidth to process the signal. To achieve an rms measurement accuracy of 1 MHz would require a S/N of at least 20.8 dB. The required signal power into a receiver with a 5-dB noise figure would be a minimum of −75 dBm.

The frequency can be measured in an *instantaneous frequency measurement* (IFM) receiver, a superheterodyne receiver, a channelized receiver, and a compressive receiver. The IFM receiver does not process simultaneous signals, so the signal whose frequency is measured must be isolated before application to the IFM receiver.

The AOA is a valuable parameter used in deinterleaving radar signals since an emitter cannot rapidly change its position. Even an airborne radar cannot significantly change its position in the few milliseconds associated with a PRI. Thus, AOA is a relatively stable parameter. Unfortunately, AOA is one of the most difficult parameters to measure. It generally requires a number of antennas and receivers, all either amplitude- or phased-matched.

A narrowband AOA system that covers the instantaneous bandwidth of the signal can be effective. For example, it is possible to measure the frequency of an incoming emitter pulse (e.g., with an IFM) and then tune narrowband receivers (i.e., superheterodyne) connected to different antennas to that frequency to measure the AOA of the next incoming pulse. One difficulty that occurs with this procedure is that it is difficult to measure frequency when several signals are simultaneously present. This difficulty is circumvented in both channelized and compressive receivers, but these solutions are expensive.

AOA measurement methods generally use either amplitude monopulse or phase interferometric methods. Amplitude monopulse methods are employed when a wide angular coverage is desired, while phase methods are more suitable for narrow angular coverage. RWRs that cover 360 degrees using quadrant antennas produce 10- to 15-degree accuracy. Interferometric systems can produce the order of 1 degree accuracy. However, when wide instantaneous bandwidth signals are involved, it is difficult to phase-match receiver channels. In some cases, phase calibration tables are used to match channels.

Two other methods used to measure AOA from a single platform are the scanning antenna and *time difference of arrival* (TDOA) methods. The scanning antenna method can provide high ES system sensitivity since it focuses a high-gain antenna beam on the emitter but generally provides a low *probability of intercept* (POI). This method is more suited for an ELINT system than an ES system, which requires fast response. TDOA is an emerging method that requires nanosecond digital integrated circuit technology for accurate measurement. Both TDOA and *frequency difference of arrival* (FDOA), which is also called differential Doppler, are currently used in operational systems employing multiple platforms.

The signal processing function of an ES system uses data that is encoded from each intercepted pulse. The data in the form of a vector ($\mathbf{S}_i$) is given by

$$\mathbf{S}_i = [AOA_i \quad f_i \quad PW_i \quad TOA_i \quad Amp_i] \qquad i = 1, n \tag{6.2}$$

where n pulses are present. This assumes that simultaneous signals have been separated so that the individual pulses can be encoded. The next step is to deinterleave the data to sort pulses into groups belonging to a particular emitter. The most reliable sorting parameters are AOA and frequency. To perform a two-dimensional sort using these parameters, the vector $\mathbf{V}_i$ is used.

$$\mathbf{V}_i = [AOA_i \quad f_i] \qquad i = 1, n \tag{6.3}$$

The sorting process then consists of collecting the vectors representing those pulses that match each other within the resolution limits of the sort. For example, in the SLQ-32, the resolution limit is 3 degrees by 10 MHz [2].

Once deinterleaving is accomplished, a number of emitter parameters can be derived from the vector set given in (6.2). The emitter's PRI can be determined by comparing successive TOA. The antenna beamwidth and scan rate can be determined from successive amplitude comparisons (Amp). Mode switching can be determined from multiple PWs. Range can often be determined from multiple AOAs.

In some systems, AOA and frequency are not sufficient to deinterleave emitters. This can occur, for example, with frequency agile emitters. Also, for low resolution systems, there may be a number of distinct emitters that fall into overlapping cells due to the coarse cell structure. The resulting ambiguities must then be resolved in an additional step of deinterleaving. This can be accomplished using either an additional primary parameter, such as PW, or a derived parameter such as PRF (or PRI) that has been determined from the first deinterleaving step. The deinterleaving process then becomes a three-dimensional sort that is more powerful than the two-dimensional sort using AOA and frequency. This three-dimensional sorting process is illustrated in Figure 6.3 using frequency PRF and PW.

The SLQ-32 uses a two-stage three-dimensional sorting process. First, AOA and frequency are used to partially separate emitters in a two-dimensional

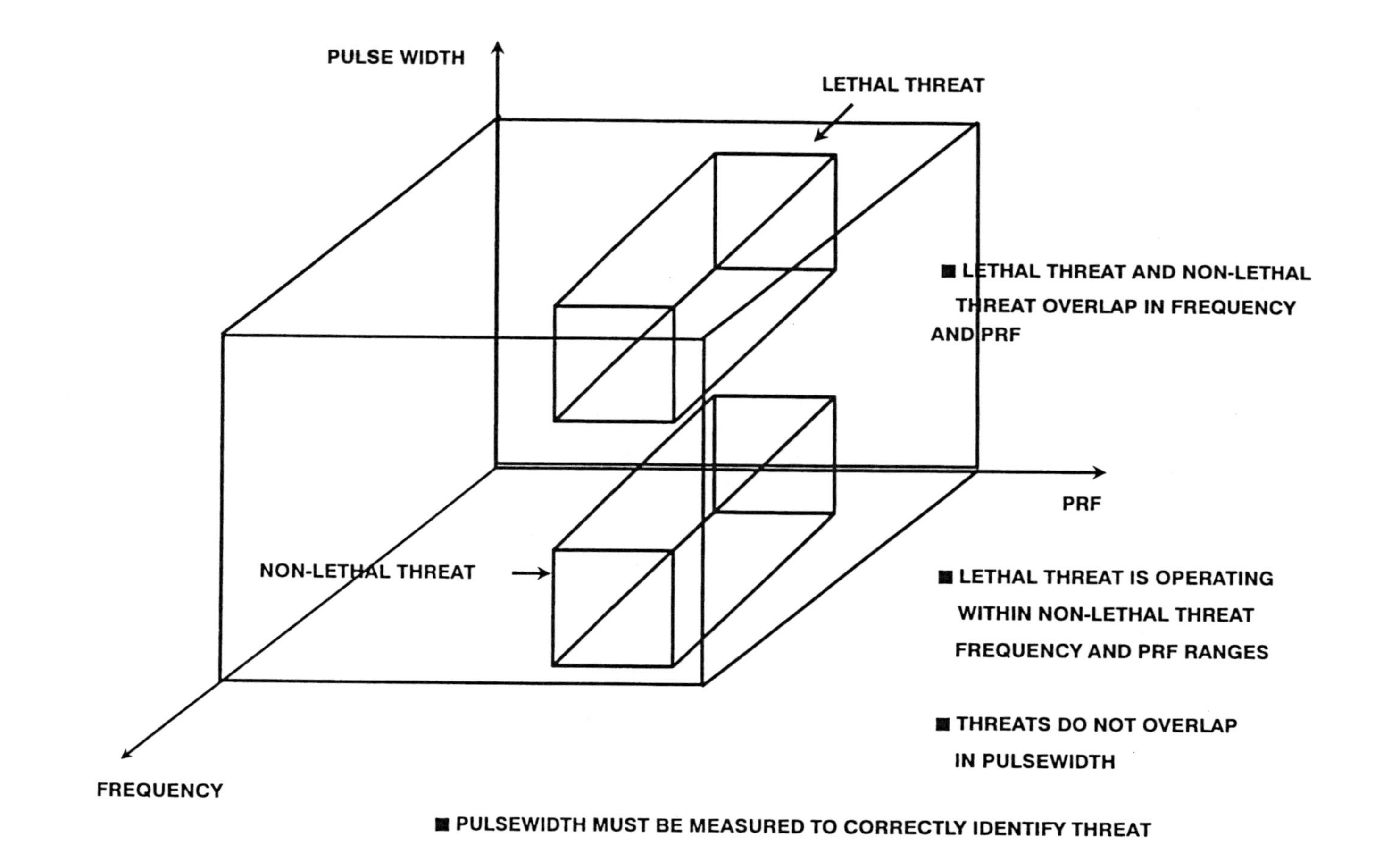

Figure 6.3 Three-dimensional sorting process.

coarse sort (3 degrees by 10 MHz). Then PRI is derived from the coarse two-dimensional sort to form a fine three-dimensional sort in AOA, frequency, and PRI (3 degrees by 10 MHz by 1 μs). The computer then processes the stored amplitude/TOA data history to determine PRF, PRF type, scan period, and scan type [2].

The next function performed is that of identifying a particular emitter. Emitters are identified by comparing the characteristics derived from the intercept functions (e.g., frequency, average PRI, PRI type, scan rate, scan type) with those from known emitters that are stored in an emitter library residing in the ES system computer. However, at times there will be more than one emitter in the library having parameter ranges that include those of the emitter being identified. In these cases, the intercepted emitter's parameters are compared with those parameters associated with other emitters in the environment to effect a match. For example, assume a threat missile is an identification candidate, one of the other emitters is a platform radar associated with the particular threat, and they both fall in the same AOA bin. Then the tentative identification is probably correct. If none of the identification candidates for the new emitter can be correlated with any of the other emitters in the environment, then the emitter is given the identification of that particular candidate having the greatest threat potential. There will also be times when an emitter has characteristics that do not match any of those stored in the library for known emitters, thereby making a positive identification impossible. In that case, the emitter is classified as either an unknown lethal or nonlethal threat based upon the emitter's frequency and scan characteristics.

In a signal processor, throughput measures the ability to absorb input data, while clock speed determines how fast each elemental operation is performed. The EW signal-processing requirements for deinterleaving estimated in the VHSIC program were for a throughput of 10^9 complex operations (i.e., two real operations) per second and a clock speed of 100 MHz. These exceeded the estimates of signal-processing capability required for communications, radar, and precision-guided weapons by one to two orders of magnitude [3].

Deinterleaving basically consists of comparing vectors to see if they match within certain limits. If E emitters with average PRF are intercepted and matched on the average at the halfway point, then the

$$\text{rate} = E^2 \cdot \overline{PRF}/2 \qquad (6.4)$$

operations per second. The matching operation generally consists of comparing the parameters of the intercepted test signal against the parameters of previously intercepted signals. This operation for a Von Neuman type processor consists of first fetching the stored parameter out of memory (RAM), comparing it with both an upper limit and a lower limit derived from the test parameter, and then returning the parameter to memory. This results in four instructions per sorting dimension (i.e., eight instructions for a two-dimensional processor). For example, assume that 500 emitters are intercepted with an average PRF of 1000 pps resulting in a pulse density of 500k pps. The resulting processing rate from (6.4) is 125 MOPs or 1000 MOPs for a two-dimensional sort. If the number of emitters were increased to 1000, then the pulse density would increase to 1M pps and the processing rate to 500 MOPs.

The ES signal processor can process only limited pulse density. This density is generally lower than the pulse density an EW receiver can intercept. Fortunately, once an emitter's signature is identified, it is generally no longer necessary to deinterleave on further pulses associated with this emitter. This is accomplished using a tracker that locks onto the signal and is able to predict the future position of these pulses. The output of the tracker is used to prevent pulses from emitters previously identified from entering the sorting process. This reduces the processing load of the ES signal processor to just that associated with examining new intercepts from different emitters than those previously identified.

Trackers can be considered as multidimensional filters that stop pulses from getting to the deinterleaving portion of the processor. They generally track the PRI and frequency of emitter and gate pulses in both the temporal and frequency domains that correspond to the tracked signal. One tracker is required for each signal to be tracked. In some cases, trackers are also part of the EA system since they isolate and condition signals for jamming operations. A generalized type of tracker that generates a blocking mask is depicted in Figure 6.4. The illustration shows how a dense RF environment can be accommodated with a modest processing capability.

The tracking concept applies generally to signals with stable PRF and frequency characteristics. Since many of the signals are of this type, it provides

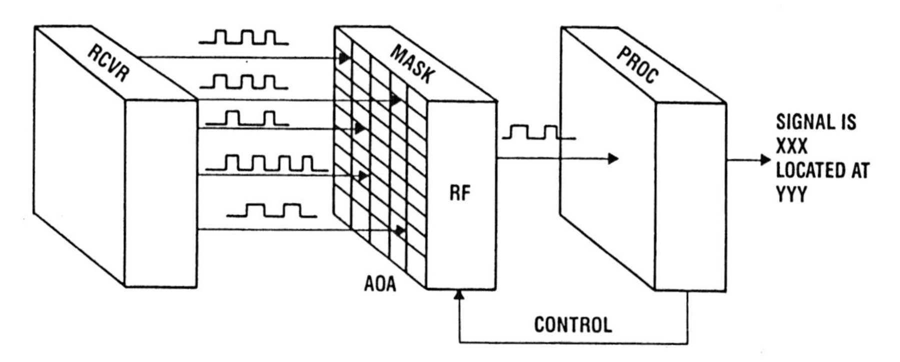

Figure 6.4 Processing in a dense RF environment using a mask.

a convenient way to reduce the pulse density that the ES signal processor has to handle. A limitation of this type of approach occurs in transient situations when the tracking loops have not settled, thereby allowing a high density of pulses to reach the tracker. This approach works best when there is a slow build-up of emitters, which lets the tracking loops converge, thereby limiting the pulse density into the signal processor.

In general, with masking-type approaches, it is necessary to revisit the emitters that have been identified to ascertain if they are still present. In a typical situation, when the ES system is associated with an EA system, the ES receiver is blanked during the period that the emitter is being jammed. During look-through periods, it is important to know whether the signal being jammed is still in operation, otherwise jamming energy may be wasted. The method employed with this procedure is to unblock the mask for critical emitters to allow them to be reprocessed. Since the maximum processing rate is known, the unblocking rate can be managed so that the processor is not overloaded.

6.2.2 Processing of Multiple Pulse Emitters

The overall RF environment seen by the RWR generally consists of a number of unsynchronized pulse-type emitters and a smaller number of CW-type emitters. The CW and high-duty factor emitters are typically separated from the pulse-type emitters in the RF portion of the receiver. As the remaining pulse emitter environment becomes dense, it can generally be assumed that the arrival of pulses is random and follows a Poisson distribution. This is represented analytically as a sequence of Poisson-distributed impulses that allows the properties of the receiver-signal processor to be determined.

Consider a Poisson impulse random process where pulses occur at a random rate of λ pulses per second. The probability that n pulses occur in a time interval τ is given by $P(n) = (\lambda\tau)^n e^{-\lambda\tau}/n!$. The probability that n pulses occur in this interval is independent of the number of pulses that occur in any other nonoverlapping interval. The expected number of pulses in a time interval τ is proportional to the length of the interval and is given by $E(n) = \lambda\tau$. The probability density function of the time random variable x, which measures the distance interior to an interval from any time to the occurrence of the next pulse, is exponentially distributed and given by $p(x) = \lambda e^{-\lambda x}$ [3].

In the analysis of receiver-signal processing combinations, two types of situations can be modeled. The first situation consists of a blocking receiver that can only process a pulse that arrives τ sec after the previous pulse, which may or may not have been processed. This results in a number of pulses being censored from the Poisson impulse process, which, in turn, results in a process called a renewal process. The renewal process is no longer a Poisson process since it was not obtained by random selection from the original Poisson process. In renewal process theory, this problem is referred to as the *paralyzable counter problem with constant dead-time* [3].

The second situation consists of a high-fidelity receiver that accurately reproduces the received pulse train but is followed by a signal processor that has a processing time τ. The result of the finite signal-processing time is to ignore certain pulses, which occur while the signal processor is processing a previous pulse. This problem is referred to as the *nonparalyzable counter problem* in renewal process theory. The distinction between the first and second models is that the first extends the blocking period whether the pulse is processed or not, while the second only blocks when a pulse is processed.

Consider the paralyzable counter model. The probability that a pulse is counted is the probability that its distance or time of occurrence with respect to the previous pulse exceeds the blocking time τ. Since the distance with respect to the previous pulse has an exponential density function (e.g., $p(x) = \lambda e^{-\lambda x}$), the probability is found by integrating from τ to ∞. The resulting probability is given by $P\,(\text{count}) = e^{-\lambda\tau}$, while the expected time between two successive processed pulses is $E(t_n) = e^{\lambda\tau}/\lambda$.

The expected time between successive processed pulses for the most general case of the nonparalyzable counter is given by $E(Y_n + V_n)$, where Y_n is the random (or constant) blocking time and V_n is the time between the end of the blocking interval and the next successive processed pulse. Because arriving pulses are Poisson distributed, it follows that Y_n and V_n are independent and V_n is exponentially distributed ($p_{v_n}(x) = \lambda e^{-\lambda x}$). The expected time between successively processed pulses then becomes $E(t_n) = E(Y_n) + 1/\lambda$, which for constant blocking time τ becomes $E(t_n) = (1 + \lambda\tau)/\lambda$. The resulting rate for processing pulses becomes $\lambda(1 + \lambda\tau)$. Note that for small values of the parameter $\lambda\tau$, the statistics of the paralyzable counter approach those of the nonparalyzable type.

As an example, consider that Poisson-distributed pulses arrive at a rate given by $\lambda = 10^6$ pulses/s. For a blocking time $\tau = 1\ \mu$s, the nonparalyzable-type configuration processes pulses at a rate of 5×10^5 pulses/s, while the paralyzable type processes pulses at the rate of 3.68×10^5 pulses/s. Table 6.1 lists the number of pulses processed for each type of processor for an input of 10,000 pulses as a function of the parameter $\lambda\tau$.

6.3 Advanced ES Systems

ES systems provide interception, identification, analysis, and location of enemy radiations for tactical use. The information they produce is used to provide

Table 6.1
Number of Pulses Processed as a Function of Blocking Time

$\lambda\tau$	Paralyzable Type	Nonparalyzable Type
0.25	7788	8000
0.50	6065	6667
1	3679	5000
2	1353	3333
3	498	2500
4	183	2000

$\lambda = 10^6$ PPS, N = 10,000 pulses.

situational awareness of the hostile threat background, warning of imminent attack, and targeting information for both the employment of EA (soft kill) and destructive suppression (hard kill) of enemy systems.

ES systems respond to threats directed at the platforms they are designed to protect. The microwave region (2 GHz to 18 GHz) is a prime frequency band since radars located in this region provide fire control data to missile and gun systems. The expansion of these radars into the MMW region (to 40 GHz and eventually to 94 GHz) as illustrated in Figure 1.4 and further into the optical region (Lidar) point to the need for expanded coverage beyond the microwave region. In addition, the communication systems that link the weapon systems through their C2 structure are also targets of ES systems. This follows from a realization that suppression of these systems is equivalent to attacking the primary sensors.

An important consideration in the design of an ES system is the sensitivity of the receiver necessary to intercept the signals of interest.

Figure 6.5 shows how the composite radar threat pulse density varies as a function of the intercept receiver's altitude and sensitivity for an environment similar to the missile- and gun-associated radars of Table 2.3. For purposes of this analysis, it is assumed that all radars are operational and homogeneously sited; PRFs are related to operating ranges; and the radar's antennas have 0-dBi average sidelobes and backlobes. The sensitivity level (–60 dBm to –80 dBm) required to pick up a majority of the emitters is apparent, as is the limitation imposed by the lower sensitivities as a function of altitude. A sensitivity of –40 dBm appears adequate for low altitudes (less than 1000 ft) but severely limits the interceptions at higher altitudes as more emitters enter the horizon limits. A sensitivity of –30 dBm appears to be inadequate, even at low altitudes of 100 ft.

An examination of Figure 6.5 shows that at high sensitivity (–80 dBm), most radars are intercepted both in their main antenna and sidelobe patterns.

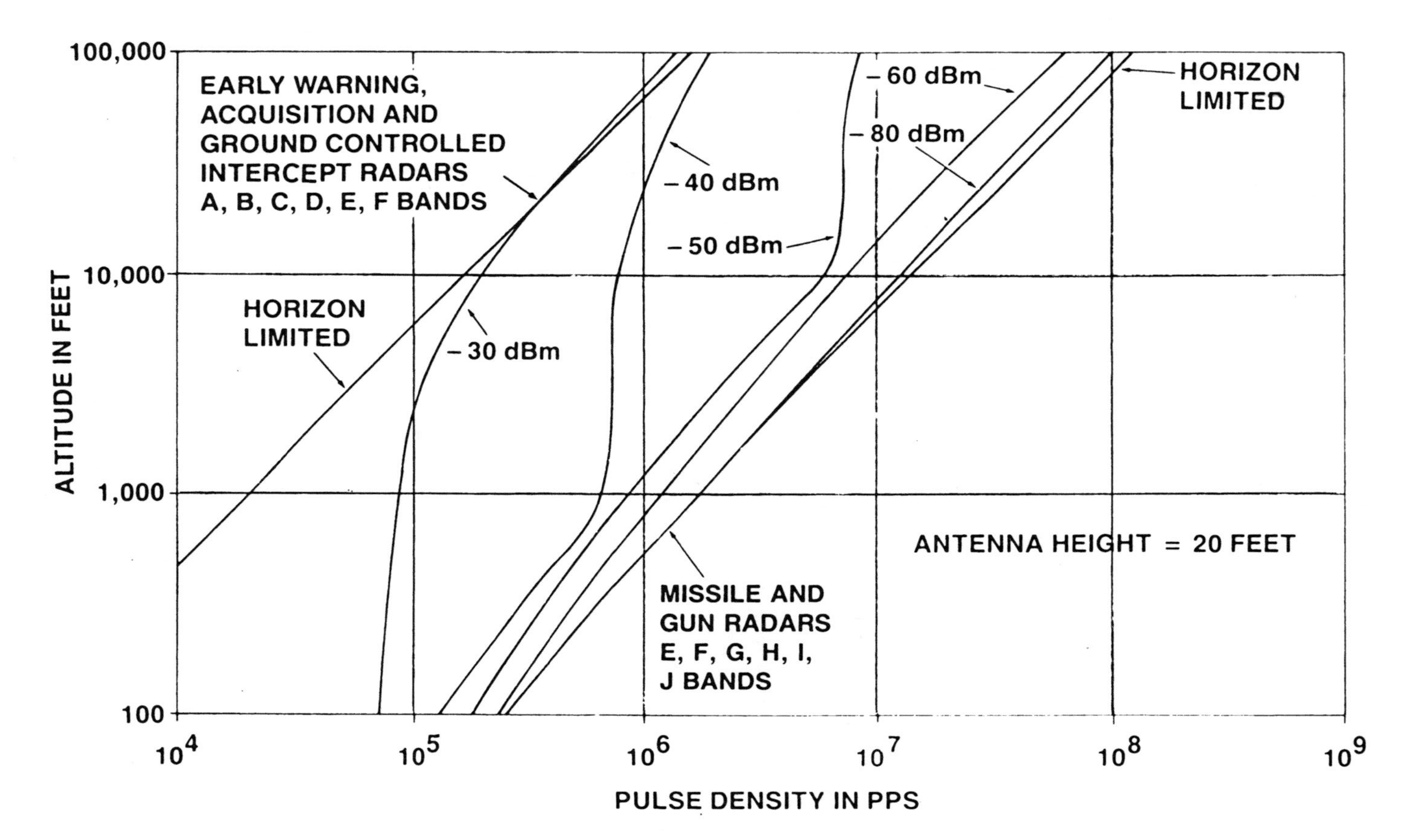

Figure 6.5 Emitter pulse density as a function of altitude and receiver sensitivity. (*Source:* [3], p. 518.)

This is then the design value to provide complete situational awareness of all radars within the field-of-view at a particular altitude. Also, DF generally dictates the interception of sidelobe responses. Another factor associated with high sensitivity is the large number of pulses (order of 10 million pulses/s) that have to be processed to identify each particular emitter received. This problem is compounded with many of the modern threats, involving pulse Doppler radars, which involve between 10k to 500k pulses/s depending upon whether an air or ground radar is involved. Note also that a lower sensitivity (order of –40 dBm) might be advantageous in some circumstances where just main-beam intercepts are important (i.e., *radar warning receivers* (RWRs)).

ES equipment designed for radar intercept falls into two broad classes: (1) RWRs operating in real time, used by aircraft, ships, submarines, and ground forces for self-protection; and (2) reconnaissance/surveillance receiver systems used to intercept, collect, analyze, and locate radar signals in near real time so as to update the situational awareness for targeting, EA deployment, early warning of enemy approach, and fusion with other sensors.

RWRs are used in military aircraft and helicopters to warn of attack by SAMs, AAMs, air interceptors, and anti-aircraft gun systems. They are also used to provide warning to tank crews of imminent missile or gun attack and by submarines on or near the surface to warn when an aircraft's surface-search radar or weapon system is illuminating the submarine. Once alerted to the type, direction, and relative priority of the threats, the crew may take evasive maneuvers and employ deceptive countermeasures, chaff, or flares, as appropriate, to foil the attack.

RWRs are generally the simplest form of ES receiver consisting of unsophisticated low-sensitivity (on the order of –40 dBm) equipment that is preset to cover the bands and characteristics of expected threats and exploits the range advantage to indicate a threat before it reaches its firing range. ES reconnaissance or surveillance receivers are generally considerably more complex than RWRs, and they are used to map enemy radar and communication installations and to monitor radio messages. The more elaborate radar ES receivers are similar in concept to RWRs, except that they generally employ more sensitive receivers to intercept radar sidelobe radiations at long ranges; have higher DF accuracy; and measure additional radar parameters such as coherency, polarization, pulse shape and artifacts, intrapulse modulation, and statistical features.

ES receivers directed against communication transmissions are similar in concept to those designed to intercept radar transmissions, except that a different approach is required to accommodate the communication signal structure. Communication systems generally operate on discrete channels, are relatively powerful, employ widebeam antennas with poor sidelobes, and use modulated

CW transmissions. Conventional radio transmissions were becoming relatively easy to jam, so modern designs use frequency-hopping transmissions over a wide bandwidth with the expectation that by the time an ES receiver has located a transmission burst, the transmission has moved on to the next random frequency. The net result of this communication signal structure at the ES receiver is a wide dynamic range and dense and overlapping signal environment. This puts a premium on DF as a method of sorting and locating communication emitters, while pulse deinterleaving and matching against a stored threat emitter file is the principal method used to sort radar emitters.

The requirements for a modern high-performance ES receiver are listed in Table 6.2. The output of the receiver is encoded in PDWs as indicated in Table 6.3, which allows deinterleaving of the received pulse set into those associated with individual emitters. After deinterleaving other parameters associ-

Table 6.2
Modern ES Receiver Performance Requirement

Frequency range	0.5–40 GHz
Signal type	50 ns pulse to CW
Sensitivity	< –70 dBm
Frequency resolution	< 2 MHz
Dynamic range	> 70 dB
Amplitude accuracy	1 dB
Bearing accuracy	Better than 1-degree rms
Pulsewidth resolution	25 ns
TOA resolution	50 ns
Probability of intercept	100%
Pulse rate	10^7 pulses/s

Table 6.3
Pulse Descriptor Words

Parameter	Bits	Resolution
Time of arrival	24	50 ns
Frequency	18	1 MHz
Polarization	16	1 degree
Amplitude	7	1 dB
Angle of arrival	9	1 degree
Pulsewidth	13	50 ns
Flags	8	
Total	96	

ated with the emitter such as PRF (multiple TOAs); scan rate, type, and antenna beamwidth (multiple amplitudes); range (multiple AOAs); and modes (multiple PWs) can be derived. The total set of parameters (measured and derived) are then compared against parameters stored in the emitter parameter list library to identify the particular emitters and their location. This information can then be used to identify the weapon complex associated with each group of emitters.

A performance measure listed in Table 6.2 that is commonly used to describe ES receivers is the *probability of intercept* (POI). This measure expresses the probability that the parameters of the ES receiving system (particularly center frequency and antenna orientation) match those of the transmitting emitter within some specified period of time. There are several important variations of this problem that occur in search intercept situations. The beam-on-beam intercept problem occurs when both the ES receiver and emitter use directional rotating antennas and interception is accomplished when the main beams of both antennas point at each other. The beam-frequency intercept problem occurs if either the emitter or ES receiver uses a rotating directional antenna and the ES receiver is scanning in frequency. Interception occurs when the rotating antenna is aligned along the receiver-emitter direction and the receiver passband encompasses the frequency of the emitter. The frequency-frequency intercept problem occurs when the emitter is hopping in frequency while the ES receiver is simultaneously scanning or stepping in frequency. Interception occurs when the emitter frequency is within the passband of the ES receiver. In general, high POI is achieved using wide-open (nonscanning) ES receivers coupled with multibeam ES antennas (high-gain antennas with parallel receivers), which provide high-gain and wide-angle coverage.

The POI problem can be put into a mathematical framework (when the scan periods are regular) by considering the equivalent problem of the intersection of two periodic pulse trains with random starting times. The width of the pulses in the pulse train are made equal to the time the emitter or receiver is matched for interception, while the period represents the scan repetition time. The analytical solution to this problem is apparently sensitive to small variations in the ratio of the periods, which can sometimes cause the trains to "lock in synchronism" for significant time periods and cause a large variation in the POI value. In practice, the conditions for 100% POI are generally clear while analytic formulations for lower POI are more useful for qualitative rather than quantitative results [4].

A generic ES system architecture that provides 100% POI is depicted in Figure 6.6. The strategy of this ES system is to divide the spatial and spectral bands into multiple subbands, thereby reducing the pulse density in each subband to a manageable level. Spiral antennas covering the 2- to 18-GHz band

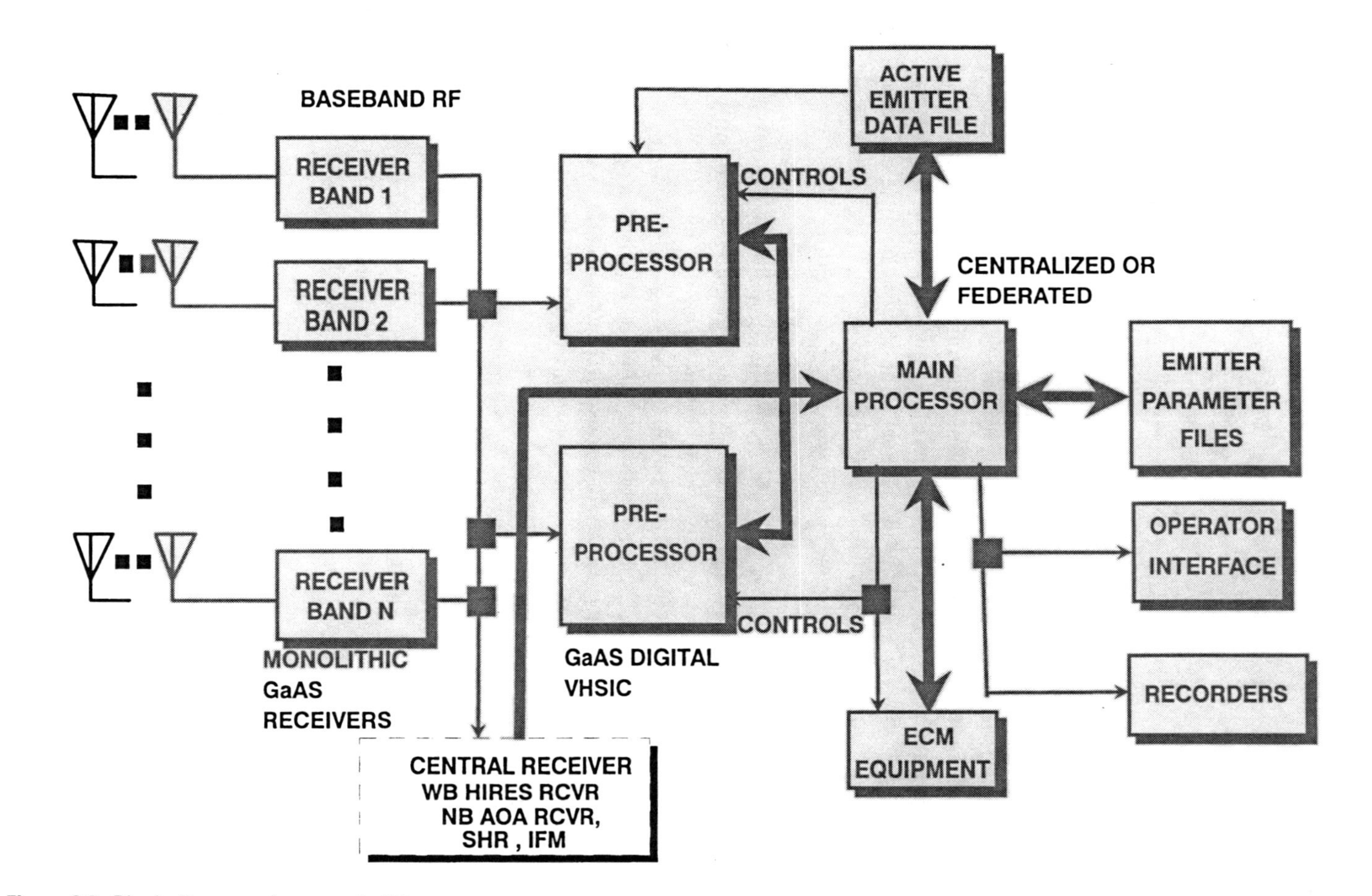

Figure 6.6 Block diagram of a generic ES system.

are often used in this type of application and have been recently extended into the MMW band. Sinuous spiral antennas providing simultaneous right- and left-handed circular polarization coverage are now available to cover all possible polarizations [5].

Monolithic microwave integrated circuit (MMIC) receivers are employed, each covering a particular sub-band. A number of receiver types are necessary to provide the encoded information required for emitter deinterleaving. A wideband high-resolution receiver provides the necessary fidelity to allow precision TOA and PW measurements. The instantaneous bandwidth of this receiver must be compatible with the largest expected signal spectral extent, which is the order of 1 GHz for modern signal modulations.

A separate receiver is necessary to allow measurement of the signal's AOA. To accomplish the AOA measurement, the signal must be isolated from other concurrent signals which is usually accomplished in narrowband superheterodyne receivers. Present methods use either amplitude monopulse or phase interferometric angle measurement techniques [3]. Recent developments in this area use TDOA techniques, which require time measurements in the nanosecond region.

Superheterodyne receivers provide several functions in an ES system. Generally, they allow specific signals to be isolated from other signals on the basis of their spectral content. This is an essential feature for DF and in capturing an accurate version of the signal so that it can be utilized in a deceptive jammer (i.e., DECM) to produce a signal that will be accepted by the victim radar. Also, superheterodyne receivers are used to search for CW, pulse Doppler, and other high-duty cycle waveforms among the complete signal set. This allows these signals, which would overload the processing system, to be extracted from the total signal set leaving just the conventional signals to be processed. The parallel output then consists of conventional signals and high duty cycle signals that can be separately analyzed and processed.

Instantaneous frequency measurement (IFM) receivers are used to provide an accurate frequency measurement generally required for pulse deinterleaving. Unfortunately, IFM receivers, which employ correlators, only respond to the strongest simultaneously occurring RF signal, thereby suppressing weaker signals in a dense signal set. Thus, it becomes essential to remove high-duty factor signals (i.e., CW, pulse Doppler) from the signal set before processing in an IFM receiver. When multiple signals are simultaneously present at the input to an IFM receiver, a flag is generated that indicates the data output of the IFM is in question.

▶ *Example 6.1*

The basic IFM receiver divides the incoming signal into two paths, one having a delay (τ_d) of known length and the other with zero delay. If the two signals are correlated, then the resulting output is given as $\cos(\omega_0 \tau_d)$. Similarly, if a 90-degree phase shift is inserted in the zero delay path, the correlated output is $\sin(\omega_0 \tau_d)$.

In a *digital IFM* (DIFM), the output of each correlator is quantized into a zero or a one. For example, if the delay (τ_d) is 1/8 ns, then the total frequency range of the sine correlator will be $f_d = 1/\tau_d = 8$ GHz. A positive correlator output indicates that the input frequency (f_o) is between 0 GHz to 4 GHz, while a negative output indicates that f_o is between 4 GHz and 8 GHz. Similarly, a cosine correlator will provide a positive indication if the input signal is between 0 to 2 GHz or 6 GHz to 8 GHz and a negative output if between 2 GHz to 6 GHz. Thus, using sine and cosine correlators permits the signal frequency to be resolved to $f_r = f_d/2^m = 2$ GHz, where m is the number of correlators. Following this procedure and increasing the delay in binary increments for successive correlators provides a digitization of the input frequency to a resolution determined by the number of correlators used in the DIFM. For example, if 9 correlators are used to process a frequency band of 8 GHz, then the output for an input signal at 7.7 GHz is 100011010. The output of the DIFM appears in a Gray code. To convert from Gray code to Binary: (1) the leftmost digit is the same and (2) the sum of $b_{n+1} + b_n + g_n$ is even, where n is measured from the right, b_n is the nth place of the Binary code, and g_n is the nth place of the Gray code. The resulting Binary code is 111101100 or 7687.5 GHz.

Write a MATLAB program that provides the Gray code, Binary code, and digital frequency output of a DIFM given the number of correlators (m) and the shortest delay segment (τ_d). Test this program for $m = 9$, $\tau_d = 0.125$ ns, and a frequency of 7.7 GHz. What is the error in the measurement? Is this within the theoretical resolution bound?

```
% DIFM Output Gray and Binary Codes
% ----------------------------------
% difm.m

clear;clf;clc;

% Input DIFM parameters and input frequency

td=input('Enter time delay in sec:  ');
fo=input('Enter frequency in Hz:  ');
m=input('Enter number of correlators:  ');
T=1e-4;

% Calculate measured frequency of analog IFM
```

```
t=0:T/1000:T;
vin=exp(j*2*pi*fo*t);
vd=exp(j*2*pi*fo*(t-td));
vo=vin.*conj(vd);
vc=mean(real(vo));
vs=mean(imag(vo));
f=atan(vs/vc)/(2*pi*td);
if vs<=0 & vc>=0;
   f=(2*pi+atan(vs/vc))/(2*pi*td);end;
if vs<=0 & vc<=0;
   f=(pi+atan(vs/vc))/(2*pi*td);end;
if vs>=0 & vc<=0;
   f=(pi+atan(vs/vc))/(2*pi*td);end;

fprintf('nfo= %-6.4e n',f);

% Find Gray code

if vs<=0;x(1)=1;
else x(1)=0;end;

for i=1:m-1;
   td1=td*2^(i-1);
   vd1=exp(j*2*pi*fo*(t-td1));
   vo1=vin.*conj(vd1);
   vc1=mean(real(vo1));

   if vc1<=0;x(i+1)=1;
   else; x(i+1)=0;end;
end;

graycode=x;
graycode

% Convert Gray code to Binary code

y(1)=x(1);

for i=1:length(x)-1;

   if y(i)+x(i+1)==1;y(i+1)=1;
   else y(i+1)=0;end;
end;

binary=y;
binary

% Find Decimal output

z=fliplr(y);
z=find(z);
```

```
zz=z-1;
zzz=2.^zz;

decimal=sum(zzz);
fr=decimal/(td*2^m)
```

```
» difm
Enter time delay in sec: .125e-9
Enter frequency in Hz: 7.7e9
Enter number of correlators: 9

fo=7.7000e+009

graycode=

  1 0 0 0 1 1 0 1 0

binary=

  1 1 1 1 0 1 1 0 0

fr=

7.6875e+009

»
```

An essential part of an ES system is the emitter parameter list stored in the computer library. Comparison against these files is the mechanism that allows identification of the intercepted emitter. Interception of an unknown emitter generally provides little tactical information. Also, identification of friendly emitters allows a parsing of the data set that becomes important when processing dense emitter sets.

Signals intelligence (SIGINT) is the process that facilitates compilation of an accurate threat data base. A distinction is made between the collection of *communications* (COMINT) and noncommunication *electromagnetic* (ELINT) data, but both come under the umbrella of SIGINT.

SIGINT data generally focuses on producing intelligence of an analytical nature that is not as time critical as ES data. The prime customers for SIGINT data are the upper echelons of military forces, which can include the commanders at national levels.

ELINT is generally performed on a regular basis in times of peace prior to a specific mission but can occur under actual war conditions or during an attack. Peacetime operations have the objective of securing the maximum possible data on the complete electromagnetic environment within those areas of interest to any one nation. Special ships, aircraft, and satellites, as well as fixed and mobile land-based ELINT facilities are employed, often operating on comprehensive reconnaissance schedules.

The basic targets for ELINT are all types of radar (land, sea, or airborne for surveillance, fire control, navigation, and other applications) that are detected, located, and identified by their radar signatures in all their operating modes (e.g., search, tracking) and their transmissions hence recorded. These recordings contain the radar characteristics such as PRF, PW, transmitter frequency, modulations, and any other parameters constituting the signature of a radar that enable it to be identified without being seen. Other electronic reconnaissance targets that are given similar attention are navigation systems, command and telemetry links, and data links.

ELINT data are used in several ways. First, there is the straightforward intelligence function where the recorded signals are analyzed for the purpose of establishing the likely function and mode of operation of each individual piece of electronic equipment. The information may also permit an assessment of the equipment's performance or that of its associated system. This may in turn allow evaluation of the state of the art attained by the equipment surveyed. In addition, the types, numbers, and locations of the detected electronic systems permit an evaluation of the other nation's strength and intentions. Furthermore, regular monitoring may reveal potential changes in strategy.

ELINT and its subsequent analysis also fulfills a number of tactical functions. As previously mentioned, the determination of an enemy's EOB is used to form a threat library for use in ES and EA equipment. Also, in tactical engagements, special ELINT missions may be mounted for the purpose of gaining data to use in planning a specific attack. A typical mission might be the determination of the numbers, activity, types, and locations of defensive search radars, acquisition radars, and weapon control radars in a particular area. These data are intended for use in determining the best mode of attack and deployment of EA equipment to suppress these radars.

Ambiguities or overlaps in the parameters used to distinguish one emitter from another are prevalent in the data base. This leads to misidentification of intercepted emitters and presents problems when deinterleaving signal sets. A method sometimes used to identify a specific emitter is known as fingerprinting and generally relies on *unintentional frequency modulation on pulse* (UFMOP). This type of signature is most prevalent in radars that use magnetron transmitters or other emitters that use pulsed oscillators [6].

UFMOP can potentially identify a particular emitter (serial number) among emitters of a similar type on the basis of its transmitter's signature. It can also be used for deinterleaving to pick a particular emitter from a dense set of pulses. However, most unintentional frequency modulation occurs on the leading and trailing edge of pulsed oscillators as they move from a quiescent to an active state. This implies a high sampling rate (the order of 16 to 32 samples per radar pulse) to capture the signature information, which in turn implies high sampling bandwidths (the order of 320 MHz to 640 MHz for a 100-ns pulse). Thus, it appears that UFMOP would be most practical for hard-to-identify emitters rather than as a generalized deinterleaving or identification tool.

Figure 6.6 is one of many possible ES system architectures. It is appropriate where 100% POI is required and made possible by the miniaturization and economy realized using MMIC techniques. Other types of proven architectures are listed in Table 6.4. The ALR-45F is typical of airborne RWR-type ES systems. It employs four-quadrant coverage spiral antennas that provide amplitude monopulse DF capability. Low-sensitivity crystal video receivers provide main-beam threat warning. The ALR-45F has been replaced by the ALR-67(V)V2, which adds superheterodyne receivers for signal analysis and detection of high-duty cycle signals. A major upgrade called the ALR-67(V)3/4 that uses a channelized receiver has recently been introduced.

The SLQ-32 is used for ship defense and consists of five different variations providing various levels of EW capability including ES and EA. The system's lens-fed multibeam array used for both the receiving and transmitter antennas creates a set of individual contiguous high-gain beams, all existing simultaneously, with each beam possessing the full gain of the array aperture. The array used for the ES subsystem provides highly sensitive, wideband, simultaneous coverage over 360 degrees in azimuth with the capability of high-accuracy measurement of emitter direction.

Each quadrant multibeam array feeds seventeen crystal video receivers in bands H to J and nine in bands D to G. The outputs of these receivers are used to provide pulse descriptors consisting of AOA, TOA, amplitude, and carrier frequency. The emitter's frequency is obtained using an IFM receiver processing the output of an omniantenna. AOA is obtained using amplitude monopulse processing between adjacent crystal video receiver ports. Emitters are coarse sorted using a combination of AOA and coarse frequency (3-degree by 10-MHz resolution). The emitter's PRI is derived from TOA measurements applied to the coarse sort. Then a fine sort is accomplished using AOA, carrier frequency, and PRI with a resolution of 3 degrees by 10 MHz by 1 μs [2].

The architecture of the SLQ-32 is similar to that of the generic ES system. Essential system requirements include quick reaction time and 100% POI to respond to the ASCM threat. However, the use of crystal video receivers limits system sensitivity and the omni-fed IFM is susceptible to overload in a dense emitter environment.

The architecture of the ALR-47 (upgraded to the ALR-76) is similar to that of the ALR-45, but scanning superheterodyne receivers are used in place of the wide-open crystal video receivers. This allows the system to provide

Table 6.4
Proven Antenna/Receiver Combinations

Example Systems	Antenna (Type)	Receiver (Type)	System Sensitivity	System Response Time	Application	Platform
AN/ALR-45F	4 spirals (wide open)	Crystal video (wide open)	Low (–45 dBm)	Fast (few seconds)	Threat warning	A/EA-6, A-7E
AN/SLQ-32	Array lens (wide open)	Unpreamplified (crystal videos and IFMs (wide open))	Low (–50 dBm)	Fast (few seconds)	Threat warning/ ESM	Many cruisers and destroyers in future
TAC 105	4 pairs of spirals (wide open)	Swept superhets (narrowband)	Moderate (–70 dBm)	Moderate (many seconds)	Threat warning/ ESM	Gun boats S - 3a
AN/WLR-11 with AS-899 equivalent (foreign navies)	Rotating dish (narrowbeam plus omni)	Preamplified IFMs (wide open)	Moderate to high (–70 to –80 dBm)	Fast (few seconds)	ESM/cruise missile defense	Many cruisers and destroyers

greater sensitivity but limits its POI to less than 100% and results in greater response time. The maximum time-to-intercept of a pulsed radar signal for a single scanning superheterodyne receiver can be shown to be given by [3]

$$t_i = \frac{D}{B_{if} \cdot f_r} \tag{6.5}$$

where D is the receiver's coverage band, B_{if} is the receiver's bandwidth, and f_r is the radar's pulse repetition frequency. Smart scanning can often reduce this time by employing a wideband search followed by a narrowband search of the frequency regions found by wideband search to contain signals.

► *Example 6.2*

A scanning superheterodyne receiver is to cover the band from 2 GHz to 18 GHz to locate pulsed radar signals whose PWs range from 0.1 μs to 30 μs with PRFs of 500 to 8000 pps. What is the maximum time to intercept on emitter? Devise a smart scan system that uses a wideband/narrowband search strategy. What is the maximum intercept time for this system? How does this depend on the number of emitters in the field?

The WLR-11 employs a rotating dish antenna covering the band from 7 GHz to 18 GHz that feeds an IFM receiver. The dish can scan at rates up to 300 rpm. The IFM receiver uses a low-noise preamplifier that provides a relatively high sensitivity of −70 dBm to −80 dBm (includes the antenna gain). This naval ES system is primarily intended to provide detection of radar-guided ASCM.

The high-speed spinning antenna in this system provides increased sensitivity, spatial filtering, and DF capability. The sensitivity is sufficient to provide sidelobe detections if the ASCM seeker is in a search mode, while the antenna dwell time allows main-beam intercept if the seeker is tracking the ship. Identification of the emitter is accomplished by measuring its frequency and PRI. The directivity of the scanning antenna basically provides a spatial filter effect excluding all returns except those from a narrow angle included within the main antenna beam. This limits the number of multiple signals that might otherwise cause the IFM to become desensitized. As with all rotating antenna systems, an omnicomparison antenna is required to protect against spurious sidelobe responses that exceed the sensitivity threshold and appear as weak but bona fide signals.

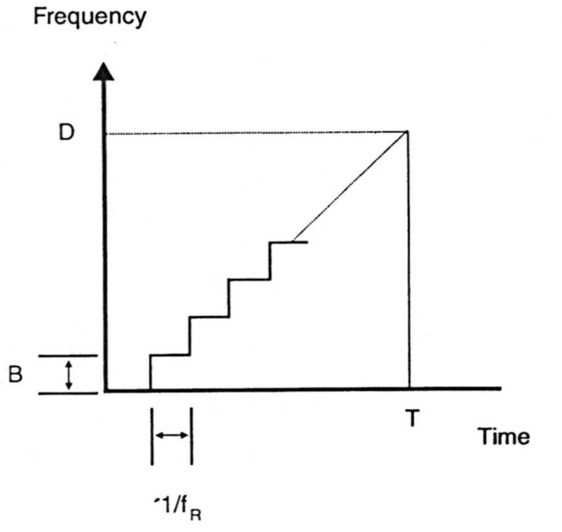

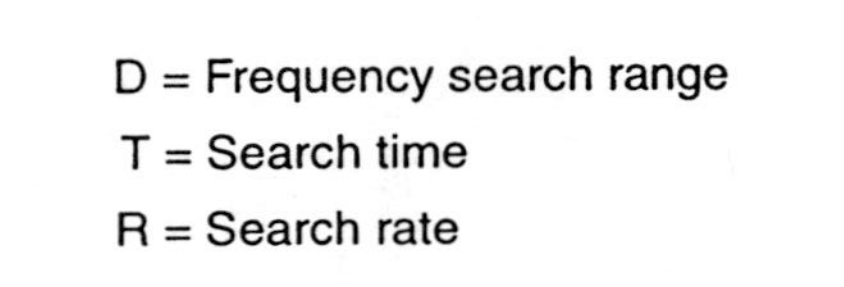

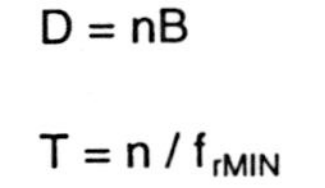

$D = nB$

$T = n / f_{rMIN}$

$R = D / T = B f_{rMIN}$

Example

B = 10 MHZ	R = 5000 MHZ / S
D = 16 GHZ	T = 3.2 S
f_r = 500 HZ	

Wideband / Narrowband search

WB = 100 MHZ	R = 50,000 MHZ / S T = 0.32 S
NB = 10 MHZ	R = 5,000 MHZ / S T = 0.02 S

Total time = 0.32 + 0.02 x n

n = Number of 10 MHZ bins with emitters

Example 6.2 Wideband/Narrowband RWR search.

6.3.1 Advanced Receiver Systems

Conventional receiving systems (i.e., crystal video, superheterodyne, IFM) generally break down as the signal environment becomes dense. One approach to a dense signal environment is to combine a number of specialized receiver techniques with each receiver designed to handle a particular subset of the overall signal set. This approach tends to break down as new threat signals are introduced. A better solution is a receiver that is robust with respect to the threat environment.

When many emitters are present, it is ultimately necessary to break down the emitter set in order to classify each signal. In the radar case, the emitter parameters available for this purpose are amplitude, PW, frequency (carrier and modulation), AOA, and pulse train characteristics. Frequency and AOA are considered the most reliable parameters for the sorting process. Another requirement is that the receiver be able to make a measurement in a time that corresponds to the shortest PW of the emitter set to be analyzed.

The process of breaking down a parameter set into subsets is generally referred to as filtering. The available filtering possibilities for the ES receiver are the frequency, spatial, and temporal domains. Filtering in the frequency domain is generally called spectral analysis and must be accomplished within a radar PW. Spatial filtering is accomplished by breaking up the antenna coverage into narrow multiple beams or by placing antenna nulls over selected emitters. Temporal filters can be formed using either analog or digital filters that respond to a certain subset of pulse train parameters. In this section, we will concentrate on spectral analysis techniques, which are key to advanced receiver designs and provide an essential parameter for emitter sorting and the efficient allocation of EA resources.

The basic structure of an ES receiver spectrum analyzer is a set of contiguous filters that cover the band of interest with individual filter bandwidths that support the shortest pulse to be analyzed. This type of ES receiver is called a channelized receiver. For example, if it is desired to cover the band from 2 GHz to 18 GHz and analyze PWs down to 0.1 μs, then 1600 filters, each with 10-MHz bandwidth, are required. This channelized receiver would have a resolution of 10 MHz, while the frequency measurement accuracy would depend on the duration of the radar pulse and the available signal-to-noise ratio. The channelized receiver is useful for resolving multiple-beam radars that transmit from the same site but at different frequencies as well as any emitters separated by more than the basic channel bandwidth (10 MHz in the example). The hardware associated with the generation of 1600 filters is obviously large, and the art of designing channelized receivers is primarily concerned with compromises resulting from the use of a lesser number of filters in a manner that approaches the effect of a full set.

There are a number of other methods available to accomplish spectral analysis that can be carried out with considerably less hardware than the brute force approach of the channelized receiver. These methods are generally related to finding the Fourier transform of the input waveform and are implemented in either the form of a compressive, an *acoustic-optical* (AO) Bragg cell, or a completely digital receiver using the FFT.

The most advanced of these high-technology receiver types is the channelized receiver that is deployed in a number of operational equipment. The critical technology necessary to build channelized receivers are GaAs MMIC devices and SAW filters. The MMIC devices provide the economy and miniaturization necessary to fit a large number of microwave components into a small volume. The SAW devices provide the sharp filter skirts that are necessary to separate signals into distinct frequency channels without producing spurious crosstalk between channels.

Recent programs using channelized receivers are the ALR-67(V)3/4 RWR, the *emitter location system* (ELS) carried by the Tornado aircraft, and the ALQ-161 used in the B-1B but scheduled for replacement. With the exception of ALQ-161, these systems make extensive use of MMIC devices.

6.3.1.1 Channelized Receiver

There are basically three different configurations of channelized receivers that are used in practice. These are the pure, band-folded, and time-shared variations. The pure channelized receiver divides the frequency range to be covered into contiguous channels equal to the final resolution required. The band-folded configuration folds a number of bands into a common sub-band. The time-shared configuration switches only those channels that are active into a common sub-band.

Figure 6.7 depicts a time-shared or fast call-type channelized receiver architecture designed to cover the current established threat range of 0.1 GHz to 18 GHz. The microwave region is split into nine segments, each 2-GHz wide. Each segment is converted to a standard 2- to 4-GHz first IF and then amplified. The segmentation is done with the aid of lowloss multiplexing bandpass filters. RF preamplifiers are generally used to provide low noise figures (e.g., $NF < 10$ dB). Conversion to the first IF uses conventional diode mixers. In most cases, a large dynamic range (e.g., 50 dB to 60 dB) is required and the mixers must be designed for low spurious responses due to higher order mixing products. This can be accomplished using four-diode doubly balanced mixer circuits.

The various local oscillator frequencies are usually produced by frequency multiplication of a stable reference frequency, which in Figure 6.7 is 2 GHz. Each channel segment has a low-noise 2- to 4-GHz IF amplifier with a typical noise figure of 3 dB and an amplifier gain of 30 dB.

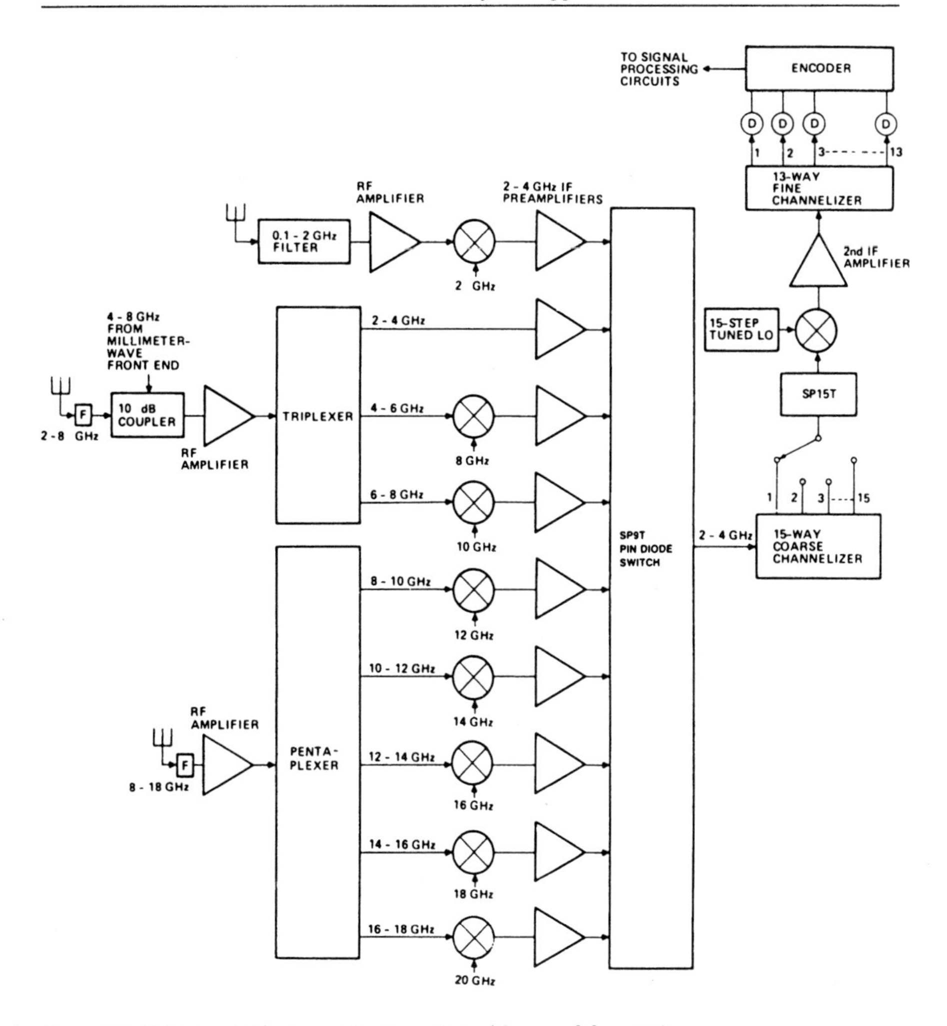

Figure 6.7 Wideband ES channelized receiver. (*Source:* [3], p. 83.)

If the channelized receiver architecture were continued, then each of the nine channel segments would require further parallel back-end channelization. However, there is often knowledge from other receivers (e.g., IFM receivers) as to which channels have received a signal. In this case, only a single back-end is required. A single-pole nine-throw switch selects the proper channel. The 2- to 4-GHz band is further segmented in 15 continuous channels that are each 133-MHz wide. This is accomplished using a coarse channelizer, which is typically implemented using a bank of bandpass filters constructed using stripline techniques. The channel is further narrowed using a single-pole 15-throw switch, which selects the desired 133-MHz segment.

The various segments are then mixed (using a doubly balanced mixer) with a local oscillator stepped in frequency by 133 MHz as 1.833, 1.967, . . . ,

3.833 GHz to produce a second IF band of 167 MHz to 300 MHz. The second IF output is amplified and feeds a bank of 13 contiguous filters each with 10-MHz bandwidths, which forms the fine channelizer function. The outputs from each filter are detected and, by comparing amplitudes, one can locate a received frequency of the order of one-tenth of a fine channel width, hence enabling the receiver to locate a threat signal frequency to an accuracy of 1 MHz. The encoder usually codes the frequency in a digital bit sequence for transmission to the signal processing portion of the ES system.

In Figure 6.7, an input port from a millimeter-wave frequency extension receiver is shown. This port accepts millimeter-wave signals in the 26- to 42-GHz and 88- to 120-GHz bands, which have been down-converted into the 2- to 8-GHz band for analysis by a conventional microwave channelized receiver.

The development of SAW filters was the significant technological development that made channelized receivers practical. A typical receiver may require from 20 to 200 of these filters. The filters in the output channels must exhibit response characteristics exceeding those of an eight-pole, 1-dB ripple Chebyshev bandpass filter. The triple-transit suppression (reflections that travel from the output to the input to the output transducers) must approach 55 dB in a typical design.

The basic SAW filter consists of input and output transducers attached to a crystal substrate. An IF signal applied to the input transducer causes a piezoelectric distortion of the crystal (usually quartz or lithium niobate) on which the wave propagates. The SAW propagates from the input transducer to the output transducer. The wave amplitude is proportional to the piezoelectric constant of the crystal, the strength of the signal, and the length of the transducer fingers, which are in the form of thin metal strips deposited interdigitally on the crystal surface. By shaping the finger lengths (called apodization), which weights their effect in the direction of propagation, the filter's effects in the time domain (and hence in the frequency domain by duality) are controlled. This results in the frequency response of the filter being directly controlled by the mask used to metalize the transducers onto the crystal. Hence, they require no tuning, are inherently stable, and have excellent manufacturing repeatability.

Currently, SAW filters can be built with center frequencies up to 1 GHz and bandwidths on the order of 0.4 of the center frequency. Projected capability is to 2 GHz and bandwidths to 0.8 of the center frequency. The resolution-determining filters in a channelized receiver are generally used to measure the frequency of the emitter signal and, hence, must be stable. This requires the use of ST-cut quartz substrates in the SAW filter that nominally have a 0 ppm/°C temperature coefficient. SAW filters that use lithium niobate substrates can be designed with complex transducers, which provide complex filter shapes but have temperature coefficients of the order of +95 to +108 ppm/°C. Another advantage of SAW filters is their particularly compact design, whereby up to 32 filters can be fabricated on a single substrate.

In some applications, there is a conflict between the desired frequency resolution and the bandwidth required to support the narrowest expected PW. A further increase in frequency resolution can be achieved by staggering two sets of contiguous filter banks in frequency. By digital-logic comparison of the detected output of the filter bank in groups of three filters around the detected signal, it is possible to divide the bandwidth of each filter into three cells [3]. The key factor in this resolution-extending design is the ability to precisely shape the filter's response through appropriate design of the SAW filter's transducer response.

A significant problem in channelized receivers is the activation of multiple channels by the occurrence of a high-power, narrow pulse. For example, for a 0.1-μs rectangular pulse, the sideband energy at the −60-dB level extends over ±3000 MHz. For a triangular-shaped pulse, which is a more realistic expectation, this reduces to about ±100 MHz. Thus, the weaker of two 0.1-μs overlapping pulse signals that are 60 dB apart in amplitude cannot be resolved unless they are separated by more than 100 MHz in frequency. This is the so-called "splatter" effect in channelized receivers.

A number of approaches that cope with this problem including guard filter bands, channel comparison, and wideband-narrowband filters have been implemented [3].

6.3.1.2 Compressive Receiver

A compressive receiver provides wideband frequency coverage with high resolution in a dense signal environment. For this reason, it is a natural candidate for ES applications against either radar or communications emitters. Its most serious competitor is the channelized receiver, which provides similar overall performance at the expense of potentially more hardware and cost [9].

Both compressive and channelized receivers generally use analog designs that are based upon SAW-based filters and delay lines. They both perform instantaneous spectral analysis within the period of a radar PW, but the compressive receiver's output occurs serially (i.e., in a time-based waveform) as compared to the parallel output of the channelized receiver. This difference results in a hardware advantage for the compressive receiver, which is most compelling in direction-of-arrival (i.e., DF) applications. In DF applications, multiple receivers (compressive or channelized) are required and the hardware and cost advantages of the compressive receiver are magnified. It has been stated that channelized receivers are impractical in wideband DF applications for this reason [10].

A number of experimental compressive receivers have been successfully built, thereby confirming both the principle and its practical application [3,10,11]. Wide-bandwidth compressive receivers are a particular challenge since they stress the state of the art in both SAW delay lines and digital output circuitry. The state of the art in both areas is about 1 GHz, which limits the maximum bandwidth of compressive receivers to this order of magnitude. This is a disadvantage in most ES applications that have to cover a frequency range of 2 GHz to 8 GHz in two or three bands. Prospects for increasing this bandwidth limitation using the present approach for compressive receivers are not promising since the bandwidth of the output digital circuitry is directly proportional to the receiver bandwidth.

The basic configuration of a compressive receiver is depicted in Figure 6.8. A chirp waveform is mixed (multiplied) with the input signal so that each Fourier component is converted into a chirp of the same frequency-time slope but with a starting frequency fixed by that of the original component. The mixed output is then processed in a chirp filter (dispersive delay line) of matched slope that correlates, or pulse compresses, the constituent chirps. The linear correspondence between frequency and time in the dispersive delay line results in the original Fourier components being translated to time displacements between the resulting pulse-compressed outputs.

The principle of the basic compressive receiver is shown in Figure 6.8. The incoming signal f_t is multiplied by a down-chirp waveform $\exp(-j\alpha t^2/2)$, where α is the slope in radians per second of the linear-FM swept local oscillator signal. This signal is applied to a dispersive delay line filter whose slope is matched to the premultiplier chirp but is opposite in sign. The impulse response of this filter is therefore given by the up-chirp function $\exp(j\alpha t^2/2)$. The output of the chirp filter is found by convolving the multiplied waveform $f_t \cdot \exp(-j\alpha t^2/2)$ with the impulse response of the chirp filter. As shown in Figure 6.8, the resulting time output g_t is proportional to the Fourier transform $F(\alpha t)$ of the input signal f_t, where the argument is a linear function of time referenced to the premultiplying chirp waveform and proportional to the chirp's slope. The other factor associated with g_t is a phase term $\exp(j\alpha t^2/2)$, which can be eliminated by post-multiplying by another chirp waveform $\exp(-j\alpha t^2/2)$ to find the Fourier transform $F(\cdot)$, or using an envelope detector to find the Fourier spectrum $|F(\cdot)|$.

In order to implement the compressive-receiver algorithm depicted in Figure 6.8 with practical chirps, finite limits must be imposed in time and frequency. For example, assume that a 500-MHz bandwidth is required and that the shortest pulse to be analyzed is 0.1 μs. One way to design the compressive receiver would be to use a premultiplying sweep that covered the 500-MHz operating band in 0.1 μs or a chirp rate of 5000-MHz/μs. Because

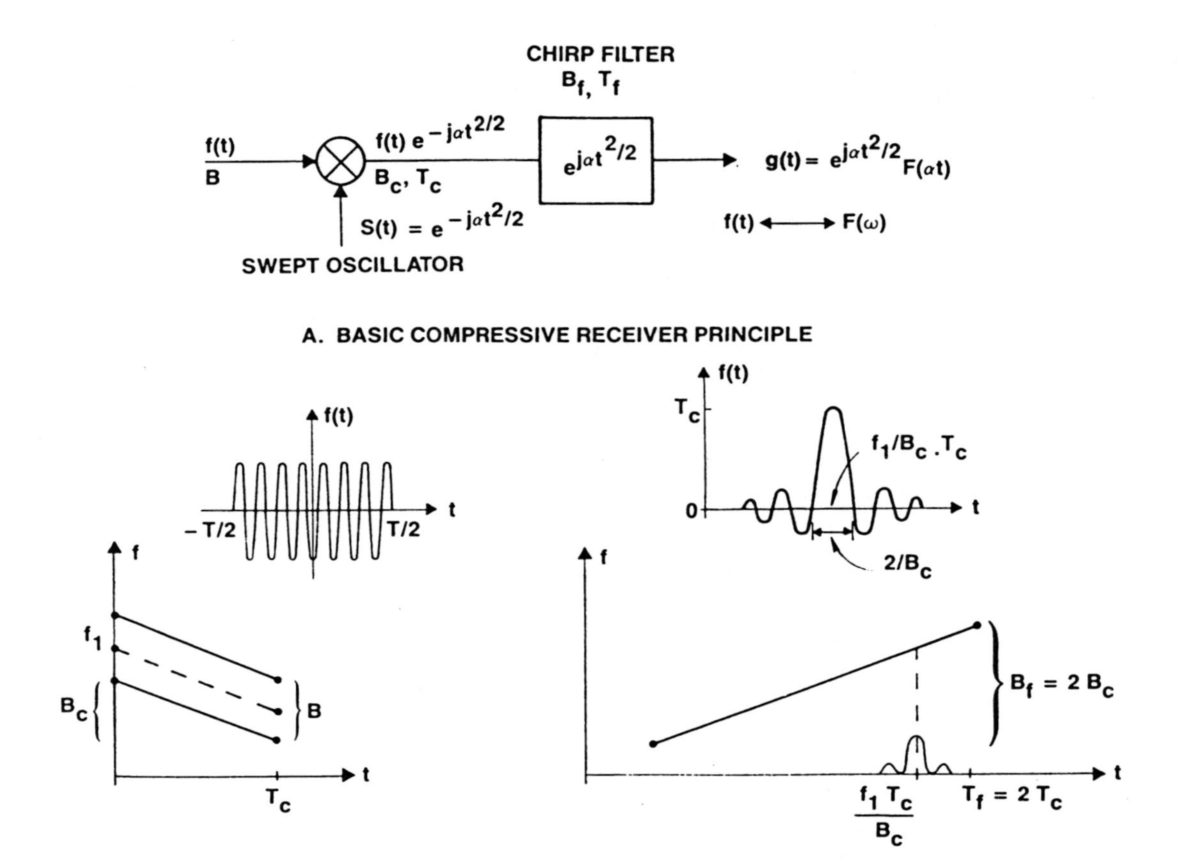

Figure 6.8 Compressive receiver operation. (*Source:* [3], p. 83.)

the input signal can be located anywhere in the coverage band, the frequency range at the chirp filter extends over twice the bandwidth, or 1000 MHz in the example. Because the magnitude of the chirp rate of the impulse response of the filter is the same as the premultiplying chirp, the filter delay must be twice the period of the premultiplying chirp, or 0.2 μs. This configuration is referred to as the $M(s) - C(l)$ type, where a short premultiplying chirp is followed by a long chirp filter or convolver. The relationships between the various waveforms and the PC outputs are depicted in Figure 6.8.

The action of a compressive receiver is identical to that of a PC radar. In the radar case, a chirp waveform received from the target is compressed to a narrow pulse using a matched PC filter. In the compressive receiver case, the chirped waveform is induced on the input signal and then compressed in a matched chirp filter. The time delay position of the compressed pulse in both cases varies in a linear manner as a function of the input frequency. In the radar case, this is an apparent range shift, which is a function of the target's Doppler frequency; while in the ES case it is a measure of the input frequency.

SAW devices make both the channelized and compressive receivers practical. In the channelized receiver, the SAW devices allow multiple, stable, shaped filters to be fabricated on a single substrate. In the compressive receiver case, the SAW devices are used as dispersive delay lines both to process the chirp waveform and to generate the premultiplying chirp local oscillator signal. The chirp waveform is typically generated by exciting a SAW dispersive delay line with an impulse function. This provides an easy method for matching the chirp waveform to the dispersive network, since both SAW devices can be similarly fabricated and hence tend to track under various environmental conditions. The alternate configuration of an active swept local oscillator and a passive SAW device is more difficult to match.

The SAW dispersive delay line tends to limit the overall frequency coverage band of the compressive receiver. The present state of the art is of the order of 500 MHz to 1000 MHz in a single channel. Parallel channels can be employed to process a wider band. In addition to device limitations, the necessity of providing a fast serial readout of frequency places severe demands on the speed and power of digital interface circuitry as the coverage increases. For example, a resolution of 1 MHz requires an output display period of 1 μs. With a 500-MHz coverage bandwidth, the output data must be sampled as a 1-GHz rate, which is a challenge to current high-speed digital circuit technology.

► *Example 6.3*

The basic compressive receiver principle is described in Figure 6.8. The receiver forms the instantaneous Fourier spectrum of the signals contained within its analysis window, allowing the separation of simultaneously occurring signals with nonoverlapping frequency spectra.

Using MATLAB, simulate a compressive receiver with a 1-MHz chirp bandwidth, an analysis window of 100 μs, and with two input signals, one at 300 kHz and the other at 100 kHz. How close can the two signals be in frequency and still be resolved?

```
% COMPRESSIVE RECEIVER USING FFT
% ----------------------------------
% comprcv.m

clear;clc;clf;

% Input Chirp Bandwidth and Pulsewidth

bw=1e6;                    % Receiver bandwidth - Hz
T= 100e-6;                 % Analysis window - secs
fo=1e5;                    % Input signal #1 frequency - Hz
f1=1.165e5;                % Input signal #2 frequency - Hz

% Input FFT Points

lfft=1024;                 % FFT points

% Set Time Base and Compute Parameters

smp= 1/(2*bw);
pts= bw*T;
t=-T/2:T/(lfft-1):T/2;

% Sample Complex Chirp

s= exp(-j*pi*bw*t.*t/T);

% Sample Input Signals

vt=exp(-j*2*pi*fo*t)+exp(-j*2*pi*f1*t);
svt=s.*vt;SV=fft(svt,lfft);

%Find Fourier Spectrum of Chirp

Y= fft(s,lfft);

% Find Power Spectral Density

Psd=SV.*conj(Y)/lfft;
A=ceil(max(abs(Psd)));
l=length(Y);
Psd1=fftshift(Psd);
```

```
f=(1/T)*(-(l-1)/2:(l-1)/2);

% Find Matched Filter Time Waveform

zh=ifft(Psd,lfft);
h=fftshift(zh);
h=h/max(h);

% Plot Analysis Window- Waveform

B=max(abs(h));

plot(t,abs(h));grid
axis([-(T/2) 0 0 B])
title(['Compressive Receiver Output']);
xlabel('Time - sec');ylabel('Amplitude');
```

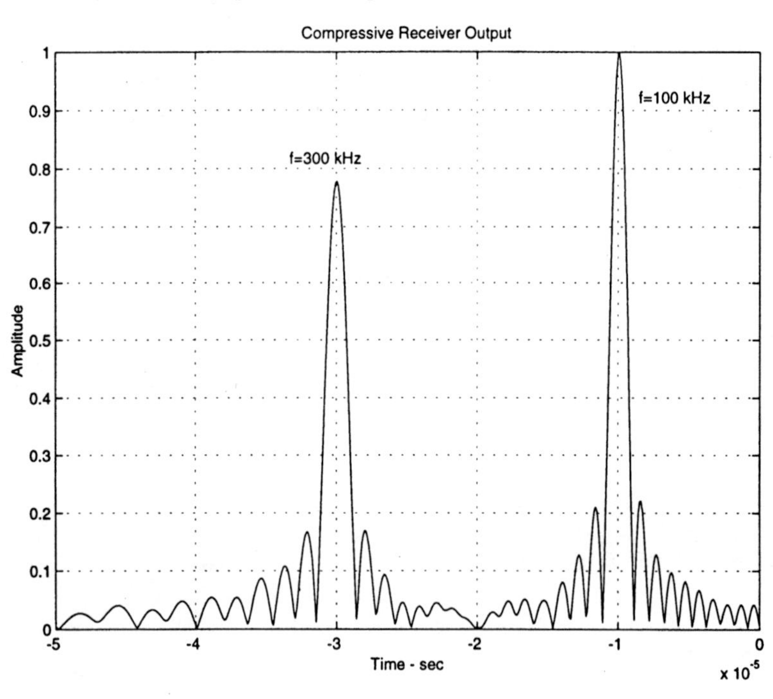

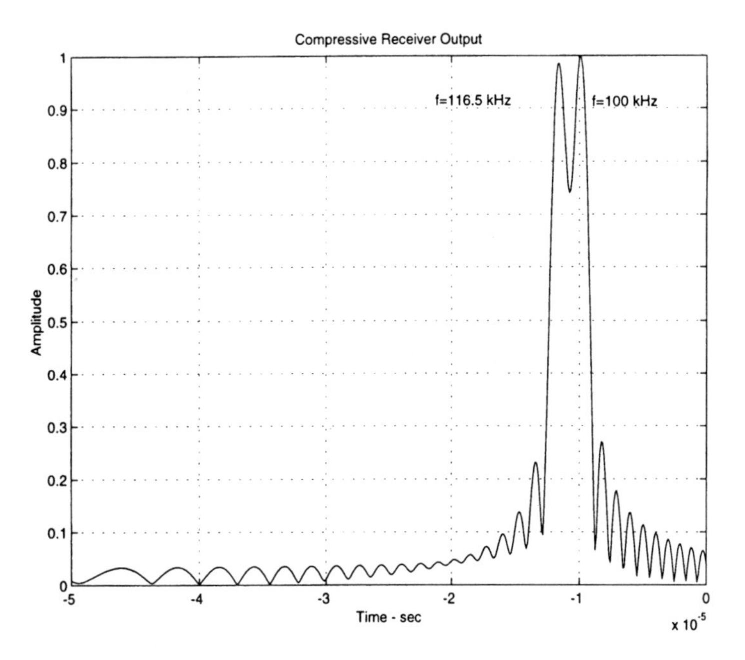

6.3.1.3 Acousto-Optic Bragg Cell Receiver

Another advanced approach to the design of an ES receiver with the capability of detecting and resolving signals in a dense emitter environment is the *acousto-optic spectrum analyzer* (AOSA) or Bragg cell receiver. The functions performed by this type of receiver are quite similar to those performed by the compressive receiver, except that the mechanization is different. The elements of the AOSA as depicted in Figure 6.9 are the laser, which illuminates the Bragg cell with coherent light; the Bragg cell, which deflects the light in proportion to the frequency of the applied signal; and a transform lens, which focuses the light onto a photodiode detector array. The spatial distribution across the photodiode array represents the instantaneous Fourier transform of the input signal across the Bragg cell aperture, while the energy distribution across the output of the photodiode array is the Fourier spectrum of the signal.

In comparison with the compressive receiver, the Bragg cell performs the function of the premultiplying chirp while the transform lens performs the function of the chirp convolving filter. By analogy, the frequency resolution is approximately equal to the reciprocal of the acoustic transit-time through the crystal. The bandwidth is limited by the acoustic attenuation and diffraction characteristics of the Bragg cell. As with compressive receivers, the time-bandwidth product is an important performance measure of the AOSA.

The key element of the AOSA is the Bragg cell that deflects the light beam in proportion to the frequency components present in the input signal.

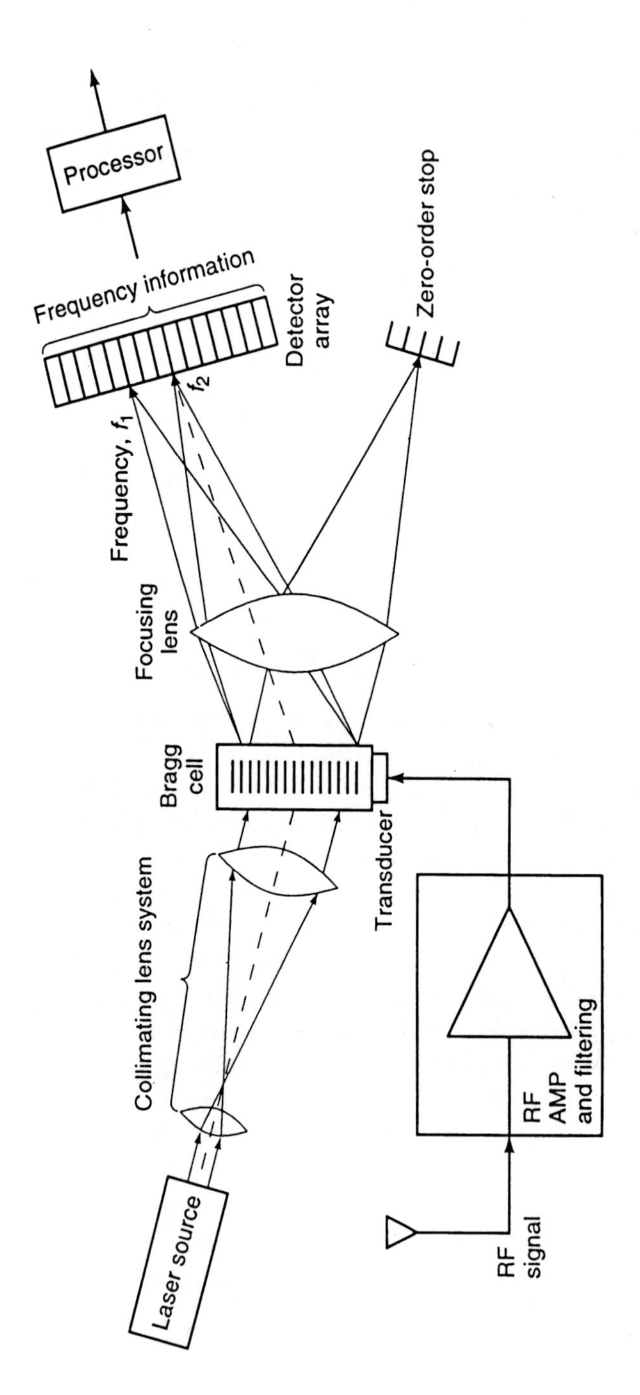

Figure 6.9 Acoustic optic Bragg cell receiver.

The basic action is an acousto-optic interaction whereby a traveling sound wave set up in an acoustic substrate (e.g., lithium niobate) causes alternate compression and rarefaction of the crystalline structure. This, in turn, affects the substrate's optical index of refraction, which sets up an optical diffraction grating whose spacing is proportional to the acoustic wavelength (proportional to RF input signal). As the sound-wave frequency is changed due to receiving a signal at a different frequency, the grating spacing is thus generated by the index of refraction. Because the angle of the deflected light is directly proportional to the frequency of the received RF signal, the basic deflection-versus-frequency phenomena is established, which is the principle upon which AO signal processing and spectrum analysis are based. An interesting side effect in the acousto-optic interaction is that the light wave is not only diffracted by the sound but is shifted in frequency as well.

The limited dynamic range (approximately 25 dB) available from present Bragg cell receivers is an area of concern in the design of advanced AOSA receivers. Ultimately, dynamic range is limited by the amount of optical power that can be delivered to the detector array, the detector sensitivity, and the third-order intermodulation products that result from operating the Bragg cell or other components beyond their linear range. Another factor that affects the AOSA's ability to handle multiple signals is the generation of in-band spurious signals resulting from sidelobes of other signals or those from the undiffracted portion of the beam.

The primary improvements in dynamic range are expected by higher Bragg cell efficiency (from 0.1% to 3% per RF watt diffraction efficiency) and increased laser power (from 10 mW to 20 mW). This is expected to increase the spurious-free Bragg cell dynamic range limitation from the current 25 dB to about 50 dB. The problem of spurious signals must be controlled through the utilization of high-quality components, appropriate weighting, and adequate amplifier compression points.

Another approach that promises a breakthrough in the dynamic range performance area is the use of heterodyne detectors rather than the present direct-detection photodiodes. With this approach, a part of the original laser signal is combined with the Bragg cell diffracted light in the photodiodes. The signal beam heterodynes with the reference laser beam, which acts like a local oscillator, causing the output to vary linearly with the RF signal input to the Bragg cell. Dynamic ranges in excess of 50 dB have been achieved using this approach.

6.3.1.4 Digital Receiver

In addition to the analog spectrum analysis techniques of the channelized, compressive, and Bragg cell receivers, it is also possible to use all-digital techniques. Real-time digital spectrum analysis is possible using special hardware configured to accomplish the FFT.

The block diagram of a basic digital ES receiver is depicted in Figure 6.10. To understand the operation of the receiver, assume that a 500-MHz frequency band must be analyzed in 0.25 μs with a resolution of the order of 5 MHz. The RF input would first be translated into I- and Q-channels to form a complex signal with 500-MHz bandwidth. The complex signal would have to be sampled at the Nyquist rate by the A/D converter at a 1-GHz rate (1000 samples per microsecond consisting of I- and Q-channel samples at a 500-MHz rate) in order to preserve the signal fidelity. A minimum of six bits would be required to allow sufficient dynamic range (on the order of 36 dB) so that sidelobes from strong signals would not obscure weak signals. During the 0.25-μs pulse, at least 125 samples would be produced in each I- and Q-channel. If a radix 2 (base 2) FFT processor were employed, this would impose a requirement for a 128-point transform since the number of points must be a multiple of two. The basic 128-point FFT algorithm can be broken down into a number of "butterfly" operations, which become a measure of the complexity of the FFT. For this example, 448 butterfly operations ($n/2 \log_2 n$) for each I- and Q-channel must be performed in 0.25 μs, resulting in 1.792×10^9 butterflies/s. Currently, this is beyond the capability of dedicated digital chips, but the rapid advancement in the state of the art in this area is expected to cure the deficiency. Note that the FFT of the example would form 128 contiguous filters, so the basic resolution would be of the order of 4 MHz.

The main objective of present digital receiver research is to achieve at least 1-GHz bandwidth with an eight-bit (48-dB) dynamic range [11]. The critical component in such a design is the input A/D converter. Figure 6.11 depicts the current state of the art (circa 1997) in such A/D converters, which shows the availability of a 3-GHz eight-bit A/D converter that could be used in an I- and Q-channel digital receiver to provide the order of 3-GHz overall bandwidth. However, FFT butterfly chips that process at this rate are not yet available. This would require that the A/D converter output be divided into 12 parallel paths to handle the sampling rate with each FFT operating at 250 MHz [11].

Although digital components are available to provide wideband digital receivers, they are expensive and currently limited to experimental use. Another approach called the sub-Nyquist sampling method has been proposed to allow wideband digital receivers to be built whereby the sampling speed does not match the instantaneous bandwidth [11]. The sub-Nyquist receiver is basically an IFM-type receiver implemented using digital techniques. However, in contra distinction to an IFM receiver, the digital version can handle simultaneous signals because of the FFT operation.

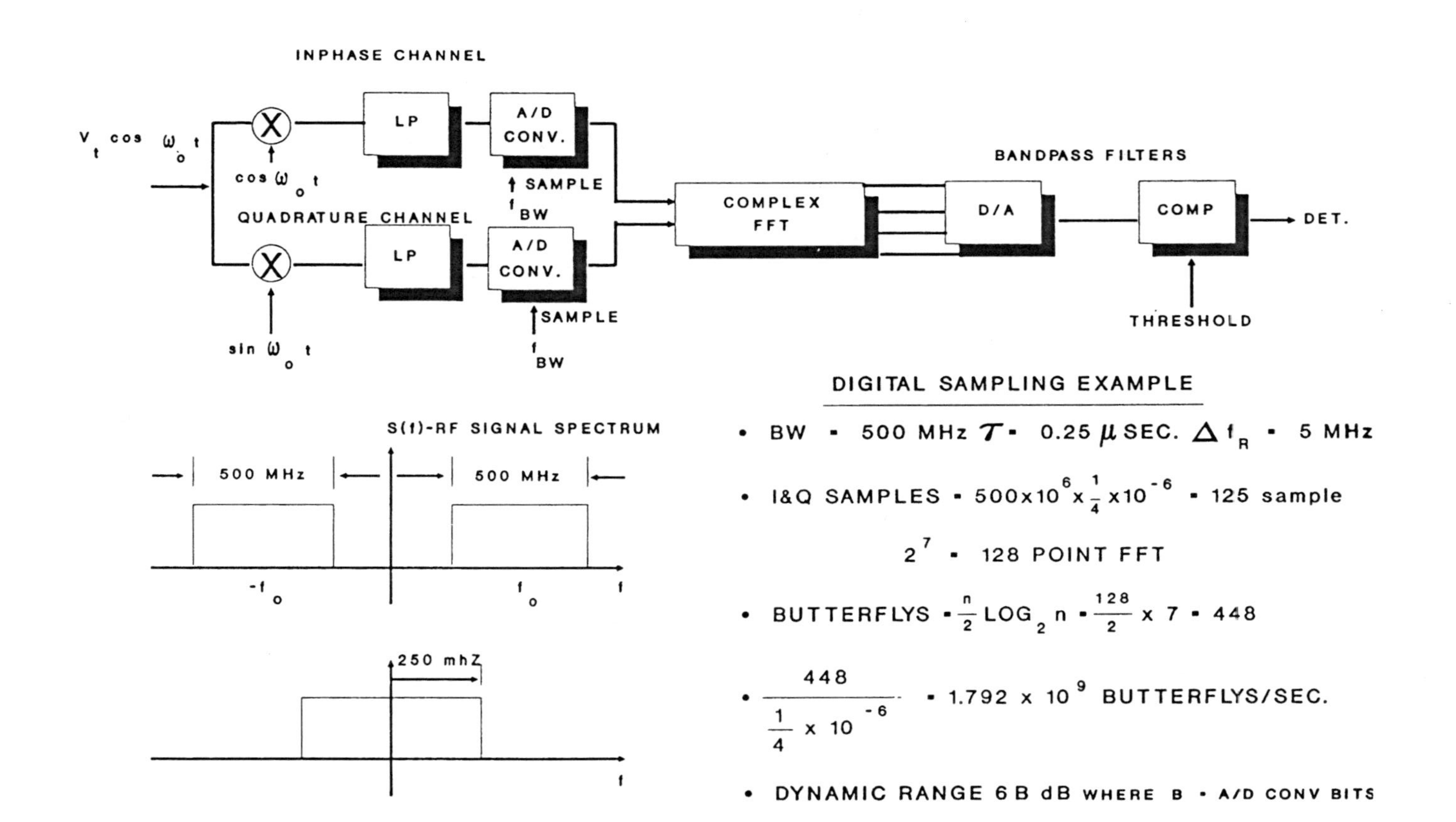

Figure 6.10 Digital ES receiver.

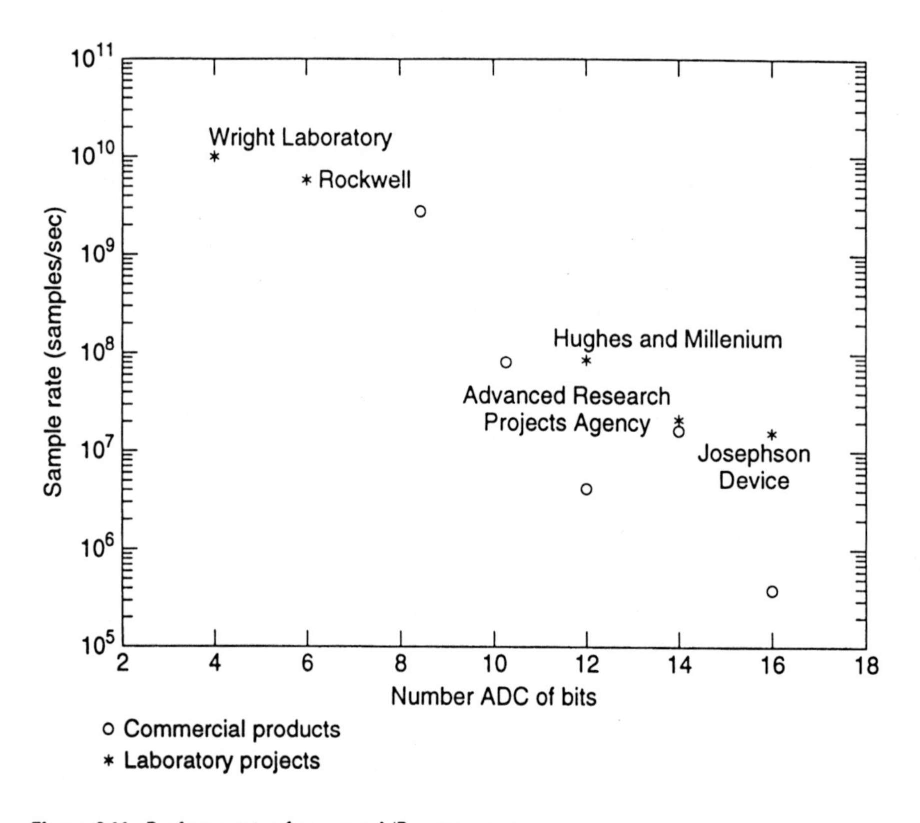

Figure 6.11 Performance of current A/D converters.

A block diagram of the basic sub-Nyquist digital receiver is depicted in Figure 6.12. In this receiver, the desired instantaneous bandwidth is 1 GHz, where the input occupies a frequency range of ±500 MHz centered about 1 GHz. The A/D converter samples each I- and Q-channel at 250 MHz, which is one-fourth the Nyquist rate. The sampling clock of the lower channel has a delay of 0.8 ns that provides a maximum capability for overall digital receiver bandwidth of 1250 MHz ($B = 1/\tau$). The overall receiver bandwidth of 1 GHz folds into a 125-MHz (half the sampling rate) band resulting in eight ambiguous bands. In other words, the FFT is only capable of resolving the input signal to 125 MHz in frequency. However, since the maximum overall capability of the receiver is 1250 MHz, each of the ambiguous outputs within the frequency range can be resolved by correlating the upper and lower channels in a phase detector. As discussed in [11], there are several anomalies in this type of design, but the basic concept generally provides the desired result.

The advantage of the sub-Nyquist sampling rate digital receiver is that higher dynamic range (12 to 14 bits) can be obtained with slower, more economical A/D converters. Moreover, available FFT butterfly chips can be

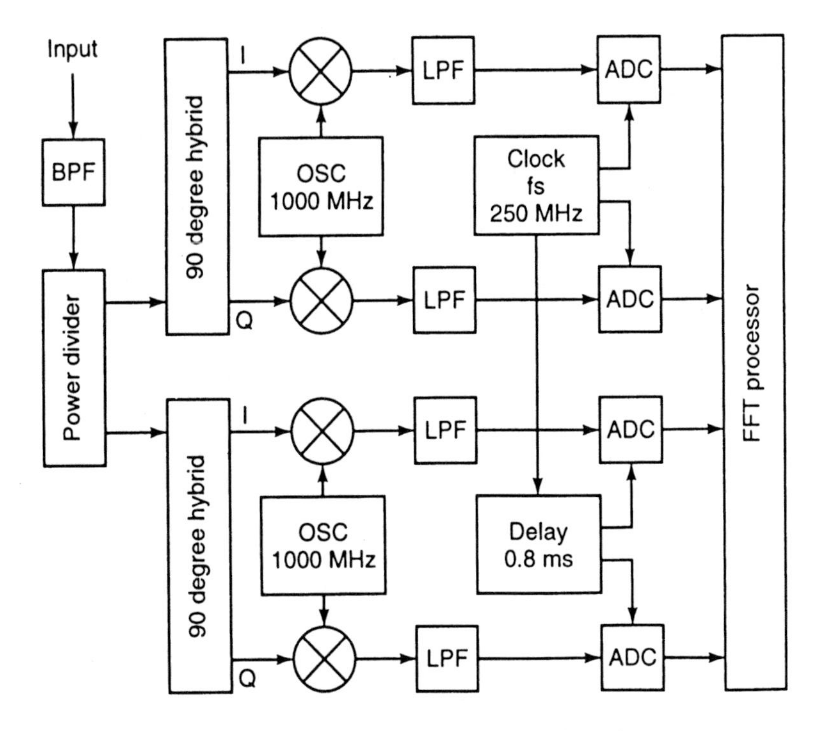

Figure 6.12 Basic sub-Nyquist sampling structure of a digital receiver.

used to implement the FFT processing. On the other hand, twice as many A/D converters are required, as is complex processing to resolve the ambiguities.

6.4 Direction Finding

DF systems provide several important functions in modern EW systems. We have already discussed the importance of measuring the emitter's bearing, or AOA, as an invariant sorting parameter in the deinterleaving of agile pulsed radar signals and in separating closely spaced communication emitters. In addition, the conservation of jamming power in power-managed EA systems depends on the ability of the associated ES system to measure the direction to the victim emitter. A function that is becoming increasingly important in defense suppression and weapon delivery systems involves the grouping of a number of spatially dispersed emitter-angle measurements to locate the emitter's position passively. This can be accomplished from a single moving platform through successive measurements of the emitter's angular direction or from multiple platforms that make simultaneous angular measurements.

The emitter identification function requires identifying and associating consecutive pulses produced by the same emitter in a two-dimensional parameter space of *angle of arrival* (AOA) and frequency. The AOA is a parameter that a hostile emitter cannot change on a pulse-to-pulse basis. However, to measure the AOA of pulses that overlap in the time domain first requires them to be separated in the frequency domain. The advanced ES receivers that accomplish this function must operate over several octaves of bandwidth while providing rms-bearing accuracies on the order of at least 1 degree with high POI and fast reaction time in dense signal environments.

The most elemental approach toward measuring AOA is to use a narrowbeam rotating antenna. As the antenna beam sweeps over the target, it essentially traces out the antenna pattern. By selecting the peak of this response, the AOA to the target can be determined. A complimentary omni-directional antenna is generally required to prevent the peak response of a sidelobe from being interpreted as the direction of the emitter. The theoretical error of this technique in radians is given by

$$\Delta\theta = \frac{\sqrt{2}\lambda}{\pi D(2S/N)^{1/2}} \tag{6.6}$$

where λ is the wavelength, D is the diameter of the antenna, and S/N is the signal-to-noise ratio [3]. Equation (6.6) shows that a large aperture high-gain antenna is required for accurate measurement. Note that either a mechanically scanning or a circular phased array-type antenna can be used to implement this technique [12].

The two primary techniques used for DF are the amplitude-comparison method and the phase-comparison or interferometer method. The phase-comparison method generally has the advantage of greater accuracy, but the amplitude-comparison method is used extensively due to its lower complexity and cost. Regardless of which technique is used, it should be emphasized that the ultimate rms angular accuracy is given by $\Delta\theta = k\theta_B/\sqrt{SNR}$, where θ_B is the antenna's angular beamwidth, or interferometer lobe width, and SNR is the signal-to-noise ratio. Thus, phase interferometers that typically use very wide beam antennas require high signal-to-noise ratios to achieve accurate AOA measurements. Alternately, a multielement array antenna can be used to provide relatively narrow interferometer lobes, which require modest signal-to-noise ratios.

As previously described, virtually all currently deployed RWR systems use amplitude-comparison DF. A basic amplitude-comparison receiver derives a ratio, and ultimately AOA or bearing, from a pair of independent receiving channels that utilize squinted antenna elements that are usually equidistantly spaced to provide an instantaneous 360-degree coverage. Typically, four or six antenna elements and receiver channels are used in such systems and wideband logarithmic video detectors provide the signals for bearing-angle determination. The monopulse ratio is obtained by subtracting the detected logarithmic signals, and the bearing is computed from the value of the ratio.

Amplitude comparison RWRs typically use broadband cavity-backed spiral antenna elements whose patterns can be approximated by Gaussian-shaped beams. Gaussian-shaped beams have the property that the logarithmic output ratio slope in decibels is linear as a function of the AOA. Thus, a digital look-up table can be used to determine the angle directly. However, both the antenna beamwidth and squint angle vary with frequency over the multi-octave bands used in RWRs. Pattern shape variations cause a larger pattern crossover loss for high frequencies and a reduced slope sensitivity at low frequencies. Partial compensation of these effects, including antenna squint, can be implemented using a look-up table if frequency information is available in the RWR. Otherwise, gross compensation can be made depending upon the RF octave band utilized.

The output of two squinted antennas using Gaussian-shaped beams can be expressed as

$$A = A_{1,2} \exp[-k(\theta \pm S/2)^2] \tag{6.7}$$

where $A_{1,2}$ are the amplitudes in channels 1 and 2, S is the squint angle between adjacent antennas, and $k = 2.776/\theta_B^2$ with θ_B equal to the 3-dB beamwidth. Expressing the ratio of the antenna outputs in decibels results in

$$R_{dB} = 10 \log(A_1/A_2) + 8.68kS\theta \tag{6.8}$$

Differentiating this ratio expresses the angular error as a function of amplitude imbalance, antenna squint, and 3-dB beamwidth as

$$\Delta\theta = \frac{\Delta R_{dB}\theta_{3dB}^2}{24S} \tag{6.9}$$

Therefore, reducing the imbalance between channels, reducing the antenna beamwidth, and increasing the squint angle increases the accuracy of an amplitude monopulse type AOA system.

Simultaneous multiple-beam systems use an antenna, or several antennas, to form a number of simultaneous beams (e.g., Rotman lens), thereby retaining the high sensitivity of the scanning antenna approach while providing fast response. However, it requires many parallel receiving channels, each with full frequency coverage. This approach is compatible with amplitude-monopulse angular measuring techniques, which are capable of providing high-angular accuracy.

Interferometer systems, which by definition use phase-comparison techniques, have the advantage of fast response, but generally use wide-coverage antennas, which result in low sensitivity. In addition, they require relatively complex microwave circuitry, which must maintain a precise phase match over a wide frequency band under extreme environmental conditions. When high-angle location accuracy is required (of the order of 0.1 to 1 degree), wide baseline interferometers are utilized with ambiguity-resolving circuitry.

Phase interferometer DF systems are utilized when accurate AOA information is required. The basic geometry is depicted in Figure 6.13, whereby a plane wave arriving at an angle is received by one antenna earlier than the other due to the difference in path length. The time difference can be expressed as a phase difference in radians as

$$\phi = \frac{2\pi d}{\lambda}\sin\theta \tag{6.10}$$

where θ is the AOA, d is the antenna separation, and λ is the wavelength in compatible units. The unambiguous *field-of-view* (FOV) is given by

$$\theta = 2\sin^{-1}(\lambda/2d) \tag{6.11}$$

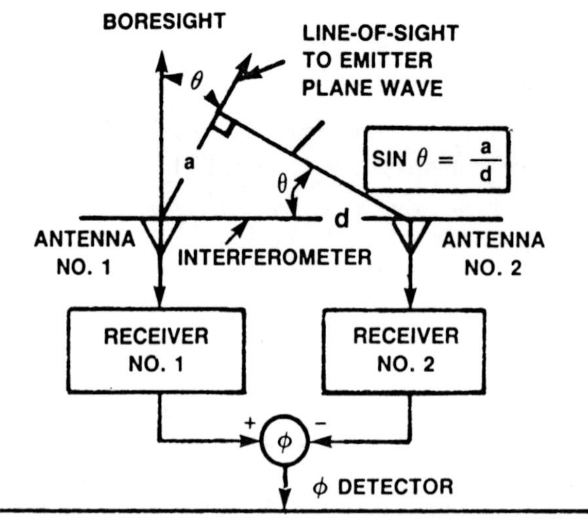

Figure 6.13 Phase interferometer principle. (*Source:* [3], p. 101.)

which for $\lambda/2$ spacing results in 180-degree coverage. The accuracy of the interferometer can be determined by differentiating (6.10) and solving for the differential AOA

$$\Delta\theta = \frac{\lambda}{2\pi d \cos\theta}\Delta\phi \tag{6.12}$$

where $\Delta\phi$ is the phase error. An examination of this equation indicates that accuracy is greatest at the antenna boresight and can be increased by increasing the spacing between antennas, which generally results in ambiguous angle measurement. This spacing must be established for the highest frequency to be received.

Interferometer elements typically use broad antenna beams with beamwidths of the order of 90 degrees. This lack of directivity produces several deleterious effects. First, it limits system sensitivity due to the reduced antenna gain. Second, it opens the system to interference signals from within the antenna's broad angular coverage. The interference signals often include multipath from strong signals which can limit the accuracy of the interferometer.

In an interferometer, the locus of points that produce the same time or phase delay forms a cone. The indicated angle is the true azimuth angle multiplied by the cosine of the elevation angle. The error in assuming the incident angle to be the azimuth angle is negligible for signals about the antenna's azimuth and elevation angle boresight. At 45-degrees azimuth and 10-degrees elevation, the error is less than 1 degree, increasing to 15 degrees for both at 45 degrees, independent of instrumentation errors. Two orthogonal arrays, one measuring the azimuth angle and the other the elevation angle, can eliminate this error. For application to targets near the horizon, the depression angle is small, thereby requiring only horizontal arrays [3].

The theoretical rms angular error caused by noise in an interferometer in radians is given by $\sigma_\theta = \Delta\alpha/\pi \cdot \sqrt{SNR}$, where $\Delta\alpha = \lambda/d \cdot \cos\theta$ is the separation between adjacent nulls. For a two-element interferometer, the spacing (d) must be $\lambda/2$ or less to provide unambiguous, or single-lobe ±90-degree, coverage. This, in effect, sets a wide interferometer (or grating) lobe that must be split by a large factor to achieve high accuracy. This, in turn, imposes a requirement for a high signal-to-noise ratio to achieve the large beam-splitting factor. For example, if 0.1-degree accuracy is required from an unambiguous two-element interferometer, then a signal-to-noise ratio of about 50 dB is required even if other sources of error are well below the required total error of 0.1 degree. Such accuracy and signal-to-noise ratio are seldom available, considering the inherent vulnerability to multipath and the low sensitivity of an interferometer system.

When high accuracy is required from an interferometer system, it is usual to employ separations greater than $\lambda/2$ to achieve this accuracy. The increased separation sets up a multi-grating-lobe structure through the coverage angle that requires less signal-to-noise ratio to achieve a specified accuracy. For example, a two-element interferometer with 16λ spacing would set up a 33-grating-lobe structure (including the central lobe) throughout the ±90-degree coverage angle. Within each of the 33 grating lobes, it would only require a signal-to-noise ratio of the order of 20 dB to achieve 0.1-degree accuracy. However, there would be 33 ambiguous regions within the ±90-degree angular coverage and also 32 nulls (where the phase detector output is zero), about which the system would be insensitive to an input signal. The ambiguities could be resolved by employing a third antenna element with $\lambda/2$ spacing, which would provide an accuracy of the order of 3 degrees with a 20-dB signal-to-noise ratio. This accuracy is sufficient to identify which of the 33 lobes contains the signal. Providing coverage in the null regions requires additional antenna elements.

Interferometers employing multiple antenna elements are called multiple-baseline interferometers. In a typical design, the receiver consists of a reference antenna and a series of companion antennas. The spacing between the reference element and the first companion antenna is $\lambda/2$; other secondary elements may be placed to form pairs separated by 1, 2, 4, and 8 wavelengths or may use one or two spacings that are related by prime numbers. The initial AOA is measured unambiguously by the shortest spaced antenna pair. The next greatest spaced pair has a phase rate of change that is twice that of the first, but the information is ambiguous due to there being twice as many lobes as in the preceding pair. A greater phase rate of change permits higher angular accuracy while the ambiguity is resolved by the measurement involving the previous pair. Thus, the described multiple-baseline interferometer provides a binary-based AOA measurement where each bit of the measurement supplies a more accurate estimate of the emitter's AOA.

Harmonic multiple-baseline interferometers use elements that are spaced at $2^n \cdot \lambda/2$, $n = 0, 1, 2, 3$. In nonharmonic interferometers, no pair of antennas provides a completely unambiguous reading over the complete field-of-view. For example, the initial spacing in the nonharmonic interferometer might be λ, while the next companion element spacing is $3\lambda/2$. Ambiguities are resolved by truth tables, and hence the accuracy is set by the spacing of the widest baseline antenna pair. Nonharmonic interferometers have been implemented over nine-to-one bandwidths (2 GHz to 18 GHz) with rms accuracies from 0.1 to 1 degree and with no ambiguities over ±90 degrees. The principal advantage of the nonharmonic over the harmonic interferometer is the increased bandwidth for unambiguous coverage.

Interferometer DF accuracy is determined by the widest baseline pair, limited by errors due to quantization of the phase difference, phase bias from inaccurate component tracking, imperfect frequency correction, and phase noise due to receiver thermal noise. Typical multi-octave microwave antennas, such as cavity-backed spirals, track to six electrical degrees, and associated receivers track to nine degrees, resulting in an rms total of 11 degrees. At a typical 16-dB signal-to-noise ratio, the rms phase noise is approximately nine electrical degrees. The quantization error can be made arbitrarily small. For these errors and an emitter angle of 45 degrees, a spacing of 25λ is required for 0.1-degree rms accuracy while a spacing of 2.5λ is needed for 1-degree accuracy. For high accuracy, interferometer spacings of many feet are required. In airborne applications, this usually involves mounting interferometer antennas in the aircraft's wingtips. Maintaining the electrical length of transmission lines over time and under extreme environmental conditions presents a significant problem. One solution to this problem involves heterodyning the frequency of the signal to a lower frequency at the antenna so that any subsequent change of transmission line length represents proportionally less phase mismatch.

The characteristics of typical airborne amplitude comparison and phase interferometer DF systems are summarized in Table 6.5. The phase-interferometer system generally uses superheterodyne receivers that provide the necessary selectivity and sensitivity for precise phase measurements.

DF can be accomplished using the Doppler shift effect, which is given by

$$f_d = (v/\lambda)\cos\theta \tag{6.13}$$

where v is the antenna's velocity, λ is the wavelength, and θ is the angle between the line of sight and antenna velocity vector. Solving for the AOA results in

$$\theta = \cos^{-1}(f_d\lambda/v) \tag{6.14}$$

This implies that both the velocity of the collecting antenna and the frequency of the source emitter be known.

A pseudo-Doppler shift can be induced onto an array of spatially displaced antennas by switching between these antennas. The pseudo-velocity is then determined by the switching pattern while the source frequency can be measured using any one of the stationary antennas. This technique has been successfully employed in the UHF and VHF bands using a circular array of antennas that generates a sinusoidal Doppler shift onto the source emitter.

Table 6.5
Direction of Arrival Measurement Techniques

	Amplitude Comparison	**Phase Interferometer**
Sensor configuration	Typically four to six equispaced antenna elements for 360-degree coverage	Two or more RHC or LHC spirals in fixed array
Theoretical DF accuracy	$DF_{ACC} \approx \frac{\theta_{BW}^2 \Delta C_{dB}}{24S}$ (Gaussian antenna shape)	$DF_{ACC} = \frac{\lambda}{2\pi d \cos \theta} \Delta \theta$
DF accuracy improvement	Decrease antenna BW; decrease amplitude mistrack; increase squint angle	Increase spacing of outer antennas; decrease phase mistrack
Typical DF accuracy	3- to 10-degrees rms	0.1- to 3-degrees rms
Sensitivity to multipath/reflections	High sensitivity; mistrack of several dB can cause large DF errors	Relatively insensitive; interferometer can be made to tolerate large phase errors
Platform constraints	Locate in reflection free area	Reflection free area; real estate for array; prefers flat radome
Applicable receivers	Crystal videos; channelizer; acoustic-optic; compressive; superheterodyne	Superheterodyne

ΔC_{dB} = amplitude monopulse ratio in decibels; S = squint angle in degrees; θ_{BW} = antenna beamwidth in degrees.

Another method for measuring AOA is the TDOA. This method uses the same configuration as the interferometer depicted in Figure 6.13. However, the time difference between the signals in the two antennas is measured instead of the phase difference. This has the advantage that the measurement is invariant with frequency. However, this method is only applicable when a reasonable baseline is involved. To examine this, consider that the TDOA between the signal in the two antennas is given by

$$\Delta t_d = (d/c)\sin \theta \tag{6.15}$$

where d is the distance between the antennas, c is the speed of light, and θ is the AOA. The resulting AOA is given by

$$\theta = \sin^{-1}(c\Delta t_d / d) \quad (6.16)$$

The sensitivity to time difference measurement noise can be determined by differentiating (6.15), resulting in

$$\Delta\theta = (c/d \cos\theta) \cdot \Delta t_d \quad (6.17)$$

The resulting time-measuring accuracy for a 1-degree angle accuracy using single and multiple platforms is given in Table 6.6.

The passive geolocation of ground emitters from airborne or space platforms is an important function useful for weapons targeting and situational awareness. The basic technique for geolocation of an emitter involves measuring AOA from multiple locations and triangulating to determine its location.

Two methods that have proven to be capable of providing highly accurate estimates of emitter location are the TDOA and the *frequency difference of arrival* (FDOA), also called differential Doppler techniques [13]. These methods are used in the *Communication High-Accuracy Airborne Location System* (CHAALS) used in the *Guardrail Common Sensor* (GRCS) to provide high location accuracy on communications and radar sensors.

A single TDOA measurement from two airborne platforms provides a hyperbolic curve containing the emitter location that is formed by the intersection of a cone with the surface of the Earth. A second measurement between a third airborne platform and either of the first two platforms provide similar type hyperbolic lines, the intersection of which provides the emitter location. The method requires a precise knowledge of the distance between the airborne platforms and a precision system clock to synchronize TDOA measurements. The *Global Positioning System* (GPS) can be used to accurately locate the platform, while a cesium clock is generally used to provide high time precision. Under certain unfavorable alignments of the airborne platforms a *geometric dilution of precision* (GDOP) occurs that can reduce the accuracy of this approach.

Table 6.6
Time Difference of Arrival Measurement Accuracy (1 Degree Angle Accuracy)

Platform	Baseline (m)	Δt_d (ns)
Aircraft wingtips	30	1.75
Ship bow/stern	100	5.8
Two A/C or ships	1000	59.2

A complementary method that can be used to determine emitter location uses the differential Doppler shift, which results when two platforms are closing on the emitter with different radial velocities. This effect can be analyzed by considering the frequency of the signal received at two stations that is given by

$$f_1 = f_0 - (r_{12} - r_{11})/\lambda T \qquad (6.18)$$
$$f_2 = f_0 - (r_{22} - r_{21})/\lambda T$$

where r_{ij} is the distance of receiver i from the transmitter at time $t = t_j$, $T = t_2 - t_1$ and λ is the wavelength. The FDOA is defined by

$$\Delta f_{\text{d}} = f_1 - f_2 = (1/\lambda T)[r_{22} - r_{21} - r_{12} + r_{11}] \qquad (6.19)$$

which defines a surface in three-dimensional space on which the transmitter must lie. If the transmitter lies on the surface of the Earth, then the intersection of the two surfaces defines a curve on the Earth on which the emitter lies. If a second measurement of FDOA is made or if another type of measurement such as TDOA or AOA can be obtained, two curves can be defined, and the intersection of the two curves provides an estimate of the transmitter location.

The use of both TDOA and FDOA generally provides greater precision than either used alone. FDOA is appropriate for a stationary emitter, while TDOA can be used for either fixed or mobile emitters.

6.5 Probability of Intercept

The POI is a measure that expresses the probability that the parameters of an ES receiving system (particularly center frequency and antenna orientation) match those of the transmitting emitter within some specified period of time. The intercept time, which is a key factor in the performance of EW systems, is defined as the time required to achieve a given POI. Modern ES systems have the objective of achieving 100% POI.

The achievement of 100% POI is generally approached using antennas that examine simultaneously all spatial angles containing emitters while employing wide-open receivers that are tuned to accept the frequencies radiated by the emitters. An example of this approach is the SLQ-32 ES system that employs a multibeam Rotman lens antenna to cover all possible spatial angles and a crystal video receiving system tuned to receive all the emitter frequencies. Many RWRs also use similar architectures to achieve 100% POI. However, ES systems employing rotating antennas or scanning superheterodyne receivers do not achieve 100% POI. The question with these systems relates to the observation time required to achieve a specified POI.

A few of the more common situations encountered with these types of systems are the beam-on-beam intercept problem, the frequency-frequency intercept problem, and the beam-frequency intercept problem. Each of these types of systems are reflected in currently operational hardware.

The beam-on-beam intercept problem generally occurs when a scanning antenna is employed to intercept an emitter with a scanning antenna. If the ES system's sensitivity is not sufficient to intercept the sidelobe radiations of the emitter, then an intercept occurs when the main beams of both antennas are pointing at each other. The parameters available for adjustment are the ES system's antennas beamwidth and scan rate. This problem assumes that a wide-open receiver such as a crystal video, IFM, or channelized receiver is utilized so that a search in frequency is not required.

The frequency-frequency search problem occurs when scanning or switched ES receivers are used to cover the emitter's expected frequency range. A classic situation is the scanning superheterodyne receiver. For CW signals, the superheterodyne receiver must dwell for a time in each frequency cell equal to the reciprocal of the receiver's bandwidth. For pulse signals, the dwell time is for the longest PRI associated with any emitter. A further dimension is added to the problem when the emitter is hopping in frequency. Smart scan techniques are used in some cases to reduce the intercept time as described in Example 6.1.

The beam-frequency problem is an extension of the first two types. Either the emitter or intercept antenna or both use a rotational, directional antenna and the receiver scans in frequency. An intercept occurs when the antennas are aligned in the receiver-emitter direction and the receiver passband encompasses the frequency of the emitter.

A further POI problem occurs in the time domain when the ES system is used as part of an EA system. With this situation, a short random look-through period is provided for the ES system to locate the emitters with the jamming disabled. When the look-through occurs, the ES system's receiver and antenna must be directed at the emitter for an intercept to occur.

A convenient way to investigate the intercept problem is by representing each activity of the receiving and transmitting systems by window functions as indicated in Figure 6.14. Each window function represents a time in which the ES system or the emitter are in a state consistent with intercept. For example, one window function could represent the time a directional antenna points at the emitter, a second window the time the emitter's directional antenna points at the intercept receiver, and a third window the time the frequency of the receiver is tuned to the frequency of the emitter. Each window

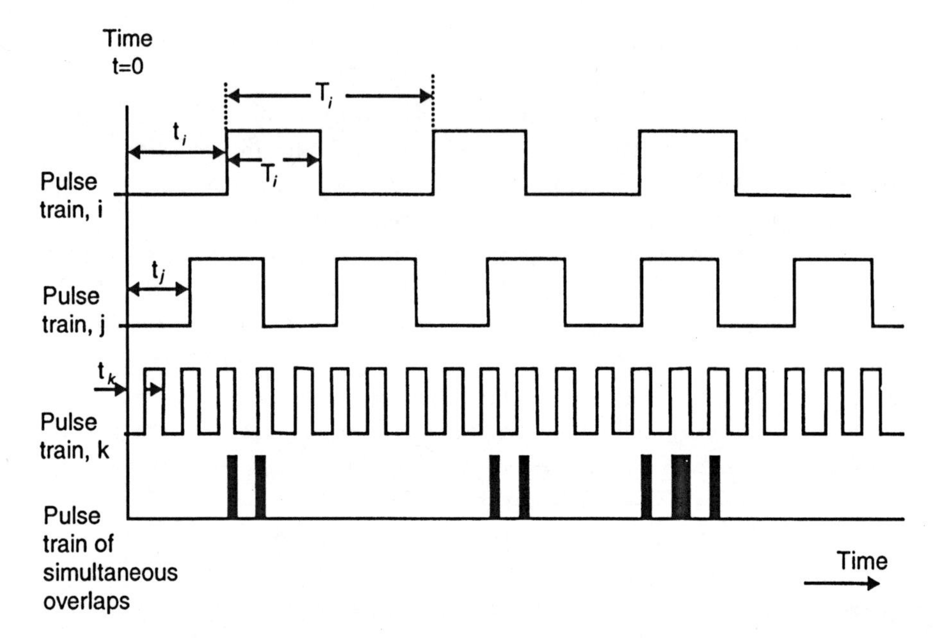

Figure 6.14 Intercept window function.

function is characterized by a window period representing the cyclic nature of the intercept phenomenon, a window width representing the time in each cycle when interception is possible, and a random starting time representing the independence between the intercept activities.

If we define the average time for an interception as T_0 sec, then the probability that a coincidence occurred in the time interval $t + \Delta t$ is given by $P(t + \Delta t)$. It is also given by the probability $P(t)$ that a coincidence occurred in the time interval t plus the probability $(1 - P(t))$ that no coincidence occurred in the time interval t multiplied by the probability of a coincidence in time interval $\Delta t(\Delta t/T_0)$. This can be expressed as

$$P(t + \Delta t) = P(t) + (1 - P(t)) \cdot \frac{\Delta t}{T_0} \tag{6.20}$$

which results in a differential equation for $P(t)$ given by

$$\frac{dP(t)}{dt} = \frac{1}{T_0}(1 - P(t)) \tag{6.21}$$

This equation can be solved as

$$P(t) = 1 - (1 - P_0)e^{-t/T_0} \tag{6.22}$$

where P_0 is the product of the coincidence probabilities of the individual window functions given by

$$P_0 = \prod_{i=1}^{m} \frac{\tau_i}{T_i} \tag{6.23}$$

Generally, P_0 is very small so that

$$P(t) = 1 - e^{-t/T_0} \tag{6.24}$$

The average window time T_0 for the two-window problem can be found as follows. The probability of coincidence for the first window is given by τ_1/T_0. The probability is also given by the conditional probability of a pulse in the first window (τ_1/T_1) times the probability of a coincidence in the second window given by $(\tau_1 + \tau_2)/T_2$ [14].

Equating the two probabilities results in

$$T_0 = \frac{T_1 T_2}{\tau_1 + \tau_2} \tag{6.25}$$

This can be extended to multiple windows as

$$T_0 = \prod_{i=1}^{m} \left(\frac{T_i}{\tau_i}\right) \Big/ \sum_{i=1}^{m} \left(\frac{1}{\tau_i}\right) \tag{6.26}$$

The time (T_j) to achieve a certain level of intercept probability can be expressed as

$$T_j = -T_0 \ln(1 - P(T)) \tag{6.27}$$

These relationships describe the intercept time and POI for well-behaved independent window functions that do not lock into synchronism. If these functions are derived from a common source (e.g., locked to a power line), then there may be long periods of coincidence or no coincidences that differ from the average values defined in the preceding equations.

▶ *Example 6.4*

Consider an intercept system that uses a 1-degree beamwidth antenna that rotates at 60 rpm to intercept a radar emitter having a 2-degree beamwidth rotating at 10 rpm. Assume a wide-open ES receiver is used (e.g., IFM). Using MATLAB, plot the probability of intercept versus the intercept time using (6.26) and (6.27). To determine the accuracy of these equations, simulate the intercept probability using the square (t, duty) function and plot a histogram of the intercept probability using the hist (y, x) function. To reduce the computer run time, use T_1 = 6.13 sec, t_1 = 0.0333 sec, T_2 = 1.05 sec, and t_2 = 0.00278 sec. How do the results compare?

```
% Probability of Intercept Simulation
% ----------------------------------
% probint.m

clear;clf;clc;

% Input Window Parameters

s=[];
T1=6.13;           % Emitter 1 Period in s
T2=1.05;           % Emitter 2 Period in s
t1=.0333;          % Intercept 1 Pulse in s
t2=.00278;         % Intercept 2 Pulse in s
df1=100*t1/T1;     % Duty Factor 1 - %
df2=100*t2/T2;     % Duty Factor 2 - %
n=1000;            % Number of Trials

% Simulate Probability of Intercept

for i=1:n;
p=[];
c=0;
r=rand;
while length(p)<1,
   t=linspace(c*T2,c*T2+t2,100);
   x=square(2*pi*(t/T1+r),df1);
   z=(x+1)/2;
   p=find(z);
   tot=c*T2;
   if tot>10000,break,end;
   c=c+1;
end;

% Find Probability of Intercept

s(i)=c*T2;
phase(i)=r;
end;
[hits,time]=hist(s,600);
cumulative=cumsum(hits);
```

```
y=cumulative/n;

% Plot Probability of Intercept

x1=time;
stairs(x1,y);
axis([0,500,0,1]);
grid;
title('Probability of Main Beam Intercept');
xlabel('Intercept Duration-seconds');
ylabel('Cumulative Probability of Intercept');
text(350,.5,['T1=',num2str(T1),'sec']);
text(350,.45,['t1=',num2str(t1),' sec']);
text(350,.4,['T2=',num2str(T2),' sec']);
text(350,.35,['t2=',num2str(t2),' sec']);
hold on;

% Calculate Probability of Intercept

POI95=.95;
To=T1*T2/(t1+t2);        % Mean Intercept Time
Tj=-To*log(1-POI95);     % Time POI=.95
tx=0:1:Tj;
POI=1-exp(-tx/To);
plot(tx,POI);
hold off;
```

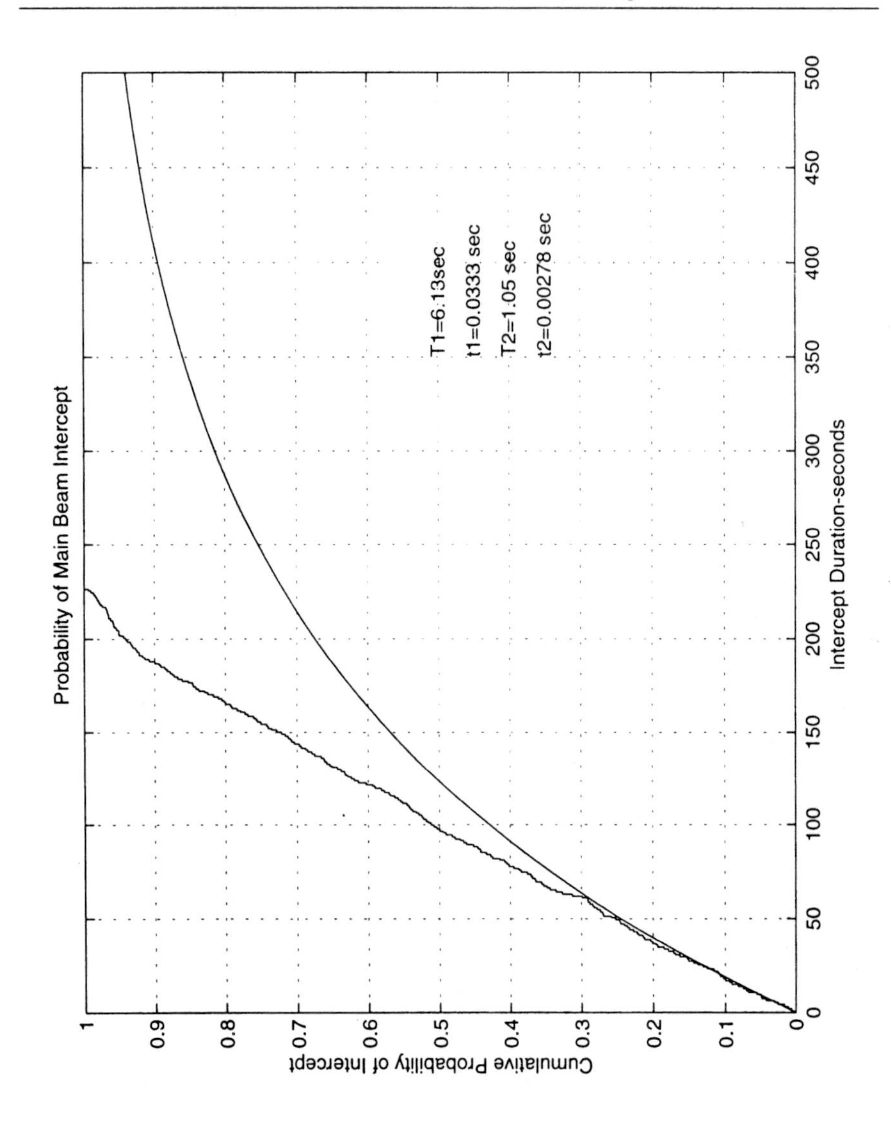

▶ *Example 6.5*

Consider an intercept system that includes both a narrowbeam rotating antenna and a scanning narrowband receiver that operates against a scanning radar emitter. Assume parameters are the same as Example 6.2 for the rotating antenna systems. Assume the narrowband scanning receiver covers 2 GHz in 100-MHz steps and that the minimum PRF of the emitter is 500 Hz. (Use T_3 = 0.043 sec and t_3 = 0.0021 sec to reduce the computer run-time). Plot the intercept probability versus the observation time.

```
% Probability of Intercept Simulation
% Using Three Windows
% ------------------------------------
% probint3.m

clear;clc;clf;

% Input Window Parameters

s=[];
T1=6.13;         % Emitter 1 Period in s
T2=1.05;         % Emitter 2 Period in s
T3=.043;         % Receiver Scan Period in s
t1=.0333;        % Intercept 1 Pulse in s
t2=.00278;       % Intercept 2 Pulse in s
t3=.0021;        % Receiver Open Time in s
df1=100*t1/T1;   % Duty Factor 1 in %
df2=100*t2/T2;   % Duty Factor 2 in %
df3=100*t3/T3;   % Duty Factor 3 in %
n=200;           % Number of Trials

% Simulate Probability of Intercept

for i=1:n;
p=[];
c=0;
r1=rand;r3=rand;
while length(p)<1,
   t=linspace(c*T2,c*T2+t2,100);
   x=square(2*pi*(t/T1+r1),df1);
   xx=(x+1)/2;
   y=square(2*pi*(t/T3+r3),df3);
   yy=(y+1)/2;
   z=xx.*yy;
   p=find(z);
   tot=c*T2;
   if tot>5000,break,end;
   c=c+1;
end;

% Find Probability of Intercept

s(i)=c*T2;
phase(i)=r1;
end;
[hits,time]=hist(s,600);
cumulative=cumsum(hits);
yz=cumulative/n;

% Plot Probability of Intercept
```

```
x1=time;
stairs(x1,yz);
%axis([0,500,0,1]);
grid;
title('Probability of Main Beam Intercept');
xlabel('Intercept Duration-seconds');
ylabel('Cumulative Probability of Intercept');
hold on;

% Calculate Probability of Intercept

POI95=.95;
To=(T1*T2*T3)/(t1*t2+t2*t3+t1*t3); % Mean Intercept Time
Tj=-To*log(1-POI95);                % Time POI=.95
tx=0:1:Tj;
POI=1-exp(-tx/To);
plot(tx,POI);
hold off;
```

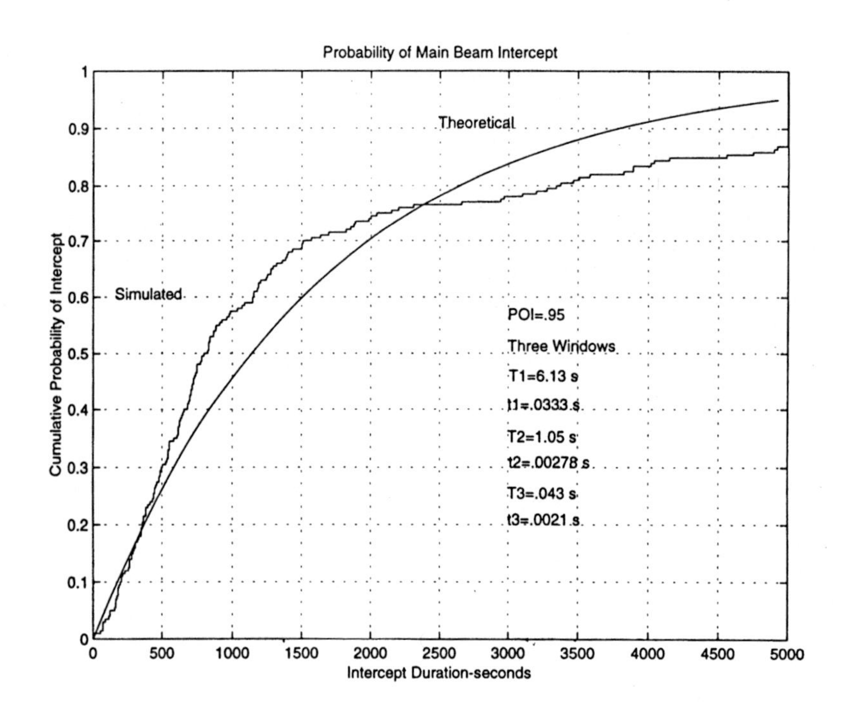

▶ *Example 6.6*

Assume a radar emitter has a 1.5-degree beamwidth that rotates at 12 rpm and that it is to be intercepted using a rotating antenna with a wide-open receiver. Using MATLAB, plot curves showing the observation time for a 90% POI versus the interceptor beamwidth with the interceptor scan rate as a parameter. From these curves determine the best interceptor scan rate and beamwidth.

```
% Intercept Time for Specified POI
% ----------------------------------
% intcept.m

clear;clf;clc;

% Input Radar Beamwidth and Pulsewidth

bw=1.5;          % degrees
scan=12;         % rate in rpm

% Calculate Window Function

T1=60/scan;      % seconds
t1=(bw/360)*T1;  % seconds

% Input Intercept Bandwidth

ibw=1:.1:10;     % degees

% Input POI

POI=.9;

% Calculate Obsevation Time

for k=1:6;
  iscan=30*k;     % degrees
  T2=60/iscan;
  t2=(ibw/360)*T2;
  To=(T1*T2)./(t1+t2);
  Tj=-To*log(1-POI);
  Tjx(:,k)=Tj';
end;

% Plot Observation Time

plot(ibw,Tjx,'k');grid;
xlabel('Beamwidth - degrees');
ylabel('Intercept Time - seconds');
title('Intercept Time for Main Beam Intercept');
text(7.5,750,['POI =',num2str(POI)]);
```

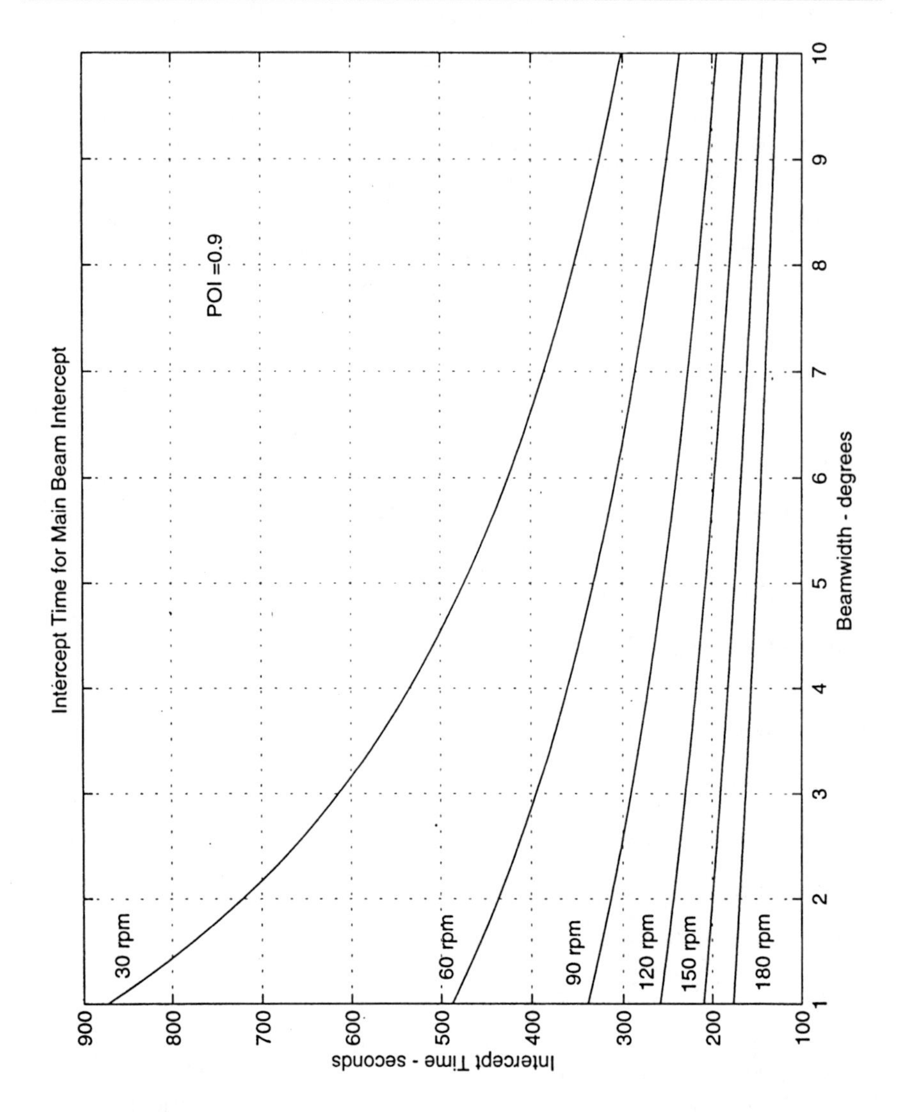

6.6 Problems

1. A RWR has a circularly polarized receiving antenna with a gain of 3 dB. The antenna feeds a scanned superheterodyne receiver with a 20-MHz bandwidth and a sensitivity (S/N = 13 dB) of –63 dBm. The superheterodyne searches a frequency band from 1 GHz to 20 GHz by stepping its 20-MHz bandwidth every 4 ms.

a. How much time does it take the receiver to scan through the 1- to 20-GHz band?

b. Given the radar with the indicated parameters, determine the range at which detection of the radar by the receiver occurs.

c. If an onboard jammer radiates 200W of noise over a 100-MHz band centered on the transmitter frequency of the radar in part b, and if the isolation between the jammer and receiver is 60 dB, at what range will the radar be detected?

Transmitter power	50 kW
Antenna gain, maximum	34 dB
Transmitter frequency	5.6 GHz
Antenna beamwidth	2 degrees
Pulse width	0.4 μs
Pulse repetition frequency	1800 Hz
Polarization	Vertical
Noise figure	6 dB
Scan axis gain	31 dB

2. A police radar is designed to measure the velocity of a moving vehicle, with a RCS σ up to a distance of 50m. The car is equipped with a radar detector whose effective antenna aperture is $k\sigma$ (k = 0.001). The antenna gain is 100 and the frequency is 15 GHz. Assuming that the car's radar detector has the same sensitivity as the radar receiver, at what distance will the radar detector provide a warning?

3. A multiband ES receiver employs an acceptance bandwidth in the search mode of 10 MHz from 100 MHz through 1 GHz and 20 MHz from 1 GHz to 18 GHz. Minimum PRFs are 200 Hz in the lower band and 500 Hz in the upper band. Find the minimum sweep time that provides 100% probability of detections of at least one radar pulse, assuming the signals are above the receiver thresholds and the receiver scans successively through the bands.

4. Design a DIFM to indicate the frequency to within ±5 MHz from 6 GHz to 12 GHz. How many correlators are required? What delay is required for the least and for the most significant bit?

5. The equivalent noise bandwidth of a wideband direct detection receiver with an RF bandwidth followed by a video bandwidth B_V is given by $B_e = B_{RF}^{1-\alpha}(2B_v)^{\alpha}$, where α is a factor depending upon the ratio $B_{RF}/2B_V$, the probability of detection (P_d), and the probability of false alarm (P_{fa}). Find the equivalent noise bandwidth for a receiver that has a passband from 8 GHz to 12 GHz and a video bandwidth of 20 MHz with an α = 0.7. How does the sensitivity of this receiver compare to a superheterodyne receiver with a 4-GHz bandwidth and one with a 20-MHz bandwidth?

6. Alberhseim's relationship approximates the signal-to-noise ratio in decibels for a given P_d, P_{fa}, and number of pulses noncoherently integrated. This relationship is given by

$$S_{dB} = -5\log(n_e) + k\left(6.2 + \frac{4.54}{\sqrt{n_e + 0.44}}\right)$$

$$k = \log(A + 0.12A \cdot B + 1.7B)$$

where

$$A = ln(0.62/P_{fa}) \qquad B = ln(P_d/(1 - P_d))$$

$$n_e = \text{number of pulses integrated.}$$

Use Alberhseim's relationship to show that the integration exponent α is given by

$$\alpha = 0.5 + \frac{4.54k}{10\log(n_e)}\left(\frac{1}{\sqrt{1.44}} - \frac{1}{\sqrt{n_e + 0.44}}\right)$$

where $n_e = B_{RF}/2B_v$. What is the exponent α for a $P_d = 0.9$, $P_{fa} = 10^{-6}$, and $B_{RF}/2B_v = 100$?

7. Using MATLAB, plot contours of the exponential integration factor α derived in Problem 6 with $B_{RF}/2B_v$ (1 to 10,000) as the ordinate, α as the abscissa, and $P_d = 0.5$ ($P_{fa} = 10^{-2}, 10^{-4}, 10^{-6}$), $P_d = 0.9$ ($P_{fa} = 10^{-6}$), and $P_d = 0.99$ ($P_{fa} = 10^{-8}, 10^{-10}$). What values of α are most prevalent for large $B_{RF}/2B_v$?

8. The parameters in Table 6.7 are measured from an RWR. Sort the data set by AOA and RF. From the sort, derive the PRF of the emitters.

9. N parameters are used to identify a particular emitter. If the measurement tolerance is ±2% of the total parameter range, what is the maximum number of sources that could be uniquely identified by the system for (a) $N = 3$ and (b) $N = 5$?

10. In an ES receiver, radars with PWs from 50 ns to 10 μs are intercepted. What is the minimum receiver bandwidth necessary to allow measurement of the emitter's PW? What signal-to-noise ratio is required to allow measurement of the emitter's frequency to 1 MHz? What signal level into a 5-dB noise figure receiver is required to provide this capability?

11. The signal from an FM-CW LPI radar sweeps 40 MHz in 1 ms and then repeats. What bandwidth receiver is necessary to intercept this waveform? What ES receiver is best suited to detect this form of signal?

Table 6.7
Data Stream for RWR

Angle of Arrival (Degrees)	RF (GHz)	Amplitude (V)	Pulsewidth (μs)	Time of Arrival (ms)
10	2	1	1	1
10	1	6	6	1.1
90	3	3	2	1.5
10	2	1.5	1	2
90	3.1	3	2	2.5
10	2	1	1	3
90	3	3	2	3.5
10	2	0.5	1	4
10	1	6	6	4.2
90	3.1	3	3	4.5

12. A PD signal with a PW of 1 μs and a PRF of 100 kHz is intercepted in an ES receiver. This signal is processed in a dedicated superheterodyne receiver and is to be filtered from lower PRF signals simultaneously received in the data. What bandwidth is required in the superheterodyne receiver to isolate the PD signal? What bandwidth is required in an eight-pulse, l-dB ripple, Chebyshev band reject filter to attenuate the PD signal by 60 dB so that it does not interfere with lower PRF signals in the deinterleaving process?

13. To reduce the intercept time, a wideband IFM receiver is used to tune a superheterodyne receiver to the signal. The superheterodyne receiver has a sensitivity of −80 dBm and the IFM receiver's sensitivity is −60 dBm. What is the overall detection sensitivity of the ES receiver?

14. An IFM receiver has a sensitivity of −60 dBm (S/N = 0 dB). It is used to measure the frequency of a signal with 0.1-μs PW. What minimum signal level is required to measure the frequency of the signal to an rms accuracy of 1 MHz? If a superheterodyne receiver with −80 dBm sensitivity is placed in front of the IFM, what signal level is required to achieve the 1-MHz rms accuracy?

15. A crystal video receiver has an RF bandwidth of 2 GHz and a video bandwidth of 20 MHz. Its tangential sensitivity (S/N = 4 dB) is measured as −40 dBm. Using the model of Problem 5 with $\alpha = 0.7$, deduce the equivalent noise figure of the receiver. What is the resulting tangential sensitivity if a *low-noise amplifier* (LNA) with 30-dB gain and 3-dB noise figure is placed before the crystal video receiver?

16. An IFM receiver operation over a 2-GHz band with a direct detection type video diode is used to measure the frequency of pulse signals with minimum

PWs of 0.1 μs. The tangential sensitivity (S/N = 4 dB) of this system is measured as −40 dBm. What input signal level is required to provide a minimum measurement accuracy of 1 MHz? If an LNA with 30-dB gain and 3-dB noise figure is placed in front of the IFM receiver, what signal level is now required to provide a minimum measurement accuracy of 1 MHz?

17. Radar signals between 2 GHz to 18 GHz are to be intercepted using a channelized receiver. If the minimum PW of these signals is 0.1 μs, how many filters or channels are required to detect the signal if a full channelized receiver is employed? What is the sensitivity (S/N = 0 dB) of this receiver with a 5-dB noise figure? If a 3-dB noise figure LNA front-end is used to drive the channels, each with 15-dB noise figure, what amplifier gain is required to provide an overall 5-dB noise figure?

18. The channelized receiver of Problem 17 is designed using three manifolds. The first folds the signals into 2-GHz-wide channels, the second into 125-MHz-wide channels, and the third into 10-MHz-wide channels (see Figure 6.7). If a band-folded configuration (all channels in each manifold combined) is used, what overall sensitivity is provided for the parameters of Problem 17? If a fast call configuration (only those channels with a signal energized) is used, what is the sensitivity of the receiver for the parameters of Problem 17? What amplifier gain is required to maintain the 5-dB noise figure? Note that in the fast call configuration, a separate mechanism is required to determine those channels to be energized while in a band-folded configuration, ambiguity resolution is required.

19. A 0.1-μs pulse with carrier frequency is applied to a channelized receiver with 60 dB of dynamic range. Over what frequency range does the channelized receiver provide an output for a rectangular pulse, a triangular pulse, and a Gaussian-shaped pulse?

20. Using MATLAB, simulate a 1-μs pulse into a channelized receiver at a carrier frequency of 100 MHz. Apply this signal to a bandpass filter with 1-MHz bandwidth tuned to 80 MHz. Plot the envelope of the output signal. Why is this called "rabbit ears"? How would you design a circuit in a channelized receiver to determine that this is a spurious signal?

21. A linear FM PC signal with 10-MHz chirp and 100-μs PW is to be detected in an ES receiver. What processing gain (increase in amplitude out of the PC network as compared to the amplitude of the uncompressed signal) is provided by this signal? If a crystal video receiver is used, what reduction in amplitude of the video output will occur versus processing an equivalent 0.1-μs pulse with the same energy as contained in the PC signal? If a superheterodyne receiver is used, what bandwidth is required to pass the PC signal?

22. Show that for the same resolution and sensitivity, a compressive receiver that uses a chirp waveform with bandwidth (BW) and pulsewidth (T)

can scan a band $\beta = T \cdot BW$ faster than a conventional superheterodyne receiver.

23. The compressive receiver shown in Figure 6.8 uses a chirp that sweeps over 1 GHz in a 0.2-μs period. How many signals at different frequencies can theoretically be resolved in the 0.2-μs analysis window? What is the duration of the video signal in the analysis window for a CW signal within the acceptance band? What bandwidth digital output circuitry is required?

24. An RWR uses quadrant spiral antennas in an amplitude monopulse configuration to determine the intercepted emitter's AOA. The antennas provide Guassian-shaped beams and cross over at the 3-dB points. What minimum signal-to-noise ratio is required to provide 10-degree angular accuracy? What is the maximum amplitude imbalance that can be tolerated while meeting the angular accuracy requirement?

25. In a two-element interferometer, what signal-to-noise ratio is required to provide 0.1-degree accuracy? What maximum phase measuring error can be tolerated for 0.1-degree accuracy? If an overall 0.1-degree accuracy is desired, how would you combine the inaccuracies resulting from the two error sources?

26. The angle measured by an interferometer is accurate only for its design frequency. Show that the error due to frequency inaccuracy is given by $\Delta\theta = \Delta f \cdot \sin\theta / f_0$, where Δf is the frequency error, θ is the angle of arrival, and f_0 is the design frequency.

27. Consider an interferometer designed with antennas in an azimuth plane. Derive an expression for the angular error caused by the assumption that the incident signals are contained in the azimuth plane when they actually have an elevation angle with respect to this plane. How would you design a system to correct this problem?

28. Design a wideband interferometer system to cover the frequency range from 1 GHz to 10 GHz. A linear array of antennas is used. The outermost antennas may be spaced as much as 3m apart. In each band, a closely spaced and widely spaced antenna pair should be used to obtain good accuracy. Overall accuracy should be 1-degree rms. Specify antenna spacing, sets to be used in each band, and the accuracy to which phase differences must be measured.

29. A TDOA system is used to measure the AOA of a pulsed signal from an emitter. Two antennas are used, separated by a distance *d.* The PW of the signal is τ. Derive an expression for the angular accuracy of the system as a function of the signal-to-noise ratio (SNR). What SNR is required for 0.1-degree accuracy with 30-m separation using 0.1-μs pulses?

30. TDOA systems sometimes use antennas located on separate platforms, thereby providing a long baseline with resulting higher accuracy. To synchronize the time measurements, a precision clock is required as is accurate measurement of the baseline *d.* Derive expressions for TDOA angular accuracy as a function of time and baseline measurement errors. Given a baseline of 1000m, what time-measuring accuracy and distance-measuring accuracy is required to provide an angular accuracy of 0.1 degrees?

31. FDOA is used to measure the angle of arrival of a ground emitter from two moving aircraft platforms. The frequency of the ground emitter is f_0 and the Doppler frequencies of the receiver signals are measured in each aircraft as f_{d1} and f_{d2} over time Δt_m. From (6.18), show that the pseudo-TOA of each signal is given by

$$t_{ai} = \frac{f_{di}}{f_0} \Delta t_m$$

Using this relation in the TDOA equations, derive an expression for the AOA using FDOA. How accurately must the emitter's frequency be known for this method to achieve comparable accuracy to TDOA measurements? Can the two methods be used in complementary ways to provide greater overall accuracy than either method alone?

32. A rotating reflector DF system is used to determine the AOA of an emitter. A wide-open receiver is used. If the beamwidth is 2 degrees and the minimum emitter PRF is 500 Hz, what is the maximum antenna scan rate for 100% POI on emitter antenna sidelobes? Using this scan rate, what is the POI provided for a main-beam intercept if the radar has a beamwidth of 5 degrees and rotates at 12 rpm?

33. In a rotating reflector DF system, why is it necessary to have a complementary omnidirectional antenna system?

34. Two DF systems have an rms accuracy of 3 degrees. The first measures the bearing of the emitter as 92 degrees, while the second measures 310 degrees. If the source is 500 km from the first DF system and 300 km from the second, find the footprint in the location of the source from the DF measurements.

35. A rotating reflector DF system has a beamwidth of 30 degrees and a scan rate of 200 rpm. Compute the beam wander if it is used against a source with 100% modulation at a 160-Hz rate.

References

[1] Ås, B, - O., "The Pilot, A Quiet Naval Tactical Radar," Bofors Electronics AB Report, 1990.

[2] Hatch, W., "Computer Use in Electronic Warfare," *Microwave J.*, Sept. 1978.

[3] Schleher, D. C., *Introduction to Electronic Warfare*, Norwood, MA: Artech House, 1986.

[4] Wiley, R., *Electronic Intelligence: The Interception of Radar Signals,* Norwood, MA: Artech House, 1986.

[5] Scherer, J., "The Dual Polarized Sinuous Antenna," *J. Electronic Defense,* April 1996.

[6] Herskovitz, D., "The Other SIGINT/ELINT," *J. Electronic Defense,* April 1996.

[7] Tsui, J., *Microwave Receivers with Electronic Warfare Applications,* New York: John Wiley, 1986.

[8] Abercrombie, H., and J. Wood, "Channelized Receiver Application Guide," *J. Electronic Defense* 1995 Suppl.

[9] Luther, R., and W. Tanis, "Advanced Compressive Receiver Techniques," *J. Electronic Defense,* July 1990.

[10] Tsui, J., *Digital Microwave Receivers: Theory and Concepts,* Norwood, MA: Artech House, 1989.

[11] Tsui, J., *Digital Techniques for Wideband Receivers,* Norwood, MA: Artech House, 1995.

[12] Rehnmark, S., and T. Burgher, "Passive, Precision Direction Finding," *J. Electronic Defense,* Aug. 1990.

[13] Chestnut, P., "Emitter Location Accuracy Using TDOA and Differential Doppler," *IEEE Trans.,* Vol. AES-18, No. 2, March 1982.

[14] Self, A., and B. Smith, "Intercept Time and Its Prediction," *IEE Proc.,* Pt. F, July 1985.

7

Expendables and Decoy Systems

Expendable EA, as the name suggests, refers to EA systems using expendables that are deployed only once for a limited time offboard the platform that they are designed to protect. The expendable nature of this type of EA makes economics an important consideration in its design. To be cost effective, the life-cycle cost of the number of expendables intended to protect a platform must be less than the cost of the platform itself.

Chaff and flares are generally the most inexpensive and effective expendables. Chaff is a form of volumetric radar clutter that is composed of distributed metalized reflectors dispensed into the atmosphere to interfere with and confuse radar operation. The chaff usually consists of a large number of dipoles that are designed to resonate at the frequencies of the radars they are attempting to confuse. Flares are designed to be effective against *infrared-* (IR) seeking missiles. They are dispensed as the missile approaches its target to capture the IR seeker's tracking system, thereby diverting the missile away from the target.

Decoys are intended to divert incoming missiles away from their intended target and are not necessarily consumed in performing their function. They are particularly effective against missiles employing monopulse seekers, which are difficult to counter by other means. Decoys can be either active or passive, but the current trend favors active decoys. This trend is supported by the recent development of *monolithic microwave integrated circuits* (MMICs), which provide both the miniaturization and low cost necessary in these systems.

Deployment of active decoys depends on both the platform and the type of decoy. In aircraft, small active decoys can be launched from chaff/flare dispensers (i.e., GEN-X). Larger active decoys are towed from the aircraft and operate either independently (i.e., ALE-50) or in concert with the onboard jamming system (i.e., IDECM). For operation with an onboard EA system, *fiber optic* (FO) links are prevalent. Other forms of airborne decoys are launched from munitions racks and use gliders or miniature UAVs to simulate strike aircraft (i.e., Tactical Air-Launched Decoy) [1].

Shipboard decoys are generally intended to divert antishipping missiles away from the ship they are protecting. The high RCS of ships dictates a high ERP for active decoys. The towed *active electronic decoy* (AED) features a 1- to 2-kW ERP active repeater that covers the I/J-band and facilitates defense against frequency agility seekers. A passive form of towed decoy called the Rubber Duck uses linked pairs of octahedral corner reflectors. However, the most common shipboard decoys employ rocket-propelled chaff cartridges. The dispenser systems resemble a matrix of mortar barrels that are loaded with explosive squibs and payload cartridges. The squib is used to propel the cartridge that in addition to chaff may contain flares, active EA decoys, or hybrid packages. Rapid blooming chaff is required for shipborne defense. It is contained in large numbers of pie-shaped plastic cassettes along with small explosives in each shell package. The chaff shell cartridges are propelled to an altitude of 100m or more to ensure that the cloud has an adequate lifetime. At altitude, explosive rod located in the center of the shell first explodes, shattering the chaff cassettes that then explode, releasing the chaff.

► *Example 7.1*

The naval scenario presented here consists of an antishipping missile directed against a ship that is defended by two jamming buoys. The missile is launched 30 km from the ship. The jamming buoys are located at (5000, 100) and (3000, 5) in an x, y coordinate system. They transmit 100W of spot noise over a 10-MHz bandwidth. The RCS of the ship is 400 m^2. The missile seeker transmits 40 kW and its receiver has a 2-MHz bandwidth. The missile seeker antenna has 30 dB gain and employs cosine square aperture weighting. Using MATLAB, plot the *jam-to-signal* (J/S) ratio as the missile engages the ship.

```
% Program Calculates J/S for Jamming with
% Decoys in Naval Scenario
% ---------------------------------------
% buoyjam.m

clc;clf;clear;

% Jamming Decoy Parameters

Pj1=100;Gj1=1; % Erp
Pj2=100;Gj2=1; % Erp
x1=5000;y1=100;% meters
x2=3000;y2=50; % meters
wl=.03;        % meters

% Ship at origin

Rcs=400;       % m^2

% Missile with cos^2 Antenna and Go=30 db
% Launch range of 30 km

Rmax=30000;Rmin=2000;
Put=40000;

% N Element Range Vector

N=1024; step=(Rmax-Rmin)/(N-1);
R=Rmax:-step:Rmin;

%  Angle Vectors fi to Ship

fi=zeros(1,N);

% Angle Vectors to Jamming Buoy #1

fi1=atan(y1./(R-x1));

% Angle Vectors to Jamming Buoy #2

fi2=atan(y2./(R-x2));

% Range to Jamming Buoy #1

R1=((R-x1).^2+y1^2).^(1/2);

% Range to Jamming Buoy #2

R2=((R-x2).^2+y2^2).^(1/2);

% Received Jamming Power from Buoy #1
```

```
Prj1=(Pj1*Gj1*gcos2(30,fi1)*wl^2)./...
((4*pi)^2*(R1).^2);

% Received Jamming Power from Buoy #2

Prj2=(Pj2*Gj2*gcos2(30,fi2)*wl^2)./...
((4*pi)^2*(R2).^2);

% Jam Power Dilution for 10 Mhz Band Width
% in 2 Mhz Receiver Band Width

Prj1=Prj1*2/10;
Prj2=Prj2*2/10;

% Zero Prj1 and Prj2 for R<=x1 and R<=x2

for n=1:N;
  if R(n)<=x1;Prj1(n)=0;end;
end;

for n=1:N;
  if R(n)<=x2;Prj2(n)=0;end;
end;

% Total Received Jamming Power

Prj=Prj1+Prj2;
Prjdbm=30+10*log10(Prj+1e-12);

% Recived Target Power from Ship

Prtar=(Put*gcos2(30,fi).^2*Rcs*wl^2)./...
((4*pi)^3*R.^4);
Prtardbm=30+10*log10(Prtar+1e-12);

% Plot Received Jammer and Target Power

plot(R/1000,Prjdbm,R/1000,Prtardbm);
title('Jam and Target Power for Buoy Jamming');
ylabel('Received Signal - dbm');
xlabel('Target Range - km');grid;
```

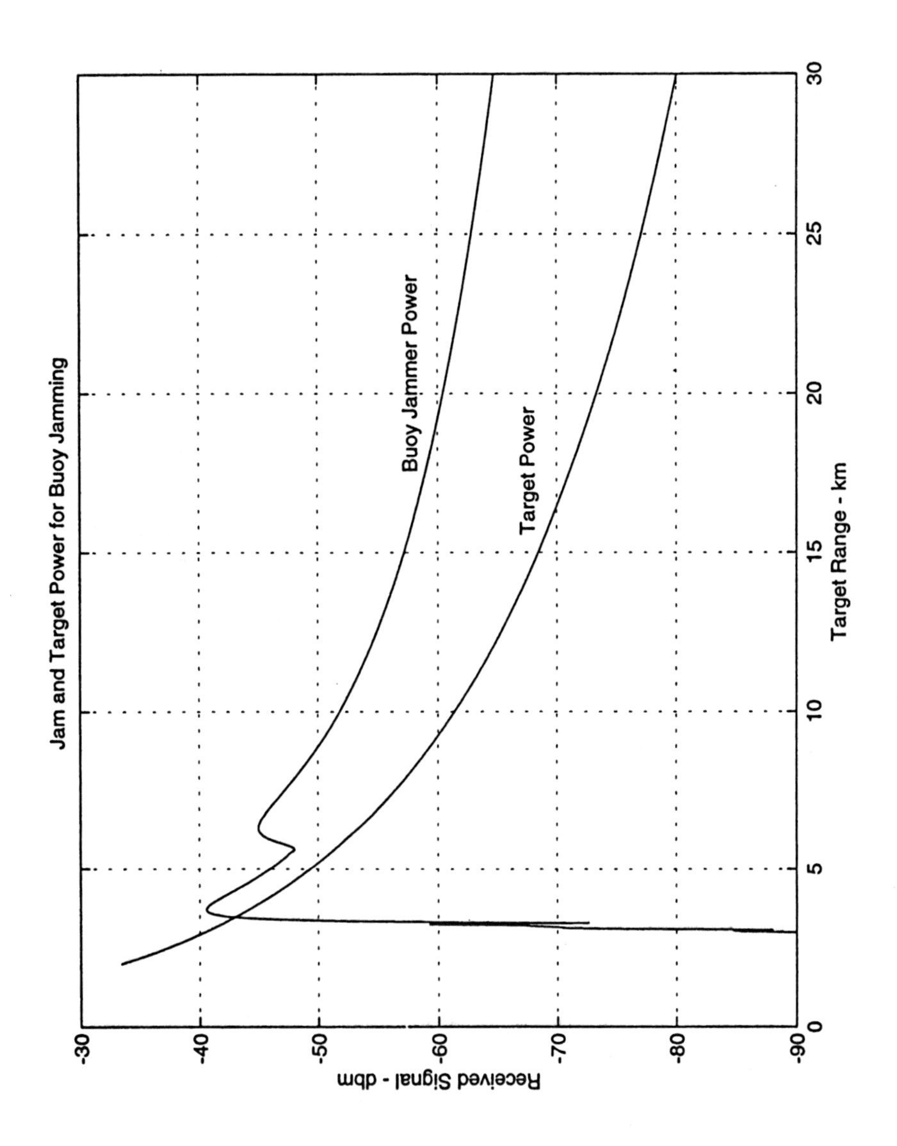

7.1 Design of Expendable EA Systems

A number of factors must be considered in the design of expendable EA systems. These may be conveniently separated into those associated with (1) the delivery vehicle, (2) payload, and (3) a deployment requirement as depicted in Figure 7.1. As can be seen from this diagram, a large number of possibilities exist for each category, but a complete description is beyond the scope of this discussion.

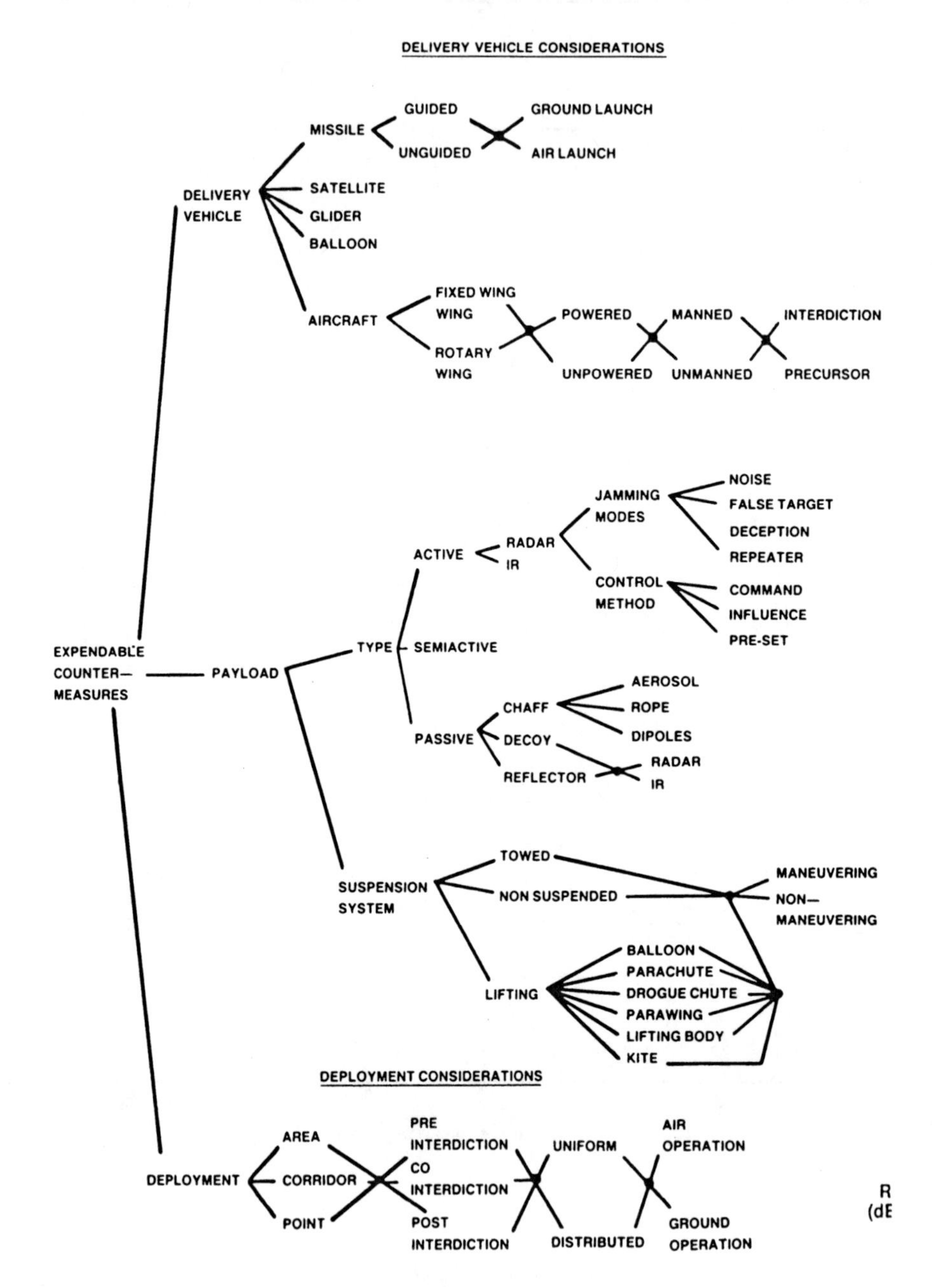

R
(dE

Figure 7.1 Expendable countermeasures design factors. (*Source:* [14], p. 101.)

Instead, we will concentrate on a few general principles followed by a more detailed treatment of the characteristics and deployment of chaff.

In general, a systematic approach must be applied to the design of expendable EA. First, it must be recognized that several forms of expendable EA and conventional onboard and stand-off EA may be simultaneously deployed in an overall system. Expendables should be deployed only when they provide an advantage over other forms of EA, either through increased performance or lower cost. General advantages of expendables are their ability to generate real angle deception and to provide jamming in close proximity to the threat emitter. This latter characteristic provides the potential for low-power active jamming, which, in turn, allows the cost-effective deployment of multiple jammers.

Time is a critical parameter that affects the design of expendable EA. First, the time over which the expendable is effective is generally a function of a number of factors including the delivery method, the suspension system, and the mission dynamics. For corridor chaff released into the atmosphere, this time may range from of the order of 40 sec for aluminum chaff to over 100 sec for aluminized-glass chaff to reach full efficacy; while residual effects are apparent well over a half-hour after being dispensed. This time is considerably reduced for chaff dispensed into a rarefied atmosphere (e.g., ballistic missile defense application) because the rate of fall varies inversely with the square root of the density of air. Active expendable jammers dispensed from an airborne platform and suspended by parachute can radiate for 5 to 10 minutes. Alternatively, an expendable jammer dropped by parachute and implanted into the ground can radiate for over an hour, injecting jamming energy into the threat emitter's sidelobes. Expendable EA can be suspended on balloons for an indefinite period of time, provided that winds are favorable. Alternatively, the balloon can be tethered if allowed by mission dynamics. Expendable EA carried by small *remotely piloted vehicles* (RPVs) can operate for some hours using either command or inertial navigation guidance.

With either passive or active expendables, the objective against radar threats is to provide a time window during which the target is shielded from radar detection and tracking. This places a premium upon proper management and timely application of the expendable resources because, when the limited number of expendables are dissipated, this form of jamming protection is no longer available. Most modern expendable EA systems employ some form of computer control to determine the most effective response so that the least amount of expendables are deployed in a timely sequence to counter the perceived threat. In future expendable EA systems, this function will be integrated into the overall platform defensive system that operates by fusion in concert with the other platform sensors and jamming resources to optimize the proper set of defensive responses at any instant of time.

Active expendable EA systems are generally more expensive than passive expendable EA systems such as chaff or retrodirective decoy reflectors and, hence, tend to be used where passive systems are not effective. This generally occurs in the lower frequency range (e.g., below 1 GHz) where chaff dipoles are no longer practical, requiring the use of long (e.g., 750 ft) nonresonant streamers called "rope" and where decoys that are proportional in size to the incident wavelength become overly large.

Active expendable jammers can use either noise- or deception-jamming techniques. Generally, noise techniques are preferred because of their simplicity. One mode of operation is for the active expendable jammer to radiate a noise-modulated or CW signal, which is automatically tuned after deployment to the radar's frequency. Another mode uses frequency-adaptive repeater transponders, which generate multiple false targets. The transponder is initially tuned to the radar's frequency as the radar beam sweeps over the target and then continually responds, generating false targets. There is no need for range or angle deception because one of the objectives of the expendable jammer is to be spotted and draw attention away from the real target. Chirp frequency modulation is sometimes applied to the transponder output to overcome Doppler velocity-discrimination modes associated with PD-, MTI-, and CW-type radars.

The major components of an active expendable jammer are the antenna, RF power oscillator or amplifier, receiver, modulation or techniques generator, and power supply. The issues concerning the antenna are a requirement for a relatively high-powered antenna in a small package combined with meeting the performance requirements of high gain, wide bandwidth, compatibility with the victim radar's polarization, and a low VSWR. Also, isolation between transmitting and receiving antennas is a significant problem, particularly when operating against high duty-cycle emitters. GaAs FET amplifiers are suitable for use in higher frequency (e.g., above 3 GHz) expendable repeater-type jammers. The issue here is whether to employ high-power designs over narrow bandwidths (e.g., 10% to 15% of center frequency) or octave-wide amplifiers with relatively low-power outputs.

Another significant issue is the power source (e.g., battery), which must be capable of reliably providing high current for long periods. Thermal batteries, so named because they need heat for activation, are one solution. The thermal battery contains an integral heat source, which when ignited, melts the electrolyte and delivers the necessary power. Also, low-temperature zinc-silver oxide batteries and lithium batteries are suitable for use in expendable jammers.

Decoys are a particular form of offboard EA that are used to increase the probability of survival of ships and aircraft by creating false-target information in enemy weapon systems. One mode of decoy operation is to saturate the enemy defense with a large number of false decoy targets, thereby causing commitment of weapons against these false targets, which dilutes the overall defense effectiveness. Another mode of decoy operation is to enhance the signature (usually both radar and IR) of either a low-valued target or a heavily defended target such that the weapon is attracted to the decoy rather than the protected target.

Decoys are often deployed using either drones or remotely piloted vehicles. These vehicles can be either expendable or recoverable. Many times they serve the dual function of deception and reconnaissance, which helps justify their relatively high initial cost. An advantage of decoys deployed on drones is that their characteristics can be made to resemble closely those of a real target in such areas as RCS and IR signature and also to maneuver like a real target. This realism makes it very difficult for EP to be effective against decoys.

Decoys often use corner reflectors or Luneburg lenses to enhance their RCSs. Below microwave frequencies (e.g., 1 GHz), these devices become too large, necessitating the use of active repeaters when low-frequency enhancement is required. For example, the maximum RCS available from a triangular corner reflector is given by $\sigma_{max} = 4\pi l^4/3\lambda^2$, where l is the length of a side. Therefore, in order to simulate a 10-m^2 cross section airborne target at 150 MHz would require a corner reflector with sides at least 5.8-ft long, which is too large to be carried by most drones. Luneburg lenses, which are most appropriate for naval applications, provide an RCS given by $\sigma_{max} = 4\pi^3 r^4/\lambda^2$, where r is the radius of the lens. To simulate a 2500-m^2 cross section destroyer at 150 MHz would require a lens of at least 19.7 ft in diameter, which would be large for a small decoy-type boat.

There are several methods that might be used to discriminate between decoys and real targets. These methods generally rely upon imperfect simulation by the decoy of realistic target characteristics. One of the more powerful methods is to examine the fluctuation characteristics of the target. The amplitude return from real targets tends to fluctuate while decoy target returns tend to remain steady. The physical mechanism that causes this effect is that real targets are complex, being composed of many scatterers of various sizes, while decoys, because of practical constraints, are generally composed of a single reflector. As the aspect angle to the complex target changes, the distances between the various scatterers change, causing the vector addition of the complex envelope to change. This effect can also be induced by changing the radar frequency from one pulse to another. Since they are point reflectors, decoys are relatively insensitive to aspect change and frequency changes over moderate bandwidths.

7.2 Chaff

Chaff is the oldest, and still the most widely used, radar countermeasure. It is generally used to protect tactical aircraft, strategic aircraft, and ships in either a corridor-laying or self-protection mode. Chaff dispensed from an aircraft at a steady rate over a fairly long period is used to form a corridor that conceals following aircraft. Self-protection involves launching relatively small quantities of chaff in controlled bursts to cause a weapon-associated tracking radar to point at the chaff rather than the protected vehicle. The design of the chaff dispensing and deployment system is quite different for each of the two modes of operation. Types of chaff distribution are depicted in Figure 7.2 (also see Figure 1.5).

In an airborne corridor-screening operation the objective is to dispense a large volume of chaff that fills a significant number of radar resolution cells in the areas that lie along the expected flight path of the penetrating aircraft. Aircraft or helicopters are ideal platforms for dispensing chaff at high altitudes because this leads to long lifetimes for the chaff clouds, which essentially float down through the atmosphere. Mean fall rates for aluminum foil chaff are 0.4 to 0.55 m/s at sea level, while the rate for aluminized glass chaff is of the order of 0.3 m/s. Because the rate of fall would be expected to vary inversely as the square root of the air density and the air density at 40,000 ft is approximately one-fourth that at sea level, the resulting comparable average fall rates at 40,000 ft would be 0.8 to 1.1 m/s for aluminum and 0.6 m/s for aluminized glass chaff. The resulting "hang time" for aluminized glass chaff dispensed at a high altitude (e.g., 40,000 ft) is of the order of from 2 to 3 hrs.

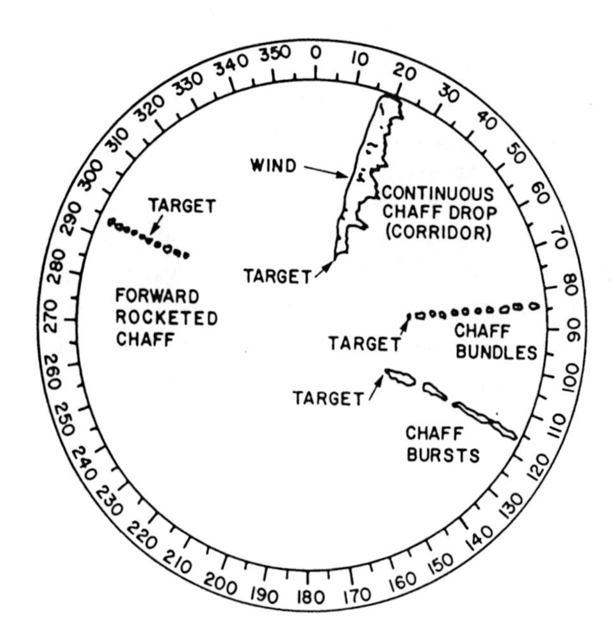

Figure 7.2 Types of chaff distribution. (*Source:* [2].)

Corridor chaff is normally launched from gravity dispensers into the atmosphere, and a typical magazine might contain 25 kg of chaff with several of these dispensed in one mission. Many systems of this type employ prepackaged chaff units containing half-wave resonant dipoles of various lengths selected to cover the frequency band of interest. Alternatively, some systems employ airborne chaff cutters, which can be used to cut long strands of stored material to form dipoles resonant to the viewing radar frequency.

One difficulty with the gravity-dispensing method is that the chaff blooms behind the aircraft leaving a pointerlike pattern on a surveillance radar display that points at the dispensing aircraft. This potentially exposes the dispensing aircraft to hostile action. One solution is to use forward launching, which is accomplished by firing chaff rockets ahead of the dispensing aircraft. A problem with this approach is that forward-launched chaff is expensive because a relatively large number of rockets must be carried to seed even a relatively small area with chaff.

A fundamental consideration in seeding chaff corridors involves the amount of chaff required to shield a penetrating aircraft. One criterion that must be met in this regard is that, within a radar processing cell, the return from the chaff must be stronger than the return from the protected target. For a radar without Doppler processing, this can be translated into an equivalent criterion, which states that the equivalent RCS associated with the chaff must be greater than the RCS of the protected target. When a radar with matched Doppler filtering is involved, the fraction of return power from the chaff that enters the detection Doppler filter and competes with the target return must be stronger than the target return. This involves consideration of the spectral densities of both the chaff and the target, which is beyond the scope of the present discussion. Although not explicitly covered in the following discussion, it is apparent that the effect of Doppler processing can be accounted for by reducing the equivalent chaff radar cross sectional area by the ratio of the chaff power accepted by the target Doppler filter to the total chaff power intercepted by a radar resolution cell.

The equivalent RCS concept introduced previously replaces the large number of elemental chaff scatterers contained within a radar resolution cell with a single equivalent scatterer, which returns the same amount of power to the radar. To calculate the size of this equivalent scatterer, we must find the volume of a radar resolution cell and multiply this by the volumetric back-scattering coefficient (η with dimensions m^2/m^3). The radar's volume resolution cell can be found by multiplying the area subtended by the radar's circular antenna beam pattern (possibly elliptical for fan beams) at a particular range by the range equivalent to the radar's processed pulsewidth. When the resolution volume cell is multiplied by the back-scattering coefficient, the equivalent RCS for chaff (assuming Gaussian-shaped antenna beams) is given by $\sigma_c = \pi\eta\theta_A\phi_E R^2 \tau c/16$, where η is the back-scattering coefficient in m^2/m^3, θ_A is the azimuth beamwidth in radians, ϕ_E is the elevation beamwidth in radians, R is the range in meters to the resolution cell, τ is the radar pulsewidth in seconds, and c is the velocity of light in meters per second.

The equivalent radar cross sectional area of the chaff is dependent on a number of radar-associated parameters and one chaff-associated parameter. The radar-associated parameters indicate that the effect of the chaff can be minimized by high resolution in range and angle. This leads to the use of three-dimensional radar (e.g., narrow elevation beam) with PC as a chaff EP technique. However, the main radar EP technique used to combat chaff is Doppler processing. While the radar designer attempts to minimize the power returned from the chaff into a radar resolution cell, the EA designer attempts to maximize this factor. The only parameter that can be controlled to accomplish this objective is the chaff back-scattering coefficient, η. We will discuss the determination of this rather complex function, but we must first digress and discuss some fundamentals of chaff physics and meteorology.

7.2.1 Chaff Fundamentals

Chaff consists of a large number of shorted tuned dipole antennas. The tuned dipoles are cut to the first resonance point (length = $\lambda/2$) since this provides the maximum RCS for the least amount of material and makes them easiest to handle. The capture area of the tuned dipole is given by $G\lambda^2/4\pi$, where G is the antenna gain. The captured signal is reflected from the short so that it essentially doubles the voltage or generates four times the power at the antenna terminals. This is further reradiated with gain G to produce an equivalent RCS equal to

$$\sigma = \frac{\lambda^2 G^2}{\pi} \tag{7.1}$$

The dipole gain varies with angle and for vertical polarization can be expressed as

$$G = \frac{2 \cdot \left[\dfrac{\cos(kl \cdot \cos\theta) - \cos kl}{\sin\theta}\right]}{\displaystyle\int_0^{\pi} \left[\dfrac{\cos(kl \cdot \cos\theta) - \cos kl}{\sin\theta}\right]^2 \sin\theta d\theta} \tag{7.2}$$

where $k = 2\pi/\lambda$. The gain is maximum normal to the antenna ($\theta = \pi/2$) where

$$G = \frac{2 \cdot (1 - \cos kl)^2}{\int_0^{\pi} \frac{(\cos(kl \cdot \cos \theta) - \cos kl)^2}{\sin \theta} d\theta} \tag{7.3}$$

The integral in the denominator equals 1.2188 for half-wave dipoles, 3.3181 for full-wave dipoles, and 1.7582 for $3\lambda/2$ length dipoles. The corresponding maximum gains are 1.64 (2.15 dB) for half-wave dipoles, 2.41 (3.82 dB) for full-wave dipoles, and 1.41 (0.56 dB) for $3\lambda/2$ length dipoles. Substituting this into (7.1) results in the maximum RCS from a single vertically polarized dipole as

$$\begin{aligned} \sigma &= 0.86\lambda^2 && \text{half-wave dipole} \\ \sigma &= 1.85\lambda^2 && \text{full-wave dipole} \\ \sigma &= 0.41\lambda^2 && 3\lambda/2 \text{ dipole} \end{aligned} \tag{7.4}$$

In a chaff cloud, the dipoles are randomly oriented with respect to the illuminating radar. An approximate relationship that expresses the RCS of a randomly oriented dipole is

$$\overline{\sigma} = (1 + \cos 2kl)^2 \cdot \frac{G^2(\theta, kl)\lambda^2}{\pi} \tag{7.5}$$

Evaluation of this relationship is accomplished in Example 7.1 and indicates the harmonic structure of the dipole response. The average RCS ($\overline{\sigma}$) for randomly oriented half-wave (typically 0.46 to 0.48λ) dipoles at resonance is generally given as $\sigma = 0.15\lambda^2$, which is a smaller amount than indicated by (7.1).

▶ *Example 7.2*

Plot the mean response for a randomly oriented dipole as a function of frequency. Assume that the chaff is resonant at a frequency of 10 GHz.

```
% Radar Cross Section of Dipole
% -------------------------------
% dipole.m

clc;clf;clear;

l=.03/4;          % Half Wave Resonance in X-band
Sig=[];Sigf=[];

% Vary angle aspect

M=8;
for Ang=pi*(1/M:1/M:1-1/M);
fGhz=[];

   % Inner loop varies frequency,kl
   % and converts frequency to Ghz

   for fMhz=5000:100:50000;
   wl=300/fMhz;
   k=2*pi/wl;
   kl=k*l;
   fGhz=[fGhz;fMhz/1000]; fMhz=[];

   % Evaluation of Integral as function of kl

   N=32;ang=pi*(1/N:1/N:1-1/N);I=0;
   for n=1:N-1;
   I=I+pi/N*((cos(kl*cos(ang(n)))-cos(kl)).^2)/sin(ang(n));
      end;

   G=2*((cos(kl*cos(Ang))-cos(kl))/sin(Ang))^2/I;
   sig=G^2*wl^2/pi;
   Sig=[Sig;sig];sig=[];
       end;

   % Convert for angle aspect

   Sig=Sig';
   Sigf=[Sigf;Sig];Sig=[];
end;

% Correction for phase between I1 and I2
% and polarization losses

korr=.5*(cos(4*pi*fGhz*l/.3)).^2;

% Plot Mean RCS

plot(fGhz,korr'.*mean(Sigf));
xlabel('Frequency - Ghz');
```

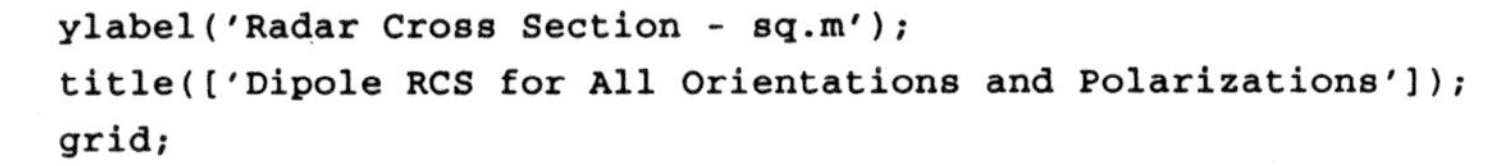

```
ylabel('Radar Cross Section - sq.m');
title(['Dipole RCS for All Orientations and Polarizations']);
grid;
```

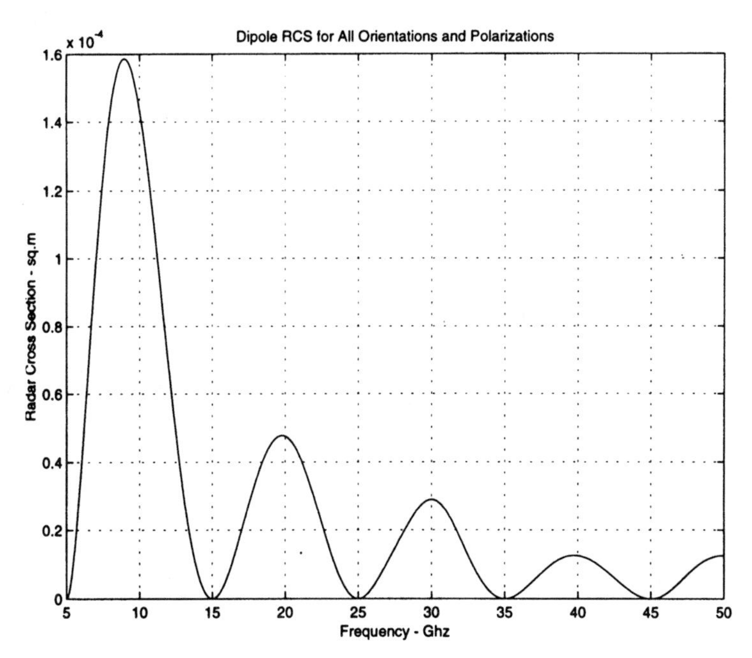

If a number of elemental chaff dipoles are combined to form a chaff cloud whose spacing between elements is of the order of two wavelengths or greater, then the average RCS of the cloud is given by $\sigma = 0.15N\lambda^2$, where N is the number of effective chaff dipoles. The bandwidth of an elemental dipole is a function of the length-to-diameter (L/d) ratio for circular dipoles and the length-to-width ($4L/w$) ratio for rectangular dipoles. The chaff bandwidth ranges from 10% to 25% of the resonant frequency for L/d ratios of 100 to 10,000, where the higher bandwidths are associated with the lower ratio. A typical design provides a bandwidth approximately equal to 15% of the center frequency for an L/d ratio of 1000. Figure 7.3 shows the RCS versus frequency for a representative design of a broadband chaff package covering the frequency range from 2 GHz to 12 GHz where various elements are stagger-tuned to cover the whole band of interest. Note that the numbers of dipoles associated with the higher frequency chaff elements are substantially greater than those associated with the lower frequency chaff elements. Also, the cross section response at any particular frequency is the sum of the responses of all chaff dipoles that are resonant at any frequency lower than the design frequency. Thus, the cross section associated with a particular group of elements cut to

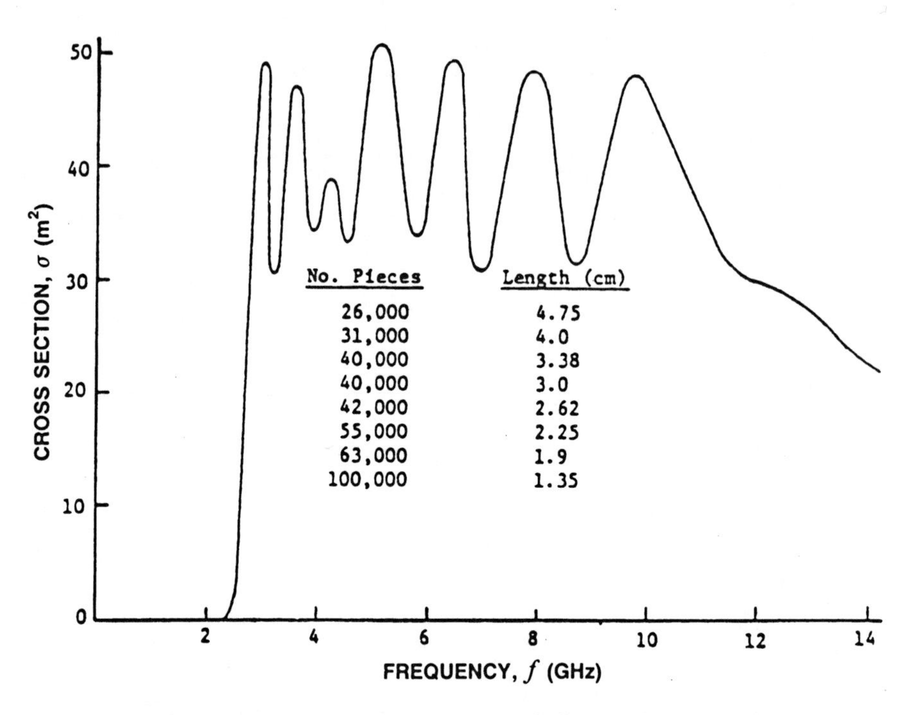

Figure 7.3 RCS of multiband chaff package. (*Source:* [14], p. 187.)

the same length (e.g., typically $L = 0.48\lambda$) is generally lower than the actual cross section at this frequency, except for the group with the longest cut lengths. Note that the weight of 25- by 50-μm aluminum foil chaff to achieve the multiband RCS of 40 to 50 m^2 depicted in Figure 7.3 is approximately 0.1 pound.

7.2.2 Chaff Shielding Effects

In predicting the theoretical response of a cloud of dipoles, the fact that each individual dipole is located in a cloud of thousands of other dipoles results in a degradation effect known as shielding. Shielding occurs when the dipole density is such that it prevents every dipole from receiving the full amount of energy incident from the radar. One extreme occurs when the dipole cloud is so dense that only the surface dipoles reflect the incident energy. This dense chaff cloud would generally have dimensions much smaller than the radar's angular resolution, resulting in an RCS equal to the geometric projected area of the chaff cloud (A_0) in a direction normal to the radar beam. At the other extreme, the chaff dipoles are widely spaced and the overall RCS of the chaff cloud is given by the summation of the elemental chaff RCSs (e.g., $\sigma = N\sigma_1$) as previously discussed. At intermediate chaff cloud densities, the RCS can be approximated by the relationship $\sigma = A_0(1 - \exp(-n\sigma_1))$, where A_o is the geometric projected area of the chaff cloud, $n = N/A_o$ is the number of dipoles per unit of cloud projected area, and $\sigma_1 = 0.15\lambda^2$. Evaluating this relation at the extremes, it is seen that for dense clouds $\sigma = A_o$, as expected, while for clouds with widely spaced dipoles $\sigma = A_0(N\sigma_1/A_0) = N\sigma_1$, which checks against our assumption that the RCS is simply the sum of the cross sections of the elemental dipoles. Note that the maximum value of RCS equals the sum of the values of the individual elements' cross sections regardless of the physical size of the chaff cloud.

The relationship for chaff RCS can be used to determine how far apart chaff dipoles must be spaced in a cloud to allow for minimum interaction. By expanding the shielding relationship in a power series and retaining the first two terms, it can be shown that the RCS is given by $\sigma = N\sigma_1(1 - 0.075/\alpha^2)$, where α is the uniform dipole packing factor expressed in wavelengths. For example, with λ spacing the chaff RCS is given by $\sigma = 0.925N\sigma$, while for 2λ spacing it is given by $\sigma = 0.981N\sigma_1$. This relationship assumes that all chaff elements are contained within a volumetric radar resolution cell, which has higher probability for a dense cloud than for a widely dispersed chaff cloud.

7.2.3 Chaff Characteristics

Because the volume and weight available on an aircraft for chaff dispensers is limited, it is of interest to translate the preceding relationship for chaff RCS into volume and weight requirements. For circular dipoles of length 0.48λ, it can be shown that the chaff volume required to provide a dispersed chaff cloud with RCS σ is $V = 2.514\sigma d^2/\lambda\rho$, where d is the dipole diameter (typically $d = L/1000$), ρ is the chaff packing density ($\rho = 0.2$ to 0.6), and compatible units are used. The volume of the chaff package can be multiplied by the density of the chaff dipoles to obtain the weight of the chaff. This expression can then be solved to provide the chaff RCS as a function of the chaff weight, resulting in $\sigma = kW/f$, where W is the chaff weight in pounds; f is the radar frequency in gigahertz; and k is a constant that depends on the chaff material density, dipole volume, and chaff efficiency (typically, a degradation in chaff cross section of 2 to 1 is used for mature chaff). The properties of various types of chaff and their RCS factor (k) are listed in Table 7.1. Table 7.1 indicates that aluminized glass chaff provides the highest RCS for a given weight, among the chaff types listed, and this coupled with its relatively slow fall time (0.3 m/s) makes it the most popular type for area- or corridor-chaff applications. The high RCS-to-weight ratio (e.g., $k = 17{,}000$) attributed to aluminized glass chaff in Table 7.1 must be tempered by the fact that the

Table 7.1
Properties of Chaff Dipoles

Type	Dimension (μm)	Density (kg/m^3)	k(m^2-GHz/lb.)
Aluminized glass	25	2550	17,000
2x1 Aluminum foil	50 × 25	2700	6,310
4x1 Aluminum foil	100 × 25	2700	3,155
V-Bend aluminum foil	200 × 12	2700	3,235
Silver-coated nylon	90	1300	2,570

small glass filaments employed are quite stiff and brittle, which may lead to a high breakage factor and a reduced value of *k* due to the decreased efficiency. Note that the dimensions for chaff given in Table 7.1 represent recent improvements in reduced size over previously reported values for aluminum foil chaff (e.g., 254 by 25 μm), which resulted in an RCS $\sigma = 3000\ W/f$ with 100% efficiency.

7.2.4 Dispensing Chaff

When chaff is first launched from an airborne dispenser, the drag on an individual chaff dipole is so great compared to its mass that it comes to air mass velocity almost instantly. The initial length of the chaff cloud is determined by how far the aircraft travels during the time between the launching of the first and last chaff particles in the cartridge. In conventional cartridges, this time is of the order of 6 ms, which results in an initial size of about 2m. The cloud tends to grow very rapidly due to aircraft-induced turbulence in a direction along the flight axis of the aircraft, reaching a value of 10m to 15m in 120 ms. During this time, the chaff cloud expands more slowly in the cross flight direction, reaching a size of about 2m to 3m in 120 ms. The resulting chaff cloud at this point is very dense and has the shape of a cylinder or ellipsoid, whose major axis is aligned parallel to the flight path of the dispensing aircraft. The actual values for the dimensions of the chaff cloud vary widely, depending upon the chaff dispenser design, the type and dynamics of the launching aircraft, and the deployment point on the aircraft.

During the initial period (e.g., of the order of seconds) after launch, the chaff cloud is relatively small and is generally contained within a volumetric resolution cell of the radar. The dipoles in the cloud at this point are closely packed so that the RCS is equal to the geometric projected area of the cloud in a direction normal to the radar beam. For example, if the cloud were a cylinder 40-ft long and 6-ft wide, the RCS could vary from 2.8 m^2 to 24 m^2 as the aspect angle to the radar changes from 0 to 90 degrees.

After the initial effects of the turbulence generated by the dispensing aircraft, the cloud continues to grow as a result of three factors. These are the distribution of the fall rates of the elements (partially due to lack of uniformity in size and coatings), the prevailing winds, and the air turbulence. In general, the growth is predominantly in the direction of the prevailing winds.

There are several other effects that result in the chaff cloud continuing to grow in volume after the initial dispensing period. First, the lifetime of the chaff cloud is limited by the length of time required for the chaff to fall to Earth. Horizontal transport of the chaff as it settles is primarily due to the wind. Thinner pieces of chaff fall more slowly and, hence, are transported through greater distances than larger pieces as they fall through a given altitude differential. This causes a dispersion effect which aids in the growth of both the width and thickness of the chaff cloud. This effect can be accentuated using chaff dipoles with multiple widths in a single chaff package. One example using a mixture of one-mil and one-half-mil glass dipoles tripled the blooming rate of the chaff. The parameters associated with the one-half-mil glass chaff at 15 min after release were 7,000-ft horizontal transport, 450-ft cloud width, and 220-ft cloud thickness.

Another effect that apparently occurs as the chaff falls, particularly with glass chaff, is a gradual separation into two clouds: one that is predominantly horizontally polarized and the other vertically polarized. This is caused by the vertically oriented dipoles falling faster than the horizontally oriented dipoles resulting in a horizontally polarized layer above the vertically polarized material. One solution to this problem is to make one end of the chaff dipole heavier than the other so that the chaff falls in slow spirals, which approximates 45-degree polarization.

Having discussed the physics and meteorology of chaff, we are now in a position to summarize the use of chaff in a corridor-screening operation. A key element in planning the chaff-laying mission is an initial determination of the volumetric resolution cells of the radar to be jammed. A desirable chaff-corridor-dispensing strategy is to launch the chaff cartridges at distances spaced by the radial or range dimension of the smallest radar volumetric resolution cell. This ensures that each radar resolution cell along the flight path of the chaff-seeding aircraft contains an initial chaff packet.

The second consideration concerns how much chaff should be dispensed in each burst. In general, the RCS of the chaff should be of the order of twice the RCS of the maximum-sized target to be protected. For example, the multiband chaff package design illustrated in Figure 7.3 might be used to protect aircraft whose RCSs were of the order of 20 to 25 m^2. However, as we previously discussed, the RCS values shown in Figure 7.3 are the maximum realizable by this chaff package within a radar resolution cell. When the chaff is initially launched, the RCS is given by the geometric projection of the chaff cloud in a direction normal to the radar beam. This cross section is generally much smaller than the maximum value, and the chaff must be allowed sufficient time to bloom before the penetrating aircraft enters the protected corridor.

As the chaff blooms, its physical area enlarges until the RCS approaches the value given by the sum of the individual chaff elements (e.g., $\sigma = 0.15N\lambda^2$). Aluminized glass chaff designed for corridor shielding might take on the order of 100 sec to reach its maximum value, while the comparable value for 25- by 50-μm aluminum foil chaff is 40 sec. The corridor screening is then effective as long as the chaff elements within the cloud lie within a volumetric radar cell. As the chaff cloud continues to expand, the individual chaff elements are dispersed into several radar volumetric cells, thereby diminishing the effective overall RCS of the chaff. In a typical situation, the effective radar cross sectional area of the chaff in a radar volumetric resolution cell might fall to 50% of its maximum value in 250 sec for aluminized glass chaff and approximately 80 sec for 25- by 50-μm aluminum foil chaff. Note that the previously specified chaff RCSs apply to horizontal polarized values, while vertical polarized values are approximately 90% for aluminum foil chaff and 50% for aluminized glass chaff.

For effectively denying information regarding strike composition and timing, a chaff corridor should originate outside the radar detection range and be uninterrupted throughout its length. The chaff RCS in the radar resolution cells should be sufficient to prevent detection of the shielded aircraft, taking into consideration any chaff discrimination capabilities of the radar. The chaff RCS initially deployed by the chaff-dispensing aircraft must be greater than the chaff RCS subsequently required for corridor use by the penetrating aircraft. This is necessary to allow for chaff cloud growth and chaff diffusion across the resolution cell boundaries. How much greater the initial chaff RCS should be compared to the chaff RCS per resolution cell at time of use would depend on several factors, such as the chaff-dispensing system, the RCS of the aircraft to be shielded within the corridor, the corridor growth and chaff fall characteristics, the length and use time required of the corridor, and the resolution cell's volume.

As an example, consider the ASR-9 radar (λ = 0.1m, θ_{az} = 1.3 degrees, ϕ_{el} = 4.8 degrees, τ = 1.05 μs) that at a range of 50 nmi has a volume resolution cell $V = 10^9$ m^3. Assume that we want to generate a chaff corridor with an RCS of 50 m^2 within this volume cell. The chaff reflectivity will then equal $\eta = 5 \times 10^{-8}$ m^2/m^3. The number of chaff dipoles contained within the radar's resolution cell can be found as $N = 3.33 \times 10^4$ dipoles. The density of dipoles within the resolution cell is then 3.33×10^{-5} dipoles/m^3 or 1 dipole in a cube 30 m on a side. The most efficient chaff is made by aluminizing cylindrical glass dipoles. The diameter of each chaff dipole is 25 μm and its density is 2550 kg/m^3. The volume of each dipole is given by $V = 0.48\pi d^2\lambda/4 = 0.377d^2\lambda$. The weight of each dipole is thus $w_1 = 961d^2\lambda$ kg. The total weight of the aluminized glass dipoles that will provide an RCS = 50 m^2 within the radar's volume resolution cell at a wavelength $\lambda = 0.1$m is then $w = Nw_1/e$, where e is an efficiency factor that allows for breakage, mistuning, mismatching, and other chaff inefficiencies. For a chaff efficiency of 50%, the total weight of chaff to fill one resolution cell would hence equal 4×10^{-3} kg.

The preceding analysis can be expressed as a relationship of the chaff's RCS with respect to the weight of chaff required to produce a given RCS at a specified frequency. This relationship is given by $\sigma_c = kw_{lbs}/f_{GHz}$, where w_{lbs} is the weight of the chaff in pounds, f_{GHz} is the radar's frequency in gigahertz, and k is a constant that depends on the properties of the type of chaff utilized. The value of k ranges from 17,000 for aluminized glass chaff to the order of 3,000 for aluminum foil chaff.

7.2.5 Rope Chaff

It is impractical to produce lightweight resonant dipoles that have a slow rate of fall for frequencies below 1 GHz. For coverage in the B and C bands, long metal, or metal-coated, fiber streamers that are very long compared to the wavelength of the frequency of interest are used. This nonresonant form of chaff, called rope, can be dispensed along with resonant dipoles to provide multiband coverage.

The theoretical cross section response of three 750-ft-long sections of one-quarter-inch-wide rope is depicted in Figure 7.4. This curve is only an approximate estimate, since the RCS is a function of the shape and orientation of the rope streamers and these cannot be defined with any degree of accuracy. Skin depth, which is inversely proportional to frequency, would cause a large loss if aluminized glass fibers were used as a material for rope. Rather, string ball (aluminum-coated glass fibers wound up like a ball of string) made with of the order of 2 aluminum-coated fibers is used. The use of multiple fibers permits a balance between the radiation resistance and the ohmic loss of aluminum-coated string balls while still having the physical properties of light weight and low fall rate.

7.2.6 Self-Protection Chaff

Chaff is used for self-protection of naval ships against radar-guided antiship missiles. The principles of chaff protection for ships are generally the same as

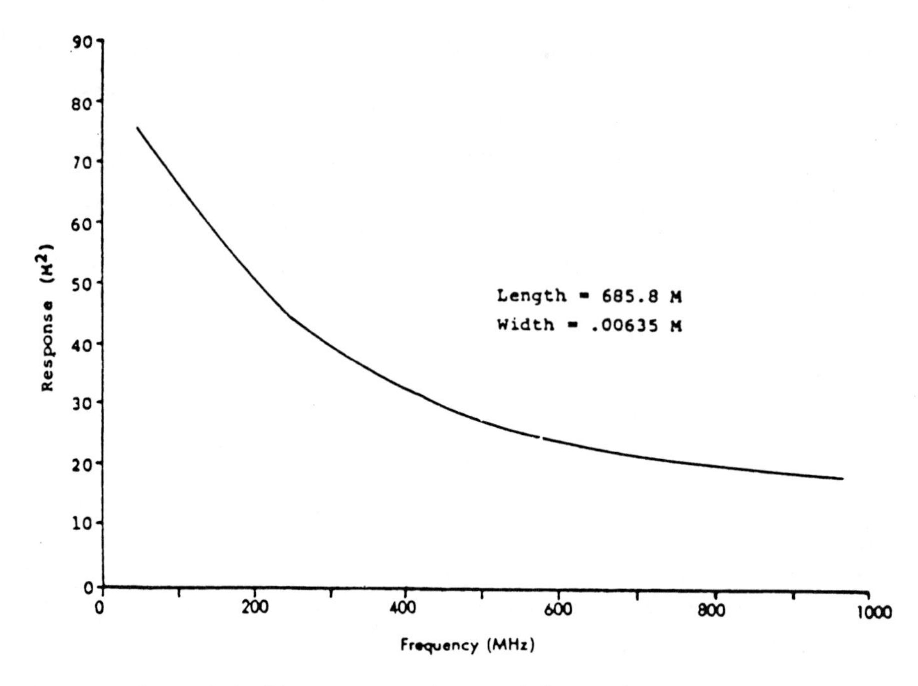

Figure 7.4 Theoretical RCS of rope chaff. (*Source:* [14], p. 193.)

for aircraft, except that the large RCS of ships (e.g., 17 dBsm to 60 dBsm) makes the timing and implementation of the naval mission more difficult.

For naval use, chaff is most commonly ejected from rocket, shell, or mortar systems. Naval rockets can contain up to 7 kg of chaff, and mortar systems typically dispense up to 3 kg of chaff from several grenades fired simultaneously. There are two main modes of use of chaff at sea. Distraction decoys are dispensed by rockets and shells at ranges up to 2 km from the vessel, and a pattern of several rockets fired in different directions is used to provide alternate targets to missiles while they are still at some distance from the vessel. The decoys may last for several minutes; and if the threat is still present, more decoys are periodically sown. The second mode, known as the dump mode, is used closer to the vessel (e.g., 1 km away), where active EA jamming is used to deny range information to the seeker and, in conjunction with chaff fired from a rocket, to lure the attacking missile away. The centroid of the chaff is very close to the ship, where the chaff cloud is dispensed at a range of about 100m to 400m. A large echoing area must be realized within a few seconds of firing the chaff near the ship. The ship then moves quickly out of, and away from, the chaff echo and the missile is lured away, thus avoiding a direct hit. The centroid mode can be achieved with rockets or mortars and is regarded as a last resort tactic, which will succeed best with vessels of relatively small RCS, such as small, fast patrol boats. In naval applications, multipath effects can be used with advantage where the free-space RCS of a chaff cloud is greatly enhanced by its proximity to the sea. Significant enhancement can be achieved with clouds up to 200m above the sea surface, depending upon the height and range of the seeker. Naval chaff systems are generally designed to achieve the required RCS of the order of 3 sec to 10 sec.

When chaff is used in self-protection applications, the dispensers must be quick-reaction devices that eject relatively small quantities of chaff in controlled bursts. This is commonly achieved using cartridges fitted with pyrotechnic squibs, where the squibs are fired electrically by a programmable control unit. Self-protection chaff cartridges typically contain 100g to 150g of chaff carried in modules of 30 cartridges. At least two modules are normally carried. Alternatively, mechanical dispensers can be used, where individual packs are ejected in short bursts from an assembly of long tubular magazines.

7.2.7 EP Against Chaff

EP against chaff generally exploits the difference between the chaff return and that of an actual target. Some of the characteristics of chaff that are different than those of a target are: (1) the lack of motion, (2) different fluctuation characteristics, (3) different polarization characteristics, (4) different echo size, and (5) scatterers distributed over a large volume rather than concentrated within a limited spatial extent.

In general, all of these characteristics of chaff are used to filter the target return from the chaff return. However, the most effective antichaff technique available to the radar is the use of Doppler filtering, which exploits the different motion characteristics of the target and the chaff. There are two basic Doppler filtering techniques that are used. The first is called MTI, which employs a PRF that provides unambiguous range coverage while using a comb filter whose nulls are tuned to the average radial motion component of the chaff. The second is called pulsed Doppler, which can use a high PRF to provide unambiguous Doppler coverage in conjunction with a bank of contiguous Doppler filters, allowing separation of the target from the chaff.

The key issue involved with the performance of these types of radars is the spectral spread of the chaff return. The determination of chaff spectra is complicated by the rotation of the chaff dipoles that occurs naturally and can be induced by certain chaff design characteristics or by air turbulence caused by the dispensing aircraft and atmospheric disturbances. Wide chaff spectra can cause MTIs to have limited performance, leading to a general preference for pulsed Doppler radar.

In an MTI radar, the techniques used in an MTD-type processor [2] are applicable. These techniques generally allow the rejection of both ground clutter and chaff (or precipitation) clutter, although these two types of clutter occupy different parts of the frequency spectrum. High-PRF PD radar is considered to be an even better method for eliminating chaff because its unambiguous Doppler response allows a clutter rejection filter to be set above the velocity response of the chaff. However, care must be exercised in radars operated in extended clutter with range ambiguities (e.g., high- and medium-PRF PD) because the spectral response at each range-ambiguous clutter cell will be different (both center frequency and spread), and the total clutter spectrum is the sum of the individual clutter spectra. This is illustrated in Figure 7.5 for four separate range ambiguities, where the chaff clutter spectrum is spread over a significantly wider range than that associated with an individual range cell.

Chaff dipoles have high aerodynamic drag that causes their velocity to drop to that of the local wind conditions a few seconds after being dispensed. This means that the horizontal velocity of the chaff will be determined primarily by the wind speed in the radar resolution cell containing the chaff. Its vertical velocity (falling rate) depends on the dipole dimensions and material, with a falling velocity of 0.3 m/s attributed to aluminized glass at sea level.

The width of the Doppler spectra for chaff can be divided into four components in a manner similar to that for precipitation clutter. The compo-

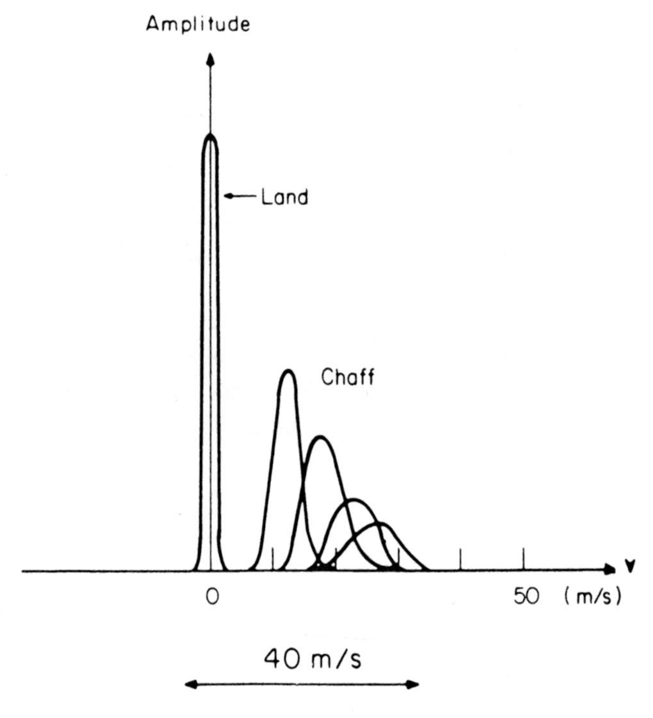

Figure 7.5 Spectrum of range of ambiguous chaff clutter. (*Source:* [15].)

nents due to wind shear and beam broadening due to the finite widths of the beam should be identical to those for precipitation. The value for turbulence is on the order of σ_{turb} = 1 m/s below 12,000 ft altitude and 0.7 m/s above that altitude. The final component of the chaff spectra is the Doppler spread due to the variation in falling velocity of the dipoles. Because chaff dipoles fall more slowly than raindrops, this component is estimated at $\sigma_{fall} = 0.45 \sin \phi_{el}$ m/s.

As with precipitation, the spectra of chaff clutter is generally determined by the wind shear component, which is given by $\sigma_{shear} = 0.42kR\phi_{el}$, where k is the velocity gradient (4 m/s -km is suggested for arbitrary azimuth angles), R is the slant range to the clutter resolution cell in kilometers, and ϕ_{el} is the two-way elevation antenna beamwidth (rad). However, in contrast with precipitation, chaff altitude is not generally limited to 20,000 ft, and spectral spreads greater than 6 m/s may be observed when chaff is dispensed above this altitude. Spectral widths for this component are depicted in Figure 7.6. Note that observed spectral widths for chaff are apparently not affected by the age of the chaff (assuming the chaff has bloomed) and, when normalized to wavelength, are independent of frequency (i.e., σ_v does not depend on frequency).

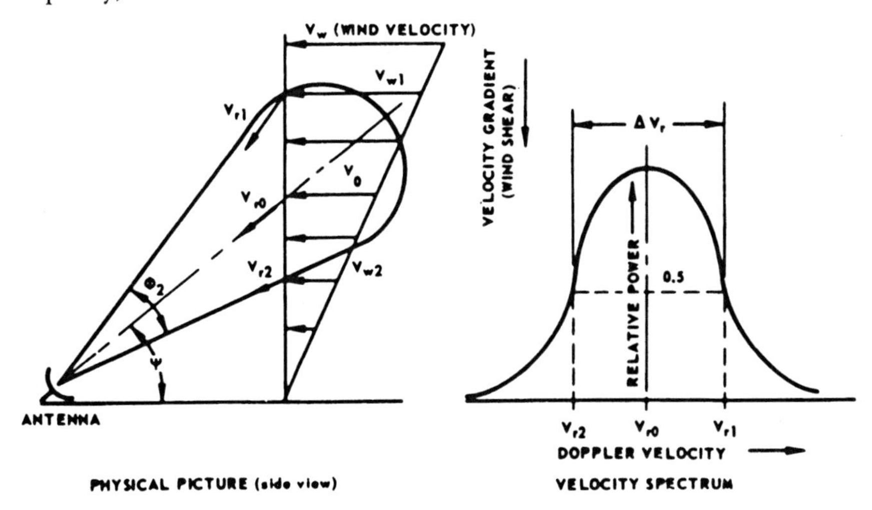

Figure 7.6 Effects of wind shear on the Doppler spectrum. (*Source:* [16].)

7.3 Infrared Missile Attack

IR missile kills have dominated attrition statistics against aircraft and helicopters since these types of missile became prevalent in the 1950s [3,4]. They also are effective against ships in naval engagements using ASCMs. They can be launched from aircraft, helicopters, naval ships, and troops using shoulder fired weapons.

IR missiles are passive, using heat energy emitted by the target in the IR spectrum as a tracking source. The hotter an object is, the more energy it emits, and also the lower the spectral wavelength of the peak of its energy emission (see Figure 7.7).

A primary countermeasure against this type of missile is an expendable decoy flare that attempts to lure the missile from its intended target by presenting a large heat source offboard the target. Flares can be effective against early types of IR missiles (e.g., Red-eye, SA-7, AA-2) that basically employ real engagement hot spot trackers that are tuned in the 1- to 3-μm region to hot engine parts in the 1300° to 2000°K temperature range. However, the flares must be dispensed at an appropriate time and in an appropriate direction to be effective. This is particularly critical in aircraft protection applications where the flares rapidly separate from the dispensing aircraft due to aerodynamic drag effects. Like all expendables, only a finite number of flares are available on the protected platform and, hence, once they are dispensed, the platform is left unprotected. This leads to a requirement for MAW that allows both the

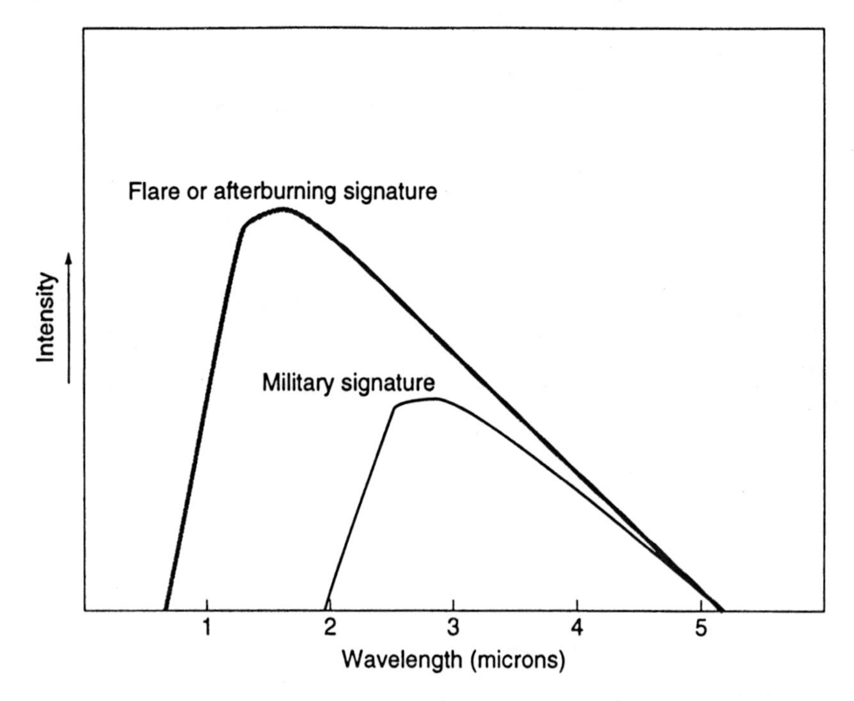

Figure 7.7 Emission of aircraft and flares. (*Source: Journal of Electronic Defense,* Jan. 1994, p. 47, Figure 1.)

direction of the missile attack and its time-to-impact to be determined before launching a flare. The characteristics of a typical magnesium flare are listed in Table 7.2.

First-generation IR missiles use uncooled lead-sulfide detectors that have a peak sensitivity at 2 μm. This limits this type of missile to stern engagements since the missile seeker must look at the hot turbine in the engine tail pipe to track the target. Cooled seekers have been introduced to make the detectors more sensitive and other detector material (such as indium antimonide in the 3- to 5-μm region and mercury-cadmium-telluride in the 8- to 10-μm region) is used to see longer wavelength radiations. The improved detection response allows advanced IR missiles to achieve all-aspect track capability. By operating at longer wavelengths (in the 3- to 5-μm region), advanced seekers can detect the engine exhausts as well as hot engine parts and thus attack at a wider range of engagement angles. Further advanced seekers operate at even higher wavelengths (up to 10 μm), providing the capability to detect the warm skin of the aircraft, in contrast to a colder sky background, at all aspect angles. Figure 7.8 depicts the IR signature of a helicopter [5].

Along with the spectral diversity in the future mix of threats, there will be variety in the techniques used by seekers to find and track the target as shown in Figure 7.9. Early seekers used a spinning reticle in front of the detector to reject background radiation and encode target position. The reticle is gradually being replaced by techniques that either sequentially scan the target with small *field-of-view* (FOV) detectors or stare at the target with a mosaic of detectors in a focal point array configuration [4]. Staring detectors essentially image the target, thereby allowing relatively easy discrimination between decoy and target. Multicolor discrimination techniques generally employ at least two sets of detectors, each peaked in response to different spectral bands. A ratio of the spectral response to the same radiator in different bands, at the same point in the seeker's FOV, is taken to determine whether the radiating object

Table 7.2
Typical Flare Characteristics

Composition	**Magnesium and Teflon**
Volume	900 cm^3
Burn time	22 sec
Peak power output	12,675 W/steradian
Power after 11 sec	6,337 W/steradian
Total energy	1.543 kW-s
Spectral region	1.8 to 5.4 μm

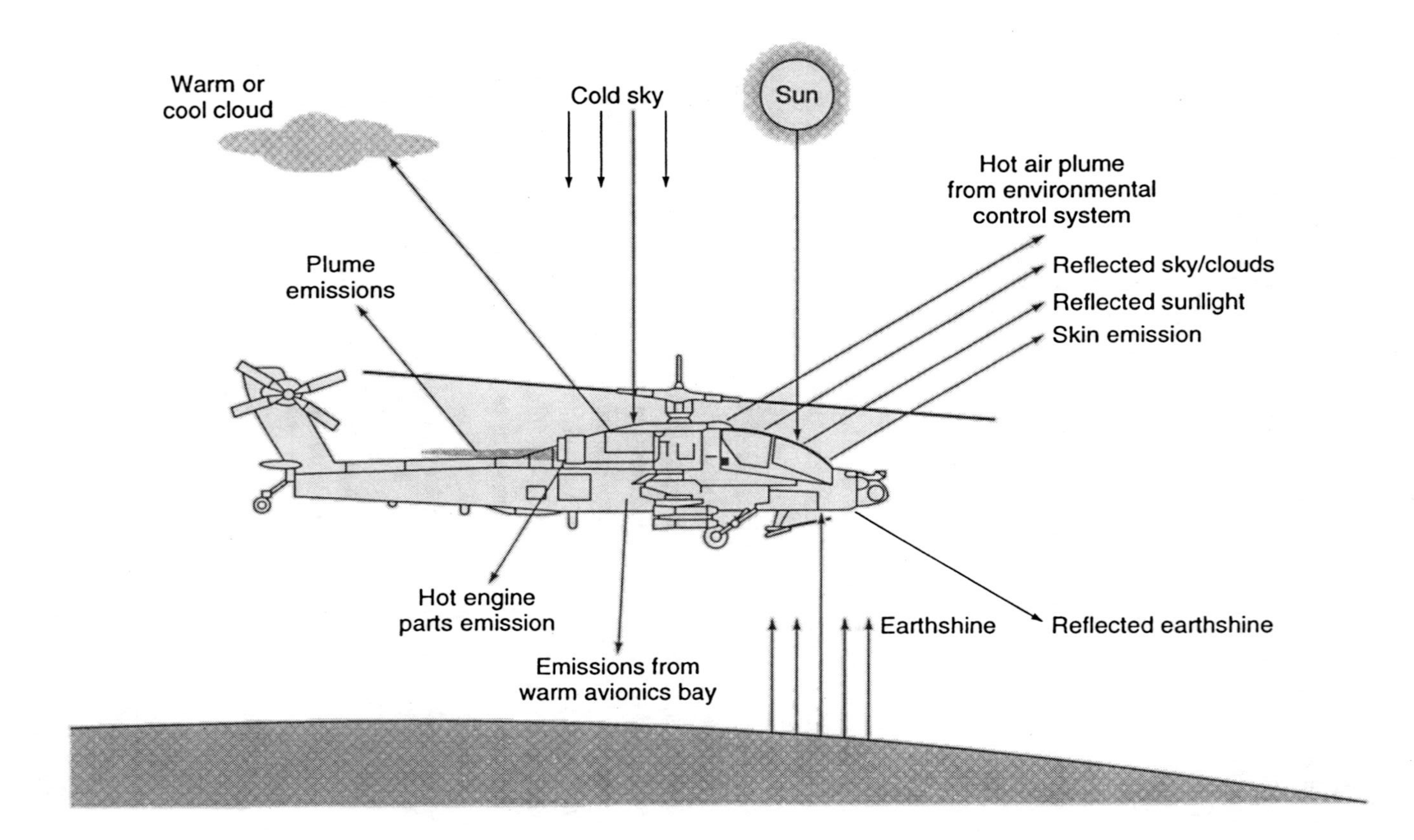

Figure 7.8 Helicopter IR signature. (*Source: Journal of Electronic Defense,* May 1991, p. 62, Figure 3.)

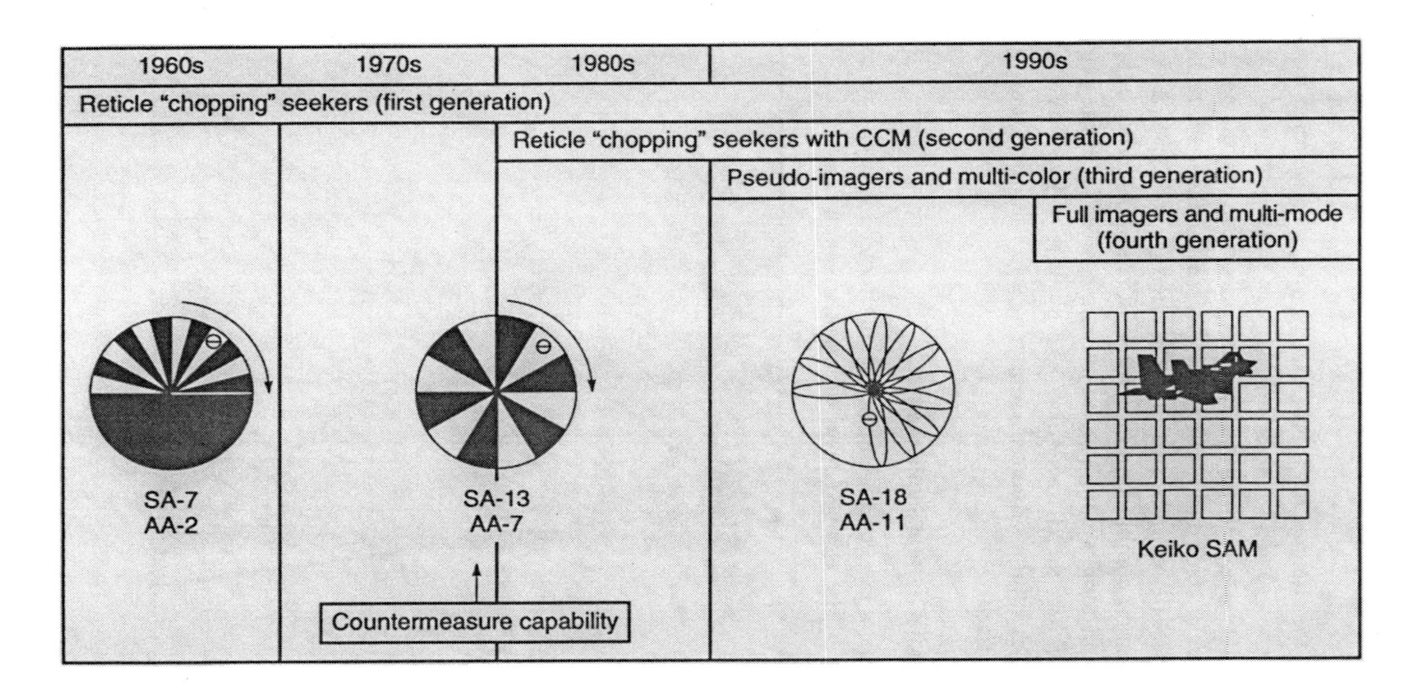

Figure 7.9 Threat seeker evolution. (*Source:* [3].)

is a decoy or an aircraft. For example, as depicted in Figure 7.10, conventional magnesium-based flares provide a response in the 1- to 3-μm region that tends to far exceed the signature of the target in this region. However, in the 3- to 4-μm region, the energies are approximately equal. Thus, when the seeker in track experiences a sudden increase in energy in the shorter wavelength band, it is a strong indication that a flare has been launched in the seeker's FOV. The importance of color discrimination is that it virtually prevents the efficient magnesium-based flares from being used as decoys. The flare designer must revert to a different combination of pyrotechnic compositions that is generally less efficient and more voluminous to overcome this EP feature [4].

Table 7.3 lists the characteristics of a number of IR missiles [3]. Many of these missiles exhibit advanced countermeasures features and provide all-aspect target attack.

7.3.1 IR Missile Seeker Fundamentals

IR missile seekers use energy emitted by the target in the IR spectrum as a tracking source. The wavelength of the peak emission is given by

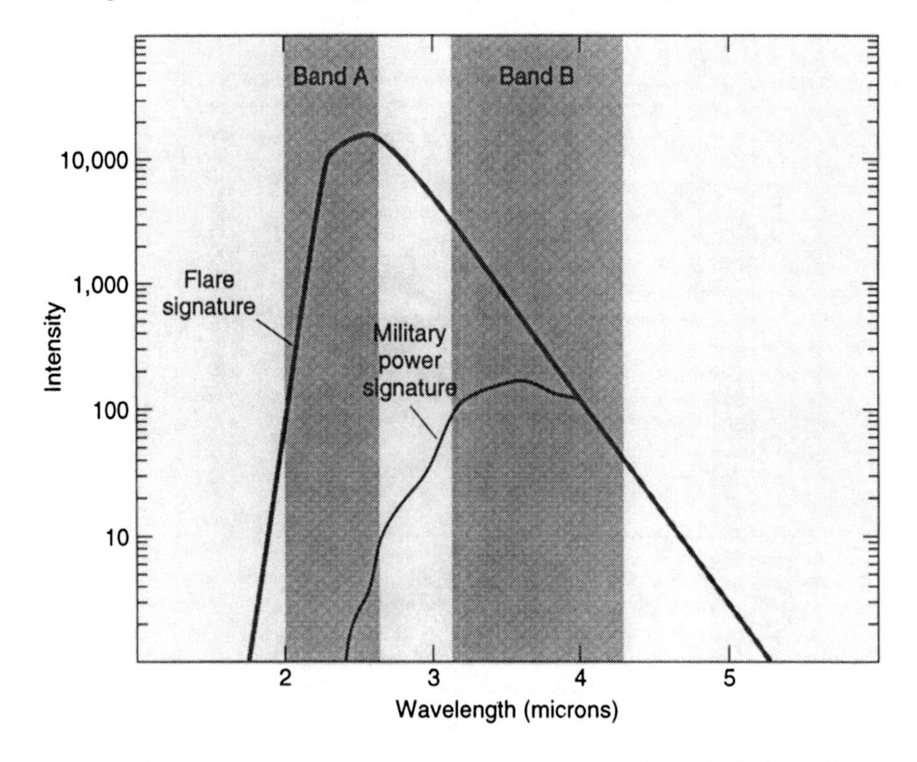

Figure 7.10 Two-color IR countermeasure. (*Source: Journal of Electronic Defense,* Jan. 1994, p. 67, Figure 4.)

Table 7.3
IR Missile Characteristics

Name	Type	Target Acquisition	Missile Guidance	Firing Range (m)	Target Altitude (m)	Speed (m/s)	Missile CCM
Strella (SA-7)	SAM	Visual	Uncooled IR	4200	2300	385	None
Grouse (SA-18)	SAM	Visual	—	5200	3500	—	Two color
Gimlet (SA-16)	SAM	Visual	Cooled IR	5200	3500	570	Two color (3.5-5.5 μm all aspect)
Redeye (FIM-43)	SAM	Visual	Uncooled IR	5500	2700	522	None
Stinger (FIM-92)	SAM	Visual	—	4500	3800	718	Cooled conical scan (4.1–4.4 μm all aspect)
Matra Mistral	SAM	FLIR, LOBL, or A/C radar	Cooled IR	6000	4500	852	Two color
Sidewinder (AIM-9)	AAM	LOBL or A/C radar	—	8000	—	—	Flare rejection
Alamo (AA-10)	AAM	LOBL or A/C radar	Terminal IR	40 km	—	—	—

$$\lambda_{max} = \frac{2898}{T} \tag{7.6}$$

where T denotes temperature in Kelvins. The temperature of the hot metal exhaust at the tail pipe of an aircraft is approximately 2000°K, resulting in a $\lambda_{max} = 1.5\ \mu m$; while the temperature of the exhaust plume is approximately 1000°K, resulting in a $\lambda_{max} = 2.9\ \mu m$. The IR signatures of aircraft afterburners, flare, and exhaust plumes are depicted in Figure 7.7.

▶ *Example 7.3*

The radiant emittance ($w \cdot cm^{-2} \cdot \mu m^{-1}$) from a black body depends only on the temperature of the surface and the wavelength. This is given by Planck's black-body radiation law

$$w(\lambda) = \frac{C_1}{\lambda^5(\exp(C_2/\lambda T) - 1)}$$

where

$$C_1 = 3.741 \cdot 10^4\ w \cdot cm^{-2} \cdot \mu m^4$$

$$C_2 = 1.438 \cdot 10^4\ \mu m \cdot {}^\circ K$$

$$\lambda = \text{wavelength in } \mu m$$

$$T = \text{temperature } {}^\circ K$$

Using MATLAB, plot the radiant emittance for temperatures T = 2000°K, 1000°K, and T = 400°K, which correspond to the temperatures of the hot metal exhaust at the tail pipe, exhaust plume, and skin of a missile.

```
% Black Body Radiation
% Plank's Black Body Radiation Law
% --------------------------------
% blkbodya.m

clear;clf;clc;

% Input Wavelengths

wl=logspace(-1.5,2);         % microns

% Input Equation Constants

C1=3.741*1e4;            % w*cm^-2*micron^4
C2=1.438*1e4;            % micron*deg K

% Compute Spectral Response

T=[6000 4000 2000 1000 ];  % deg Kelvin
for i=1:4;
   Tx=T(i);
   a=exp(C2./(wl*Tx))-1;
   W=C1./(a.*wl.^5);
   Wx(:,i)=W';
end;

% Compute Spectral Response for Low Temperatures

wl1=logspace(0,2);        % microns
T1=[500 300];

for j=1:2
  Tx1= T1(j);
  a1=exp(C2./(wl1*Tx1))-1;
  W1=C1./(a1.*wl1.^5);
  Wx1(:,j)=W1';
end;

% Plot Spectral Response

loglog(wl,Wx,'k');grid;
xlabel('Wavelength - microns');
ylabel('Spectral Power Density - w/cm^2-micron');
axis([.01 100 1e-5 1e5]);
title('Power Spectral Density Of Black Body Radiator');
hold on
loglog(wl1,Wx1,'k');
```

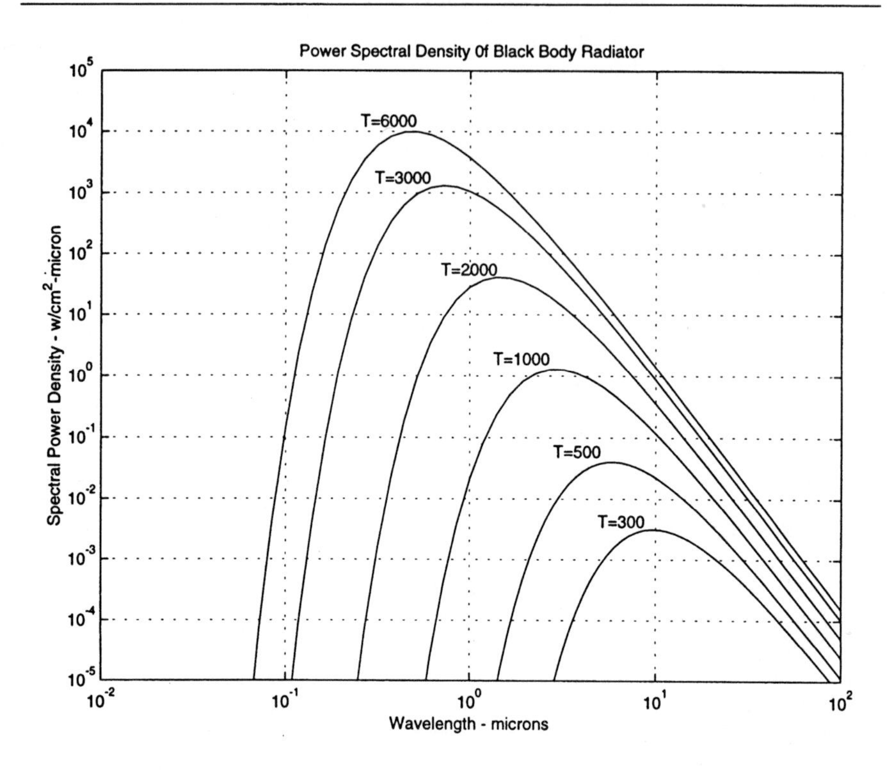

Figure 7.11 is a block diagram of an IR spin-scan seeker that uses a spinning reticle in front of the detector to chop the signal into a series of pulses. Phase information embodied in the chopped signal indicates the angular direction of a point target from the missile's boresight, while its amplitude indicates the magnitude of the error. Distributed targets such as clouds or surface reflections tend to provide less modulation than point targets and, hence, are partially eliminated.

A reticle with half the disk transmissive, as shown in Figure 7.12, tends to impart a steady signal on extended sources, such as clouds, while retaining the pulse signal from point sources. The target signal can then be extracted using a simple filter tuned to the reticle scan rate. Detector processing is illustrated in Figure 7.13, which shows the output error signal. The error signal is fed back into servos controlling the seeker, causing it to center the missile on the target.

Virtually all AAMs use proportional navigation guidance laws [3]. To fly such a flight path, the missile seekers line-of-sight rate relative to the target must be zero. When this is accomplished, the missile will ultimately collide with the aircraft target, providing it has a positive differential velocity with respect to the aircraft. Target maneuvering and missile velocity changes modify the look angle required to fly the proportional navigation path. The missile

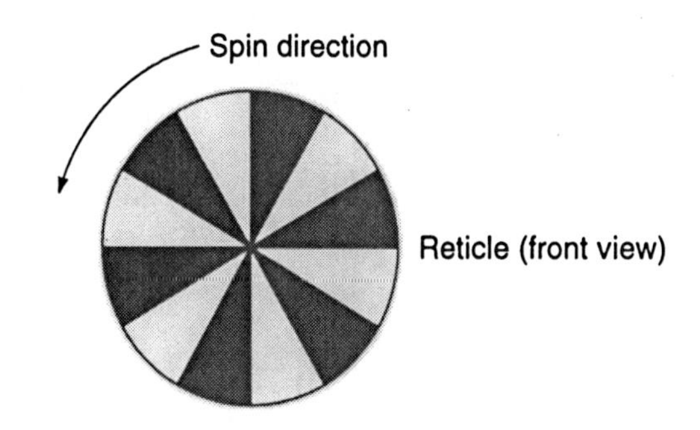

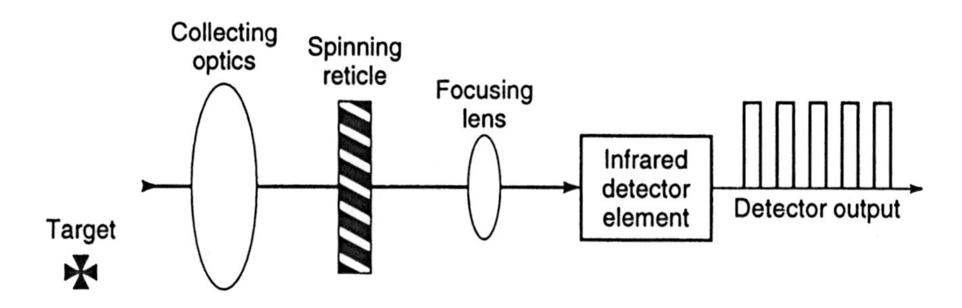

Figure 7.11 Block diagram of a spin-scan rejection. (*Source: Journal of Electronic Defense,* Jan. 1994, p. 48, Figure 2.)

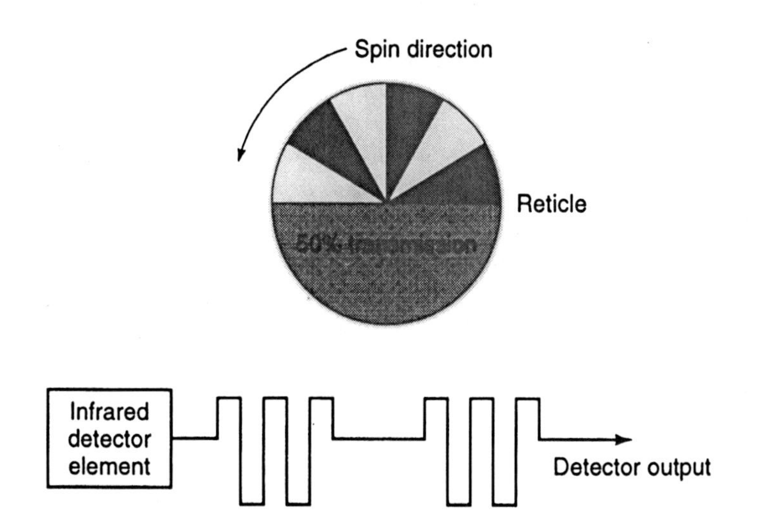

Figure 7.12 Spin-scan reticle with cloud rejection. (*Source: Journal of Electronic Defense,* Jan. 1994, p. 48, Figure 3.)

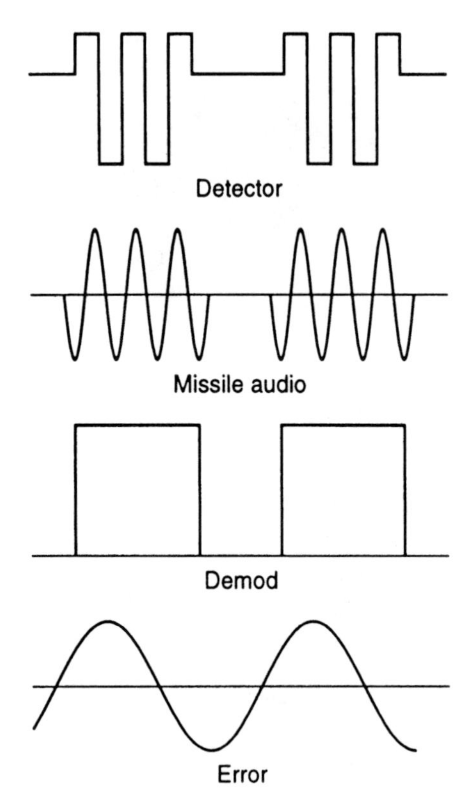

Figure 7.13 Spin-scan detector signal processing. (*Source: Journal of Electronic Defense*, Jan. 1994, p. 49, Figure 4.)

guidance control commands the missile to fly a path driving the tracking rate to zero, which ensures the implementation of a proportional navigation guidance law.

Center-spun spin-scan seekers, such as that shown in Figures 7.11 and 7.12 are relatively insensitive when the target is in the center of the seeker scan because point targets tend to illuminate all spokes evenly, thereby eliminating the pulse signal output of the detector. An error signal is generated only when the target falls away from the center of the reticle, causing the missile to fly an undulating path toward the target. Most spin-scan seekers use uncooled lead-sulfide detectors that are tuned to 2 μm and are easily countered by flares.

Flares provide the missile seeker with a hotter target than the aircraft, causing the seeker to track the flare. The typical magnesium-teflon flare burns with peak emissions in the 2-μm range. Since the flare energy emission is greater than that of the target aircraft, the missile seeker detector is dominated by the flare energy and transfers lock to the flare as long as the flare achieves a greater emissivity while the flare and target are both in the missile's FOV.

A typical conical scan seeker is depicted in Figure 7.14. The reticle is fixed and does not spin. Instead, a secondary mirror is tilted and spun. This causes the target image to be scanned in a circular path around the outer edge of the reticle. As shown in Figure 7.15, when the target is centered in the seeker scan, the detector gives an output similar to that of the spin-scan seeker. However, as the target leaves the center, the output of the detector is a *frequency modulated* (FM) square wave. The depth of FM is directly proportional to the amount the target is displaced from the center of the seeker scan. The signal is processed through an FM discriminator whose output is an AM signal with amplitude proportional to the amount of FM present. The error signal is then demodulated to produce the target-tracking error signal.

The conical scan optics are usually designed to spin the target very close to the edge of the reticle. This generates the greatest amount of FM for a given target-tracking error and, thus, a more sensitive and tighter tracking loop. The center of the reticle is only used for acquisition. Flare resistance is built in because flares tend to drop off the reticle much faster than the spin-scan reticle.

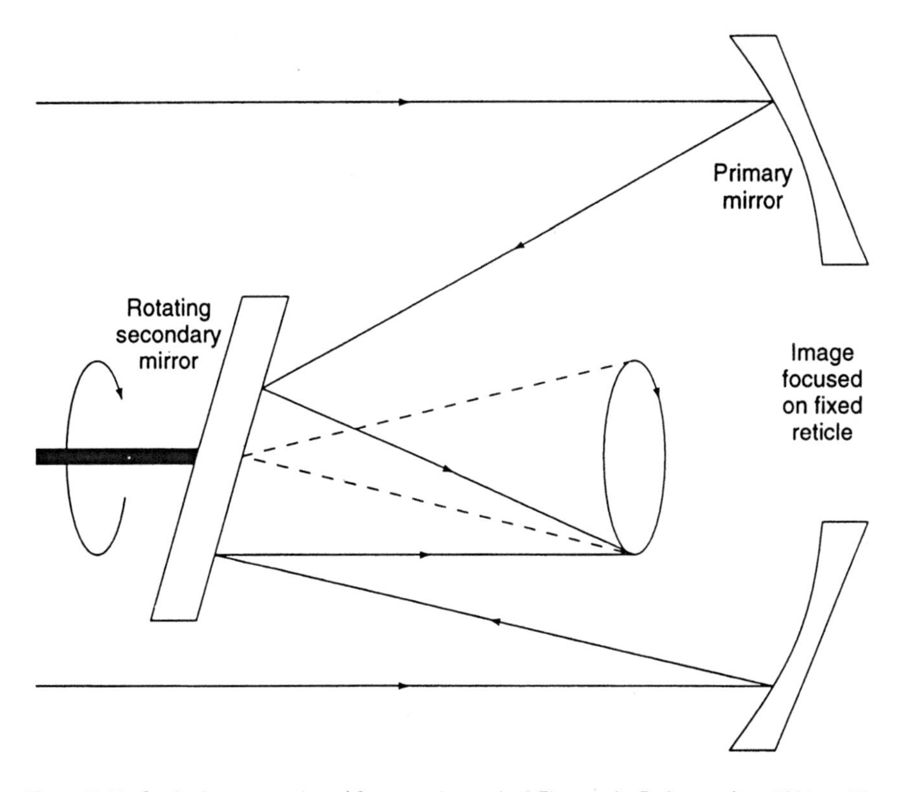

Figure 7.14 Conical scan seeker. (*Source: Journal of Electronic Defense,* Jan. 1994, p. 50, Figure 6.)

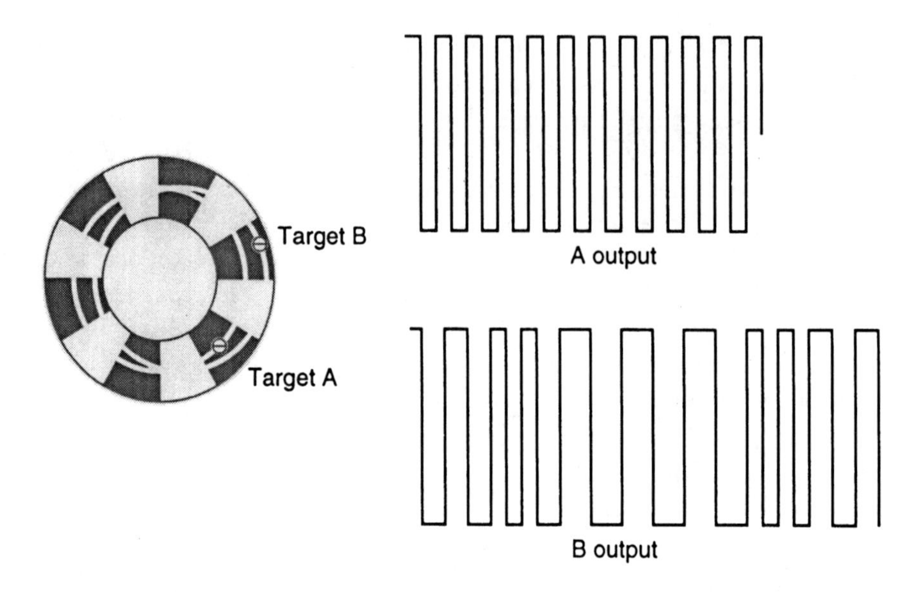

Figure 7.15 Conical scan operation. (*Source: Journal of Electronic Defense,* Jan. 1994, p. 50, Figure 7.)

Imaging seekers use a focal plane array of detector elements that detect energy from the scene, thereby building a TV-like spatial map. The image can be built from a line of detector elements across which the scene is scanned or from a two-dimensional array of detectors that stare at the scene. Imaging seekers are inherently resistant to flare-type countermeasures. They will not be readily decoyed by flares because the point-source flare simply does not look like an extended aircraft with an exhaust plume. Processing software within the seeker can select a vulnerable aim point within the target rather than simply track a hot spot in the aircraft exhaust.

The most difficult aspect of imaging seekers concerns the associated data-processing requirements for such a device. For example, a typical 128 × 128 element focal plane array has 16,384 pixels. If the staring array is sampled 100 times per second and each pixel is digitized to 10 bits, this represents a data rate greater than 16 Mbits/s. The number of computer operations necessary to perform temporal, spectral, and spatial filtering and track IR targets can easily exceed one billion operations per second, a formidable challenge for any tactical computer.

Pseudoimaging seekers do not build an image as in a true imaging seeker. However, they use spatial information from the scan to isolate the target from the background, thereby limiting the instantaneous FOV. An example of a pseudoimaging seeker using a rosette scan pattern for electronic FOV gating is depicted in Figures 7.16 and 7.9.

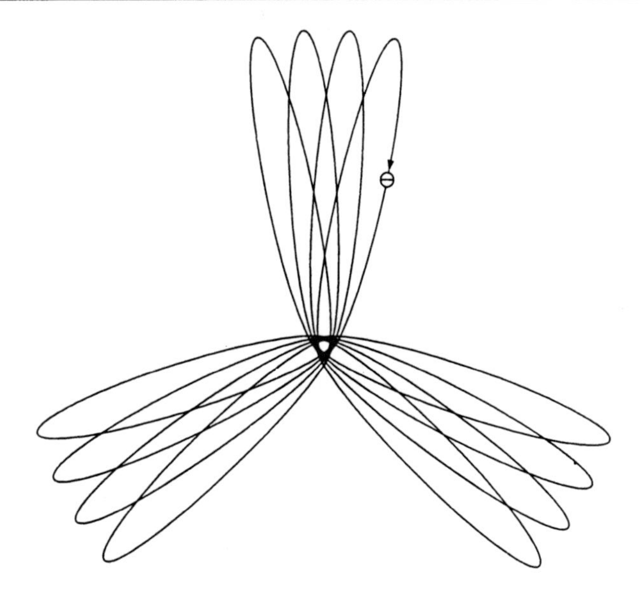

Figure 7.16 Pseudoimaging rosette scan pattern. (*Source: Journal of Electronic Defense,* Jan. 1994, p. 67, Figure 12.)

7.3.2 IR Missile Detection Range

The detection range of an IR missile depends on several factors that include the target's temperature, the atmosphere transmittance, the seeker optics, and the IR detector's sensitivity. Each of these factors are discussed in the following.

The main targets of the IR missile are the exhaust nozzle and plume of the jet engine that powers the aircraft. Temperatures of the tail pipe can range from 1000°K to 2000°K, while the plume may have temperatures between 700°K to 1000°K. When the aircraft's after burner is operational, the plume temperature can approach that of the hot tail pipe. A secondary target is the aerodynamic heating of the aircraft skin whose temperature can reach 300°K to 500°K. The temperatures of the target establish the wavelength of the emissions that peak in the 1- to 5-μm region for the aircraft's tail pipe and plume and the 8- to 12-μm region for the aircraft's skin.

From the Stefan–Boltzmann Law, the power density of the radiation in W/cm^2 is given by

$$w = \epsilon \sigma T^4 \tag{7.7}$$

where

ϵ = emissivity = 0.9 for a gray body

$\sigma = 5.67 \cdot 10^{-12} \text{W} \cdot \text{cm}^{-2} \cdot {}^{\circ}\text{K}^{-4}$

T = Temperature in °K

If we assume a tail pipe temperature of 800K with a nozzle area of 2500 cm^2, then the resultant power is 5225W. Assuming a Lambert radiator, the power density is $5225/\pi$ = 1663 W/steradian. This estimate is somewhat high since measured values for a B-52 in the 2.0- to 2.6-μm region are in the range of 600 W/steradian for a tail aspect and 200 W/steradian as seen from the front of the aircraft. The values are less for aircraft with engines mounted within the fuselage with curved air intakes. Values for a fighter aircraft range from 300 W/steradian for a rear aspect to a negligible amount for the forward aspect.

For aircraft skin heating, the temperature can be approximated as $T = T_0(1 + 0.164M^2)$, where M is the Mach number. Applying this to a Mach 2 aircraft with T_0 = 250°K results in a temperature of 414°K. The emissivity of the aluminum skin is about 0.05. Assuming a radiating area 10 m^2 results in a power density of 255 W/steradian. The wavelength for maximum emission is 7 μm from (7.6).

Transmission of IR radiations through the atmosphere is generally possible in five specific IR bands: 1.5 μm to 1.8 μm, 2 μm to 2.5 μm, 3.5 μm to 4.2 μm, 4.5 μm to 4.8 μm, and 8.5 μm to 12.5 μm. Other bands suffer from a high degree of atmospheric attenuation. IR radiation suffers from rain and fog attenuation over all bands. The attenuation coefficients generally decrease with altitude as shown in Table 7.4.

IR detectors are generally characterized by a parameter D^* that is the detectivity measured with a bandwidth of 1 Hz and reduced to a detector acceptance area of 1 cm^2. The units of D^* are $\text{cm} \cdot \text{Hz}^{1/2} \cdot \text{W}^{-1}$. D^* is the quantity generally given in detector specification sheets and increases as the

Table 7.4
IR Attenuation in Two Bands (dB/km) for 23 km Visibility

	Band	
Altitude	**3.5 μm**	**10 μm**
200 m	0.88	0.57
2 km	0.37	0.17
5 km	0.16	0.05
10 km	0.05	0.04

detector is cooled, which reduces the detector's internally generated noise. For example, photo-conductive lead sulfide detectors have D^*s at 3 μm of $4 \cdot 10^{11}$ (77°K). $2 \cdot 10^{11}$ (195°K), and $2 \cdot 10^{10}$ (295°K). This detector is sensitive from about 1.5 μm to 3 μm at room temperature. It is sensitive from about 1.5 μm to 4 μm when cooled to liquid nitrogen temperatures (77°K). To obtain the longer wavelength response, which is a better representation of the spectral IR signature of aircraft targets, liquid-nitrogen cooling is added to most nonimaging IR guidance systems.

In modern guidance systems, indium antimonide has become the preferred detector material. It requires cooling to liquid-nitrogen temperatures but is sensitive from about 3 μm to beyond 5 μm. This matches jet engine plume radiation at 4.3 μm, which comes primarily from hot CO_2 gas. This cooled detector has a D^* of about $9 \cdot 10^{10}$. Other detector materials used in newer missiles are mercury-cadmium-telluride, gallium arsenide, and iridium silicide [3].

It can be shown that the detection range of an IR missile in centimeters is given by

$$R = J^{1/2}\tau_a^{1/2}\left[\frac{\pi}{2}D_o NA \tau_o\right]^{1/2} D^{*1/2}\left[\frac{1}{(\omega\Delta f)^{1/2}(S/N)L_s}\right]^{1/2} \qquad (7.8)$$

where

J = Radiant target intensity, W/steradian

τ_a = Atmosphere transmittance

D_o = Optics diameter

NA = Numerical aperture ($D_o/2 \cdot$ focal length)

τ_o = Optics transmittance

D^* = Detector sensitivity cm $\cdot$ Hz$^{1/2}$ $\cdot$ w^{-1}

ω = Field of view, steradians

Δf = Noise bandwidth, Hz

S/N = Signal-to-noise voltage ratio

L_s = System and processing losses

The first term depends on the target, the second on the atmosphere, the third on the optics, D^* depends on the detector, while the last term depends upon the system design. Note that the atmospheric transmittance is generally a function of range, requiring an iterative solution of (7.8).

► *Example 7.4*

Using MATLAB, plot the range of an IR missile against an aircraft as a function of altitude. Consider detection of the plume at 3.5 μm and the aircraft skin at 10 μm. Atmospheric attenuation is given in Table 7.4. Optics and detector parameters for each band are given in Table 7.5.

Table 7.5
IR Seeker Parameters

	3.5 μm	10 μm
J (W/steradian)	1000	150
D_0 (cm)	7.5	13
NA	1.9	1.7
τ_0	0.7	0.6
D^* (cm $Hz^{1/2}$ W^{-1})	10^{11}	10^{10}
ϕ (degrees)	1.5	2.5
Δf (Hz)	100	100
S/N (dB)	20	20
L_s (dB)	8	8

```
% Range of IR Missile Seeker
% ---------------------------
% irrng.m

clear;clf;clc;

% Input Seeker Parameters [ Band 1  Band 2 ]

J=[1000 150];       % Target Radiant Intensity - w/sterandian
Do=[7.5 13];        % Optics Diameter - cm
NA=[1.7 1.7];       % Numerical Aperture
ao=[.7 .6];           % Optics Transmittance
Ds=[1e11 1e10];     % Detector D*
phi=[1.5 2.5];        % Optics FOV - degrees
df=[100 100];       % Noise Bandwidth - Hz
snr=[30 30];        % Voltage Signal-to-Noise Ratio - dB
Ls=[8 8];           % System and Processing Loss - dB

snrx=10^(snr(1)/20);Lsx=10^(Ls(1)/10);
fovx= (phi(1)*pi/180)^2;                    % FOV in steradians
snr1=10^(snr(2)/20);Ls1=10^(Ls(2)/10);
fov1=(phi(2)*pi/180)^2;

% Compute Range With No Atmospheric Attenuation
```

```
tarfac=J(1)^.5;
optfac=(pi*Do(1)*NA(1)*ao(1)/2)^.5;
detfac=Ds(1)^.5;
sysfac=(((fovx*df(1))^.5)*snrx*Lsx)^1/2;
Ro=tarfac*optfac*detfac/(sysfac*1e5); % Range in km

% Input Atmospheric Attenuation vs Altitude

alt=[200 2000 5000 10000]; % Atmospheric Attenuation -dB/km
att=[.88 .37 .16 .06];

% Compute Range with Atmospheric Attenuation
% Using Newton's Method

for j=1:4;
R(1)=Ro;
atn=att(j);
a=atn/20;

for i=1:10;
   x=.4343/R(i)+a;
   y=log10(R(i))-log10(Ro)+a*R(i);
   dR=y/x;
   R(i+1)=R(i)-dR;
   if abs(dR)<1e-6;break;end;
end;
Rx(j)=R(i);
end;

% Plot Range vs Altitude
% Smooth Data using Polynomial Fit

z=0:10000;
f=polyval(polyfit(alt,Rx,3),z);
plot(z,f);grid;
xlabel('Altitude - meters');
ylabel('Range - km');
title('Range of IR Seeker With Altitude');
axis([0,10000,4,14]);
hold on

% Compute Range for Second Band

alt1=[200 2000 5000 10000];
att1=[.57 .17 .05 .04];

tarfac1=J(2)^.5;
optfac1=(pi*Do(2)*NA(2)*ao(2)/2)^.5;
detfac1=Ds(2)^.5;
sysfac1=(((fov1*df(2))^.5)*snr1*Ls1)^.5;
Ro1=tarfac1*optfac1*detfac1/(sysfac1*1e5);
```

```
for k=1:4;
   R1(1)=Ro1;
   atn1=att1(k);
   a1=atn1/20;

   for l=1:10;
      xx=.4343/R1(l)+a1;
      yy=log10(R1(l))-log10(Ro1)+a1*R1(l);
      dR1=yy/xx;
      R1(l+1)=R1(l)-dR1;
      if abs(dR1)<1e-6;break;end
   end;
   Rx1(k)=R1(l);
end;

fx=polyval(polyfit(alt1,Rx1,3),z);
plot(z,fx);
```

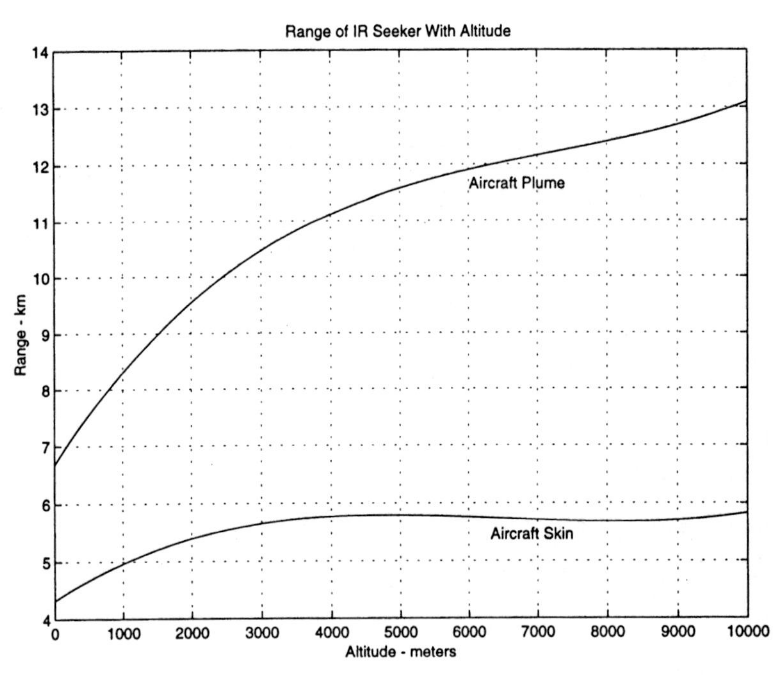

7.3.3 IR Missile Seeker Counter Countermeasures

IR missile designers have added advanced features that detect the presence of flares and reject them as targets, allowing the seeker to continue tracking the real target. *IR counter countermeasures* (IRCCM) built into advanced seekers generally operate in two phases. The first is a detection device or switch that determines whether a flare is in the seeker's FOV and activates the second phase, which is the action the missile seeker takes to reject the flare.

There are several different switches and response techniques. A device that is capable of decoying one type of advanced IR missile might be totally ineffective against another advanced IR missile type. There are a variety of techniques that can be used to detect flares. These include rise time (temporal), two-color (spectral), kinematic, and spatial switching or trigger techniques.

An IR missile using a rise-time switch monitors the energy level of the target it is tracking. A sharp rise time in the received energy indicates a flare in the seeker's FOV. Fighter aircraft have an IR signature that is much larger in the stern than on the beam or forward quadrant. Flares are designed to generally provide twice as much energy as the target they are trying to protect. Thus, a flare-to-target ratio of 2:1 in the rear and as much as 10:1 in the beam or forward quadrant may be expected. A seeker with a temporal discriminator would switch on the IRCCM if the seeker detected an energy increase above a preset threshold within a preset time limit. For example, a threshold of a 2.5:1 energy increase within 40 ms might permit flare detection in all aspects while ignoring the relatively slow energy rise from after-burner ignition. The IRCCM would be switched off when the received energy dropped to a preset value (e.g., 2:1), indicating the flare had left the seeker FOV.

A missile seeker with a two-color switch that samples the energy level in two different bands is shown in Figure 7.10. A non-after-burning target would generally have more energy in the 3- to 5-μm band than in the 1- to 3-μm band. Thus, a sudden jump in the 1- to 3-μm-band energy compared to that in the 3- to 5-μm band indicates a flare in the seeker's FOV. The seeker could use two different detectors to monitor the energy levels in the two bands (i.e., lead sulfide in the 1- to 3-μm band and indium antimonide in the 3- to 5-μm band) or use a single detector with different bandpass filters on the reticle spokes.

Kinematic switches exploit the effect that flares separate very quickly from the dispensing aircraft due to their high aerodynamic drag. In a beam-aspect engagement, a missile seeker that transfers track from the target to the flare will have a large, sudden change in the line-of-sight rate due to the rapid deceleration of the flare. As the seeker rapidly moves to track the flare, the kinematic switch senses the rapid change in relative motion between the missile and target and initiates the IRCCM response. Kinematic discriminators tend to have difficulty in head-on and stern engagements, with the small line-of-sight changes between the target and the flare.

A spatial switch uses the physical separation between the flare and the target as the discrimination mechanism. As the flares separate to the rear of the aircraft, the missile seeker sees the target on the forward side of the FOV and the flare on the rear side of the FOV. Once two hot spots on opposite sides of the FOV are distinguishable, the switch to the IRCCM mode operation is made.

The seeker's response to the switch is to reject the flare or to limit its effect on the target track. Generally, as long as the flare remains in the seeker's FOV, the missile seeker is not tracking the target or is tracking in a degraded mode. Most IR seekers have a FOV less than 2.5 degrees. This results in the flare remaining in the FOV for a long time at long ranges and a short time at short ranges. Several different response techniques are used, either alone or in combination, to defeat the flare.

Coast or memory techniques cause the seeker to reject its detected track data and maintain its motion relative to the target, waiting for the flare to leave the seeker's FOV. The missile will continue to reject track data until the flare leaves the FOV or the switch times out, returning the seeker to its normal track mode. If the switch times out while the flare remains in the FOV, the seeker usually transfers lock to the flare.

Another response technique causes the seeker gimbals to drive the seeker in the direction of target motion. Pushing the seeker forward causes the flare to depart the FOV faster than with the coast mode, minimizing the time the missile is not tracking the target. If the amount of forward bias is too great, the seeker may be pushed forward of the target, causing both the target and the flare to depart the FOV. In this mode, the seeker may be pushed ahead without affecting the missile flight path or the missile may begin to lead the target in concert with the seeker push ahead. Push-pull response techniques are appropriate when the flare and target are on opposite sides of the seeker's FOV. This technique drives the seeker toward the coolest IR target in the FOV.

In sector attenuation, an attenuation filter is placed across part of the seeker FOV. This causes the seeker to be less sensitive to objects in that FOV. If the target being tracked is in the center of the FOV, then placing an attenuator in the quadrant below and to the rear of the target should reduce any energy received from a flare. If the attenuated flare energy is below that of the unattenuation target energy, the seeker will continue to track the target.

Electronic FOV gating is used in conjunction with a noncircular seeker scan such as the rosette scan shown in Figures 7.9 and 7.16. At some time after the flare is dispensed, the trigger and flare will no longer be in the same lobe of the scan. By computing the relative motion of the target, the missile is able to determine on which lobes the target should appear. Information from all other lobes is ignored, allowing the missile to retain track on the target.

7.3.4 Missile Approach Warning

Missile approach warning (MAW) is a necessary function for countering missile attacks on tactical aircraft. MAWs generally warn aircraft that they are under attack by an approaching missile and provide directional and time-to-impact threat information necessary to counter the attack by either dispensing decoys or employing active countermeasures. MAWs are intended to supplement RWRs, which alert the crew of the presence of hostile fire control radar and missiles carrying active radar seekers, but they are impotent against weapons employing IR/EO passive sensors.

MAWs may be either active or passive. The former use a pulse Doppler radar to detect and track missiles by means of their skin return, while the latter respond to either the IR or *ultraviolet* (UV) signature of the exhaust plume from the missile's rocket engine. The two types of sensors may be combined in a hybrid system, offering the best combined benefits of each [6].

Radar provides the advantage of providing all-weather range and range-rate data that enables the calculation of time-to-impact data. Disadvantages involve the difficulty of extracting the small missile RCS target from competing clutter and providing all-aspect accurate AOA data. An additional criticism of an active radar is the possibility of its being used as a beacon for an antiradiation-type missile. In addition, LPI waveforms and antenna shielding are necessary when radars are used on stealth aircraft. Discrimination between missiles and rapidly closing aircraft is another difficult problem for active radar MAWs.

Passive MAWs employ either IR or UV sensors to detect a missile's exhaust plume at some point prior to burn-out. These sensors generally provide relatively accurate angular data (i.e., order of 1 degree), but range must be estimated on the basis of signal strength. Present systems (e.g., AAR-34 and AAR-44) use detection arrays that are mechanically scanned. Characteristics of these systems are listed in Table 7.6 in addition to those of more advanced systems using staring IR and EO detectors [7]. The advantage of scanning systems is that they use relatively few detectors while providing a wide FOV. These systems are generally susceptible to false alarms. Staring systems, which are analogous to phased array radar, use a mosaic of detectors to generate an image of the target scene. Staring IR systems are tuned in the 4- to 5-μm region to look for early threat launch warning by detecting plume radiation and using two-color, spectral-spatial-temporal clutter rejection techniques.

The AN/AAR-47 missile warning system employs optical sensors to detect the IR signatures of approaching passive SAMs. The emission that accompanies a SAM depends on the type of missile and the portion of its trajectory under observation. At launch, the emission from the booster rocket plume can be

Table 7.6
Comparison of Some Passive MAWs

	Scanning IR	Staring IR	Electro-optical
Representative	SAWs AAR-34, AAR-44	Fly's Eye	AAR-47
Acquisition cost	Moderate	High	Moderate
Development risk	Low, mature technology	High, emerging technology	Low, mature technology
Sensor	Multispectral IR FPA with rotating scanning mirror	Multispectral IR FPA	Photomultiplier tubes, selective filtering, staring mode
Advantages	Large FOV	Continuous coverage within FOV	Less atmospheric attenuation, continuous coverage within FOV, no cooling needed
Disadvantages	Less than 100% duty cycle, requires cooling, has moving parts	Requires cooling, very high data rates	Requires four sensors for 360-degree FOV

expected to produce an energetic, highly visible signature with the components in the optical and mid-range IR portions of the spectrum. After the boost phase, there may be secondary firings of steering rockets. As the missile approaches its target, the booster and steering rockets may be exhausted, but portions of the missile skin surrounding the engine and even the leading edges of the missile itself emit IR signatures. The AAR-47 detects the UV region of the IR spectrum of these missile plumes.

UV detectors are designed to detect the burning of solid rocket propellants. This can be accomplished with low false alarms in the "solar blind" region between 0.22 μm and 0.28 μm, because the Earth's ozone layer absorbs most of the solar radiation in this band [8]. Atmospheric transmittance of solar blind UV below the ozone layer is limited by scattering due to other atmospheric constituents and absorption by the surface ozone (a common urban pollutant). Thus, at altitudes below the ozone layer lies a band at which energy is, to a limited extent, transmitted with extremely low solar background levels. It is also known that there are few phenomena in the Earth that produce large amounts of UV radiation. Thus, in spite of significant attenuation, lower atmospheric events such as the burning of solid rocket propellants, which produce solar blind UV radiation, are detectable above the extremely low background. Further, since the UV radiation lies on the steep side of the blackbody curves (see Figure 7.7), small changes at high temperatures produce large changes in UV emissions. UV detectors, therefore, provide the capability of detecting the positive contrast of the rocket booster against the extremely low solar blind background.

The greatest advantage of a passive missile warning system is its passivity. Depending on the aircraft and its mission, that alone may be reason to justify its choice. Disadvantages of a passive MAW include performance degradation in bad weather, lack of accurate range data, poor time to intercept estimation, and an inability to detect an incoming missile after the rocket engine has burned out.

7.3.4.1 MAW Using Pulsed Doppler Radar

A MAW radar has the advantage of providing outputs of target range and range rate, enabling the system to calculate time-to-intercept and thus permitting optimum deployment of countermeasures and time of aircraft evasive maneuvers. Radar also provides the potential to discriminate between approaching missiles and aircraft and to accomplish noncooperative target recognition. A radar-based MAW is effective in all-weather functions at any altitude but can act like a beacon [9, 10].

The design of pulse Doppler MAW faces two main challenges posed by the small-target RCS (0.01 to 0.1 m^2), the strong ground clutter encountered at low altitudes, and limited space available for antennas in a tactical aircraft. One major task is to achieve a high probability of detection (i.e., 0.9 to 0.95) coupled with a low probability of false alarm at a range that provides adequate time-to-intercept for effective deployment of decoys and possibly for evasive maneuvers. The main impacts of this consideration are on the average transmitter power and the required antenna aperture. The other major challenge is to provide the subclutter visibility (i.e., the order of 80 dB to 90 dB) necessary to reject the large ground-clutter signals, particularly at low altitudes that might mask the desired target or cause false alarms.

The basic search radar detection range equation applicable in a noise-limited background is given by [9]

$$R = \left[\frac{P_A A_r t_s \sigma}{4\pi \Psi_s k T_s S_A L_s}\right]^{1/4} \tag{7.9}$$

where

R = Detection range over solid angle Ψ_s, m

P_A = Average transmitted power, W

A_r = Effective antenna receiver aperture, m^2

t_s = Time to search solid angle Ψ_s, s

σ = Average radar cross section, m^2

Ψ_s = Solid angle searched with uniform energy, steradians

k = Boltzman's constant, $1.38 \cdot 10^{-23}$ W/(Hz – °K)

T_s = Effective system noise temperature, °K

S_A = Effective integrated received signal energy-to-noise spectral density ratio E/N_o

L_s = Total system losses including pattern-propagation factor

The search radar equation is primarily independent of frequency except that a number of parameters (i.e., σ, T_s, L_s) may vary with frequency.

The range that must be achieved by the MAW is dependent on the warning time required to perform the necessary protective actions, which is of the order of 3 sec to 5 sec. Table 7.7 lists the detection ranges required to provide a 5 sec warning for various missile velocities [9]. The high-speed missile (Mach 5) becomes the design driver with a required detection range of 10.2 km, which is 1.7 times that of the more commonly encountered missiles (Mach 2.5). From (7.9), this translates into a power-aperture product that is of the order of nine times that of the lower speed missile. If the 5.9-km design requirement of the medium speed missile (Mach 2.5) is adopted, then the warning time is reduced to about 3 sec for the high-speed missile. This warning time may be adequate in many applications.

Table 7.7
Detection Range Requirements for 5-sec Warning

Missile Velocity		Detection Range Requirements (km)			
		M 0.95 A/C		M 0.6 A/C	
Mach	m/s	Nose	Tail	Nose	Tail
1.5	516	4.2	1.0	3.6	1.6
2.5	860	5.9	2.7	5.3	3.3
5.0	1720	10.2 (6.1)	7.1 (4.2)	9.6 (5.8)	7.6 (4.5)

( ) - Requirements for 3 seconds warning

The RCS of threat missiles varies both as a function of aspect angle and frequency. A good design value for frequencies below 2 GHz is between −10 dBsm and −15 dBsm. The RCS of small diameter missiles (e.g., 3 inches) is very frequency sensitive, providing enhancement of 10 dB at 430 MHz and reduction of about 20 dB at 5 GHz. Larger diameter missiles (greater than 8 in) are less sensitive to frequency and fall within the −10- to −15-dBsm range for aspects to 30 degrees off the missile nose [9].

A typical MAW pulse-Doppler radar design is presented in [9]. The parameters for this design are tabulated in Table 7.8. The performance of this system is presented in Table 7.9 for a 90% cumulative detection probability and a Mach 0.95 aircraft. Note that the reaction times are generally greater than 3 sec for all attack aspects and greater than 6 sec for the most common missiles. The Doppler filter bandwidth is set at 40 Hz to accommodate missiles with 16-g acceleration.

Table 7.8
Typical MAW Pulsed Doppler Radar

Transmitter frequency	1.5 GHz
Transmitter average power	500W
Noise figure	3 dB
Total system losses	10 dB
Antenna effective aperture	40 in^2
Antenna main/back lobe ratio	35 dB
Receiver dynamic range	110 dB
Subclutter visibility	70 dB
Doppler filter bandwidth	40 Hz
Pulse repetition frequency	25 kHz
Pulsewidth	10 μs
Angle coverage	All aspect

Table 7.9
MAW Pulsed Doppler Radar Performances

Missile Velocity	**Nose Attack**		**Tail Attack**	
(Mach)	**Detection (km)**	**Time to Go (sec)**	**Detection (km)**	**Time to Go (sec)**
1.5	7.7	9.2	9.7	51.3
2.5	7.2	6.0	8.4	15.9
5.0	6.2	3.0	7.0	50

▶ *Example 7.5*

The parameters of a typical MAW pulse Doppler radar are given in Table 7.8. Using these parameters and assuming a Swerling 3 type target, a search frame time of 1 sec, a probability of detection of 0.98, and an aperture efficiency of 70%, plot the detection range for antenna diameters from 1 in to 12 in. What minimum antenna aperture is required to achieve the minimum detection range of 6 km?

```
% Missile Approach Warning Pulsed Doppler
% Detection Range vs Antenna Diameter
% ------------------------------------------
% mawant.m

clear;clc;clf;

% Antenna Diameter Range

Dmin=1 ; Dmax=12;  % inches
D=Dmin:Dmax;

% Enter Radar Parameters
% Swerling Case 3 Target

Pa=500;        % Trans Ave Pwr - w
ts=1;          % Search Frame Time - s
RCS1=.1;       % Radar Cross-Section - sq. m
RCS2=.0316;    % Radar Cross-Section - sq. m
ang=4*pi;      % Search Coverage - steradians
k=1.38e-23;    % Boltzman's Constant - w/(Hz-Ko)
NF=3;          % Receiver Noise Figure - dB
Sa=18;         % SNR for Pd=.9;Pfa=1e-8 - dB
Ls=10;         % System Losses - dB
eff=.7;        % Antenna Aperture Efficiency

F=10^(NF/10);Sa=10^(Sa/10);Ls=10^(Ls/10);

% Find System Noise Temperature

Ts=293*(F-1)+150;

% Find Antenna Aperture

Ar=eff*(pi/4)*(D*2.54/100).^2;

% Compute Missile Detection Range

R1=(Pa*Ar*ts*RCS1/(4*pi*ang*k*Ts*Sa*Ls)).^.25;
R2=(Pa*Ar*ts*RCS2/(4*pi*ang*k*Ts*Sa*Ls)).^.25;
end;
```

```
% Plot Range vs Antenna Diameter

plot(D,R1/1000,D,R2/1000);grid;
ylabel('Range - km');
xlabel('Antenna Diameter - inches');
title(['MAW Pulsed Doppler Detection Range']);
```

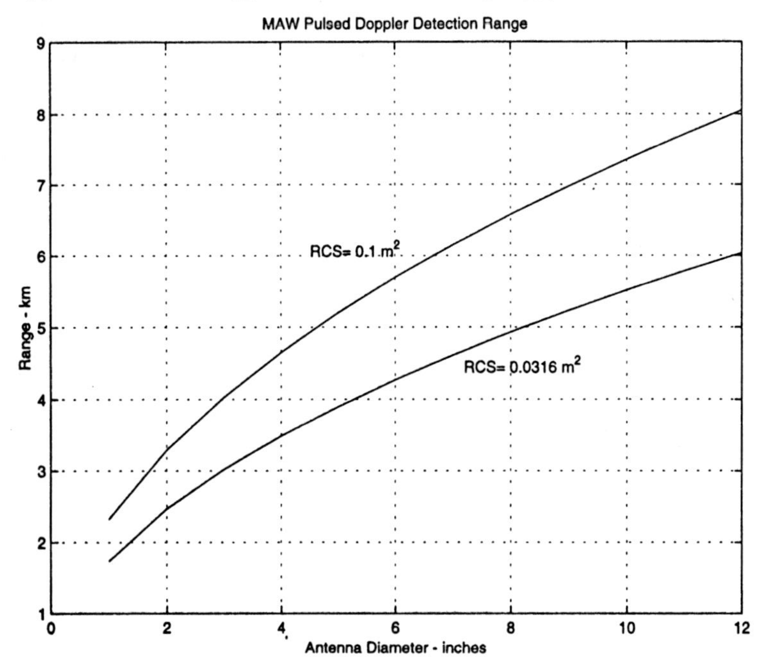

Radar clutter effects vary as a function of aircraft speed, altitude, and the approach angle of the attacking missile. The most frequently encountered case is a tail-chase missile attacking a high-speed low-altitude aircraft. Strong ground clutter returns are received from the main lobe, sidelobes, and backlobes of the tail-looking MAW antenna. These signals can be characterized as main-lobe, sidelobes, and altitude-line clutter, each of which has unique characteristics and affects the design in a different manner.

Main-lobe clutter is usually the strongest, owing to the power gain of the antenna's main lobe. For an aft-looking antenna, the main-lobe clutter appears to be moving away from the aircraft, producing negative Doppler that can be separated from the missile return (which is moving toward the aircraft and producing positive Doppler) by Doppler filtering. For a nose-mounted antenna, the clutter produces positive Doppler, but the combined closure of the missile and aircraft produces a much greater positive Doppler, allowing missile detection by Doppler discrimination.

Altitude line clutter for an aircraft in horizontal flight has zero Doppler shift since it is neither approaching or receding and, thus, can be discriminated from targets of interest. Altitude line clutter is usually received through sidelobes, which have low-relative gain, but can be very strong at low altitudes over water or other surfaces that produce specularlike reflections.

Sidelobe clutter is generally much weaker than either main-lobe or altitude-line clutter. However, for an aft-looking antenna, backlobe clutter produces positive Doppler that competes with the positive Doppler produced by missiles that are closing on the aircraft at relative speeds less than the aircraft speed. Thus, for this situation, the clutter must be attenuated by providing low antenna backlobes that reduce the clutter magnitude below that of the missile return.

An estimate of the subclutter visibility requirements can be made by estimating the magnitude of the clutter return. The area of the clutter patch at low altitudes is given by $A_c = \theta_{AZ} \cdot R \cdot R_\tau$, where θ_{AZ} is the azimuth antenna beamwidth, R is the range to the patch, and R_τ is the range gate length [2]. From Table 7.7 at 6-km range, the clutter patch area is $A_c = (20/7 \cdot 2.54) \cdot 6000 \cdot 1500 = 10^7\ m^2$. The equivalent RCS of the clutter is then $\sigma_c = \sigma^\circ \cdot A_c$, where σ° is the backscattering coefficient equal to 0.01 for mountainous and urban return. The signal-to-clutter is given by $S/C = \sigma_t/\sigma_c$, where σ_t is the target RCS of 0.1 m^2. The resultant S/C is then 10^{-5}.

A low false alarm rate of less than 1/hr is desired [9]. To search 4π steradians in 1 sec requires the order of $4\pi/(\lambda/D)^2 = 10$ beam positions/s. Since there are three range gates per beam position, this results in 30 range gate false alarm positions per second. The Doppler filter bank for each range gate must span a velocity differential of Mach 5 (1720 m/s) to detect both nose and tail attacking missiles. Each Doppler filter has a velocity range of $v = (\lambda \cdot B_d)/2 = 4$ m/s, resulting in the need for 1720/4 = 430 Doppler filters per range gate. Hence, in 1 sec, there are 12,900 possibilities for a false alarm; and in 1 hr, there are approximately 10^7 to 10^8 possibilities for a false alarm requiring a $P_{fa} = 10^{-8}$. The signal-to-clutter ratio required for detecting a Swerling Case 3 target with $P_d = 0.9$ and $P_{fa} = 10^{-8}$ is 18 dB. Hence, the required subclutter visibility for the MAW described by Table 7.5 is 50 + 18 = 68 dB, which can be rounded to 70 dB.

The 70-dB subclutter visibility can be achieved by Doppler filtering in situations where the missile is closing on the aircraft at speeds greater than the missile is closing on the ground. The phase noise from the transmitter (i.e., "transmitted clutter") and from the stable local oscillator in the MAW receiver must be such that random modulations do not spread clutter components into the Doppler filter band [2]. This phase noise must be of the order of −120 dBc [9]. Also, timing of range gates and PWs must have stability in the picosecond region over the Doppler filter processing time.

In tail chase situations, the backlobes of the antenna induce positive Doppler components on the clutter that range in magnitude up to the forward velocity of the aircraft. When the missile is closing on the aircraft at speeds less than twice the forward velocity of the aircraft, then its Doppler signature must directly compete with the backlobe clutter. This situation requires low antenna backlobes to allow detection of the missile. Since signals are both transmitted and received through the backlobes, a reduction of gain in the backlobes equal to the $SCV/2 = 35$ dB with respect to main-beam gain provides the necessary clutter attenuation for missile detection.

Receiver dynamic range is also important since the strong main-lobe and altitude-line clutter signals are present in the receiver prior to Doppler filtering. Any nonlinearity in the receiver generates intermods between the various clutter components that spread the clutter spectrum into the Doppler filter passband [2]. A dynamic range of the order of 110 dB is required in the MAW design to allow missile detection at low aircraft altitudes.

The operating frequency of the MAW affects several aspects of its design. First, lower frequencies (i.e., in the UHF band) suffer because antennas with sufficient aperture to provide missile detection cannot provide the main-lobe-to-sidelobe gain ratios necessary to meet the subclutter visibility requirements. Also, the resulting large antenna beamwidths at low frequency do not provide the necessary AOA accuracy required to counter the threat. Higher frequencies (i.e., in X/Ka Band) result in a complex antenna with narrow antenna beams that must be scanned rapidly through the coverage volume to detect the threat. In addition, small diameter missiles (i.e., of the order of 3 in to 5 in) experience an RCS reduction at high frequencies. Also, weather effects are prevalent at these higher frequencies. The results of a trade-off study considering these issues and others concluded that the L band (1 GHz to 2 GHz) was the optimum operating frequency for a MAW pulse Doppler radar [9].

Another issue that affects MAW pulse Doppler radar design is that of false alarms generated by aircraft entering the MAW radar's FOV. The false alarms are due to jet engine modulation of the radar return by the aircraft engine that generates Doppler components that correspond to those of missiles. This can generally be controlled by comparing the velocities derived from the Doppler against those derived by differentiations of the range measurements. If they provide the same result, the target is a missile; otherwise it is an aircraft.

7.3.4.2 IR Missile Countermeasures

Current EO/IR countermeasure systems fall into two categories: pyrotechnic IR flares and onboard jammers. The flare, by its very nature, has certain limitations. Specifically, since it is expendable and represents a diminishing resource to the user, it must be dispensed judiciously. Early use of flares against first-generation threats (see Figure 7.9) involved dispensing manually after visual threat sighting by the air crew. However, as IR missile systems have become smaller, combined with smokeless thrust systems, this is no longer the case. Decoy flares are generally ignited by an impulse cartridge and expended from an onboard dispenser. They contain a variety of chemical components (i.e., primarily magnesium powder) that create heat when exposed to the atmosphere, thus attempting to deceive the threat missile system by presenting a more attractive IR signature than the target aircraft.

A modern dispenser such as the ALE-47 provides three modes of operation. Automatic mode dispensing evaluates sensor and avionics inputs, selects an appropriate response, and initiates the countermeasure with no external action. Semiautomatic mode sequencing evaluates sensor and avionics inputs but requires crew activation to dispense flares. Manual-mode aviation launches six preprogrammed, cockpit selectable responses.

Unfortunately, advanced IR missile seekers employ the capability to reject the common magnesium-based flare (see Figure 7.7). Current research in flares are directed at modifying their signature (one approach uses hydrocarbon fuel) to more closely resemble the aircraft they are trying to protect. In addition, signature suppression of the aircraft itself that allows the flares to operate more effectively is possible.

An aircraft has four major types of heat source: exposed engine parts, the exhaust plume, structure warmed by aerodynamic friction, and cockpit glazing that reflects sunlight. In the 2- to 3-μm band exploited by first-generation IR homing missiles, which remain in widespread use, hot metal is the predominant source. Energy emanating from the engines (including the exhaust plume) in the 2- to 2.4-μm band is typically 250 to 600 W/steradian, compared with only about 1 to 2 W/steradian for skin heating and, therefore, accounts for more than 99% of the total signature. Skin heating becomes more important in the 3- to 5-μm band employed by advanced missile seekers but generally represents only about 20% of the total (288 W/steradian for a high-speed aircraft with engines contributing 1000 W/steradian). The exhaust plume is a major source of energy in the 8- to 14-μm band, where long wave thermal imagers operate [4].

The *Passive Infrared Radiation Engine Suppression* (PIRES) system was successful in reducing the IR signature of a C-130 by 90% in the 2- to 3-μm band and 80% in the 3- to 5-μm band. The system consisted of an exhaust transition duct, suppresser mixer section, and a cowling enclosure [11].

In addition to flares, onboard pulsating IR systems are used to confuse and jam the guidance systems of some basic IR seekers. Onboard *IR countermeasures*

(IRCM) generally use arc lamps or graphite elements to emit IR pulses. The pulsed modulations are such that the signal generated interacts with the signal generated by the IR seeker's reticle. The result generates false guidance commands to the missiles' aerodynamic control surfaces. For aircraft with large IR signatures onboard omnidirectional active IRCM systems may not be able to cover their signatures.

An example of an integrated IRCM system designed to cope with advanced threats is depicted in Figure 7.17. This system employs passive MAWs to accurately determine the direction of the threat. It then deploys a directional, multispectral IRCM pulsed signal to jam the IR missile seeker. Successful jamming can be detected using the MAW to determine if the missile is directed away from the protected aircraft. Once this is accomplished, the active IRCM system is available to service the next priority threat. If the IRCM is ineffective, a signal is sent to the dispenser that launches decoy flares [12].

Another more advanced approach is the use of *directed energy countermeasures* (DIRCM) against advanced IR missile threats including imaging seekers. The system will use a passive-type MAW consisting of an IR staring array sensor to detect and locate the threat missile. Once detected, the MAW hands over to a tracking laser system that both tracks the missile and provides an optical path for projecting a high-intensity IR laser beam at the approaching missile. The laser beam is intended to damage the seeker optics and IR detector, which in effect blinds the missile. Earlier versions of this type of system use the laser in a deceptive jamming mode to cause the missile seeker to break lock [12].

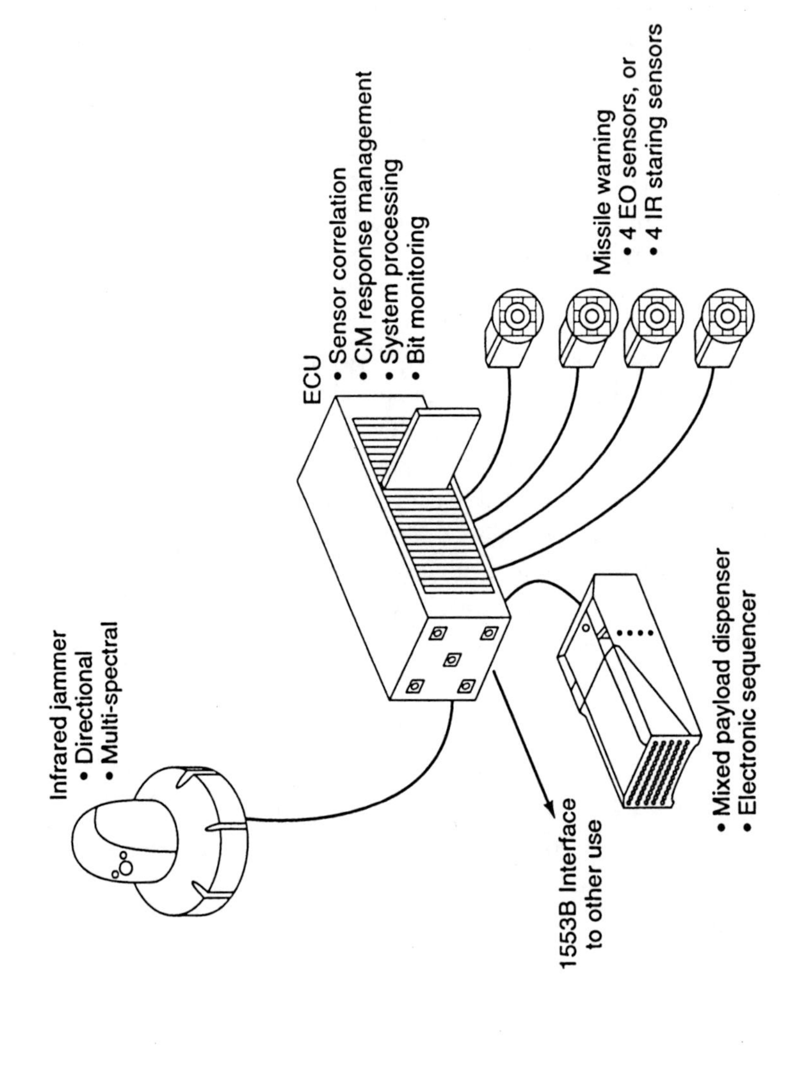

Figure 7.17 Integrated IRCM system. (*Source: Journal of Electronic Defense,* May 1992, p. 55, Figure 1.)

7.4 Problems

1. Chaff falls at a rate of 0.45 m/s at an altitude of 13,000m. Air pressure varies with altitude as shown in Table 7.10. Find the total time required for the chaff to reach sea level. If the average wind velocity is 20 knots, how far will it travel horizontally? If a second group of chaff falls at a rate of 0.35 m/s at 13,000m, how far will it travel horizontally?

Table 7.10
Pressure Variation with Altitude

Altitude (m)	Pressure (Millibars)
0	1016
1000	900
2000	796
4000	618
6000	479
8000	369
10,000	258
13,000	169

2. A chaff package consists of 10,000 pieces at 3 cm, 12,000 pieces at 2.4 cm, and 15,000 pieces with a length of 1.9 cm. Plot the RCS of this chaff package by modifying the MATLAB program dipole.m.

3. Consider an S-band PC radar (λ = 0.1m, θ_{AZ} = 1.5 degrees, ϕ_{eL} = 2.4 degrees, τ_e = 0.5 μs) at a range of 50 nmi. Assume that a chaff corridor is generated with an RCS of 50 m^2 in each resolution cell. What chaff reflectivity (η) is required? How many chaff dipoles are required within each radar resolution cell? How far is each dipole separated in the chaff cloud? What weight of aluminized-glass chaff is required to create a corridor that is 20-nm long, 50-nmi wide, and extends 20,000-ft high?

4. Consider a radar with the same parameters as given in Problem 3 except that it now uses a CSC^2 antenna with ϕ_{eL} = 20 degrees. What weight of aluminized glass chaff is now required to provide the required corridor chaff coverage?

5. The spectrum of chaff is primarily determined by wind shear effects and its rms value is given by $\sigma_{shear} = 0.42 \cdot k \cdot R_{km}\phi_{eL}$, where k = 4 m/s-km. What Doppler spectral width is indicated for the parameters of Problem 3?

6. Assume that an adaptive MTI system is used to allow moving targets to be detected in the chaff cloud. The power transfer function of an MTI filter is given by $|H(f)|^{2n} = [2\sin(2\pi fT/2)]^{2n}$, where n is the number of cancelers and $T = 1/PRF$. The normalized spectrum of the chaff is given by $S_f = \exp(-f^2/2\sigma^2_{\text{shear}})$. The chaff attenuation is found as

$$CA = \frac{1}{2^{2n-1}} \int_0^\infty |H(f)|^{2n} S_f df$$

Integrate the expression for CA and find the amount by which the chaff is reduced using the parameters of Problem 3 for a single-, double-, and triple-MTI canceler and a $PRF = 1000$ Hz. For an optimal speed target (velocity at peak of MTI response), what is the signal-to-chaff power ratio for a 5-m^2 target using a triple-MTI canceler?

7. An MTD processor uses a bank of Doppler filters following the MTI canceler. The MTI is not velocity adapted, so in general, the chaff's spectrum falls within the MTI's passband. The sensitivity of the Doppler filter containing the chaff spectrum (this depends on the chaff's velocity) is reduced until the chaff response falls below the noise level. Since the target has a different velocity than the chaff, it is contained within a Doppler filter not desensitized, thereby allowing it to be detected. For the parameters of Problem 6, how many Doppler filters are required in the Doppler filter bank such that one Doppler filter just straddles the chaff?

8. A PD radar is generally more effective than an MTI when used to detect moving targets in a chaff background. In the high-PRF form of this radar, the PRF must be at least twice the value of the highest frequency component associated with the chaff since this allows a filter to be unambiguously placed to extract the target from the chaff. What minimum PRFs are required at X-band (λ = 3 cm), S-band (λ = 10 cm), and L-band (λ = 30 cm) to achieve this type of operation for wind velocities up to 50 knots with wind shear parameters as given in Problem 5?

9. In high-PRF PD radar, range is highly ambiguous. Since chaff is distributed throughout the full range extent of the radar, its return in a particular range cell r will result from returns at ranges $r + nR_u$, where R_u is the radar's unambiguous range. The target range is also generally ambiguous, but its return results from a single range point. Develop an expression for the equivalent RCS of chaff for a high-PRF PD radar. What are the spectral components of the chaff returns from the multiple range cells (see Figure 7.5) that fold into a single range cell?

10. A drone having a RCS of 0.5 m^2 is to carry a repeater to simulate a total RCS of 5 m^2 at a frequency of 3 GHz. Omnidirectional antennas are to be used for both transmitting and receiving. Find the required electronic gain if the repeater is gated with a duty cycle of 0.35. Assume 45-degree slant polarization for the repeater.

11. A repeater jamming buoy is used to divert an ASCM from a ship. The ship has an RCS of 2,500 m^2, while the ASCM seeker operates at 10 GHz and uses a 30-kW peak power transmitter and a 25-dB gain antenna. The repeater jammer uses omnidirectional antennas and is 45 degree slant polarized. How much peak power must the jammer provide to capture the ASCM seeker (at least 10 times the ship return) at a range of 2 km?

12. Calculate the maximum RCS of a triangular corner reflector where the sides of each corner are 30-cm long at a frequency of 10 GHz. At what angle from the maximum response is the response 6-dB down? What diameter Luneburg lens provides the same maximum response as the triangular corner reflector? What diameter lens is required to provide the 6-dB down response? What is the advantage of the Luneburg lens?

13. Consider an aircraft that tows a 1-m^2 spherical radar reflector behind itself using a 100-m-long line in order to divert a fighter launched missile. Assume that the output of a linear repeater jammer aboard the aircraft is directed against the spherical reflector to increase the magnitude of the deceptive signal to the equivalent of 50 m^2. What electronic gain is required in the repeater assuming that it uses 10-dB gain antennas and is 45-degree slant polarized? What sidelobes are required for the repeater antenna to prevent the repeater from acting as a beacon?

14. Assume that a 10-GHz linear repeater towed behind an aircraft provides 80-dB isolation between transmit and receive antennas. What maximum effective RCS can be generated by the repeater, assuming it uses 10-dB gain 45-degree slant polarized antennas?

15. An expendable noise-jammer decoy is used to protect an aircraft strike group from an AAA Fire Control tracking radar. The aircraft has an RCS of 5 m^2. The characteristics of the radar are given in Table 7.11. What jamming ERP density (W/MHz) is required if the expendables are deployed 600m from the radar to provide a 1000- and 5000-m burn-through range?

Table 7.11
Tracking Radar Parameters

Parameter	Value
Peak power	200 kW
Pulsewidth	0.2 μs
S/N for target detection	13 dB
Antenna gain	36 dB
Antenna sidelobes	−10 dB
Jammer polarization loss	3 dB

16. An aircraft emits 600 W/steradian toward an IR detector operating in the 2- to 2.6-μm region. The distance to the detector, which is aimed at the aircraft, is 10 nmi. The detector effective aperture is 20 cm^2. Average atmospheric attenuation is 0.8 per nmi along the propagation path. Find the power (in dBm) intercepted by the detector.

17. In Problem 16, what detector D^* ($cm - s^{1/2}/W$) provides a 10-dB signal-to-noise ratio for a bandwidth of 100 Hz?

18. A ship has a total surface area of 8000 m^2. Its surface temperature is 30°C. Find the integrated radiation per solid angle in the 8- to 14-μm region, assuming the average emissivity is 0.8.

19. An aircraft radiates 500 W/steradian to the rear in the 3- to 4-μm region. The seeker of a heat-seeking missile is tracking the aircraft at a range of 1 km. Find the intercepted energy in the missile optics that uses a 7.5-cm-diameter lens. The atmospheric transmittance is 0.95 and the IR dome transmittance is 0.90.

20. At a wavelength of 10 μm, 80% of the IR radiation is transmitted through a 1-nmi path of air at sea level with a water vapor content of 9.2 g/m^3. What fraction of this radiation is transmitted through a similar path 30-nmi long?

21. The exponential rate of attenuation of transmitted radiation with distance R (km) is given by

$$I / I_0 = \exp(-\alpha R)$$

where α is the extinction coefficient. Table 7.12 gives extinction coefficients in the 8- to 12-μm band. Compare the ability of an IR seeker in this band to detect signals in the clear as compared to light fog, heavy fog, light rain, and heavy rain over a distance of 4 km.

Table 7.12
Extinction Coefficients for Weather Conditions

Weather Conditions	α (km^{-1})
Clear	0.08
Haze	0.105
Light fog	1.9
Heavy fog	9.2
Light rain	0.36
Heavy rain	1.39

22. A helicopter employs a scanning IR missile warning system. The missile is launched at a range of 5 km from the helicopter and during its powered phase radiates 100 W/steradians in the 3- to 5-μm spectral band. The scanning IR system has a 1.44-mrad FOV, a numerical aperture of 1.78, and a bandwidth of 10 kHz. The transmittance is 0.48 over the 5-km path. What detector D^* is needed to provide a signal-to-noise ratio of 50 for detection at the warning system's maximum range?

23. What is the maximum rate at which the missile warning system of Problem 22 can scan? How long does it take to scan a 15- by 6-degree sector? If the missile has a velocity of 600 m/s, what distance does it move during the time spent scanning the sector?

24. Consider a helicopter and a cargo aircraft. Their engine temperatures are 750°K, while their body temperatures are 300°K. For the helicopter, the projected body areas are 3000 cm^2 for the engine and $6 \cdot 10^5$ cm^2 for the body. For the cargo aircraft, it is 5000 cm^2 for the engine and $2 \cdot 10^6$ cm^2 for the body. The radiance can be approximated as $L_b = \Delta\lambda L(\lambda_{mid})$ while L_{mid} is approximated as $L_{mid} = C_1/(\lambda^5 \exp(C_2/\lambda_{mid} T))$, where L is expressed in W/cm^2-steradians. Assuming that the radiances for the two heat sources add, compute the radiant intensity (W/steradian) for the helicopter and cargo aircraft in the 2- to 2.8-μm band (2.4-μm midband) and the 3- to 5-μm band (4-μm midband).

25. Using the results of Problem 24, show that the radiant intensity in the 2- to 2.8-μm band is less than 1 W/steradian and, hence, it is safe to ignore this component.

26. Assume that a laser operating in the 2- to 2.6-μm band is used to jam an IR missile seeker. The missile seeker is designed using a 20-in^2 diameter detector that provides a received power of −56 dBm from a 600-W/steradian target at 10 nmi. The transmittance is 0.8/nmi. The laser must provide at least a power of −46 dBm to capture the seeker. The laser operating as a radar is used to track the missile seeker with 0.5-m radian accuracy. What power must the laser radiate to accomplish this mission?

27. The far field of an antenna (lens) is given by $R = 2D^2/\lambda$, where D is the diameter of the lens. What is the far-field range of the laser jammer of Problem 26?

28. Assume that the temperature of a decoy flare is 2000°K, while the temperature of an aircraft is 1000°K. Using the approximations for radiance (L) given in Problem 24, compute the radiance for the flare plus aircraft in the 2- to 2.6-μm band and the 3- to 5-μm band. Do the same for just the aircraft. Use the results to explain how a two-color missile seeker discriminates against the flare.

29. A MAW radar for helicopters is to be designed at UHF (425 MHz). The system is to provide 360-degree azimuth coverage by spacing three 120-degree separations. The antennas provide 30-degree elevation coverage. The RCS of the missile is enhanced to 1 m^2 due to resonance effects. What antenna effective aperture is required to implement this design? What PRF provides unambiguous Doppler for a Mach 5 missile? What is the PW for a 25% duty cycle? What Doppler bandwidth accommodates a 16-g missile acceleration?

30. What is the required peak power of the MAW in Problem 29 to provide 10-km detection range? Assume a Swerling case 3 target, a 3-dB noise figure, 10-dB system losses, and 0.7 aperture efficiency.

References

[1] Herskowitz, D., "Elements of EW Expendables," *J. Electronic Defense,* Dec. 1993.

[2] Schleher, D. C., *MTI and Pulsed Doppler Radar,* Norwood, MA: Artech House, 1991.

[3] Papke, N., et al., "Electro-Optical/Infrared (EO/IR) Countermeasures Handbook," Naval Air Systems Command, Arlington, VA, 1996.

[4] Deyerle, C., "Advanced Infrared Missile Counter-Countermeasures," *J. Electronic Defense,* Jan. 1994.

[5] Schwind, G., "Infrared Countermeasures for Helicopter Applications," *J. Electronic Defense,* May 1991.

[6] Hewish, M., A. Robinson, and G. Turbe, "Airborne Missile-Approach Warning," *International Defense Review,* May 1991.

[7] Hershovitz, S., "A Little Warmth, A Little Light," *J. Electronic Defense,* April 1990.

[8] McNeil, T., and J. Bachelor, "The Expanding Battlefield: DC to Light and Beyond," *J. Electronic Defense,* Feb. 1991.

[9] Black, A., "Pulse Doppler for Missile Approach Warning," *J. Electronic Defense,* Aug. 1991.

[10] Russell, W., "AN/ALQ-161 Tail Warning Function," *J. Electronic Defense,* Aug. 1991.

[11] Hewish, M., "Surviving a SAM Attack," *International Defense Review,* June 1996.

[12] Keirstead, B., and T. Herther, "Aircraft Survivability in the 21st Century—ATIRCM," *J. Electronic Defense,* May 1992.

[13] Neri, F., *Introduction to Electronic Defense Systems,* Norwood, MA: Artech House, 1991.

[14] Schleher, D. C., *Introduction to Electronic Warfare,* Norwood, MA: Artech House, 1986.

[15] Barton, D. K., *Modern Radar System Analysis,* Norwood, MA: Artech House, 1988, p. 137.

[16] Nathanson, F. E., *Radar Design Principles,* McGraw-Hill Book Company, 1969, p. 207.

8

Directed Energy Weapons and Stealth Technology

In this chapter, we discuss two relatively new *electronic warfare* (EW) approaches: *directed energy weapons* (DEW) and stealth technology. Both areas have the promise of revolutionizing certain aspects of EW. DEWs represent a new way to exercise offensive EW against a variety of threats, while stealth technology can eliminate the need for active self-defense EA jammers.

Propagation limitations generally focus the practical effectiveness of DEW toward damaging the sensitive microwave or IR detectors used in the sensors that guide missiles to their targets. The ability of DEWs to operate at the speed of light and their unlimited magazine capability make them a very attractive weapon.

Present DEWs use either lasers or *high-powered microwave* (HPM) generators. Lasers generate very narrow beams and one of the considerations involves how these beams are focused on the critical sensor front-ends. HPM systems use wider beams and, hence, are easier to aim. However, they generally have difficulty penetrating into the electronics with in-band energy through the antenna and receiver front-end (the "front door") and must leak through cables and seams (the "back door") in the electronic equipment to be effective. HPMs also are effective in so-called *high-energy radio frequency* (HERF) guns that penetrate into relatively unprotected computers, thereby disrupting their operation.

Stealth technology attempts to control the signature of a military vehicle or weapon so that it is not visible to conventional sensors. The primary signatures are RF (radar and EM energy radiated from the vehicle) and EO/IR (laser and infrared); although acoustic, visible, and magnetic signatures are also important.

Since radar is the primary all-weather sensor for missile systems, a large amount of stealth work is aimed at reducing the RCS. The most effective method for RCS reduction is geometrical shaping to divert the reflected wave away from its source. This works best in the microwave region. However, at lower frequencies (below the UHF region), a resonant effect occurs due to creeping waves around structures, which negates the stealth characteristic. At high frequencies (in the millimeter-wave region), surface roughness becomes a problem and tends to increase RCS.

Successful application of stealth technology obviates the need for an onboard EA system that may in fact make the protected vehicle more vulnerable. The low RCS does, however, make support jamming more effective. A further consequence of stealth is a reduction in the effectiveness of the radar's MTI system. However, as more powerful and sensitive radars are developed, the use of an onboard EA system may again become necessary to allow stealth vehicles to survive.

8.1 Directed Energy Weapons

The traditional role of EW as a defensive action has recently been expanded to include offensive actions that prevent the enemy's use of the *electromagnetic* (EM) spectrum through actions that include damage, disruption, or destruction of his EM capability. DEW represent one method for implementing this new offensive capability.

DEW include *high-energy lasers* (HEL), *charged particle beams* (CPB), *neutral particle beams* (NPB), and HPM. These weapons are generally characterized by speed-of-light attack, which is helpful in defeating targets such as theater ballistic missiles before they can deploy defense-saturating submunitions. They also provide the capability for rapid retargeting of multiple systems (e.g., satellites, intercontinental ballistic missiles, SAMs, AAMs, or aircraft). HEL systems have the potential for revolutionary impact in both strategic defenses and tactical battlefield applications.

HEL systems of interest are based upon six different types of lasers: gas dynamic, pulsed high-energy electrical molecular and atomic, excimer, chemical optically pumped gas, solid state, and free electron. Improvements in DE systems are anticipated due to the development of new approaches for energy transfer, efficient mixing of chemical reactants, scaling of present systems to higher power design of waveforms, and more efficient propagation of beams through the atmosphere. Improvements in HPMs are expected due to the more efficient and compact microwave sources and the design of antennas that can operate above the traditional voltage breakdown limit.

NPBs and CPBs deliver energy at a significant fraction of the speed of light. The energy from NPBs and CPBs is deposited beneath the surface of the target adding another dimension to the kill mechanism/countermeasure considerations. Expected results include warhead detonation and structural breakup at higher energy levels and electronic upset at lower levels.

HPM travel at the speed of light and may damage electronic systems by exposing their components to unwanted and unanticipated large electric fields or current spikes. These fields can cause temporary system malfunction or may result in permanent damage of equipment [1].

To be useful to the military, a HEL system has to negate a target of interest. An enhanced system would determine the effectiveness of the action, move to another target, and repeat the procedure. To do this, the system has to be able to track one or more targets, produce a laser beam, point it at the selected target, and provide enough energy on target to complete the mission. Development of a successful HEL system presupposes knowledge of the beam-target interaction physics and information about viable countermeasures. The HEL system has the capability to serve as a long-range lethal weapon for target destruction or mission abortion. The system can destroy/disable satellites, ballistic missiles, or aircraft. At lower powers, it can also be used for tracking, discrimination, or soft kill. Critical military parameters include a power greater than 20 kW CW and energy greater than 1000 joules pulsed at laser wavelengths between 0.3 μm to 30 μm [1].

Supporting technologies for laser DEWs must provide the capability to track a target or targets, aim the beam at a target for a specified time period, evaluate the damage, and then move rapidly to another target. In this way, multiple targets can be encountered and neutralized.

The laser beam-pointing and control techniques require advanced servo systems, integrated optics programming, adaptive optics, active focusing, alignment, and tracking techniques. The target acquisition tracking, kill assessment, and rapid beam slewing technologies required for different lasers and particle beams have different characteristics because of the different locations, atmospheric conditions, and scenarios and ranges for which the systems are designed. Aim point and jitter control better than 0.25-μrad rms is necessary to control a 0.5-m spot at 250 km. Slew capability rates greater than 0.5 rad/s with accelerations of 0.5 rad/s^2 are necessary to accomplish retargeting to less than 1-degree accuracy in under 2 sec [1].

DEW have great potential in enhancing aircraft survivability in the areas of susceptibility and vulnerability reduction against missile threats. Within susceptibility reduction, DEW systems have potential in preventing detection and targeting and also can aid in hit avoidance by deflecting, blinding, or causing the incoming missile to break lock and finally, where necessary, to destroy the missile itself. An additional approach might be to defeat the fusing system of the incoming missile.

Both laser and HPM weapons could be used to avoid detection and targeting. In the case of HPM, the localization of air defense radar and C3 systems to the order of kilometers would allow neutralization by destroying their electronic components. Antisensor laser weapons could be used to defeat optical detection and targeting systems. The laser weapons would use several lasers of different wavelengths to interrogate the environment, detect hostile EO sensors, and invoke a weapon beam tuned to the acceptance band of the sensor to destroy or damage it. Lower power lasers could also temporarily jam the sensor. In both laser and HPM cases, precise knowledge of the threat location is required to bound the countermeasure within practical limits [2].

When an aircraft has been detected, targeted, locked-on, and the missile fired, the emphasis has to shift to defeating the in-flight missile. One possibility is the use of DEW against the ground (or hostile aircraft) tracking and command guidance system. One of the more critical threats that could be addressed by DEW is the shoulder-launched "fire-and-forget" type of IR guided missiles. In most cases, such missiles require lock-on prior to launch; they do not have autonomous reacquisition capability. Given an adequate MAW system (see Section 7.3.4), it is quite conceivable that the missile can be defeated in flight. One approach is to use an RF weapon (directed from the aircraft under attack or counter launched) to defeat the guidance electronics. For optical or IR seekers that are out-of-band with respect to the RF of the HPM system, a method must be found to couple the RF energy into the attacking missile. In many cases, these "back-door" mechanisms exist; however, they are notoriously unpredictable and statistically diverse, differing by orders of magnitude from missile to missile, even those of the same class, depending upon the missile's maintenance situation.

Another approach is to use a laser to attack the threat in its seeker band. For highly maneuverable aircraft, it may suffice to simply blind the missile and assume it can be avoided. For slower high-value aircraft (e.g., C-17, AWACS, JSTARS) blinding may not be sufficient; the threat may still coast in close enough to activate its fuse and cause damage. In this case, a smart jammer is needed to identify the incoming missile; then a deceptive jamming signal is sent to break lock, thereby deflecting the missile from its course. A multiwavelength laser is currently under development to accomplish this countermeasure. Its advantage over current systems (i.e., ATIRCM) is that the laser waveform is not tailored to a specific threat and, hence, covers a wide class of missile seekers including antiship missile seekers [2].

The *Advanced Threat Infrared Countermeasures* (ATIRCM) system counters advanced EO/IR missile threats using a MAW system to cue dispensers and directable jammers. Once detected, the MAWs hand over to a pointer/tracker system, which both tracks the missile and provides an optical path for projecting a high-intensity IR laser beam at the incoming missile. The laser, which operates out-of-band or near-band, is intended to cause a plasma spark (0.1 ns to 50 ns) within the seeker head near the detectors. The plasma enhances jamming or blinding of the seeker head by pitting or scoring optics, creating debris, scattering within the optical head, or upsetting the electronics within the seeker. Present experiments generate a laser signal approximately 40 dB above the incident radiation. This has demonstrated break lock using a 1.06-μm laser [3].

In addition to plasma effects, the use of lasers to damage the IR detectors through overheating is also a viable attack mechanism. To accomplish thermal degradation or damage to detectors requires a long (> 200 ns) pulse of in-band radiation [3].

The ultimate susceptibility denial technique is provided by a DEW system that can destroy any approaching missile, thereby creating a defensive shield around the aircraft. Unless it can be depended upon to induce premature fusing (a problematic capability), an RF weapon is not a likely candidate for a direct destruction system. However, laser weapon technologies (both chemical and solid state) are reaching the point, for large aircraft that can afford the volume and weight penalty, a self-defense laser weapon is entering the realm of practicality using weapon lasers that would operate in the multikilowatt range [2].

In the design of DEW, it is prudent to consider the effect of protective measures against these types of weapons. Some of the protective techniques suggested for aircraft protection consist of in-band laser goggles to protect crew eyes and aircraft/sensors, hardening of aircraft structures against thermal energy, shielding of electronic equipment against RF effects, and polishing of surfaces to reduce laser energy coupling. Many of these same techniques can be used to protect missile seekers against DEWs [2].

8.1.1 High-Power Microwave Weapons

HPM weapons, also known as RF weapons and *ultrawideband* (UWB) weapons, usually comprise an electrical or explosive prime power source, an RF generator, and an antenna beam director. HPM weapons are characterized by frequencies in the 10-MHz to 100-GHz region, narrow- and wideband radiations that yield multiple simultaneous frequencies, and power levels between 100 MW and 100 GW. HPM output power is 20- to 30-dB higher than that of current EW jammers. Applications include SEAD, disrupting enemy communications and platform self-protection against AAMs, SAMs, and antiship missiles.

HPMs travel at the speed of light, are generally insensitive to weather, require only coarse pointing, and have the potential to negate target electronics in fractions of a second. Electronics can be burned out even when the target system is turned off. In general, the susceptibility of electronics to HPM increases as the scale size of the electronics decreases, making modern electronic systems potentially the most vulnerable. Hardening against HPM threats is possible but difficult to implement and maintain against a wide range of potential threats. It is more cost effective to harden during initial system development and often impractical to retroharden. HPM weapons are generally nonlethal to humans (compared to projectiles), affecting only the target's electronics. They have the potential to present a greater threat than conventional weapons due to the following characteristics:

- A higher probability of hit compared with projectiles (because the spreading RF beam can irradiate the entire target, it requires far less pointing and tracking accuracy);
- An instantaneous time of flight (speed of light) for fast engagements in nearly all weather conditions;
- A large image zone compared with the typical store of conventional projectiles and missiles.

HPM weapons are intended to upset any system that depends on electronic signals for its operation. The fluence or measure of time-integrated power density, expressed in energy per unit area (joules/cm^2), specifies the threat level of the HPM radiation to electronic equipment. At relatively high-energy fluences, HPMs burn out sensitive semiconductor components in electronic systems. Components most at risk are semiconductors with small junction areas such as microwave or IR detectors used in the front-end of radar, communication systems, RF or IR missile seekers, and low-power integrated circuits logic chips and control electronic systems such as used in ground vehicles for ignition control. Very high energy fluences might even detonate missile warheads, bombs, or artillery shells [5].

At lower energy fluences, microwaves can bombard electronic circuitry and trigger spurious signals that might jam or temporarily disable an electronic device. Semiconductor components can fail when absorbed microwave energy leads to excessive heating. The small thermal time constant of low-power semiconductor devices generally require only the order of 1 μs to heat up the junction. HPM weapons that provide the highest burn-out threat for a given source energy are pulse devices with burst durations of a microsecond or less [5].

HPM energy can penetrate electronic systems through the "front door" or the "back door." Front-door coupling refers to energy that enters through the antenna of systems containing a transmitter or receiver. Back-door coupling denotes energy that leaks into systems through cables, interconnecting wires, seams in their enclosures, or even nonmetallic enclosures (i.e., fiberglass, carbon epoxy, plastic, or kevlar) themselves.

When HPM is directed against a computer system, it is called a HERF gun. Many computers are designed without any protection against HPMs and hence are particularly vulnerable. Low fluences can disrupt computer and memory circuitry, causing erratic operation and thereby disabling the computer. The HERF gun can be aimed at personal computers, main frames, computer networks, or even at buildings containing computers. It is considered an important element of information warfare since it has the potential to disable a complete computer network while bypassing the firewalls constructed to provide computer security against the usual hacker-type attack [8].

For electronic systems that employ antennas, front-door coupling is greatest at the design frequency of the antenna. A system's vulnerability to such coupling can often be estimated from its antenna and receiver characteristic. Special attention must be given to radar systems that generally employ *transmit-receive* (TR) and limiting devices to protect their receiver from transmitter radiations. Too high a microwave pulse actuates this device and quenches the HPM signal, while a signal lower than this can potentially damage the receiver front-end. Back-door coupling is a much more complex phenomenon.

The enclosures of most electronic systems serve to support them structurally as well as to shield them to some degree against *electromagnetic* (EM) interference. Microwave signals penetrating the enclosure and also coupled-in through the cabling and power lines tend to induce signals on lengths of wire housed within the enclosure that connect modules and components. The extent to which microwaves couple to this wiring depends on the amount of energy that enters the compartment and on the resonance characteristics [5].

Measurements of the coupling to wires in enclosures with small dimensions show strong resonances at different frequencies. Coupling is strongest at the resonant frequency of enclosure openings that correspond to the size of the opening. Coupling falls steadily for frequencies below aperture resonance with occasional small peaks at the resonant frequencies of internal wiring. Above aperture resonance, the coupling displays a more gradual decrease with narrow resonances due to the many EM modes of the enclosure cavity.

If the incident HPM beam is intense enough, air in the enclosure apertures and seams may ionize and become highly conductive, shielding the openings from further penetration by the microwave energy. The vulnerability of any particular system is determined not only by shielding and coupling to interval wires and cables, but also by the response of various subsystems to voltage and current pulses induced on their various parts. These responses can often be tested by direct injection of ultrashort pulses (with gigahertz bandwidth) into subsystems to simulate pulses from microwave radiation [5].

8.1.1.1 Propagation Limitations

The attack of electronic systems with HPM ultimately depends upon the ability of the microwave beam to penetrate through the atmosphere. Propagation is limited by dielectric breakdown, diffraction, and attenuation. The extent of these effects depends upon the intensity, frequency, and PW of the microwave beam and on various atmospheric conditions.

The limited dielectric strength of air sets an upper boundary on the intensity and fluence of a microwave beam. The breakdown varies with frequency, pulse length, air pressure, water vapor in the air, and the density of seed electrons that are needed to start the avalanche breakdown process. If seed electrons are available, breakdown of air at atmospheric pressure occurs at densities of about 10^5 to 10^6 W/cm^2 for microwave pulses that last longer than several nanoseconds.

Even a perfectly collimated microwave beam spreads out as a spherical wave in the far field of an antenna (i.e., at a distance greater than $R = D^2/2\lambda$, where D is the antenna's aperture). The effect of this spreading is minimized using high frequencies and large-antenna apertures. High frequencies require tight antenna tolerances (better than $\lambda/16$) and do not couple to some targets as effectively as lower frequencies, while practical antenna diameter are limited to about 10m to 20m for stationary devices and 3m to 4m for mobile weapons [5].

Microwaves in the atmosphere are absorbed by water vapor, oxygen, and precipitation. Resonance absorption peaks occur for water vapor at 22 GHz and 185 GHz and for oxygen at 60 GHz and 118 GHz. Across tactically useful weapon ranges of 1 km to 100 km and at low altitudes, attenuation within several gigahertz of these resonance frequencies may be unacceptably high. Attenuation due to precipitation increases with frequency. Over 10 km, a 3-GHz beam is attenuated by about 0.01 dB in light rainfall (1 mm/hr) while a 30-GHz beam suffers a 15-dB loss. Choice of a design frequency for an HPM weapon varies with the need for long-range or all-weather operation.

► *Example 8.1*

Using MATLAB, plot the density (W/m^2) of a microwave beam for various frequencies between 1 GHz to 30 GHz at ranges of 10 km and 100 km. Assume a 3-m antenna is used with a peak-power density of 10^5 W/cm^2 in the near-field. Use the attenuation factors in the RGJMAT program (modified for one-way transmission) for atmospheric attenuation and precipitation of 1 and 4 mm/hr. What is the best frequency range in clear weather and in rain? What average generator power is required assuming a 30% overall efficiency?

```
% Directed Energy Power Density Levels
% ------------------------------------
% direng.m

clear;clc;clf;

% Enter System Parameters

D=3;              % Antenna Diameter - m
Pd=1e5;           % Near Field Power Density - w/cm^2
R1=10;            % Range 1 - km
r=[1 4];          % Rain Rate - mm/hr

% Enter Frequency Range

f=1:30;           % Frequency - GHz

% Compute Power in Near Field

Pnf=Pd*pi*(3*100)^2/4;

% Compute Antenna Gain

wl=3e8./(f*1e9); % Wavelength - m
G=(pi*D./wl).^2;

% Compute Power Density in Far Field

P=Pnf*G/(4*pi*(1000*R1)^2);

% Include Effects of Atmosphere

% Frequency Points - GHz

fx=[.4 1.3 3 5.5 10 15 22 35];

% Attenuation ( Two-Way ) - dB/km

att=[.01 .012 .015 .017 .024 .055 .30 .14];

fac=R1*interp1(fx,att,f,'spline');

Px=P.*10.^(-5*fac);
```

```
% Plot Power Density

semilogy(f,P,f,Px);grid;
xlabel('Frequency - GHz');
ylabel('Power Density -  w /  m  ^2');
title(['Maximum Power Density  ( R =',num2str(R1),' km )']);
axis([ 0 30 1e-60 1e10]);
hold on;

% Include Effects of Precipitation

att1=[0 .0003 .0013 .008 .037 .083 .23 .57];

fac1=R1*interp1(fx,att1,f,'spline');

Px1=Px.*10.^(-5*fac1*r(1));
Px2=Px.*10.^(-5*fac1*r(2));
semilogy(f,Px1,f,Px2);
```

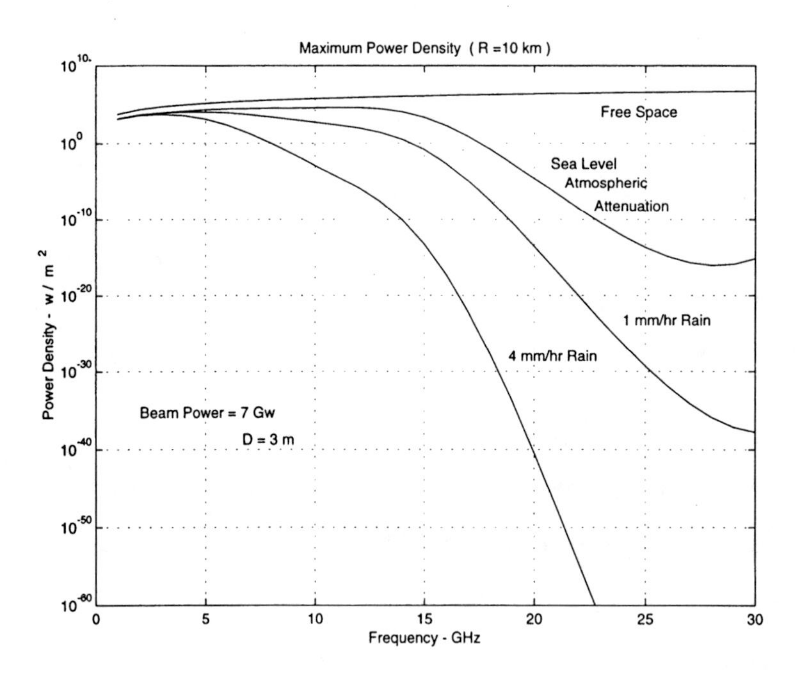

▶ *Example 8.2*

A shipborne DEW uses a high-powered microwave beam to burn out the crystal mixer of the seeker of an attacking ASCM. The ASCM operates at a frequency of 10 GHz and has a 23-dB gain antenna.

The seeker has a 2-dB front end loss between its antenna and crystal mixer. The crystal mixer (IN23C) has a burn out rating of 12-W peak for a 0.6-μs pulse width with a 0.002 duty factor. The DEW system uses a 3-m parabolic antenna with 70% efficiency. Slant polarization (45 degrees) is used to accommodate both horizontally and vertically polarized ASCMs. Using MATLAB, plot the DEW peak transmitter power for ranges from 1m to 10,000m under clear atmospheric conditions and with 1- and 4-mm/hr precipitation. What pointing and frequency accuracy is needed for the DEW system?

```
% Directed Energy Power for Seeker Burn-Out
% ------------------------------------------
% skburn.m

clear;clf;clc

% Input Range

R=500:10000;     % Range - m

% Input Parameters

D=3;             % Dish Diameter - m
eff=.7;          % Antenna Efficiency
G=23;            % Seeker Antenna Gain - dB
f=10;            % Frequency - GHz
rx=[0 1 4];      % Rain Rate - mm/hr
atm=.024;        % Atm Attenuation - db/km
atr=.037;        % Rain Attenuation - db/km
Pr=12;           % Mixer Burn-Out - w
Lx=2;            % Seeker Front End Loss - dB
Ly=3;            % DEW Polarization Loss - dB
Lx=10^(Lx/10);Ly=10^(Ly/10);

G=10^(G/10);
wl=3e8/(f*1e9);   % Wavelength - m

% Calculate Power Level

for i=1:3;
r=rx(i);
Gt=eff*(pi*D/wl).^2;
Lt=10^(-5*(atm+(r*atr)));
Pt=Pr*(4*pi*R).^2*Lt*Lx*Ly/(Gt*G*wl^2);
Ptx(:,i)=Pt';
end;

% Plot Power vs Range
```

```
plot(R,Ptx);grid;
xlabel('Range - m');
ylabel('Transmitter Power - w');
title('DEW Power for Seeker Burnout');
text(7500,3.25e7,'r=4 mm/hr');
text(7500,1.25e7,'r=1 mm/hr');
text(8000,.25e7,'clear');
```

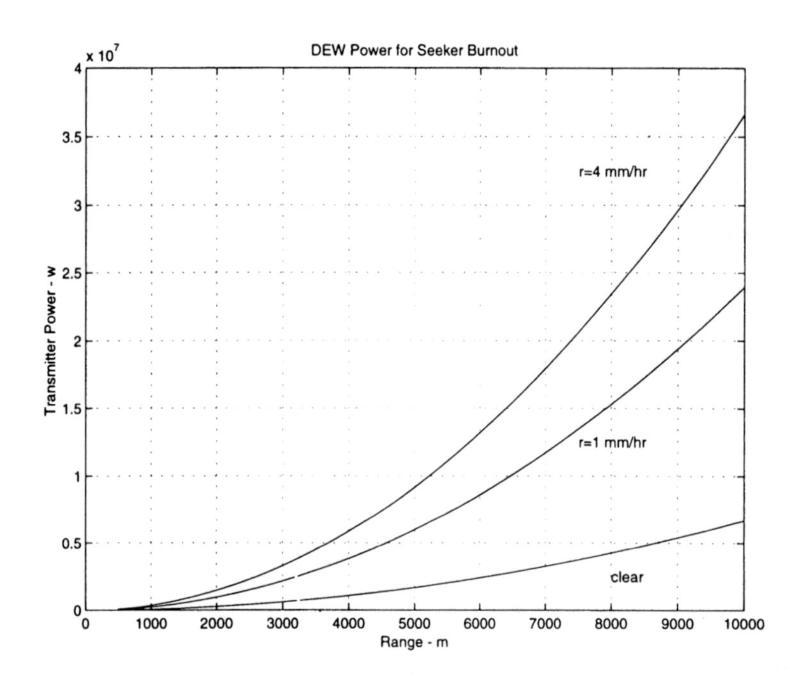

8.1.1.2 Beam Generation

Many varieties of microwave sources exist. HPM sources include the traditional magnetron and klystron as well as newer devices such as the virtual-cathode oscillator (viractor), gyrotron, free-electron laser, and beam plasma generator. All microwave sources work on the principle of converting the kinetic energy of an electron beam into the electromagnetic energy of a microwave beam. The basic components of HPM generators are a power supply, an electron beam generator, a resonant structure that converts the electron beam energy to microwave energy (e.g., a cavity), and a means for extracting and radiating the microwave beam.

The potential peak powers that can be generated by microwave tubes are depicted in Figure 8.1. In general, peak power decreases with increasing frequency. Klystrons have produced over 1 MW CW and 100 MW pulsed. Peak powers of 5 GW have been reported for magnetrons [5]. Viractors that generate microwave energy at centimeter wavelengths are leading candidates for HPM

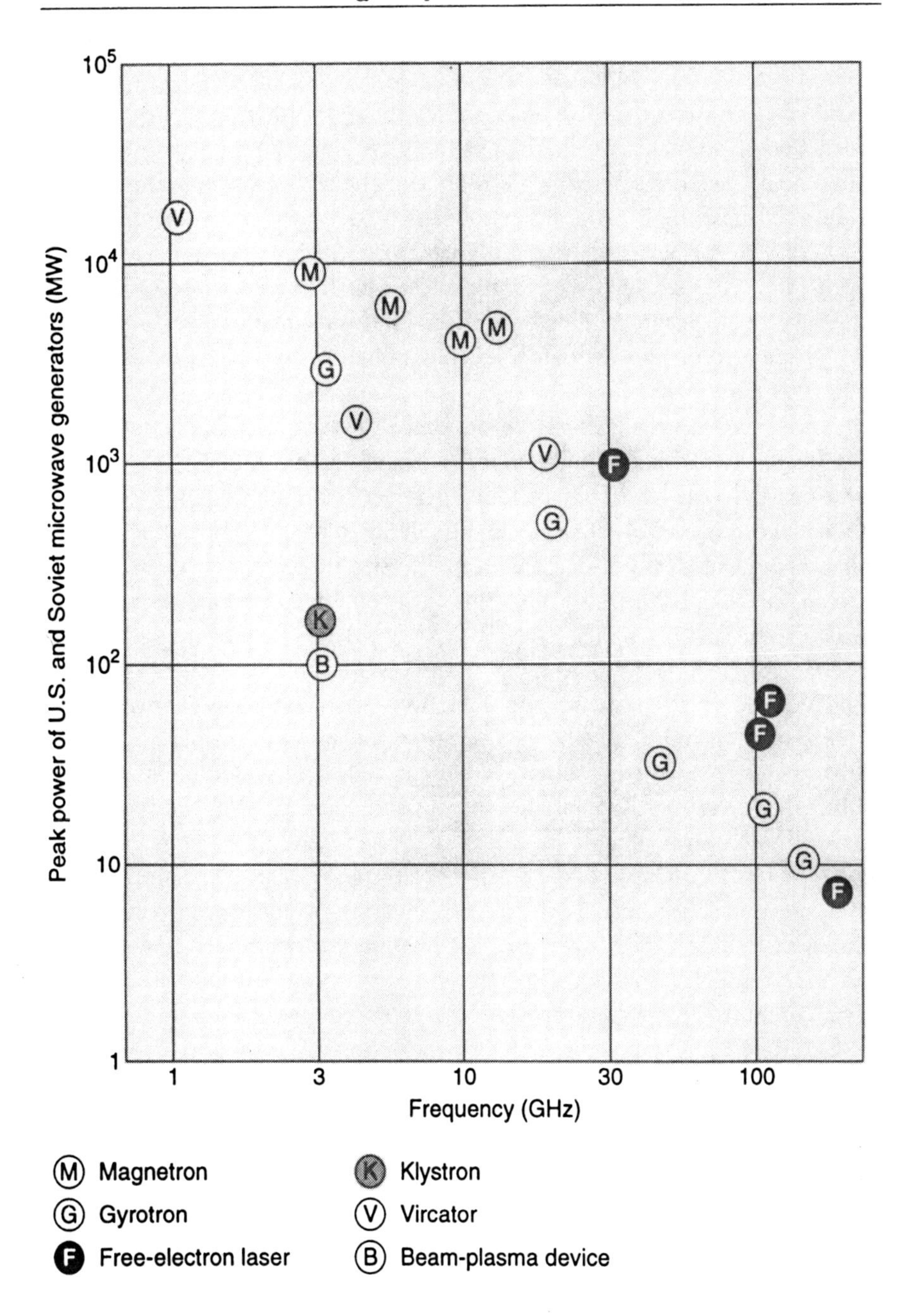

Figure 8.1 High peak-power generation of microwaves. (*Source: IEEE Spectrum,* March 1988, p. 52.)

weapons. They are simple and do not require bulky magnets. Their power output is not limited by restrictions on electron beam intensity. They are both broadband and tunable, while their natural frequency domain lies in a range where many electronic systems are most vulnerable. However, viractors have lower efficiencies than many other sources.

Gyrotrons operate in the millimeter-wave region, have narrow bandwidths, and are capable of CW operation at very high powers. These characteristics make them prime candidates for creating fusion plasmas. Gyrotrons can also be operated in a pulsed mode. Existing pulsed devices have produced peak powers of more than 7 GW.

Free-electron lasers (FEL) can produce microwave energy in the millimeter-wave region. FELs have produced peak powers of more than 1 GW. Both gyrotrons and FELs are driven by high-energy electron beams. The accelerators that produce such beams are generally too bulky to be used for mobile applications. Newer accelerators have reduced the size of these devices, making it feasible for high-power gyrotrons and FELs to be based on mobile platforms [5].

Because the susceptibilities of various electronic systems peak in different frequency bands, it would be important for a tactical microwave weapon to be either broadband or tunable over a decade or more. Beam-plasma devices, which generate broadband microwave energy (100 MW with 100-ns pulse width), have the most potential as broadband sources; while the viractor is potentially the most tunable [5].

Like radar, microwave weapons could employ either mechanical or electronic steering. Electronically steered phased arrays would be more capable of addressing multiple targets. They also present the potential for coherently combining multiple sources, allowing a number of modestly powered sources to be combined in space to provide a large microwave energy. Note that in HPM systems, it is the system's ERP that is important; for phased arrays using coherent combining, the ERP is given by $n^2 P_E G_E$, where n is the number of elements while $P_E G_E$ is the ERP per element.

The main volume and weight associated with a HPM weapon is in the power supplies and other conditioning equipment. Generally, pulse power sources are preferred over CW sources. First, short bursts are more likely than CW to couple into electronic hardware. Second, many sources are operated in a pulse mode (e.g., viractor) or generate greater pulse density in a pulsed mode.

On the other hand, since practical antenna sizes are limited and because the dielectric strength of air limits the power density of the antenna, there is a limit to how much microwave energy a weapon can emit in a single pulse with a width chosen to maximize its effectiveness against electronic hardware. For a 1-μs pulse, this limit is about 10^5J per pulse for a mobile 3-m antenna and about 10^6J per pulse for a mobile 10-m antenna. Allowing for a 5% to 20% overall efficiency of the power supply and microwave tube, the primary energy required per pulse would be 0.5 MJ to 2 MJ for a portable device and 5 MJ to 20 MJ for a stationary device. These energies are consistent with pulsed-power supplies used in particle accelerators and fusion devices. These supplies generally consist of capacitor-inductor networks with carefully timed switches designed to allow low-power pulses into the device to be compressed in several stages into a brief high-power pulse. A pulse power supply of 0.5 MJ to 2 MJ would occupy about a volume of 3 to 10 m^3 while that for a 5- to 20-MJ weapon would be 10 times larger [5].

A more compact supply can use an explosive magnetic-flux compressor. In this device, energy from a capacitor charges an inductor, which is compressed by explosive detonation. The capacitors are discharged through the inductor, resulting in a current flow of hundreds of thousands of amperes and an intense magnetic field. The result is the emission of a short duration intense electromagnetic pulse of the order of 500-ns duration. In this design, about 20% of the chemical energy is turned into the electrical energy of a pulse. Peak power as great as 18 MJ have been produced in a volume of a few cubic meters. Because the detonation destroys much of the device, its use is limited to applications where only a single pulse of HPM energy is required [5,6].

Fluence levels, power densities, and field strengths available from a HPM source of 10-GW power, 100-ns pulse width and radiated through a 100-m^2, 50% efficient antenna at a frequency of 1 GHz are listed in Table 8.1. These values represent a reasonable level for a HPM weapon using available technology. Projected developments might increase power densities by a factor of 100 and fluences by 10^4 [5].

8.1.1.3 HPM Effect on Electronic Equipment

An electronic component sensitive to microwave burnout is the point contact microwave detector diode that has a burnout level as low as 1 μJ. An upper

Table 8.1
Nominal Fluence, Power Density and Electric Field Strength 10-GW/100-ns/1-GHz Source With a 100-m^2 Dish Reflector Antenna With 50% Aperture Efficiency

Distance	Fluence	Power Density	Electric Field Strength
100m	28 J/m^2	560 MW/m^2	460 kV/m
1 km	0.28 J/m^2	5.6 MW/m^2	46 kV/m
5 km	11 mJ/m^2	220 kW/m^2	9 kV/m
10 km	2.8 mJ/m^2	56 kW/m^2	4.6 kV/m
32 km	270 J/m^2	5 kW/m^2	1.4 kV/m

bound may be put on the effectiveness of HPM in disabling electronic equipment through front-door in-band energy by considering the microwave energy collected by a 3-m antenna that is delivered to a single unprotected microwave detector diode. The single pulse fluence that would burn out such a device is 10^{-8} mJ/cm^2. A bound may also be placed on the vulnerability of electronic systems to back-door coupling by considering the microwave fluence needed to cause bit errors in unshielded computers. This fluence is of the order of 10^{-4} to 10^{-5} mJ/cm^2 at 1 GHz and is greater at higher frequencies [7].

For electronics under microwave illumination, rectification is usually the principal mode by which microwave energy is coupled into the system's electronics. Generally, this occurs by a signal entering a victim amplifier or digital circuit through interconnecting signal or power cables and sometimes is enhanced by parasitic resonances. Any nonlinear circuit devices produce a wideband video pulse that propagates through the electronic system, upsetting the normal data transmission and storage functions. In some cases, if the video pulse is great enough, damage may result to electronic components. Upset of electronic equipment can also occur due to intermodulation components that result in the nonlinear circuits of the electronic system. A major failure mechanism for low-power semiconductor devices is caused by junction overheating due to intense short pulse heating of small junctions within the semiconductor.

Energy levels for burnout of electronic equipment devices is listed in Table 8.2. Device burnout is generally proportional to junction size resulting in the smaller scale components (e.g., GaAs MESFETs and MMIC) being able to tolerate the least energy. Power levels for device burnout are typically near 100W for 1-μs pulses at frequencies of a few gigahertz.

Table 8.3 indicates the power and energy levels for upset on electronic devices. Irradiation of unshielded computers and microprocessors indicates

Table 8.2
Electronic Device Burnout Thresholds

Component Class	**Energy (J)**
GaAs MESFET	10^{-7}–10^{-6}
MMIC	7×10^{-7}–5×10^{-6}
Microwave diodes	2×10^{-6}–5×10^{-4}
VLSI	$2 \times 10^{-}6$–2×10^{-5}
Bipolar transistors	10^{-5}–10^{-4}
CMOS RAM	7×10^{-5}–10^{-4}
MSI	10^{-4}–6×10^{-4}
SSI	6×10^{-4}–10^{-3}
Operational amplifiers	2×10^{-3}–6×10^{-3}

Table 8.3
Upset Levels for Electronic Devices

Type	Power (W)	Energy (J) @ 1 μs
Operational amplifier	0.0009	$9 \cdot 10^{-10}$
TTL device	0.008	$8 \cdot 10^{-9}$
CMOS device	0.001	10^{-9}
Voltage regulators	0.09	$9 \cdot 10^{-8}$
Comparator (output switches)	0.008	$8 \cdot 10^{-9}$
VHSIC (pulsed exposure)	0.1	10^{-7}

higher susceptibility in frequency ranges corresponding to the clock frequency, thereby indicating that older computers would be susceptible to lower frequencies. Unshielded computer susceptibility reflecting bit errors has been measured at fluences as low as 10^{-7} to 10^{-8} mJ/cm^2 at 1 GHz. The fluence level required for bit errors is higher at higher frequencies [7].

System designers can take a number of steps to reduce the risk of HPM disruption or damage. System subsystems and cables can be shielded. Filters can reduce out-of-band coupling. Subsystems can be connected by fiber optic cables. Overvoltage or overcurrent protection devices can arrest large pulses. Sensitive components, such as MOS logic chips, can be replaced with hardened functional equivalents. The duty cycle of sensitive transmitters and receivers can be reduced. Redundant circuits can be incorporated. Figure 8.2 illustrates a computer room hardened against HPM attack [8].

8.1.2 High-Energy Lasers

HELs represent another class of power sources used in DEWs. These weapons, due to their highly directional beams, have the potential to operate as a long-range lethal weapon for target destruction or mission abortion. The system can destroy/disable satellites, ballistic missiles, or aircraft. At lower powers, laser weapons have the capability of actively degrading, disrupting, or destroying *electro-optical* (EO)/IR sensor systems whether permanently or only temporarily.

HELs generally operate in the IR and visible frequency bands (0.3 μm to 30 μm) and generate significant energy levels. Of particular interest are the chemical hydrogen fluoride HF/DF and *chemical oxygen iodine lasers* (COIL) that are capable of providing multimegawatts of power, although current capability is of the order of 100 kW average power. The HF/DF laser is not suitable for endo atmospheric use because the laser's radiation is strongly absorbed by the atmosphere.

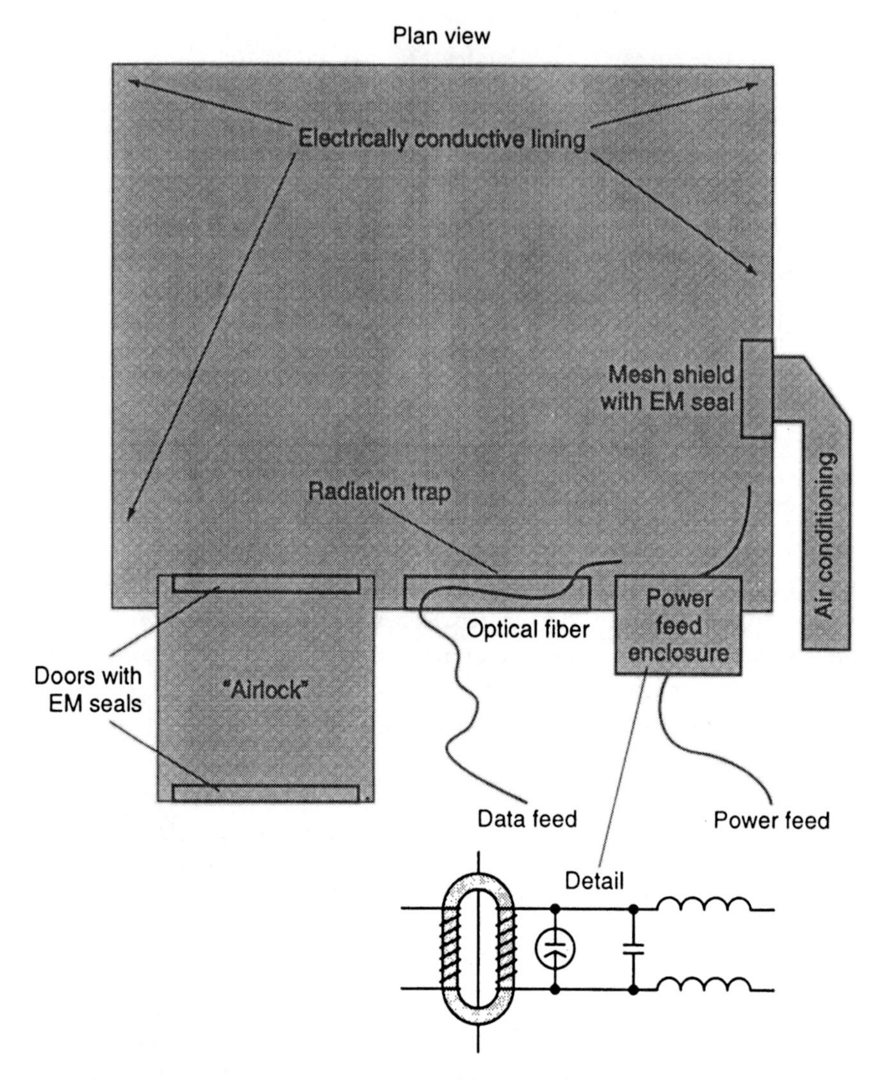

Figure 8.2 Computer room hardened against EM attack. (*Source:* [8].)

Solid-state lasers in which the active lasing elements are embedded in a host matrix and stimulated by external sources such as flash lamps, or more recently diode lasers, have not reached the average power levels of the chemical devices, but in many ways are most suited to laser countermeasure systems, intended to defeat the sensitive sensors and guidance systems of potential threats. To destroy targets, power levels greater than 20 kW CW and energy levels greater than 1 kJ are generally required [9].

The ability of lasers to operate in a military environment is well known. Lasers have become commonplace in range finders, target designators, laser radar, laser gyros, and numerous other applications. These low-power systems, although not a direct threat, assist in the acquisition, identification, and illumination of targets for various weapon systems. Advanced HEL systems have shown their ability to provide a hard kill defense capability, without the need for follow-up munitions. Between the low-power laser support system and the high-power hard kill laser lies the moderate-power tactical soft kill useful against EO/IR sensors. An example of a system of this type consists of a transportable, closed-cycle, electrically discharged, pulsed CO_2 laser operating at a wavelength of 10.6 μm that can deliver up to 1 kJ per pulse at a 10-Hz repetition rate through a 50-cm telescope [10].

The damage caused by a pulsed laser includes [9]:

- Indirect mechanical damage resulting from the intense heat of the laser beam and pressure from expanding laser plasma that can deform or perforate targets;
- Direct thermal damage caused by the laser beam that can melt, vaporize, or pyrolyze a target;
- Ionization damage caused by X-rays radiated by the hot laser plasma at the target surface, which can disrupt electronic components;
- Thermomechanical damage that is a combination of thermal and mechanical damage occurring at moderate intensity levels.

How effective a laser beam may be in causing mechanical damage can be described by an impulse-coupling factor; the total impulse produced divided by the total beam energy intercepted. This factor is a function of beam irradiance, pulse duration and wavelength, target surface material and finish, and ambient air pressure. A pulsed CO_2 laser operating at a wavelength of 10.6 μm and transmitting an intensity at the target of 10^7 W/cm^2 would deliver a three times greater impulse to a painted aluminum target in a vacuum than an unpainted one. The paint would increase the target's absorption of the laser beam energy. At high intensities, the impulse-coupling factor for metal targets in a vacuum ranges from about 1 to 8 dyne second impulse per joule [9].

To punch a hole in a 3-mm-thick aluminum plate in a vacuum, the laser would have to deliver a fluence of 5000 J/cm^2. However, components of weapon systems such as semiconductor materials, optical sensors, solar cells, and radomes may be damaged mechanically by laser-beam fluences as low as 10 J/cm^2 [9].

Direct thermal damage of metals by laser beams may be accomplished by heating them to their melting point. However, ablative material that may be used to protect targets from laser radiation cannot be melted, but must be vaporized or pyrolyzed.

For intensities too low to generate an absorbing plasma at the target surface, the burn-through time (the time required to heat the back surface of the target to a liquid point) is a function of the thickness of the target, beam intensity at the surface, and effective surface absorptivity.

At sufficiently high intensities, laser irradiation generates an absorbing plasma layer at the target surface. This layer can absorb almost all the incident laser radiation and reradiate up to half of it back onto the target as UV black body radiation. The UV radiation is highly absorbed by the target surface [9].

Hence, at high intensity, the fraction of incident beam energy absorbed by the target cannot be described by the absorptivity of the target surface. A thermal coupling coefficient is used to summarize the combined effect of surface vaporization, ionization, plasma absorption, heating, expansion, and reradiation. The thermal coupling of laser radiation at 10.6 μm to an aluminum target increases from 0.025 at low intensity to 0.14 in a vacuum to 0.37 in air. At very high intensities, laser-supported detonation waves may form in the plasma and propagate away from the target surface, thereby shielding the target from further energy transfer [10].

In the lower atmosphere, laser-supported detonation waves can cause plasma decoupling at moderate intensities. Enhanced thermal coupling has also been observed, but the pulse shape must be carefully optimized as a function of target altitude and orientation if plasma decoupling is to be avoided [9].

Thresholds for thermal damage in IR detectors depend upon damage mechanisms, irradiation time, beam diameter, laser wavelength, optical and thermal properties of the materials used in detector construction, and the quality of the thermal coupling to the heat sink. For short irradiation times ($\tau < 10$ μs), the laser damage threshold (in W/cm^2) varies inversely with the pulse width (τ^{-1}). For intermediate radiation time (10 μs $< \tau <$ 10 ms), the irradiance varies inversely with the square root of the pulse width ($\tau^{-1/2}$). For long irradiation times, the requisite irradiance approaches a constant. For short irradiation times, the energy density required to raise the surface of the detector to the melting point is inversely proportional to the absorption coefficient and directly proportional to the specific heat and the increase in surface temperature necessary to melt the detector material [28].

For the IR detector materials used in IR missile seekers, the damage irradiance for a pulse width of 100 ns is 10^6 to 10^7 W/cm^2. This corresponds to a fluence level of 0.1 to 1 J/cm^2. In general, thin-film detectors have higher absorption coefficients and damage fluence levels are lower (0.1 J/cm^2) [28].

For each proposed mission, there are a variety of conceivable laser weapons differing in choice of platform, lasing material, pumping methods, and energy source. Many systems are implausible, mainly due to the propagation phenomenon associated with lasers.

Airborne and ground-based lasers must operate at wavelengths where atmospheric absorption and scattering are not severe. Deuterium-fluoride and carbon-dioxide lasers have been proposed for endo atmospheric deployment, and iodine lasers are attracting increasing interest. Such lasers combine high specific energy as well as low atmospheric absorption that make them suitable for both airborne and terrestrial applications. Applicable missions include air defense, bomber and tactical fighter defense, and antisatellite operations.

▶ *Example 8.3*

An airborne HEL operating in the 3.5-μm band is used to damage the detector of an incoming IR missile. The damage threshold for the detector is 10^7 W/cm^2, the FOV is 10^{-8} steradians, and the optical transmission factor is 0.7. Using MATLAB, plot the laser power to damage the detector for ranges of 500m to 5 km and altitudes of 200m, 2 km, 5 km, and 10 km (see Table 7.4). What pointing accuracy is required for the HEL system to be effective?

```
% High Energy Laser Anti-IR Missile System
% ----------------------------------------
% lasdmg.m

clear;clc;clf;

% Input Range

R=500:5000;  % Range - m

% Input Seeker Parameters

Id=1e7;        % Seeker Damage Threshold - w/cm^2
Ac=10;         % Optics Aperture - cm^2
Ad=1e-4;       % Detector Area - cm^2
FOV=1e-8;      % Seeker Field-of View - steradians
ao=.7;         % Optics Transmission Factor

% Input Atmosphere Attenuation Factors

alt=[200 2000 5000 10000];  % Altitude - m
attn=[ .88 .37 .16 .05]; % Attenuation dB/km

% Laser Power Collected by Detector

Pdet=Ad*Id;      % Detector Power - w

% Find Laser Power to Damage Detector
```

```
for i=1:4;
   att=10.^(-attn(i)*R/1e4);
   Pl=(Pdet*FOV*(R*1e2).^2)./(ao*att*Ac);
   Px(:,i)=Pl';
end;

% Plot Required Laser Power

plot(R,Px/1000,'k');grid;
xlabel('Range - m');
ylabel('Laser Power - kw');
title('Laser Power to Damage Seeker IR Detector');
```

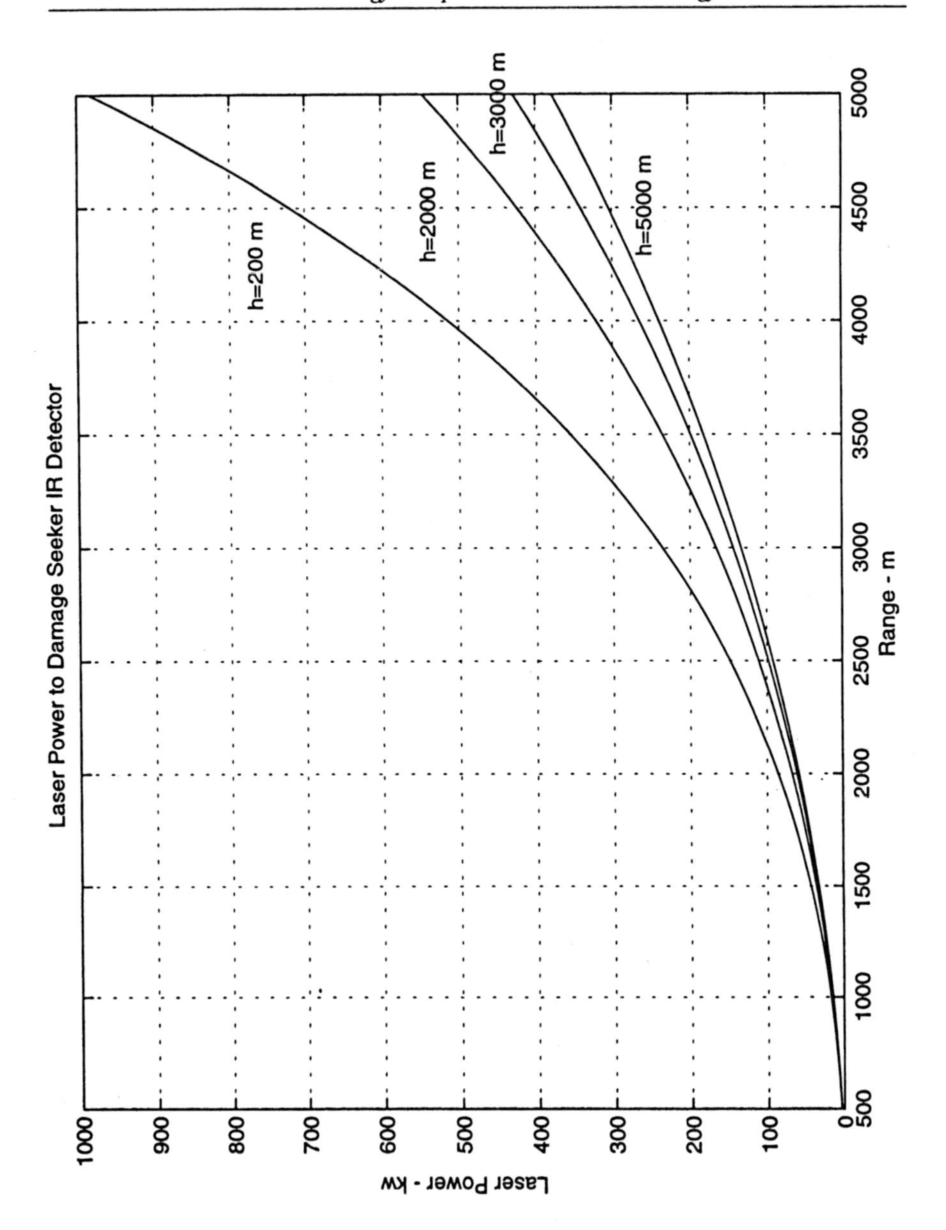

8.1.2.1 Laser Atmospheric Propagation

The most serious limitations of laser DEW involve their use in applications where they must propagate through the atmosphere. These include not only the attenuation of the laser beam in the clear atmosphere but penetration through clouds and weather. For example, a CO_2 laser radiation travels only 7m in a nimbostratus rain cloud before losing 63% of its power [9].

The atmosphere is composed of a mixture of gases that contain a variety of particles that differ widely in size and composition. Some of the gases are strong absorbers of laser radiation in certain spectral regions. Water vapor and CO_2 are the most serious absorbers in the portion of the visible and IR regions where most lasers suitable for DEW operate. Absorption by atmospheric gases is a resonant phenomenon with well-defined strong absorption bands separated by spectral windows in which absorption is small. Attenuation by particle scattering is a broadband phenomenon that is a relatively slow function of frequency. The most common particle scatterers in the optical region are haze, fog, rain, and dust. Since the composition of both the gas mixture and the particle content composing the atmosphere varies with altitude, laser transmittance through the atmosphere is a function of the altitude of the beam. General atmospheric transmittance is plotted in Figure 8.3 against wavelength [11].

The fraction of beam energy that can be delivered to the target is limited by diffraction, atmospheric absorption, atmospheric attenuation and scattering, thermal blooming, plasma breakdown, and turbulent refraction. Diffraction refers to beam spreading, causing the beam to diverge in angular radians approximately proportional to the wavelength of the light divided by the diameter of the collimating mirror or lens. As a result, the peak intensity of the beam decreases as the inverse square of the range from the laser. If the mirror is optically imperfect, then the actual beam divergence exceeds the diffraction-limited angle, causing a further reduction of beam intensity at the target.

If a high-power laser operates at a wavelength where absorption is high, then the air in the beam is heated by the energy it absorbs from the beam.

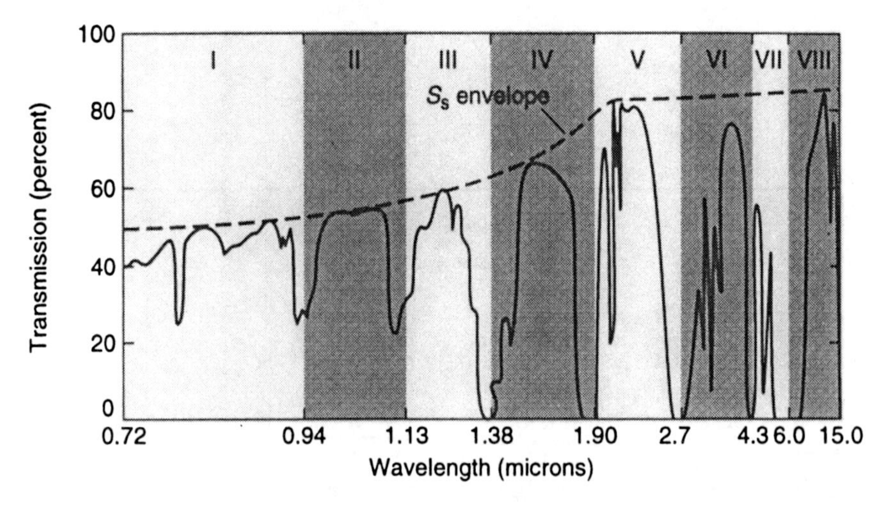

Figure 8.3 General atmosphere transmittance curves versus wavelength. (Skolnik, M., *Radar Handbook,* ©1970 McGraw Hill. Reprinted with permission.)

This causes the air to expand, leaving a low-density region of low refractive index along the beam axis. Refraction will then cause the beam to diverge, reducing its irradiance at long range. This phenomenon is called thermal blooming. The beam can also diverge as a result of scattering due to the reflection of light from aerosol particles. As with HPM beams, a high-powered laser beam can also cause dielectric breakdown of the air, resulting in a plasma that absorbs subsequent radiation through the path [9].

Even at low-beam irradiance, the beam can diverge because of refraction from inhomogenities in densities caused by atmospheric turbulence. Turbulence can also cause beam wander as can thermal blooming in the presence of the wind. Any potential change in the direction of the beam must be considered since it may be significant due to the narrowbeam dimensions at optical wavelengths [9].

8.1.2.2 Laser Beam Control

Laser energy weapons generally consist of the laser itself, beam processing that cleans the beam and neutralizes jitter, adaptive optics, beam path conditioning, and beam pointing and control. To be effective, the laser beam must be pointed at the target and trained on it long enough to cause damage. This requires a target-acquisition and tracking system and a beam-pointing and focusing system. Ideally, there should also be a damage assessment system to determine when one target has been successfully damaged so that the next target can be engaged.

Target acquisition and tracking is best performed by passive EO/IR sensors (i.e., FLIRS, IR search-and-track) that sense the blackbody radiation that the target emits or by a laser radar (lidar) that emits coherent optical radiation and detects the reflected radiation. Passive IR trackers may use a mosaic array of semiconductor photo-optical detectors that detect warm targets against a cooler structured background. This may be difficult if the target is a small satellite but is more practical when the target employs a rocket engine or heats up as it passes through the atmosphere (i.e., AAMs, antishipping missiles, and ballistic missiles). The sensitive broadband detectors are somewhat vulnerable to illumination by IR laser radiation from the general direction of the target.

Information about a target's angular position obtained by an active or passive optical tracker may be combined with pointing reference information provided by an inertial system to generate pointing commands for the transmitting optical system. The latter can range from simple, rigid cassegrain telescopes to large, adaptive segmented mirrors, each segment of which may be individually tilted for fine pointing [9].

Many laser weapons require large transmitting apertures to minimize diffraction, thermal blooming (nonlinear beam divergence within the atmosphere), or excessive thermal stress on the transmitting optics. Fabricating a rigid mirror that can maintain better than a quarter-wavelength surface tolerance over a large aperture is not feasible when the mirror is subjected to operational thermal and mechanical stresses. Therefore, adaptive optical systems are generally required whenever large transmitting apertures are needed.

An adaptive flexible mirror can be designed to provide near-perfect optical collination. However, the beam from this optical system would be distorted by the atmosphere resulting in less than optimum focusing of the laser beam. If a sounding beam can be transmitted and sensed, it is possible to distort the mirror, forming the beam in such a way that it is almost perfectly focused at the target. This mode, which is called optical path adaptation, can partly compensate for optical inhomogeneity in the laser cavity as well as in the atmosphere.

Application of HEL weapons in a destructive mode involve the heating of the target's surface to produce a hole. Figure 8.4 shows the time required to melt through a 2- to 3-mm-thick aluminum sheet (effective absorptivity 0.076) by a 5-MW laser with a 4-m aperture as a function of parameter q, which is the ratio of the actual-to-theoretical beam divergence angle. Note that this time is inversely proportional to the square of the range to the target. However, it is also possible to damage certain targets by overheating them. For example, use of a CO_2 laser having an average power of 220 kW and producing a power density of 1.4 W/cm^2 can overheat an intelligence-gathering satellite at 100-km altitude. Such a laser could be based on a large aircraft or land vehicle. The power to attack satellites at much higher altitudes becomes prohibitively difficult due to its large power requirement.

8.1.3 Charged Particle Beam (CPB) Weapons

CPB weapons are composed of an accelerator whose output is coupled to a beam director, which *projects* an intense, energetic electronic beam to a target at very nearly the speed of light. Upon striking the target, the electrons pass through its skin, depositing their energy internally. This penetration is the key element to the lethality of a CPB weapon. When a CPB strikes a missile or aircraft target, significant damage results. Even when the beam does not directly strike the target, the ionizing radiation generated by passage through the atmosphere has been demonstrated to cause transient upset in essential electronic components.

The endo-atmospheric CPB effort has focused on developing methods to deliver energy to a target over ranges that are militarily significant. Originally,

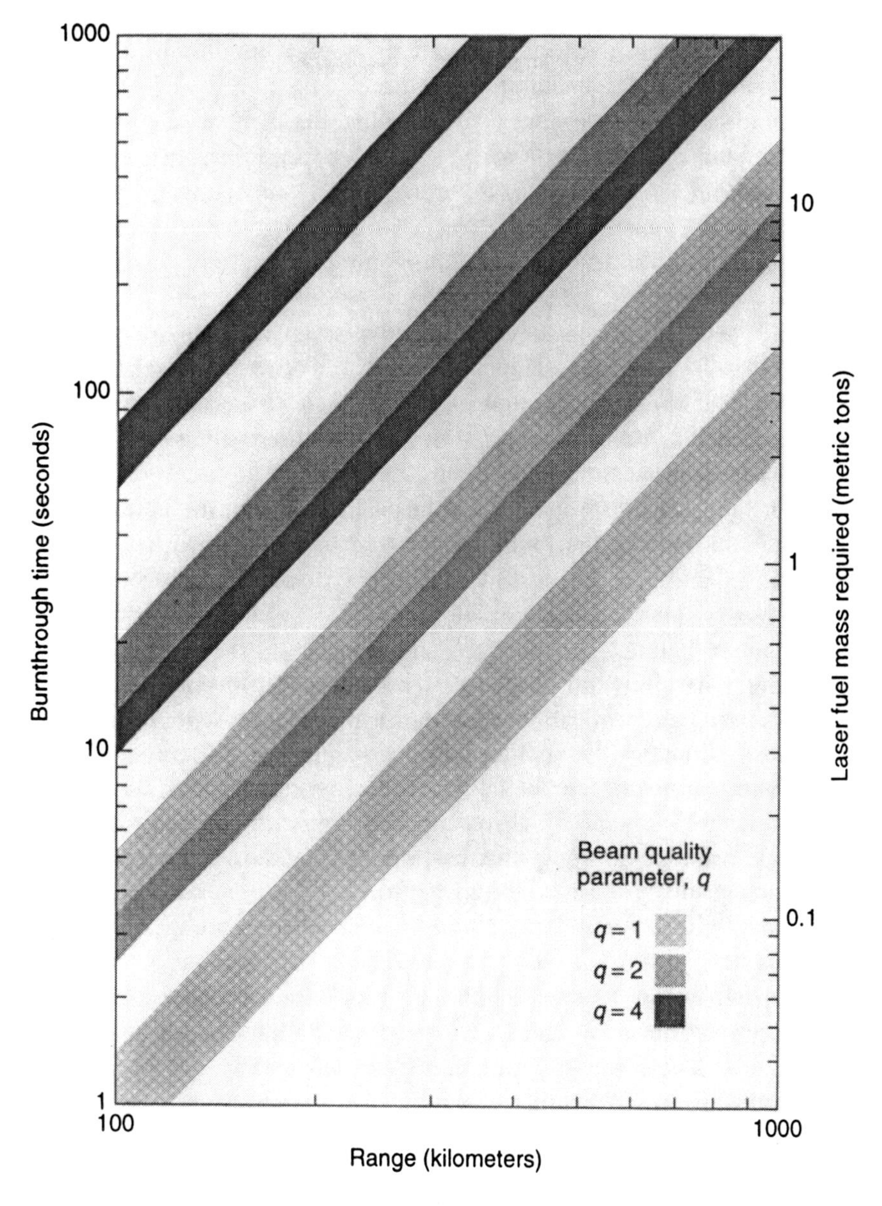

Figure 8.4 Time required for laser beam to melt an aluminum target (5 MW, 4-m aperture). (*Source: IEEE Spectrum,* March 1988, p. 52.)

endo-atmospheric CPB research was primarily of interest for *antiship missile defense* (ASMD). The principal advantages of using CPBs for ASMD include near speed-of-light engagement, instantaneous catastrophic kill capability, unlimited magazine capacity, reduced requirement for storage of energetic materials, and insensitivity of the beam to weather conditions. The robust beam parameters make shielding impractical. Reduction of engagement time greatly improves the engagement rates possible that is increasingly important as attacking missiles become more agile, faster, and stealthier, thereby reducing defense reaction times. For ASMD, propagation requirements drive the beam parameters. The possibility exists that a beam with sufficient energy fluence to propagate can also cause detonation of the high explosive used in missile warheads.

CPB technology is in the developmental stage. The primary emphasis of research has been on the design of compact, lightweight accelerators and propagation of the beam through the atmosphere. Issues concerning beam-target interaction, beam steering, and beam-produced radiation effects have also been investigated to a lesser extent.

For long-range propagation within the atmosphere, the beam is broken into short pulses. The first pulse heats the air through which it passes. The heated air expands, creating a reduced density channel through which the next pulse passes with reduced losses. The second pulse then extends this channel for subsequent pulses. This process is continued until the last pulse deposits lethal energy deeply within the target. To date, stable propagation of a single pulse has been demonstrated, but multipulse propagation requires advances in accelerator technology. For applications that require only short range (less than several hundred meters), single pulses of less energetic electrons can be used.

Accelerators capable of delivering both high-current and high-particle energy are required to apply this technology. Reduced accelerator size and weight are a primary research objective. Other applications such as *short-range air defense* (SHORAD) and tactical missile defense from mobile platforms are complementary to the ASCM application. Since the propagation range scales approximately as the mass of air through which the beam must travel, high-altitude propagation is extended and airborne applications may be attractive if a dedicated aircraft were acceptable. Aircraft self-protection applications with this technology require significant advances in size and weight reductions to make this application feasible [12].

8.2 Stealth

Signature control technology (commonly called stealth) is a critical technology area related to the survivability of military weapon platforms and the weapons themselves. Signature control makes detection by an adversary more difficult through concealment from sensor systems. Each weapon system has a blend of many different signatures with technical approaches tailored to control these signatures. Each case is a series of compromises and optimizations beginning with conceptualization of the system and sustained through operational mission execution. Technology changes have resulted in abilities to reduce/control signatures to levels that are more than two orders of magnitude lower than the basic signature. In general, RF energy signatures receive the largest share of military attention because radar is the most prevalent sensor, particularly for operation at long ranges under a wide variety of environmental conditions. IR signature technologies also receive substantial attention due to the sensor's ability to detect the ubiquitous emission of gas turbine engines and other heat sources that supply power for military systems. Visible, laser, acoustic, and magnetic signatures are of interest in proportion to their applications [13].

The minimum level of RF signature reduction for military superiority is given in Table 8.4. These also apply to any externally mounted avionics including radomes, antennas, windows, and EA equipment. The signature reduction uses critical materials such as graded resistive films, *radar absorbent material* (RAM), loaded cores and fibers, loaded planar or variable graded absorbers, and ceramics. Moreover, geometric shaping is used to divert radar return reflections away from their direction of arrival. In addition to RF, visible, IR acoustic, and magnetic signatures are of interest in proportion to their application. Signature control of these areas are listed in Table 8.5.

Table 8.4
RF Signature Reduction for Military Superiority [13]

Frequency Range	Bandwidth (% f_0)	Signature Reduction
1 MHz–2 GHz	15	5 dB
2 GHz–18 GHz	>15	15 dB
18 GHz–1000 GHz	>5	5 dB

Table 8.5
Signature Reduction for Military Superiority [13]

Technology	Critical Parameters	Critical Materials
IR	IR signature 0.7–20 μm	Materials reflectivity
Visual	90% Solar glint reduction	—
Laser	Reflectivity <5% for 0.3–10.6 μm	Optical coating
Magnetics	50% reduction	Coatings, paint
Acoustic	6-dB attenuation, f = 10 Hz–100 MHz	Materials, coating

An illustration of the military effectiveness of stealth is depicted in Figure 8.5. In this example, a stealth fighter engages a conventional fighter in a head-on engagement at 30,000 ft with both aircraft traveling at Mach 0.9. Both fighters have missiles with 100 km range, but the stealth fighter can move to firing range undetected if its RCS is sufficiently low. The conventional fighter's RCS is 5 m^2, while the stealth's fighter has an RCS that varies down to 0.01 m^2. Three curves are shown. One curve shows where both aircraft have long-range radars that detect a 5-m^2 target at 125 km. The second curve is where both fighters have short-range radars that detect a 5-m^2 target at 50 km. The third depicts a curve for the conventional fighter that has a radar with 125-km detection range, while the stealth fighter has a 90-km detection range radar, both against a 5-m^2 target. The shaded area shows a region where either fighter can fire the first missile shot. The curves show that the stealth

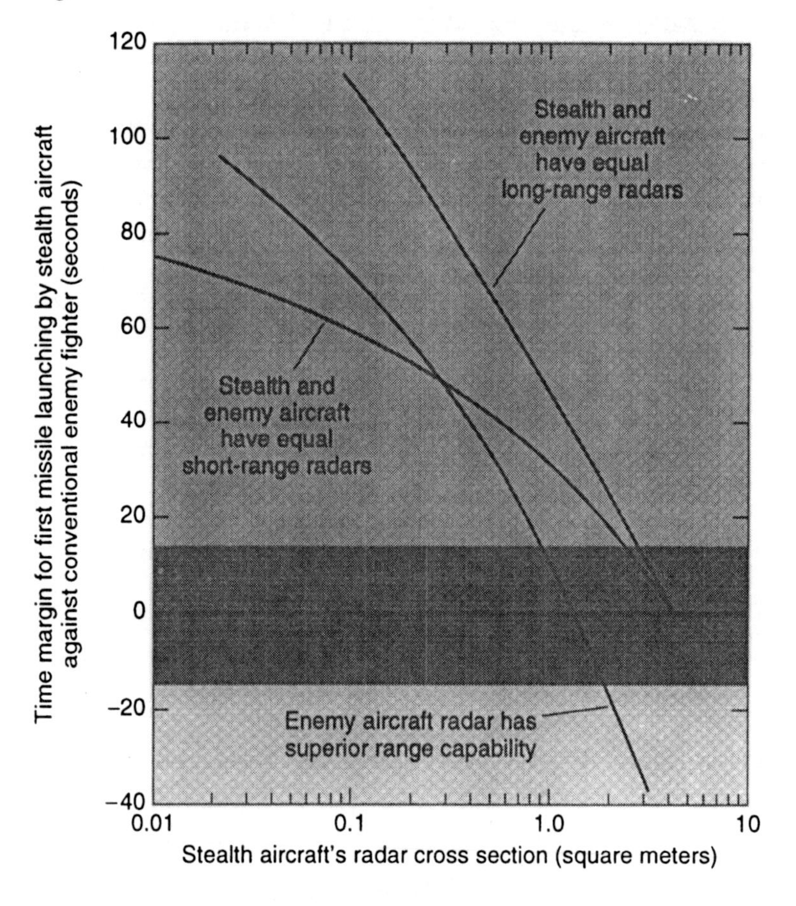

Figure 8.5 Advantage of Stealth fighter over conventional fighter. (*Source: IEEE Spectrum,* April 1988, p. 30.)

fighter with an RCS of less than 0.1 m^2 could target the conventional fighter for more than 60 sec before the conventional fighter can fire its missile, hence giving the stealth fighter a significant military advantage over the conventional fighter [14].

Although the military requirements for low observable platforms and weapons are clear, there is much controversy about both the technical and cost effectiveness of stealth. RF stealth technology matured in the production of the B-1B bomber whose head-on RCS was reduced from 10^2 in the B-1A to 1 m^2 in the B-1B. The radar range equation [15] is given by

$$R_{nm} = 129.2\left[\frac{P_{t(kW)}\tau_{\mu s}G_tG_r\sigma F_t^2F_r^2}{f_{MHz}^2T_sS_0C_BL}\right]^{1/4} \tag{8.1}$$

where

R_{nm} = Detection range in nautical miles
P_t = Peak transmitter power in kW
τ = Transmitter pulsewidth in μs
$G_{t,r}$ = Transmitter and receiver antenna power gains
σ = Average radar cross section in square meters
$F_{t,r}$ = Antenna pattern-propagation factor
f = Frequency in MHz
T_s = Effective system input noise temperature in Kelvins (°K)
S_0 = Average single-pulse signal-to-noise power ratio from Marcum–Swerling curves
C_B = Receiver matching loss relative to a matched filter
L = Detection system power loss factor

which indicates that the free-space detection range is proportional to the RCS raised to the one-quarter power ($\sigma^{1/4}$). Hence, a tenfold decrease in RCS translates to a 56% decrease in free-space detection range, which reduces the reaction time for a weapon system by about half. If the aircraft is flying at low altitude in heavy clutter, the MTI improvement factor required for detection is increased by 10 dB. This may render many radars designed to detect conventional aircraft inoperable [16].

The RCS reduction technology employed in the B-1B might be termed "moderate stealth" since it does not in general render the aircraft undetectable.

This technology also applies to the F-16, the Eurofighter 2000, and the JAS 39 Gripen and recognizes that the head-on RCS of a fighter aircraft is dominated by a few components: the inlet and engine compressor face, the radar antenna, the bulkhead and cockpit hardware, and the weapons complement [17]. The use of nonstructural RAM, reflective canopy treatments, offset radar antennas and engine inlet treatments can reduce the RCS by a factor of 10 or more. This reduces the detection range by half and also makes EW and decoys more effective.

Some aircraft may be more amenable to this treatment than others. For example, the MIG-29 and Su-27 might be hampered by their short, straight intake ducts, which are aerodynamically simple but difficult to shield on an EM basis. The F-16, by contrast, has almost complete line-of-sight inlet blockage from the same altitude, head-on aspect [17].

A more complete stealth treatment requires an aircraft specifically designed with a low signature. Examples of aircraft designed for stealth operation include the B-2, F-117, YF-22, and F/A-18 E/F aircraft and the Tier III–Darkstar UAV. The RCS of these aircraft are in the 0.001- to 0.1-m^2 range. Stealth in these vehicles is achieved primarily by geometric shaping in addition to using composite low RCS materials, judicious placements of engines, avionics and weapons, and RAM coatings.

The penalties of stealth aircraft design are a substantial cost increase and heavier aircraft relative to a conventional aircraft design. Low-observable aircraft tend to be larger and heavier to a degree that depends upon the specified mission. The reason for this is that the low-observable aircraft carries all its fuel internally. It needs weapon bays, which become larger as more weapon types have to be carried and that form structural voids and cut-outs that are hard to incorporate in a high-g fighter or a carrier-based strike aircraft.

The engine installation also drives weight and cost upward. Long, serpentine inlet ducts and two-dimensional nozzles are heavier than simple nonstealth designs, absorb more internal volume, and may cause thrust losses. Also, exhaust system weight and durability have been issues in every known stealth program [17].

Surface treatments, the avoidance of gaps and stops, and the specialized design and construction of normally routine items such as doors, apertures, and antennas also add to the cost of design, production, and maintenance. On the other hand, a DECM jammer is not generally needed to defend stealth aircraft against current radar. This can be seen by examining the burn-through range equation for a self-protection jammer

$$R_T = \left[\left(\frac{P_T G_T L_P}{4\pi S_A} \right) \left(\frac{\sigma}{P_J G_J} \right) \left(\frac{B_J}{B_R} \right) \right]^{1/2} \tag{8.2}$$

where

R_T = Range of target
$P_{T,J}$ = Power of radar transmitter, jammer
$G_{T,J}$ = Gain of radar, jammer antennas
L_P = Polarization loss
S_A = Minimum discernible signal
σ = Average radar cross section
$B_{J,R}$ = Bandwidth of jammer, radar receiver

An examination of this equation indicates that the burn-through range is proportional to the square root of the target's RCS and that the jammer ERP to maintain a specified R_T is directly proportional to the target's RCS. Therefore, a lower RCS results in shorter detection and burn-through ranges combined with minimum jammer ERP (which approaches zero in limit) requirements. The cost of a low-observable aircraft is offset somewhat by eliminating the active electronic jammer. On the other hand, a low-observable aircraft requires better passive detection systems to make the best tactical use of stealth.

▶ *Example 8.4*

The effect of an RCS reduction in the signature of a target impacts on the range it can be detected, the burn-through range in a jamming environment, the jammer power required to protect the target, and the volume within which it can be detected. Using MATLAB, plot the relative effect a reduction in RCS has on each of the preceding factors. What effect does a tenfold reduction in RCS have on each of these parameters?

```
% Effect of RCS Reduction
% ------------------------
% rcsred.m

clear;clc;clf;

% Input RCS Reduction

RCS=0:.01:1;

% Find Reductions

R=RCS.^.25;             % Dection Range
Rb=RCS.^.5;             % Burn-Through Range
J=RCS;                  % Jammer Power
```

```
V=RCS.^.75;          % Detection Volume

% Plot Effects

Px=[R' Rb' J' V'];
plot(RCS,Px,'k');grid;
xlabel('Relative RCS');
ylabel('Parameter Reduction');
title('Effect of RCS Reduction');
```

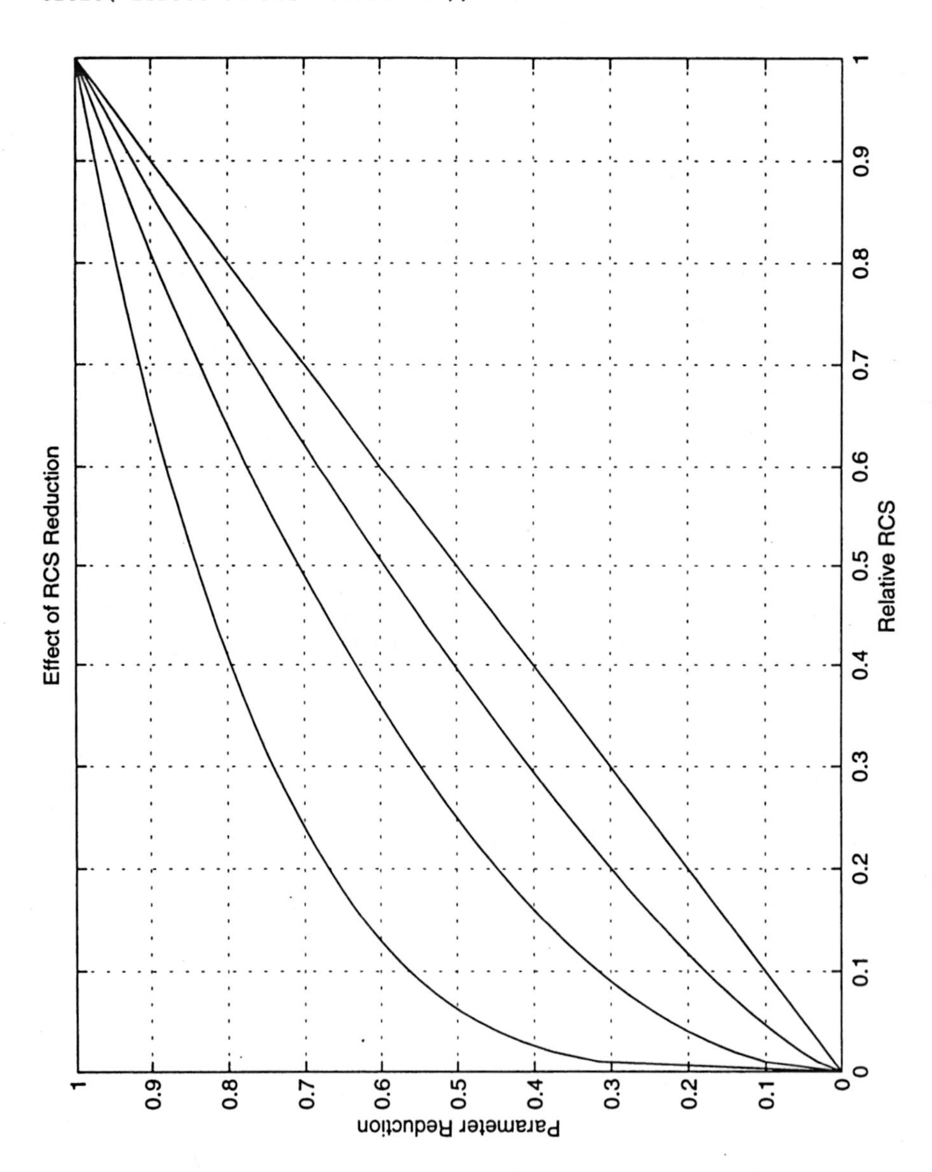

Other important examples of stealth vehicles are the ALCM and low-observable naval ships. The AGM-129A advanced cruise missile incorporates low-observable design features to reduce its RCS. It employs a faceted nose followed by a near-constant diameter for most of its length, tapering toward three relatively small control surfaces on the far end. The ALCMs faceted nose comprises four surfaces that are highly aft-swept from a sharp point that are intended to deflect radar reflections away from their angle of incidence. A conformal antenna is employed for missile guidance. The ALCM stealth design features are concentrated on the forward-and-above aspect from which a cruise missile is most likely to be detected. The upper surface of the nose is a half-cone, and features such as the inlet, exhaust, and rudder are shielded by the underbody. The range of the ALCM is greater than 250 nmi and it cruises at Mach 0.9 [18].

An example of a low-observable ship is the Swedish Navy's Visby Class Stealth Corvette. The primary responsibilities of the ship are *mine countermeasures* (MCM) and *antisubmarine warfare operations* (ASW). The design of the Visby has been directed to minimizing the optical and IR signatures, above water acoustic and hydroacoustic signature, underwater electrical potential and magnetic signature, pressure signature, RCS, and actively emitted signals. Stealth provides delayed detection, identification, and target acquisition by hostile forces that enhance both the first strike capability and survivability. A conventional non-stealth-designed Corvette would typically be detected by hostile forces at a range of 50 km in a normal environment and 25 km in a jammed environment. The state-of-the-art Corvette ship has a detection range of 13 km in rough seas and 22 km in a calm sea without jamming. In a jammed environment, the stealth ship would be detected at 8 km in a rough sea and 11 km in a calm sea [19].

A study has been conducted that assesses the vulnerability of the B-2 bomber, whose mission is the destruction of mobile ICBMs, to counter stealth concepts [20]. Potential counter-stealth concepts studied are listed in Table 8.6. The basic constraints in the study were to look beyond simple detection vulnerability to ask whether a prospective counter-stealth system can:

- Detect the B-2 at all altitudes and flight profiles with suitable coverage to defend a large area;
- Track the B-2 accurately enough and long enough to direct a weapon system against it;
- Guide a weapon and fuse it so that its warhead will explode close to the aircraft.

Table 8.6
Counter-Stealth Concepts

Acoustic systems	Upgrade existing systems
Bistatic systems	OTH radar
Special waveforms	Passive coherent detection
Balloon radar	Radar shadow detection
Bistatic reflectors	Hybrid bistatic radar
Corona detection	Detection of aircraft emissions
Correlation spectroscopy	Impulse radar
Cosmic rays	Towers and nets
Differential absorption	Advanced airborne surveillance
IR AWACS	Radar wake detection
IRST	Radiometrics
Land mines	Ultra wideband and radar
Magnetic disturbance	Polysaturation Doppler
Space-based radar	High-frequency radar

The constraints focus on whether terminal threat sensors and weapons can effectively engage a stealth target. These sensors generally operate in the EM region where the stealth signature reduction is most effective. These constraints are based upon the functioning of an air defense system that combines three inter-related independent functions—surveillance, fire control, and hard kill. If each function is carried out with 80% probability, then the overall probability of kill is about 50%. In addition, air defenses must be able to survive direct attack, resist countermeasures, and function despite the variations in the environment that include severe weather effects [21].

It is generally agreed that lower frequency radars can counter two aspects of stealth technology: geometrical shaping and RAM. Stealth shaping is designed to reflect incoming radar waves away from the direction of the source. However, when the radar wavelength is close to any dimension of the target being illuminated, resonance occurs between the direct reflections and other waves that creep around it, causing strong signals in the absence of a direct mirrorlike reflection. Resonance occurs from components as well as from the whole aircraft. For example, a gun muzzle may be resonant when illuminated by a 3-cm I-band wavelength. The E-2C AEW radar operates at UHF (400 MHz) with a 75-cm wavelength, so large components, such as aircraft fins and wing tips, fall within the resonance region [21].

VHF radars operate in the frequency range between 160 MHz to 180 MHz with wavelengths of 165 cm to 190 cm. At these wavelengths, major components of large aircraft such as wings and fins may be resonant. In addition, RAM using passive cancellation techniques, which generally are used to reduce reflections from critical edges in a stealth vehicle, are not as effective as at higher frequencies. Although these radars may be capable of detecting stealth targets, they have poor angular resolution and multipath problems that preclude them from detecting low-altitude aircraft. Another factor is that newer type stealth designs, such as those used in the B-2, have dimensions measured in meters that drive the resonance frequency into the HF region [21].

Bistatic radar (separate receivers and transmitter) was examined as a counter-stealth concept. The theory is that the stealth aircraft's shape is designed to deflect received radiation away from its source. Thus, it may send the main lobe of the reflection in another direction to be picked up at full strength by a bistatic radar receiver. However, most stealth vehicles are designed so that these high-power lobes are concentrated at a few angles where the lobes are generally narrow in width. Additionally, the reflections are moving in position with the aircraft, so a large number of receivers would be required to accomplish a high probability of detection [20].

In the study, it is conceded that *over-the-horizon* (OTH) radars have the capability of detecting the B-2. OTH radars operate in the *high-frequency* (HF) region between 3 MHz to 30 MHz with wavelengths of 10m to 100m [20]. These radars detect most targets by resonance effects and are not affected by most known forms of RAM. There have been several reported detections of stealth vehicles by these radars, particularly by the Australian Jindalee. The counter to this apparent contradiction is that OTH radars are not accurate enough to allow any other known type of sensor to detect and track the B-2 [20].

Another possibility for detecting stealth targets is to use a radar in the millimeter-wave region. The theory here is that the reflecting surfaces on the aircraft will appear rough as the wavelength approaches millimeter wavelengths. However, millimeter radars that operate in the atmosphere are primarily short-range sensors due to the high attenuation of the radiations by water vapor and oxygen. There is little available data on this conjecture.

EO/IR systems might be able to detect stealth vehicles because they must generate the same amount of heat as a conventional aircraft. The effectiveness of passive IR decreases rapidly with range and is affected by atmospheric moisture. This is why most IR area-surveillance systems are designed to protect point targets against threats that are normally not dangerous in adverse weather [20].

Airborne IR sensors, operating above most atmospheric moisture, would be more effective against medium- and high-altitude stealth targets. IR sensors have been considered as adjuncts to the radar on future versions of the AWACS (IRAWACS), but they would be used for target discrimination, raid assessment, and countermeasures protection rather than as a primary detection system [20].

Adequate detection range is the major problem for EO/IR sensors against stealth vehicles. The IR signature of an aircraft varies with speed, altitude, and exhaust temperature. Also, the major heat sources on the aircraft are visible from the rear, which is the least useful quadrant from an intercept basis. The exhaust unit of most stealth aircraft is designed to dissipate the exhaust plume, which generally is one of the greatest contributors to the IR signature [20].

Coherent laser radar, operating at 10.6 μm (CO_2 laser), has been used to detect aircraft wakes and air turbulence. It, therefore, offers the possibility of detecting the stealth aircraft by either its wake or the turbulence it creates as it moves through space. However, a more direct method might be to paint the stealth aircraft with a laser signal that in effect would be similar to an enlarged IR signature. There are several difficulties to this approach that include the problems of directing the laser beam at the stealth aircraft and of the large attenuation of laser frequencies due to absorption by carbon dioxide, oxygen and water molecules [20].

8.2.1 Stealth Fundamentals

The reduction of the RCS of a vehicle to escape detection falls into a general class of techniques called *low observables,* or more formally *signature control technology.* An aircraft is considered to be a low observable if its characteristics detectable by a radar can be reduced tenfold. For example, reducing the RCS of a fighter-bomber from 10 to 1 m^2 decreases its free-space detection range by a factor of about 2, which correspondingly reduces the range within which an opposing air defense system has only half as much time to engage the low-flying, fast-moving aircraft.

Low-observable technology, carried to its ultimate limit, is called *very low observable* or *stealth technology.* Stealth technology is defined as making an aircraft virtually impossible to detect and intercept. Stealth technology has been applied to a number of aircraft and other vehicles and must be considered by EW designers for military application against modern radar.

The question as to what constitutes a stealth target is difficult to answer and will not be pursued here. Rather, we will consider several fundamental principles that apply to detection of radar targets with low-RCS signatures. We will also discuss RCS reduction techniques and radar techniques that may provide some capability against stealth targets.

In the free-space radar detection equation that determines the clutterfree range at which a target can be detected, the RCS (σ_t) of the target has the same effect on the detection range as the transmitter power, pulse length and antenna gains; that is, $R = k\sigma_t^{1/4}$. Generally, the RCS of the target is a complex quantity that depends on the target size, shape, and radar properties such as: (1) the radar frequency, (2) the aspect angle at which the radar views the target, and (3) the radar polarization. For bistatic radar, the RCS is a function of the viewing angles and polarizations of both the transmitter and the receiver.

▶ *Example 8.5*

Using MATLAB, plot the free-space detection range versus RCS of two radars: one designed to detect a 1 m^2 target at 100 nmi and the other at 300 nmi. Note the appreciable free-space detection range of these radars even against very low RCS targets. However, even though these radars have the requisite sensitivity, it becomes increasingly difficult to detect low RCS targets in a clutter background.

```
% Target Range as Function of RCS
% --------------------------------
% tarrng.m

clear;clc;clf;

% Input Radar Parameters

Rx=[100 300];     % Detection Range - nmi
RCS=1;            % RCS for Specified Range - m^2

% Find Range vs Target RCS

Rcs=logspace(-3,1);

for i=1:2;
R=Rx(i)*Rcs.^.25;
Rxx(:,i)=R';
end;

% Plot Target Range

loglog(Rcs,Rxx,'k');grid;
xlabel('Radar Cross Section - m^2');
ylabel('Range - nmi');
title('Radar Range as Function of RCS');
text(.9,300,'X   300 nmi');
text(.9,100,'X   100 nmi');
text(.001,10.5,'\downarrow');
text(.002,10.5,'\downarrow');
text(.01,10.5,'\downarrow');
text(.02,10.5,'\downarrow'); text(.25,10.5,'\downarrow');
text(1,10.5,'\downarrow');
text(5,10.5,'\downarrow');
```

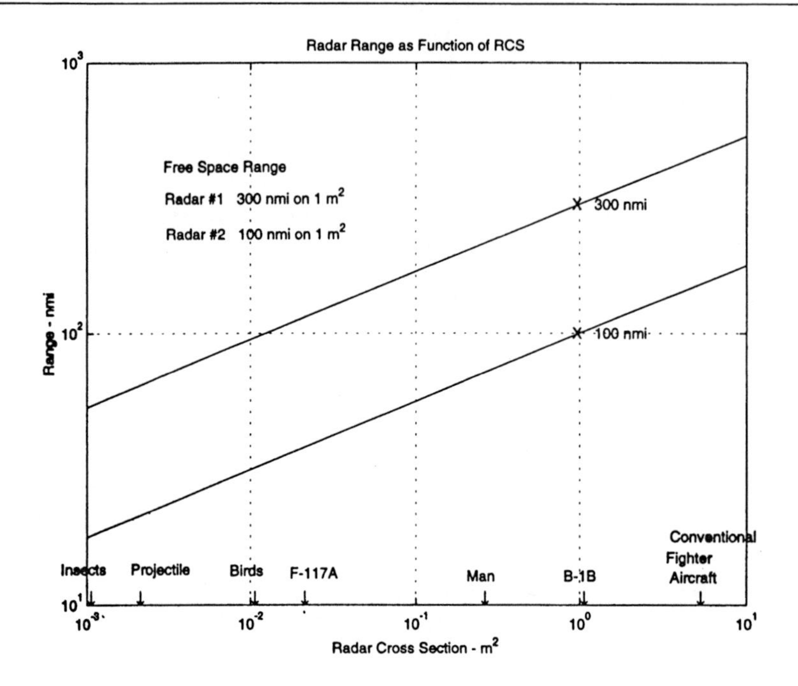

The RCS of simple geometric shapes can be predicted with reasonable accuracy, and extensive lists of formulas are available for these cases. Of this class of target, the perfectly conducting spherical target stands out because its RCS is independent of aspect angle and when its radius (*a*) is much greater than the incident radiation's wavelength ($a/\lambda > 1$), the RCS is only a function of the sphere's surface area given by $\sigma = \pi a^2$. However, the cross sections of most complex targets do not follow simple rules and must be measured to determine their RCS.

Figure 8.6 depicts the RCS of a scaled Boeing 737 aircraft. The RCS patterns of Figure 8.6 were measured at X-band with a 1/15 scale model of the 737, and hence the RCS values are much lower than for the actual aircraft (i.e., they have not been adjusted to predict values observable on the actual aircraft at longer wavelengths). The patterns in the nose-on region are dominated by the re-entrant returns from the jet engine intake ducts. Specular flashes occur within the region 15 degrees on either side of the nose due to the leading edge of the wings. A narrow specular flash occurs broadside to the aircraft whose magnitude provides the maximum aircraft RCS. A pair of broad intense lobes occur in the region 95 degrees to 105 degrees from the nose, which are probably due to the engine cowlings. The lowest RCS values are in the tail region 120 degrees to 150 degrees from the nose. To determine the detection characteristics of such a complex target, we customarily assign a mean (also the median is sometimes used) RCS value and a probability distribution

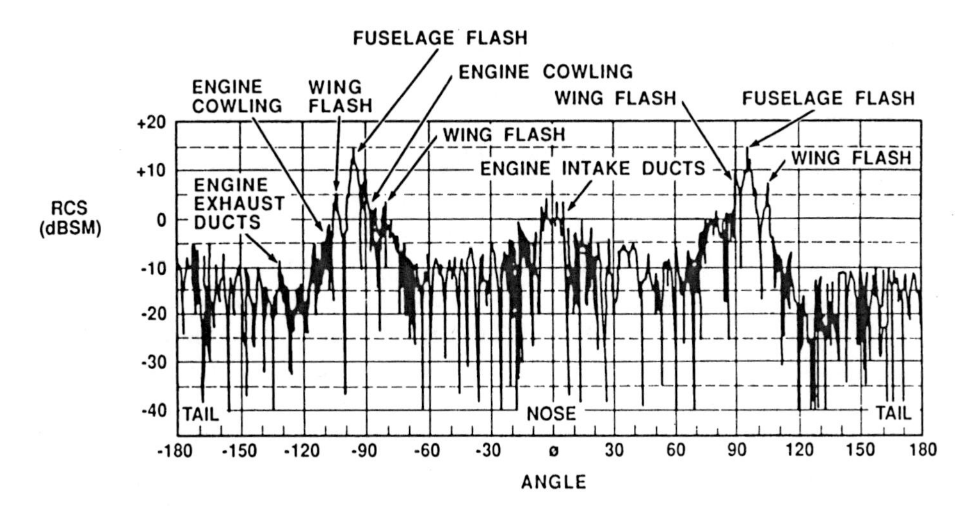

Figure 8.6 RCS of a scaled Boeing 737 aircraft measured at 10 GHz. (*Source:* [30].)

(e.g., Rayleigh) to describe the target. The radar's detection range is then determined for the mean RCS and averaged over the probability distribution to determine the overall detection probability of the radar. Extensive curves of Swerling-type targets are available for this purpose.

Given the complexity of practical radar targets, it is interesting to explore what can be done to reduce their RCS. In general, there are only four ways to reduce the radar echo of an object:

1. Shaping;
2. Radar absorbing materials;
3. Passive cancellation;
4. Active cancellation.

Each involves a different methodology and exploits different aspects of the overall problem. In applying these techniques, remember that the radar energy captured by the target must be either scattered, absorbed, or canceled. If it is scattered, the strategy is to direct it away from the radar receivers, whether monostatic or bistatic. Cancellation is practical only if there are a few primary scatterers with characteristics that are well known over the frequency range of interest.

Of the four methods for RCS reduction, shaping is the most useful. Shaping presumes that the radar threat will appear within a definable and limited cone of direction. Radar absorbing materials are typically used to complement shaping techniques or when shaping is not a useful or implementable option. A deep reduction in target RCS is possible using passive cancellation, but only over narrow aspect angle sectors and for narrow bandwidths. Active cancellation depends on the addition of active signal sources on the target and requires careful control of the reradiated energy. Of the four, only shaping and RAM have apparently been of any tactical use.

The objective of shaping is to orient target surfaces so that they reflect an incident wave in directions other than back at the radar receivers. This principle is difficult to apply to an existing vehicle, such as the scaled 737 aircraft with its RCS depicted in Figure 8.6. However, the result in reducing the average RCS would be apparent in the figure if the large specular returns could be eliminated from the RCS pattern. In new aircraft design, the shaping method has apparently been applied to both the F-117A and B-2 aircraft. As shown in Figure 8.7, the F-117A planform consists of 18 straight-line sections arranged at four main orientations: one angle is used for the leading edges of the wing and tail; another for their trailing edges; wing and tail tips are aligned roughly fore and aft, as are the fuselage sides immediately behind the wing; and the rear section of the wing tip and the aft fuselage are at a common angle. Edges tend to be strong radar reflectors. A radar signal hitting parts of the aircraft such as the wing's leading edge behaves like a light beam hitting a reflective surface. If the radar beam arrives at an angle, it departs at a similar angle in a direction away from the radiating source.

The F-117A fuselage is skinned in RAM to reduce direct reflections. Other features include: (1) engine inlets aligned to the wing's forward edge and grided to prevent engine compressor reflection, (2) a windshield coating to prevent reflecting from objects within the cockpit (such as the pilot's headgear), and (3) a 30-degree canted upper fuselage to reflect the radar wave out of the FOV [22,23].

The B-2 represents a more advanced stealth design than the F-117A. A planform with only four angles minimizes the number of directions in which the radar signals are reflected. The planform for the B-2 apparently consists of 12 straight-line sections, each of which is angled in one of four directions [24]. It employs large-dimension straight-line segments consisting of two forward wing sections that meet at the nose plus a rear zigzag profile whose edges conform to the limited edge of angles. The B-2 is constructed of composite material that has an inherently low RCS. Additional features include: (1) curved engine inlet ducts, (2) imbedded engines to provide a reduced IR and radar signature and baffles to produce a low-acoustic signature, and (3) an imbedded LPI radar located under the wing leading edge and a high-performance ESM system (ALR-50) [23,24].

RCS reduction via shaping is effective only if we are willing to accept higher echo return signals over some aspect angles to gain lower echo signals

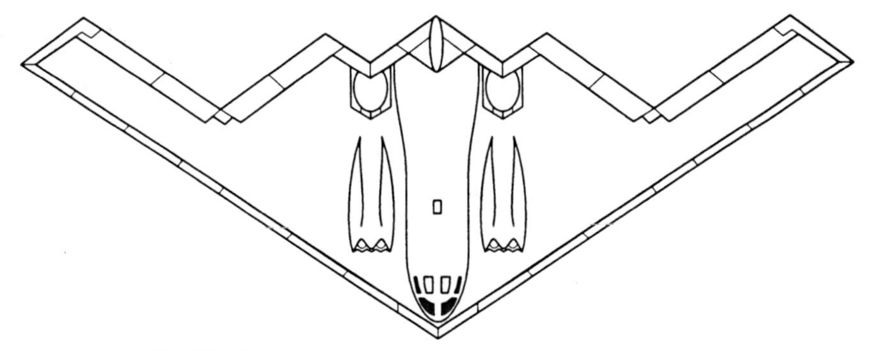

B-2 Planform
- 12 straight line segments
- Each angled in one of 4 main directions

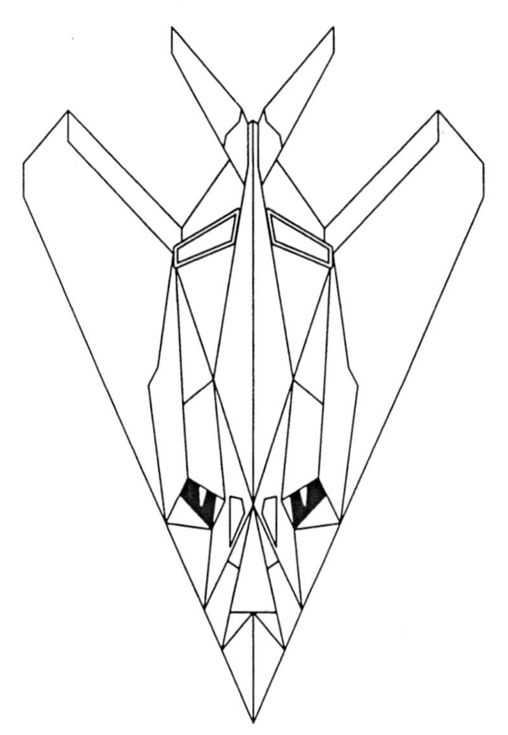

F-117A Planform
- 18 straight line segments
- 4 main orientations
- Faceted fuselage

Figure 8.7 RCS reduction using target shaping. (*Source:* [23] [24].)

over critical threat sectors since reorientation of the surfaces of the radar target for favorable reductions over the specified sector is inevitably accompanied by an increase in RCS over other sectors.

The principles of shaping are illustrated in Figure 8.8, which shows the reflection from a tilted flat plate (which is analogous to reflections from the straight line segments of stealth aircraft). The maximum RCS of the plate is given by

$$\sigma_{max} = \frac{4\pi A^2}{\lambda^2} \tag{8.3}$$

where A is the area of the plate and λ is the wavelength. The reflections in an orthogonal axis of length (L) consist of a main lobe whose width is λ / L and sidelobes that number $8L/\lambda$. The sidelobes are analogous to those produced by a uniformly illuminated planar aperture. To illustrate the frequency dependence, a 6-in square flat plate produces a maximum RCS at 10 GHz of 30 m^2 with a lobe 12-degrees wide and an RCS of 0.3 m^2 at 1 GHz with a lobe 120-degrees wide. If we consider the main B-2 wing (about 80 ft to 90 ft), the RCS flash from the reflection is 1/15-degree wide at 10 GHz and 0.7-degree wide at 1 GHz. The maximum RCS from these flashes will be reduced at a rate of 6 dB per octave (relative to the main lobe) as the direction of arrival moves away from a perpendicular to the leading edge of the wing. This example illustrates the power of shaping in producing low RCS against monostatic microwave radar whose AOAs are several degrees away from a perpendicular to the straight line segments. It also illustrates that if the straight-line segments are sufficiently long; an impractically large number of multistatic receivers would be necessary to intercept the reflections from a stealth aircraft in the microwave region.

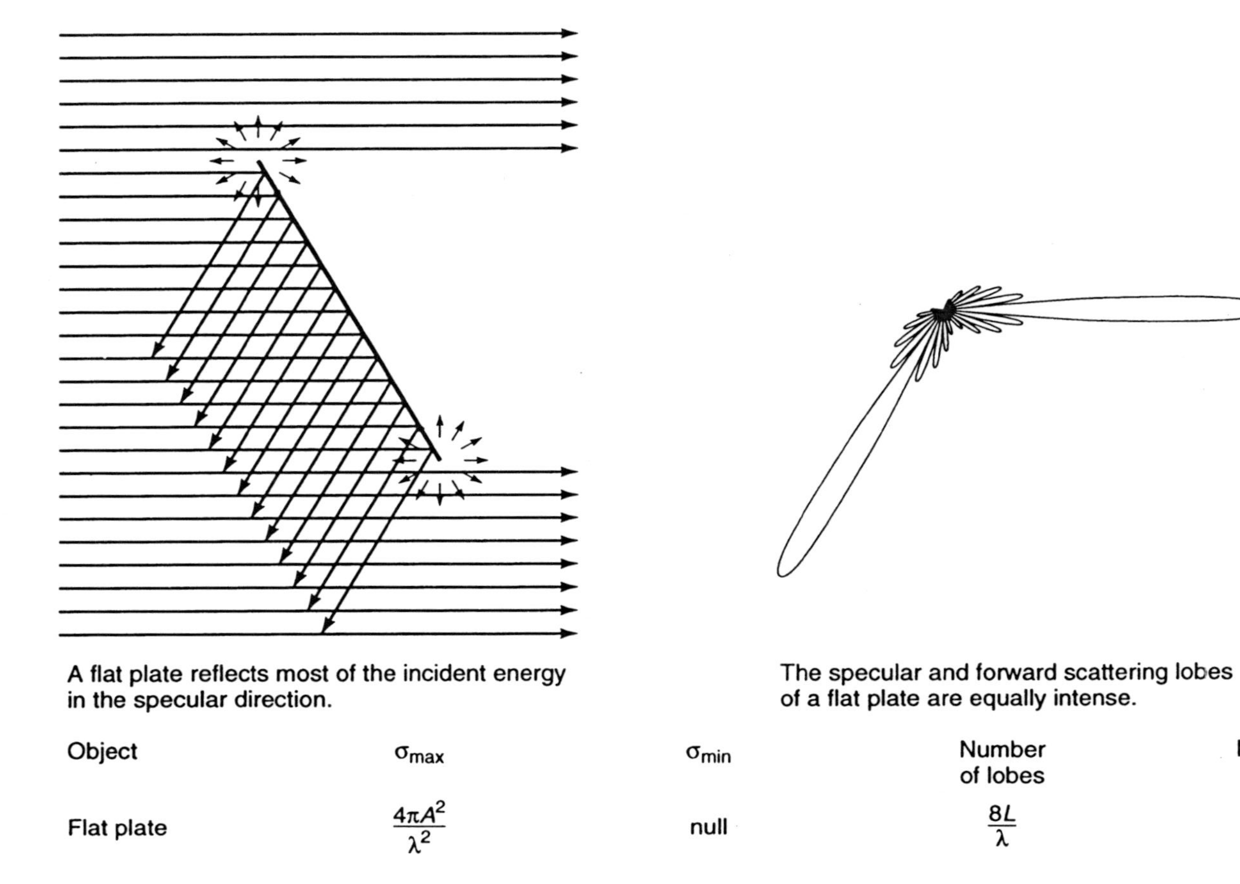

Figure 8.8 Reflections from a tilted flat plate. (*Source:* [30].)

▶ *Example 8.6*

One of the most effective methods for reducing the RCS of a target involves shaping. To demonstrate this, use MATLAB to plot the RCS of a flat plate as a function of the angle with respect to the perpendicular to the plate for frequencies of 10 GHz, 3 GHz, and 100 MHz. Use a plate 10-m long and 0.15-m wide. Note the sharp specularlike return that occurs at normal incidence and the reduced RCS of approximately 10^{-2} m^2 that occurs at 30 degrees that is typical of the angle used in stealth aircraft. Also note that as the frequency decreases, the magnitude of the specular spike decreases and the lobes become much wider.

```
% RCS of a Flat Plate
% -------------------
% rcsflt.m

clear;clc;clf;

% Flat Plate Parameters

l=10;    % Length - m
w=.15;     % Width - m
f=[10 3 1 .1];  % Frequency - GHz

wlx=3e8./(f*1e9);  % Wavelength - m

% Angle from Perpendicular

phi=0:.01:90;
phir=phi*pi/180+1e-12;

% Find RCS of Flat Plate

A=l*w;
for i=1:4;
wl=wlx(i);
sigo=4*pi*A^2./wl.^2;
k=2*pi./wl;
fac=k*l*sin(phir);
z=sin(fac)./fac;
RCS=sigo*(z.^2).*(cos(phir).^2);
RCSX(:,i)=RCS';
end;

% Plot RCS of Flat Plate

subplot(221);
plot(phi,10*log10(RCSX(:,1)));grid;
axis([0 30 -40 60]);
xlabel('Angle - deg');
ylabel('RCS - dbsm');
```

```
title('Flat Plate RCS')
text(15,30,['f = ',num2str(f(1)),' GHz']);

subplot(222);
plot(phi,10*log10(RCSX(:,2)));grid;
axis([0 30 -40 40]);
xlabel('Angle - deg');
ylabel('RCS - dbsm');
title('Flat Plate RCS');
text(15,20,['f=  ',num2str(f(2)),' GHz']);

subplot(223)
plot(phi,10*log10(RCSX(:,3)));grid;
axis([0 30 -40 30]);
xlabel('Angle - deg');
ylabel('RCS - dbsm');
title('Flat Plate RCS');
text(15,10,['f=  ',num2str(f(3)),' GHz']);

subplot(224)
plot(phi,10*log10(RCSX(:,4)));grid;
axis([0 30 -40 20]);
xlabel('Angle - deg');
ylabel('RCS - dbsm');
title('Flat Plate RCS');
text(15,0,['f=  ',num2str(f(4)),' GHz']);
```

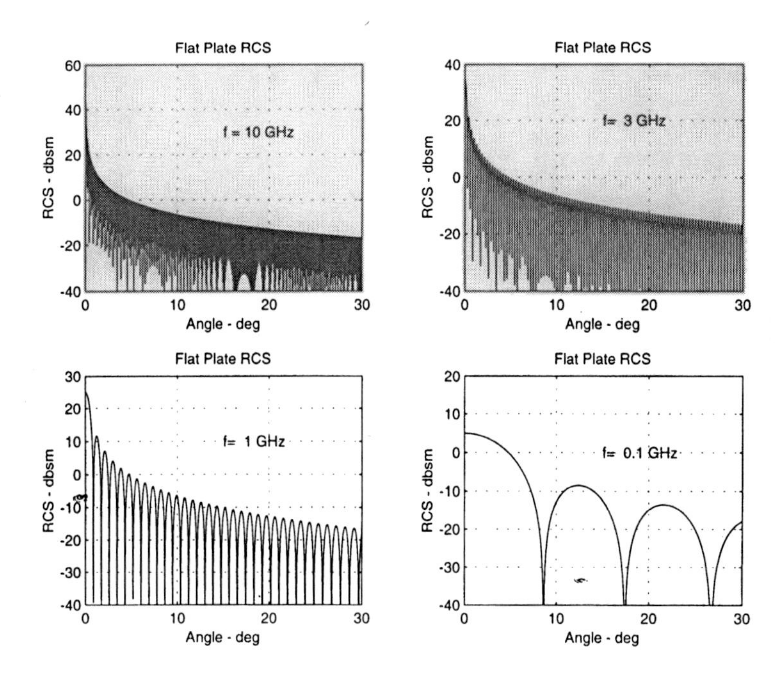

► *Example 8.7*

When a flat plate is viewed at an angle away from the perpendicular, its maximum RCS in the principal plane is given by

$$\sigma_{max} = \frac{w^2}{\pi \tan^2 \phi}$$

where w is the width of the plate and ϕ is the angle from the perpendicular. From Example 8.6, note that σ_{max} is independent of frequency. Plot this relationship using MATLAB for widths of 0.0375m, 0.075m, and 0.15m. What is the width required to provide a 10^{-2} m^2 and a 10^{-3} m^2 RCS at 30 degrees?

```
% Maximum RCS of Flat Plate as Function
% of Angle from Perpendicular
% ------------------------------------
% angfp.m

clear;clc;clf;

% Input Plate Width

wx=[.0375 .075 .15];        % Plate Width - m

% Input Angle

phi=0:.1:90;
phir=phi*pi/180+1e-12;

% Find Maximum RCS with Angle

for i=1:3;
   w=wx(i);
   Rcs=w.^2./(pi*(tan(phir)).^2);
   Rcsx(:,i)=Rcs';
end;

% Plot RCS vs Angle

semilogy(phi,Rcsx);grid;
axis([0 90 1e-4 1e4]);
xlabel('Angle - deg');
ylabel('RCS - m^2');
title('Maximum RCS of Flat Plate');
```

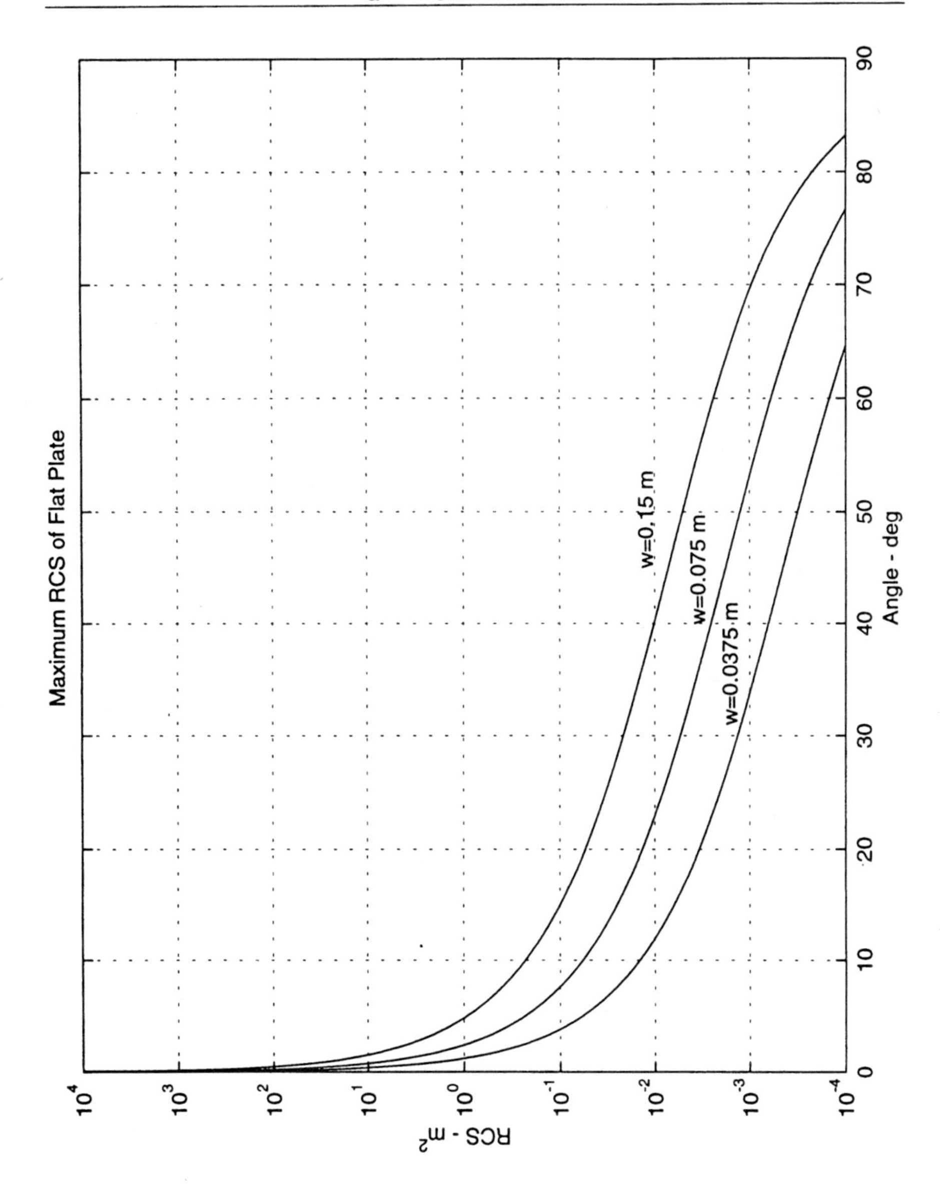

As the name implies, *radar absorbers* depend on the absorption of incident energy, thereby reducing the amount reflected back to the radar. Although commercial RAMs of several different varieties have been available for decades (e.g., for anechoic chambers), they are not necessarily suitable for tactical applications. RAM that is useful for tactical applications makes use of the energy exchange properties of carbon and certain magnetic iron compounds. When illuminated by an RF radar signal, the molecular structure of these materials is excited and the excitation converts the RF energy into heat. In this way, the material absorbs a portion of the incident signal, thereby reducing the signal available for reflection back to the radar [25].

Although the principle involved with RAM is relatively simple, its implementation is more complex, especially as applied to high-performance aircraft, where weight and the ability to withstand the stresses of the flight envelope are at least as important as the RAM's electromagnetic capabilities. In general, four types of RAM are used in practice: broadband absorbers; narrowband or resonant absorbers; hybrids, which combine features of the two; and surface paints [25].

Broadband absorbers offer a consistent performance over a range of frequencies and are more useful for aircraft where the need is for a RAM that is integral with airframe components. This might take the form of a honeycombed sandwich with absorber-impregnated foam applied to the individual cell walls in such a way as to present logarithmically graded resistance to the incoming signal. Another option for the sandwich filling might be absorbent foam with a pyramidal configuration. For example, a honeycomb arrangement using fiberglass for the skins and comb and carbon black-silver powder as the absorber is estimated to achieve 95% absorption against signals in the 2.5- to 13-GHz frequency range with a material thickness of 2.5 cm and a weight of about 310 g per 930 cm^2 of surface area [25]. This material would be suitable for subsonic vehicles, but Kevlar-based materials are generally used in modern RAM applications.

The second major type of RAM is the narrowband, resonant absorber in which an elastomer such as rubber is loaded with absorber, backed by a metal reflector, and applied parasitically to the surface needing protection in sheet or molded form. Designed for specific threat frequencies, such material can be tuned for peak performance over a narrow range of frequencies by varying the thickness of the sheet, which would ideally be equal to one-quarter of the incoming signal's wavelength. The resonant absorber works using both absorption and quasi-optical techniques that generate equal-amplitude reflections to cancel out the natural reflectivity of the material. Absorbers of this type may be targeted against a broader range of threats using several layers of material, with each layer tuned to be effective against a particular frequency [25].

Although narrowband absorbers offer the advantage of a relatively small weight penalty, their performance is degraded if the threat signal does not strike them at exactly 90 degrees. The ability to cope with other angles of incidence appears to be the major impetus behind the development of hybrid absorbers that combine features of both narrowband and broadband materials to produce a performance that, although varying with the angle of incidence, betters their individual performances over a spectrum of situations.

Another RAM alternative is radar-absorbent paint that forms a surface finish loaded with an absorber-like iron ferrite. Iron Ball paint of this type is believed to be particularly effective in the suppression of re-emissions generated by surface-current effects. When a structure is illuminated by a radar, electromagnetic currents are generated on any of its surfaces that are conductive. This, in turn, leads to a retransmission effect when the skin acts as an antenna. Even rivet heads, if not properly suppressed, materially contribute to a vehicle's RCS. More serious is the surface-current effect on the radar's visibility of such areas as wing and tail leading and trailing edges together with any external antennas that it may carry.

RAM can be selectively applied to an aircraft to substantially reduce its RCS, particularly in areas that produce high specular returns. Locations where RAM is effective in reducing the RCS of an aircraft are depicted in Figure 8.9. Note that many of the suggested RAM locations in Figure 8.9 correspond to those elements identified in Figure 8.6 as producing large RCS returns at particular aspect angles.

Antennas can contribute significantly to a vehicle's RCS. In general, the RCS of an antenna consists of two components. One, due to reflections from the antenna and antenna support structure, is termed the *structural mode response.* The second component, due to reradiation of energy captured by the antenna, is termed the *radiation mode response.* Because the illuminating radar transmitter and receiver are normally at the same point, the RCS in the direction

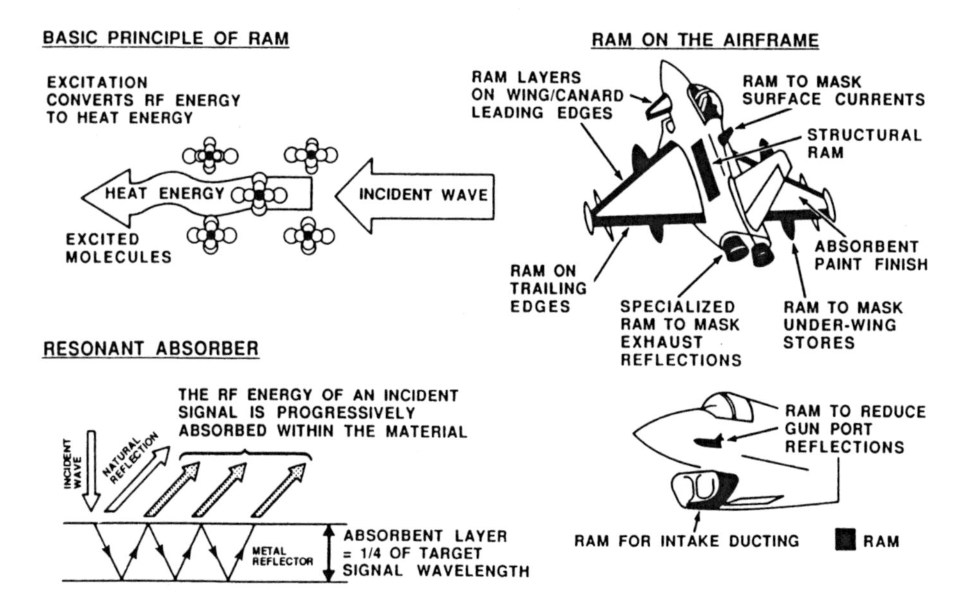

Figure 8.9 RAM application to airframe. (*Source:* [16].)

of the illuminating radar is usually of most concern. Also, the out-of-band response often can be more troublesome than the in-band response.

The RCS of the radiation mode can be estimated as RCS = $\rho G \lambda^2 / 4\pi$, where ρ is a term that depends on the reflection coefficient and efficiency of an antenna with an effective capture area of $A = G\lambda^2/4\pi$. The RCS is highly dependent on polarization. Some antennas such as the dipole basically present no backscatter response for an incident orthogonal polarization. Others like the horn or reflector antenna can look totally reflective for an orthogonal incident polarization. In general, the RCS of an antenna can be quite large when the gain and reflection coefficient of the antenna are large, as would be the case for a high-gain antenna with poor impedance match.

An antenna with interesting RCS characteristics is the phased array. The ability of the antenna to scan electronically allows orientation of the antenna aperture plane at a significant angle (e.g., on the order of 20 degrees to 30 degrees) with respect to incoming radiation. Thus, the radiation mode component of the antenna's RCS is minimized, except when the antenna is electronically scanned in the direction of the target. A similar concept has been described, whereby the phased array antenna is generally aligned in the direction of the incoming radiation, but nonreciprocal phase shifters are used. This causes the transmitting and receiving antenna patterns to point in different directions, except at boresight (perpendicular to the antenna's aperture). Thus, in general, the radiation mode RCS is that associated with the antenna sidelobe pattern rather than the main-beam pattern. In a 10- to 20-GHz phased array with a 12-in^2 aperture, the RCS was reduced from 0.15 to 0.60 m^2 to 0.004 to 0.015 m^2 using this technique.

A range of mean RCS for typical conventional radar targets is depicted in Figure 8.10 [26]. Also shown is the volume of a perfectly conducting reference sphere that exhibits the indicated RCS. The sphere is intended to show that the RCS of an object is roughly proportional to 2/3 power of the object's volume.

8.2.1.1 Rebalancing the Radar Equation

The basic thrust of current low-observable technology is to cause a mismatch between the radar, which was designed to detect conventional targets, and the target, which is now much smaller in RCS than the radar was designed to handle. The variation of free-space detection range $R = k\sigma^{1/4}$ with RCS favoring the radar since a sixteen-fold reduction in RCS reduces the detection range only by a factor of 2. This indicates that the RCS of a high-altitude stealth vehicle that attempts to escape detection must be very small if it is to be effective. In addition, the energy radiated by the radar can be increased

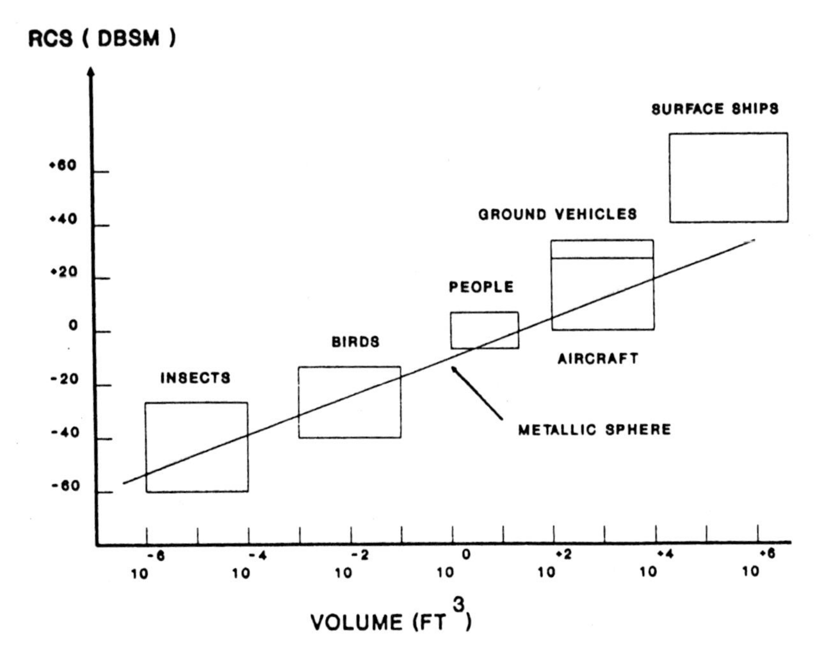

Figure 8.10 Radar cross section of conventional targets. (*Source:* [30].)

(e.g., using PC to maintain the radar's resolution) to compensate for the reduced RCS and bring the radar-target detection equation into balance.

Parameters available to rebalance the free space radar equation are identified in equation form as

$$R = k[P_t \tau A_r t_s \sigma]^{1/4} \tag{8.4}$$

where the annotations identify P_t — Peak power; τ — Pulse width (pulse compression); A_r — Antenna size; t_s — Search time.

This equation indicates that if each of the identified four factors were increased by a factor of 5.6, then this would compensate for an overall reduction of RCS of the order of 1000:1 (i.e., a target of 0.001 m^2 could be detected if the radar was designed initially to detect a 1-m^2 target). This approach is consistent with the counter-stealth vulnerability study that concluded the most effective (cost and technical) approach was to increase the capability of existing monostatic radar.

EA equations can also be evaluated for a rebalanced radar. The burn-through range for a self-screening jammer is given by

$$R_{BT} = \Big[k \overbrace{\sigma}^{\text{Stealth target RCS}} \Big/ \underbrace{P_j G_j}_{\text{Jammer ERP}} \Big]^{1/2} \tag{8.5}$$

where P_jG_j is the required jamming ERP. If the target RCS is reduced by a large factor (e.g., 1000), then it can be concluded that little or no jammer ERP is required to defend the target. However, when the radar equation is rebalanced, the burn-through equation is given by

$$R_{BT} = \left[k \frac{\overbrace{P_T G_T}^{\text{Radar ERP}} \tau B_j \sigma}{\underbrace{P_j G_j}_{\text{Jammer ERP}}}\right]^{1/2} \tag{8.6}$$

(where τ = Radar pulse width (pulse compression); B_j = Jammer bandwidth; σ = Stealth target RCS)

An examination of this equation indicates that the required jammer ERP is now very similar to that required for operation against a conventional radar.

Equation 8.5 is also appropriate for stand-off jamming against stealth targets, except that the constant k is different. This equation indicates that stand-off jamming is more effective with stealth targets than with conventional targets due to the considerable RCS reduction in the stealth RCS. However, when the radar equation is rebalanced for stand-off noise jamming, it becomes

$$R_{BT} = \left[k \frac{\overbrace{P_j G_j}^{\text{Radar ERP}} \tau B_j \sigma}{\underbrace{P_j G_j}_{\text{Jammer ERP}}} \underbrace{\left(\frac{G_T}{G_{SL}}\right)}_{\text{Sidelobe ratio}}\right]^{1/4} \tag{8.7}$$

(where τ = Radar pulse width (pulse compression); B_j = Jammer bandwidth; σ = Stealth target RCS)

An examination of this equation indicates that the jammer ERP is similar to that required to defend a conventional target, except that the ERP may have to be increased due to the use of higher PC ratios.

For repeater jammers, both the repeater gain (transmit, receive antenna gain, and electronic gain) and the maximum jammer peak power are proportional to the target RCS that must be defended [15]. To defend against a conventional radar with a stealth target, only minimal (if any) repeater gain or peak power is required. However, when the radar equation is rebalanced, the required gain remains reduced in direct proportion to the RCS reduction (since it is determined by the target's RCS), but the peak jammer power is given by

$$P_{REP} = k_1 \overbrace{P_T G_T}^{\text{Radar ERP}} \underbrace{\sigma}_{\text{Stealth target RCS}} \tag{8.8}$$

This equation shows that the peak repeater jamming power becomes similar to that for a conventional target.

The conclusions drawn from the prior discussion are that EA systems for stealth vehicles operating against conventional radars are generally not required. In fact, they may compromise the stealth characteristics of the vehicle. Stand-off jamming becomes more effective using stealth vehicles while not increasing the vulnerability of the target. Repeater jammers, if used with stealth vehicles, can be implemented with low-power solid-state transmitters against conventional radars. Decoys become more effective when used with stealth targets against conventional radars. The use of expendables to protect the integrity of the low-observable signature is indicated against conventional radars.

When more powerful radars are built (i.e., the radar equation is rebalanced), the situation changes and EA systems will be required aboard stealth vehicles. Also, against conventional radar, some form of low-frequency or millimeter-wave jammer may be required due to the increased RCS of the stealth vehicle at these frequencies.

8.2.1.2 MTI Considerations for Stealth Targets

One problem with increasing radar energy to detect small RCS targets is that clutter returns are increased in the same proportion as the radar's power (assuming that the radar's resolution is maintained). This may not only present a problem in rejecting the increased clutter return (i.e., increased subclutter visibility), but small targets (e.g., insects and birds) are now detected because they exhibit RCSs in the same range as the stealth targets (e.g., −20 to −30 dB/m^2; see Figure 8.10).

This problem is illustrated by an experiment run to determine the effect of clutter on the detection of low-RCS targets [27]. A sensitive S-band surveillance radar (P_p = 500 kW, τ = 0.8 μs, PRF 1040 pulses/s) using a double-canceler MTI with a 35-dB improvement factor was set up in a heavy ground clutter environment that also included bird and insect targets. Figure 8.11

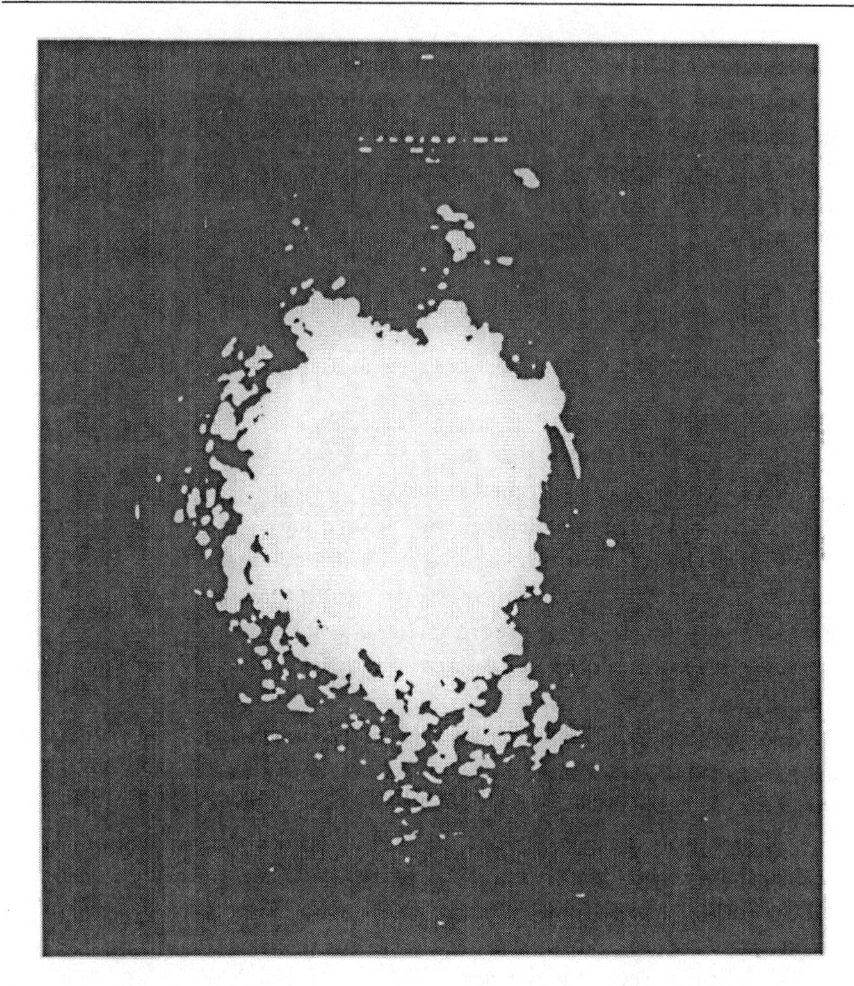

Figure 8.11 Clutter returns for MTI surveillance radar with low-RCS targets: normal video ground clutter return that extends out to about 30 nmi. (*Source:* [27].)

depicts the normal video ground clutter return that extends out to about 30 nmi. Figure 8.12 depicts the same return and MTI processing with an estimated sensitivity to detect targets with 0.1-m^2 cross section in the MTI passband. In Figure 8.13, the estimated sensitivity was extended further (by reducing STC and increasing integration gain) so that 0.004-m^2 targets could be detected in the MTI passband. The extensive clutter residuals and angel targets appearing on the display indicate the difficulty in detecting low-RCS targets on the order of –20 to –30 dB/m^2 in heavy clutter.

The performance of the MTI system depicted in Figures 8.11–8.13 illustrates a critical issue encountered in the detection of low-flying, low-observable targets. The rejection of birds, insects, and other slow-moving clutter

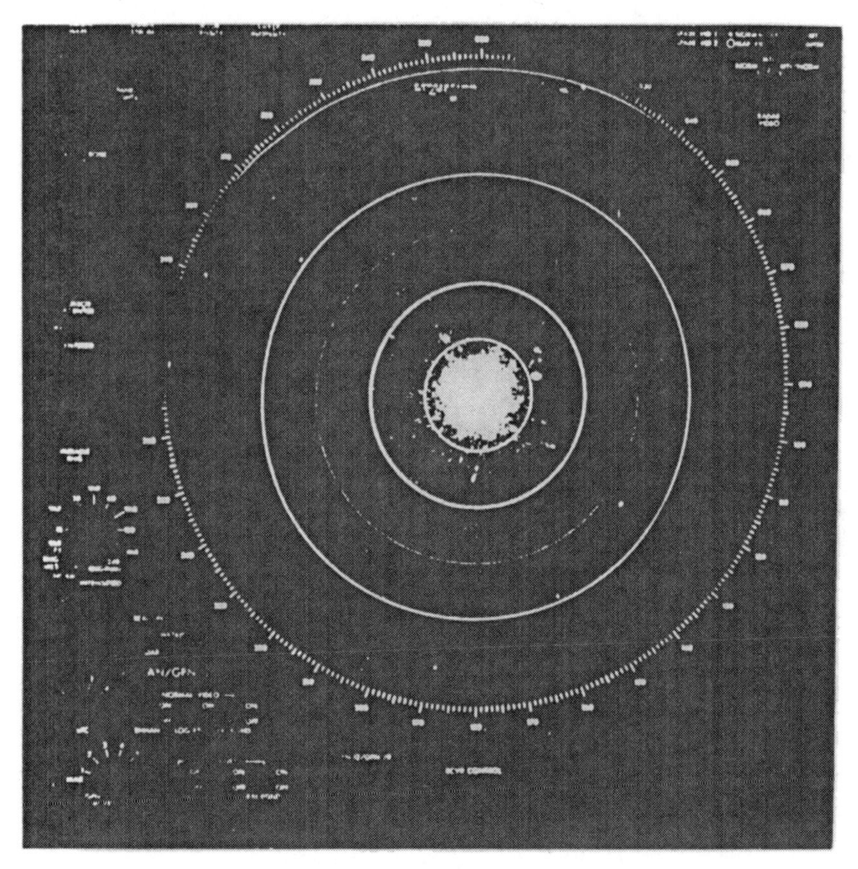

Figure 8.12 Clutter returns with an estimated sensitivity to detect targets with 0.1m^2 RCS in their MTI passband. (*Source:* [27].)

sources requires that the MTI's blind speed be high enough to create a clutter notch in the velocity region occupied by such angle clutter. For example, the S-band MTI used in the GPN-21 has a first blind speed on the order of 100 knots. If a 30-knot rejection notch were used, then the Doppler visibility (i.e., ratio of Doppler region not occupied by clutter to total interval) would be 0.4. However, using a double-canceler MTI filter in the GPN-21 rather than a sharp cut-off filter further compounds the problem. The double-canceler response is equal to $|\bar{H}(\omega)|^2 = 16\sin^4(\omega T/2)$ and the MTI gain (response to uniformly distributed velocity target) $\overline{G} = 6$. Thus, at 30 knots (the 3-dB point), the response for a uniformly distributed velocity target (possibly aliased from a higher velocity region) is 6/8 or −1.25 dB reduced from that of a low-RCS clutter target moving at this radial velocity. This points to the desirability of using a sharp cut-off MTI filter to reject birds, insects, and other slow-moving clutter sources and explains the large amount of clutter targets visible

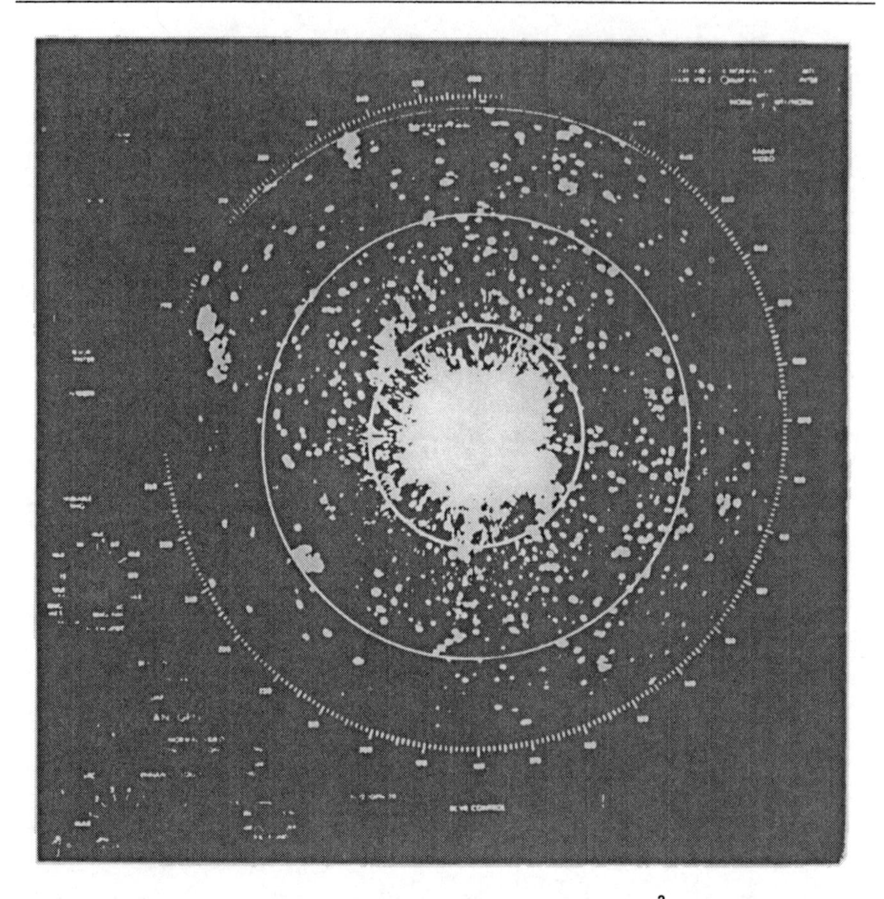

Figure 8.13 Clutter returns with sensitivity increased so that $0.004m^2$ targets could be detected in the MTI passband. (*Source:* [27].)

in Figure 8.13. In addition, use of a lower transmitter frequency (i.e., L-band or UHF) further reduces the response to birds and insects.

Next, consider the same MTI design operating at L band (f = 1.3 GHz). The first blind speed is then 230 knots and the Doppler visibility is increased to 0.74. The 3-dB response is increased to 69 knots, and the response at 30 knots for angle clutter is reduced to −8.2 dB with respect to the response for a uniformly distributed velocity target. A further decrease in the tansmitter frequency to UHF would increase the performance against birds, insects, and other slowly moving clutter sources. When moving clutter sources are removed, system stability becomes the critical factor in determining the performance of an MTI to detect low-flying, low-observable targets.

The quality factor that determines the ability of an MTI radar to detect a low-RCS target in a clutter background is its subclutter visibility or equivalent MTI improvement factor. The subclutter visibility of a radar is ultimately limited by the frequency and phase stabilities of its reference master oscillator and transmitter. Hence, the detection of low-RCS targets in a cluttered background generally depends on how stable the radar can be made.

▶ *Example 8.8*

The free-space variation of radar detection range with RCS is given by $R = k\sigma_t^{1/4}$. The variation of detection range in a clutter-dominated environment is given by

$$R = \frac{\sigma_t}{\sigma_0 \theta R_\tau S_0} = k\sigma_t$$

where σ_t is the target RCS, σ_0 degrees is the backscattering coefficient, θ is the radian azimuth beamwidth, R_τ is $c\tau/2$, τ is the radar PW, and S_0 is the single pulse signal-to-clutter power ratio (i.e., 13.2 dB for $P_d = 0.9$, $P_{fa} = 10^{-6}$). Plot the effect of a relative RCS reduction on detection range in both clutter and in the clear. Assume the value of $\sigma°$ remains constant for all grazing angles experienced by the radar. Note that the variation of detection range in clutter with RCS is more stringent than the free space variation.

```
% Effect of RCS Reduction
% in Free Space and Clutter
% ------------------------
% clutred.m

clear;clc;clf;

% Input RCS Reduction

RCS=0:.01:1;

% Find Reductions

R=RCS.^.25;                 % Free Space Dection Range
C=RCS;                      % Clutter Detection Range

% Plot Effects

Px=[R' C'];
plot(RCS,Px,'k');grid;
xlabel('Relative RCS');
ylabel('Detection Range Reduction');
title('Effect of RCS Reduction on Detection Range');
```

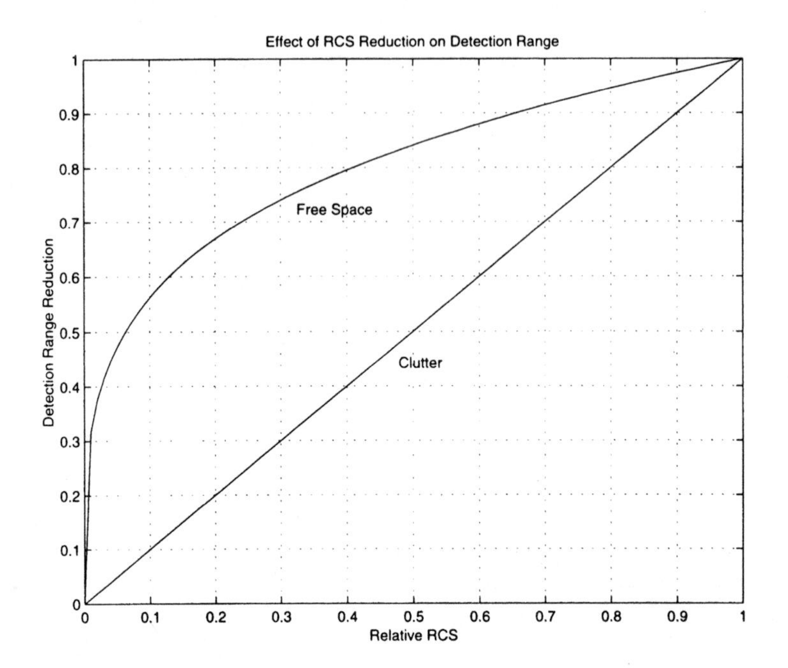

As a further example, the free-space detection range and subclutter visibility requirements for an ASR-9 radar are tabulated as a function of the target's RCS in Table 8.7. From Table 8.7, we can see that useful free-space detection ranges are provided on targets as small as of the order of –30 dBsm. Note that the ASR-9 is only modestly powered compared to many military-type radars of the same class.

The MTI in the ASR-9 provides of the order of 45- to 50-dB subclutter visibility. This would restrict the capability of the ASR-9 to the detection of

Table 8.7
Detection Range and Subclutter Visibility Requirements Against Low-Observability Targets

$\sigma(m^2)$	Free-Space Range (nmi)	SCV (dB)
1	62.6	32.7
0.1	35.2	42.7
0.01	19.8	52.7
0.001 ♦	11.1	62.7
0.0001 ♦ *	6.2	72.7

Note: ASR-9 parameters; ♦, RCS of birds; and *, RCS of insects: σ° = –30 dB.

low-observable targets immersed in ground clutter (low-level penetrators) of the order of −20 dBsm RCS. The ultimate capability of an MTI radar to detect targets immersed in clutter depends on the stability of the transmitter, receiver, and signal-processing elements. In the ASR-9, this limitation is of the order of 60 dB, which is considered representative of the current technology.

The detection of low-observability cross-section targets must also consider that the RCS of birds is of the order of −20 to −40 dBsm at S band and insects of the order of −40 to −70 dBsm. Also, weather clutter at this band is of the order of 0 to −20 dBsm. Elimination of these spurious targets becomes more difficult as the RCS of the desired targets becomes comparable, requiring judicious design of the MTI's velocity response and selection of an appropriate transmitter frequency.

8.3 Problems

1. A HERF gun is used to disrupt computers. The HERF gun operates at 1 GHz, provides a 100-kW 10-μs pulse with a 1-kHz PRF, and uses a 1-m 50% efficient antenna. At what range can it disrupt a computer that is susceptible to 10^{-7} mJ/cm^2 of radiation? Assume a 17-dB loss (equivalent to a wire grid with 0.1-λ spacing) due to penetration of the microwave energy through a building wall. Does this HERF gun meet the microwave radiation safety standard of 10 mW/cm^2?

2. A satellite ELINT receiver is designed to detect the sidelobe radiations (−30-dB sidelobes) of a 100-kW radar operating at 10 GHz using a 1-m 50% efficient antenna and a 1-μs pulse width. The satellite orbits at 100-km altitude. What satellite antenna aperture is required to provide a receiver signal of −70 dBm? Assume an 85-ft-diameter antenna with 70% efficiency is available for use as a ground-based DEW. What peak power transmitter is required to burn out the front-end of the receiver that uses an LNA with a burn-out rating of 25W? What is the resulting power density at the DEW antenna? Is this within the breakdown limits of air at sea level?

3. The breakdown in air for a high-powered microwave beam occurs at densities of 10^5 to 10^6 W/cm^2. Two high-powered operational radars are the FPS-108 Cobra Dane (P_t = 15.4 MW, 95-ft-diameter aperture, f = 1.175 GHz to 1.375 GHz) and the FPQ-6 (P_t = 3 MW, 29-ft-diameter aperture, f = 5.4 GHz to 5.9 GHz). Power densities are highest in the near field of the radar's antenna ($R = 2D^2/\lambda$). How do the power densities of these powerful radars compare to the maximum available power densities limited by air breakdown?

4. At what distance is it safe (10 mW/cm^2) to be illuminated by the main beam of the radars described in Problem 3?

5. For the HPM system described in Example 8.1 (P = 7 GW, D = 3m) assume that the antenna has sidelobes of 0 dBi. At what distance is it safe to approach the HPM system (i.e., where is the power density at a level of 10 MW/cm^2)?

6. The ASR-9 airport terminal surveillance radar transmits 1.2 MW of peak power with an antenna gain of 33.5 dB. The frequency is 2.7 GHz to 2.9 GHz. At what distance is it safe (10 mW/cm^2) to be illuminated by the main beam of this radar?

7. Laser radars often operate in the near field of the optical system whose range is defined as $R = 2D^2/\lambda$, where D is the diameter of the lens. For the CO_2 laser system (λ = 10.6 μm), described in Section 8.1.2, that uses a 50-cm-diameter telescope, what is its near-field range? Laser range finders operating at 0.904 μm use a 12.7-cm-diameter aperture. What is its near-field range?

8. Show that the power received by a laser radar operating within its near-field range is given by

$$P_r = \frac{P_T \sigma_T D^2 \eta}{4R^2}$$

where

P_T = Laser pulse power

σ_T = Target backscatter reflectivity

D = Optics receiving lens diameter

R = Target range

η = Atmosphere and optics transmissivity

9. A ground-based HEL is deployed against a surveillance satellite designed to detect the exhaust plume of an ICBM. The plume has a temperature of 1400K that provides a radiance of $7 \cdot 10^4$ W/m^2-μm at a wavelength of 2.1 μm over a 1-μm band. The plume radiates isotropically with an estimated power density of 3 MW/steradian. The satellite orbits at 1000-km altitude. The IR system uses a 20-cm-diameter collection lens. What is the power level of the signal collected by the lens of the IR system? The HEL employs a 50-cm-diameter telescope to focus the laser energy on the IR satellite. What laser output power is required to produce a jam-to-signal ratio of 20 dB at the IR system? What laser power level is necessary to damage the IR detector (a power density of 10^7 W/cm^2 on a detector of 10^{-4} cm^2 collection area)? What pointing accuracy is required for the HEL system?

10. The ground-based CO_2 laser (λ = 10.6 μm) described in Section 8.1.2 generates 1 kJ per pulse at a 10-Hz repetition frequency through a 50-cm-diameter telescope. What is the minimum pulse width that can be employed by this system to avoid air breakdown (10^7 W/cm^2)? At sea level what is the maximum range at which this system could damage a tactical IR missile's detection system (*Ac* = 10 cm^2, FOV = 10^{-8} steradians, power at detector = 10^3 W)? What is the corresponding range if the detector is located in a satellite overhead, thereby reducing the atmospheric attenuation?

11. A CO_2 laser (λ = 10.6 μm) is used to destroy a drone target by punching a hole through the drone's fuel tank. If the CO_2 laser was located on a mountain top (5000-ft altitude), how much laser power is required to produce a fluence of 5000 J/cm^2 (necessary to burn a hole in a 3-mm-thick aluminum plate) through a 30-cm telescope at 2-km range? Assume the laser is focused on the target for 10 sec.

12. A ground-based HEL is used to disable an intelligence-gathering satellite orbiting at 100-km altitude. The satellite's silicon photocells will shatter at a fluence of 10 J/cm^2. How long would a laser producing 1 kJ of energy at a 10-Hz pulse repetition rate through a 30-cm-diameter telescope have to be trained on the satellite's solar cells to produce this effect?

13. A satellite is orbiting at 100 km. It is desired to overheat the satellite by illuminating it with an energy flux ten times that of solar radiation or 1.4 W/cm^2. How much power is required using a CO_2 laser (λ = 10.6 μm) that illuminates the satellite through a 30-cm-diameter telescope?

14. A fluence of 10^{-5} J/cm^2 from a laser at visible wavelengths can cause retinal damage of an unprotected eye. A fluence of 1 J/cm^2 at IR wavelengths can damage the cornea of the eye. A typical laser range finder operates at 1.06 μm, transmits 75 mJ/pulse, has a PW of 25 μs, a repetition rate of 20 pps, an aperture of 2.5 cm, and a maximum range of 10 km. Is this laser range finder eye safe? The CO_2 laser described in Section 8.1.2 transmits a 1-kJ pulse through a 30-cm telescope at a 10-pps repetition rate. Is this laser eye safe?

15. Figure 8.5 depicts the advantage of a stealth aircraft over a conventional aircraft in terms of the differential time the stealth aircraft has to launch its missile. The radar detection ranges on a 5-m^2 target are listed in Table 8.8.

Using MATLAB, develop the curves given in Figure 8.5 for stealth RCS of 0.001 to 10 m^2.

Table 8.8
Radar Detection Ranges for Conventional and Stealth Fighters ($\sigma = 5$ m^2)

Case	Conventional	Stealth
I	125 km	125 km
II	50 km	50 km
III	125 km	90 km

16. Target resonance enhances the RCS of a target when the wavelength of the radar approaches the periphery of a target feature. Illustrate the resonance effect by using MATLAB to plot the RCS of a sphere with radius a. The formula for the spheres RCS is given by

$$\frac{\sigma}{\pi a^2} = \left| \sum_{n=1}^{\infty} \frac{(-1)^n(2n+1)}{f_n(x)[xf_{n-1}(x) - nf_n(x)]} \right|^2$$

where $x = 2\pi a/\lambda$ and the spherical Bessel functions $f_n(x)$ can be found from the recursive relationship

$$f_n(x) = \frac{2n-1}{x} f_{n-1}(x) - f_{n-2}(x)$$

where $f_0(x) = 1$ and $f_1(x) = (1/x) - j$. The number of terms needed to evaluate the series is given by

$$N = 8.53 + 1.21x - 0.001x^2$$

Note that x is the circumference of the sphere in wavelengths. What is the magnitude of the resonance enhancement of the RCS?

17. A ship made of steel has a length of 300m, a beam of 45m, and an average height of 22.5m. To minimize RCS, the ship is made in the form of an ellipsoid, half of which is below the surface. Estimate the RCS of this ship from the bow, beam, or stern aspects. What would the RCS of this ship be if it were covered with 98% efficient RAM material [29]?

18. An offset horn-fed radar antenna operating at 3 GHz uses a parabolic reflector 3 ft in diameter. It is illuminated by another radar operating at 10 GHz that is aligned to its boresight. Assume that the horn feed is followed by a preselector that acts as a short circuit except in a 500-MHz band about the operating frequency. What is the effective RCS of the reflecting antenna?

What would its RCS be if the illuminating radar was 15 degrees off boresight? What would the RCS be if the frequencies of the illuminating and reflecting antennas were interchanged?

19. The TPS-70 radar that operates at a frequency of 2.9 GHz to 3.1 GHz uses a 215- by 100-in flat plate antenna. What is the effective RCS of this antenna when illuminated at boresight by a radar operating at 1.2 GHz? What is the RCS when illuminated 30 degrees off boresight?

20. Two black boxes and a shelf in an aircraft cockpit form a square corner reflector 15 cm on a side. At a wavelength of 3 cm, find the maximum contribution of this corner reflector to the radar cross section of the aircraft [29].

21. The ASR-9 radar is designed to detect a 1-m^2 target at a free-space range of 50 nmi. It transmits a 1-μs 1.2-MW pulse at a PRF of 1200 pps and has a scan rate of 75 degrees/s. How would you modify the parameters of this radar to detect a 0.001-m^2 target at 50 nmi?

22. The MTI improvement factor is defined as the ratio

$$IF = \frac{(S/C)_i}{(S/C)_0}$$

where $(S/C)_0$ is the signal-to-clutter ratio after MTI processing. Modify the equation given in Example 8.8 to include the effects of the MTI improvement factor. What is the variation with reduced RCS on the improvement factor requirements?

23. The MTI improvement factor of the ASR-9 radar has been measured as 45 dB. Assume that this is composed of a component due to transmitter stability (IF_s) of 48 dB and a filtering component (IF_f) of 48 dB. These combine as

$$IF = \frac{1}{IF_s^{-1} + IF_f^{-1}}$$

Using the results of Problem 22, determine what stability and filtering improvement factors are needed to allow the ASR-9 to detect 0.001-m^2 targets when it was designed to detect 1-m^2 targets.

24. Assume that a stand-off noise jammer is designed to provide a 15-nmi burn-through range on the ASR-9 radar that is designed to detect a 1-m^2 target at 50-nmi range with no jamming. It requires a jamming power of 2000 W/MHz for a 50-nmi jammer stand-off range. What noise jamming power is required to shield a stealth target with 0.001-m^2 RCS to 15 nmi?

What noise jamming power is required to shield the stealth target to 5 nmi burn-through range?

25. A repeater jammed radar exhibits a burn-through range ($J/S = 1$) of 3 nmi. The repeater is designed to provide a magnification factor of 10 ($J/S = 10$) against a 1-m^2 target and has a peak power of 1 kW. If a stealth target of 0.01 m^2 is to be defended by the repeater jammer, what is the burn-through range for the stealth target? If the 3-nmi burn-through range is maintained, what peak repeater jammer power is required?

References

[1] The Military Critical Technology List, Section 4, Directed and Kinetic Energy Systems Technology, Aug. 1996.

[2] Marquet, L., "Aircraft Survivability and Directed Energy Weapons," *Fall 1995 Aircraft Survivability Newsletter,* Surviac, 1995.

[3] Bogner, A., "ATIRCM Overview," *ATEDS Conf.,* Naval Postgraduate School, Feb. 1996.

[4] Knowles, J., "Early Morning DEW: Directed Energy Weapons Come of Age," *J. Electronic Defense,* Oct. 1996.

[5] Florig, H., "The Future Battlefield: A Blast of Gigawatts," *IEEE Spectrum,* March 1988.

[6] Herskovitz, D., "Killing Them Softly," *J. Electronic Defense,* Aug. 1993.

[7] Taylor, C., and N. Younan, "Effects from High Power Microwave Illumination," *Microwave J.*, June 1992.

[8] Kopp, C., "Hardening Your Computing Assets," *Computer Magazine,* 1996.

[9] Callahan, M., "Laser Weapons," *IEEE Spectrum,* March 1982.

[10] "Pulsed Laser Vulnerability Test System," Aircraft Survivability and Directed Energy Weapons," *Fall 1995 Aircraft Survivability Newsletter,* Surviac, 1995.

[11] Skolnik, M., *Radar Handbook,* New York: McGraw-Hill, 1970.

[12] Miller, J., and E. Nolting, "Charged Particle Beam Weapons, Aircraft Survivability and Directed Energy Weapons," *Fall 1995 Aircraft Survivability Newsletter,* Surviac, 1995.

[13] The Military Critical Technology List, Section 16, Signature Control Technology, Aug. 1996.

[14] Adams, J., "How to Design an Invisible Aircraft," *IEEE Spectrum,* April 1988.

[15] Schleher, D. C., *Introduction to Electronic Warfare,* Norwood, MA: Artech House, 1986.

[16] Schleher, D. C., *MTI and Pulsed Doppler Radar,* Norwood, MA: Artech House, 1991.

[17] Sweetman, W., "The Future of Airborne Stealth," *International Defense Review,* March, 1994.

[18] Sweetman, W., "Radical Design of Stealth ACM," *Janes Defense Weekly,* March 1990.

[19] Swedish Navy Web Page.

[20] Bond, D., "USAF Study Asserts That Soviet Defenses Would Be Ineffective Against B-2 Bomber," *Aviation Week and Space Technology,* Oct. 30, 1989.

[21] Sweetman, W., "A Needle in the Haystack—Assessing Stealth Countermeasures," *International Defense Review,* March 1994.

[22] Dornham, M., "USAF Display of F-117A Reveals New Details of Stealth Aircraft," *Aviation Week and Space Technology,* April 30, 1990.

[23] Richardson, D., "Is Stealth Misleading," *Interavia,* Oct. 1989.

[24] Gething, M., "American Stealth Emerges," *Defense,* 1989.

[25] Streetly, M., "Hiding from Radar," *Interavia,* Nov. 1988.

[26] Knott, E., "Radar Cross Section," in E. Brookner (ed.), *Aspects of Modern Radar,* Norwood, MA: Artech House, 1988.

[27] Brown, R., E. Starczenski, and M. Wicks, "Clutter Considerations in Modern Radar Systems," RADC-TR-88.252, Rome Air Development Center, Rome, NY, Oct. 1988.

[28] Pollock, D., *The Infrared & Electro-optical Systems Handbook, Vol. 7 Countermeasures Systems,* Bellington, WA: SPIE Optical Engineering Press, 1993.

[29] Hosington, D., *Electronic Warfare/Electronic Combat,* 1994.

Appendix A

Radar Jamming Modeling and Analysis Tool

RGJMAT calculates the search radar detection range for conventional, frequency agility, and PD radar under various jamming and enviromental conditions. Both steady and fluctuating Swerling-type targets are considered in the presence of self-screening and up to five stand-off jammers, including the effects of standard atmospheric and weather attenuation.

RGJMAT is a MATLAB-based program that consists of three variations, depending upon the input-output data requirements of the user. These variations are selected by opening MATLAB files rj.m for the basic version, rjx.m for additional output data, and rjv2.m for a tableau type input. All versions print out the radar's detection range with atmospheric and weather attenuation and with jamming for all five Swerling-type targets and two additional variations due to frequency agility effects. The directory in which the basic file is run must also include five other MATLAB files: dB_ratio.m, ar_atten.m, det_fact.m, jam_eff.m, and soj_eff.m. All versions provide a bar chart of the detection ranges with and without jamming for the five Swerling types and for the two frequency agility cases.

The capabilities of the three versions are listed in Table A.1. The rj and rjx versions must be edited before use to enter the appropriate radar, target, weather, and jamming parameters. If these programs are run repetitively, it is appropriate to save to a special file (e.g., rjasr-9). The rjv2 version is arranged so that the default values appears in the input tableau that must be overwritten to enter the parameters for a new radar.

Table A.1
Input and Output Parameters for RGJMAT

	Version		
	rj	rjx	rjv2
Input:	Edit program	Edit program	Tableau
Output parameter printout:			
Radar	x	x	
Target		x	
Weather		x	
Jammers		x	
No Jam Range	x	x	x
Jam Range	x	x	x

A.1 Run Instructions

To run RGJMAT, first install MATLAB files rj.m, rjx.m, rjv2.m, dB_ratio.m, ar_atten.m, det_fact.m, jam_eff.m, and soj_eff.m in the active MATLAB directory (e.g., MATLAB 5). For versions rj.m and rjx.m, one must enter the radar, target, weather, and jamming parameters by editing these files—specifically, by clicking *file, open, rj* (or *rjx*), *open* in succession. The parameters of the subject radar are then entered as follows by overwriting the default values.

- trans_pwr_radar — radar peak power in kW
- freq_radar — radar frequency in MHz (400 MHz to 240 GHz)
- pulse_width — radar pulsewidth in μs
- GTDB_radar — radar transmitter antenna gain in dB
- GRDB_radar — radar receiver antenna gain in dB
- GRDB_*sl*_radar — radar receiver antenna sidelobe gain in dBi
- noise_fig_radar — noise figure of radar receiver in dB
- loss_radar_dB — total radar losses in dB
- PRF — radar pulse repetition frequency in Hz
- prob_det — probability of detection
- prob_false_alarm — probability of false alarm
- BW_dop — Doppler filter bandwidth in Hz (if non-Doppler radar enter the PRF value)
- BW_fa — frequency agility bandwidth in MHz

• Azimuth_bw	azimuth beamwidth in degrees
• az_rate	azimuth scan rate in degrees/sec
• RCS	radar cross section in m^2
• el_tgt_deg	target elevation in degrees
• target_length	target length in m
• rain_fall	rain fall rate in mm/hr
• SSJ	self-screening jamming (1 for yes, 0 for no)
• SOJ	stand-off jamming (1 for yes, 0 for no)
• SSJ_pwr_jam	self-screening jam power in W
• SSJ_gain_dB	antenna gain of self-screening jammer in dB
• SSJ_bw	bandwidth of self-screening jammer in MHz
• SSJ_loss_dB	jammer losses with self-screening jammer in dB

The parameters for stand-off jamming are entered in vector form where the length of the vector is equal to the number of stand-off jammers.

• SOJ_pwr_jam	stand-off power of jammers in W
• SOJ_gain_dB	antenna gain of stand-off jammer in dB
• SOJ_bw	stand-off jammer bandwidth in MHz
• jam_range	range of stand-off jammer with respect to radar in nmi?
• jam_height	height of jammer in ft
• SOJ_loss_dB	jammer losses with stand-off jamming in dB

When the rjv2 variation is used, it is not necessary to edit the MATLAB files. A question-and-answer tableau will appear that allows the default values to be overwritten.

To run RGJMAT type rj, rjx, or rjv2 at the MATLAB prompt and press *enter.* With the rj, rjx, and rjv2 versions, the output data appears on the display screen in addition to the bar chart figure. To record the graph, press *print* on the figure menu. The numerical data can be recorded by closing the figure and clicking *print* on the *file* menu (be sure to clear the session before starting). The figure can also be recorded in Word by typing *notebook* at the MATLAB prompt, which transfers the program to Word, and then typing rj (or rjx) followed by *control, enter.*

A.2 Program Discussion and Notes

The RGJMAT program uses Blake's search radar equation to calculate the detection range. Barton's and Alberhseim's methods are used to calculate the various detection factors associated with the Swerling target models [1].

For pulse Doppler radar, the number of effective coherent pulses is calculated by the ratio of the PRF to the Doppler bandwidth. The number of noncoherent pulses integrated is determined by dividing the total number of hits by the number of effective coherent pulses. For a non-Doppler radar, the Doppler bandwidth is made equal to the radar's PRF so that all pulses are noncoherently integrated.

Frequency agility has two effects in the program. First, the jamming bandwidths (both self-screening and stand-off) must be adjusted so that they are larger than the frequency agility bandwidth of the radar. For target detection, the number of independent hits is calculated as $n_e = 1 + 2\Delta f \cdot l / c \leq n$ where l is the target length and Δf is the frequency agility bandwidth. The fluctuation loss of the Swerling 1 and Swerling 3 targets are reduced in proportion to the number of independent hits [1].

Atmospheric attenuation is obtained via a look-up table of coefficients stored in the program. Values are available over a frequency range from 400 MHz to 240 GHz. Logarithmic interpolation is used to find intermediate values of the coefficient.

The relationship used to find the jamming range is

$$R_j = \frac{R_c}{\left[1 + \frac{1}{L_a}\left(\frac{R_o}{R_{sso}}\right)^4 + \sum_{i=1}^{n}\left(\frac{R_o}{R_{sooi}}\right)^4\right]^{1/4}} \tag{A.1}$$

where

$$R_c = R_o \cdot 10^{-D_i(n)/40} \tag{A.2}$$

and R_o is the clear range where $S/N = 1$

$$R_o = 129.2\left[\frac{P_{t(\mathrm{kW})} G_t G_r \tau_{\mu_s} \sigma}{f^2_{(\mathrm{MHz})} T_s L_R}\right]^{1/4} \tag{A.3}$$

R_{sso} is the range with just self-screening jamming and $S/J = 1$

$$R_{\text{sso}} = 4.8116 \times 10^{-3}\left[\frac{P_{t(\text{kW})} G_t G_r \tau_{\mu_s} \sigma}{p_{j(\text{W/MHz})} G_{\text{j}} L_{\text{j}}}\right]^{1/2} \tag{A.4}$$

and R_{ssoi} is the range for each individual stand-off jammer

$$R_{ssoi} = (R_{\text{j}} R^{*}_{ssoi})^{1/2}\left(\frac{G_R}{G_{Ri}}\right)^{1/4} \tag{A.5}$$

where

R^{*}_{ssoi} = R_{sso} evaluated at $p_{\text{j}} G_{\text{j}} L_{\text{j}}$ for the stand-off jammer

p_{j} = Jammer noise spectral density (W/Hz)

G_{j} = Jammer antenna gain in direction of radar

G_{Rj} = Radar antenna gain in direction of jammer

R_{j} = Range between the radar and jammer

L_{j} = Loss due to attenuation between point where the jammer transmitter power is measured and the radar antenna terminals

$D_{\text{i}}(n)$ = The appropriate detectability factor in decibels for the appropriate Swerling-type P_{d}, P_{fa}, and the number (n) of integrated pulses

The loss term due to atmospheric and rain attenuation is

$$L_a = 10^{L_{\text{dB}}/20} \tag{A.6}$$

where

$$L_{\text{dB}} = k_{\text{a}} R_{\text{E}}(1 - e^{R/R_{\text{E}}}) + r k_{\text{aR}} R \tag{A.7}$$

k_{a} = Standard atmospheric coefficient (dB/km)

k_{aR} = Weather coefficient (dB/km)

r = Rain rate (mm/hr)

R_{E} = Range parameter calculated from ϕ_{c}

$$\phi_{\text{c}} = \phi_{\text{t}} + \frac{2.5 \cdot 10^{-4}}{\phi_{\text{t}} + 0.028} \tag{A.8}$$

where ϕ_{t} is the target's elevation angle. The range parameter in kilometers is then

$$R_E = \frac{3}{\sin(\phi_c)} \tag{A.9}$$

where the preceding relationship uses an approximation due to Barton.

Since the value of range depends on the attenuation and vice versa, a nonlinear equation must be solved to obtain the range. This is accomplished using Newton's method.

The range equation is

$$F(R) = \log(R_j/R_c) + \frac{1}{4}\log(1 + a/L_a + b) = 0 \tag{A.10}$$

$$a = \left(\frac{R_o}{R_{sso}}\right)^4 \qquad b = \left(\frac{R_o}{R_{sso}}\right)^4 \tag{A.11}$$

The derivative is

$$F'(R) = (1/R) - \frac{aL'_a}{4L_a^2(1 + b) + aL_a} \tag{A.12}$$

where

$$L'_a = 2.303L_a\left[\frac{1}{R} + \frac{k_{aR}}{20} + \frac{k_a}{20}e^{-R/R_E}\right] \tag{A.13}$$

The range is found from

$$R_N = R_{N-1} - \frac{F(R_{N-1})}{F'(R_{N-1})} \tag{A.14}$$

with a starting value of

$$R_{N-1} = R_{sso} \tag{A.15}$$

for self-screening jamming and

$$R_{N-1} = R_{soo} \tag{A.16}$$

for standoff jamming. The program iteratively steps through the recursive algorithm until the differences in the successive range estimates are less than 0.1 km.

A.3 Examples

RGJMAT can be used to determine the detection range of a number of search radar that use various combinations of frequency agility, Doppler processing, mixed coherent and noncoherent integration, and different transmitter and receive functions under varying environmental and jamming conditions. One difficulty is in estimating the detection losses associated with the radar and jamming system. A default value is provided and can be used if more accurate information is not available.

The following examples illustrate the program's capabilities.

▶ *Example A.1*

The TPS-43 is a stacked beam search radar with the given parameters. Determine its detection range against a 1-m^2 target under clear weather conditions and with stand-off jamming. The jammer parameters are also given. The rms sidelobe level is estimated at 10 dBi for this relatively complex antenna system. Note that the system has a frequency agility that requires 200-MHz jammer bandwidth. Also, the radar does not provide coherent integration that requires the Doppler bandwidth be set to the PRF value.

Radar Parameters	**Stand-off Jammer Parameters**
P_t = 4000 kW	P_j = 2000W
f = 3000 MHz	G_j = 5 dB
PRF = 250 pps	B_j = 200 MHz
PW = 6.5 μs	R_j = 50 nmi
G_t = 36 dB	L_j = 7 dB
G_r = 40 dB	H_j = 20,000 ft
θ = 1.1 degrees	G_{sl} = 10 dB
θ = 36 degrees/s	Number = 5
NF = 4.5 dB	
L = 12 dB	
ϕ_t = 0.4 degrees	
RCS = 1 m^2	
B_{dop} = 250 Hz	
FA = 200 MHz	

```
% Range Calculations in a Jamming Environment
% -----------------------------------------
% rj.m
%
% This program calculates radar ranges in a jamming environment.
% It works with both Stand-off jamming and self-screening jamming
% for steady and Swerling type targets with frequency agility,
% coherent integration and standard atmosphere/rain attenuation
%
% Code translated from BASIC ''RGJMAT.BAS''
%
% Last edited: 28 August 1997.

%%%%%%%%%%%%%%%%%%%%%%%%%%%%%%%%%%%%%%%%%%%%%%%%%%%%%%%%%%%%%%%%
%%%%%%%%%%%%%%%%%%%%%
%
clear
clc
echo off
%
% Set user variables.  The user should change the following
% variables as needed for the simulation they are running.
% Descriptions of the variables follow.
%
% Radar related:
%
trans_pwr_radar = 4000;     % Transmitter power of radar - kw
freq_radar = 3000;          % Radar's frequency - MHz
pulse_width = 6.5;          % Radar's pulse width - microsecs
GTDB_radar = 36;            % Transmitter gain of radar - dB
GRDB_radar = 40;            % Receiving gain for radar - dB
GRDB_sl_radar = 10;         % Side-lobe of radar - dBi
noise_fig_radar = 4.5;      % Noise figure of receiver - dB
loss_radar_dB = 12;         % Radar losses - dB
PRF = 250;                  % Pulse repetition freq - pps
prob_det = 0.90;            % Probability of detection  < 1
prob_false_alarm = 1e-6;    % Probability of false alarm < 1
BW_dop = 250;               % Doppler filter BW - Hz
BW_fa = 200;                % Frequency agility BW  - MHz
Azimuth_bw = 1.1;           % Azimuth beamwidth - degrees
az_rate = 36;               % Azimuth scan rate - degrees/sec

% Target related:
%
RCS=1;                      % Radar cross section - m^2
el_tgt_deg = 0.4;           % Elevation of target - degrees
target_length = 15;         % Length of target - m

% Select jamming type (enter 1 for yes, 0 for no; can select
% none, one, or both):
%
```

```
SSJ = 0;                    % Selfscreening Jamming? boolean
SOJ = 1;                    % Stand-off Jamming?  boolean

% Enter characteristics for selfscreening Jamming.  If not
% selected, values won't be used.
%
SSJ_pwr_jam = 10;           % SSJ power of jammer - w
SSJ_gain_dB = 0;            % Gain of SS jammer in dB - dB
SSJ_bw = 20;                % Bandwidth of SSJ - MHz
SSJ_loss_dB = 7;            % Losses with SSJ - dB

% Enter characteristics for stand-off Jamming.  If not, selected,
% values won't be used.
%
% Each array must be equal and there must
% be a value for each stand-off jammer you
% wish to simulate
%
SOJ_pwr_jam = [2000 2000 2000 2000 2000];   % SOJ power  - W
SOJ_gain_dB = [5 5 5 5 5];                  % Gain of SOJ - dB
SOJ_bw = [200 200 200 200 200];             % SOJ Bandwidth - Mhz
jam_range = [50 50 50 50 50];               % Jammer Range -Nmi
jam_height = [20000 20000 20000 20000 20000];% Height of jammer - ft
SOJ_loss_dB = [7 7 7 7 7];                  % SOJ Losses - dB

% Weather
%
rain_fall = 0;          % Rain fall rate - mm/hr

% Print radar, target , jamming and Weather Parameters

fprintf('Radar Parameters \n\n');
par1=[trans_pwr_radar;freq_radar;pulse_width;PRF;GTDB_radar;GRDB_radar;
noise_fig_radar];
fprintf('  Pt(kw)       f(Mhz)     PW(1e-6*s)  PRF(pps)    Gt(dB)
Gr(dB)      NF(dB)\n');
fprintf('%-14.1f',par1);fprintf('\n\n');
par2=[loss_radar_dB;BW_dop;BW_fa;Azimuth_bw;;az_rate];par3=
[prob_det;prob_false_alarm];
fprintf('Lr(dB)    Dop Bw(Hz)  FA Bw(MHz)  Az Bw(deg)  Az Rate(deg/s)   Pd
Pfa\n');
fprintf('%-17.2f',par2);fprintf('%-14.1g',par3);fprintf('\n\n');

%%%%%%%%%%%%%%%%%%%%%%%%%%%%%%%%%%%%%%%%%%%%%%%%%%%%%%%%%%%%%%%%%%%%%%%
%%%%%%%%%%%%%%%%%%%%%%%
% STOP EDITING.  ONLY PROGRAM FOLLOWS
%%%%%%%%%%%%%%%%%%%%%%%%%%%%%%%%%%%%%%%%%%%%%%%%%%%%%%%%%%%

%%%%%%%%%%%%%%%%%%%%%%%%%%%%
%
% Begin Main
```

```
%

% Convert radar gain, loss, noise to ratios
% loss_radar = db_ratio(loss_radar_dB);
GT_radar = db_ratio(GTDB_radar);
GR_radar = db_ratio(GRDB_radar);
noise_factor_radar = db_ratio(noise_fig_radar);

% Calculate system input noise temp
%
sys_noise_temp = 290 * (noise_factor_radar - 1) + 150;  % in K

% Calculate range without jamming, FA, etc (S/N = 1)
%
range_index = 129.2 * ((trans_pwr_radar * pulse_width * GT_radar
* GR_radar * RCS) / ...
(freq_radar^2 * sys_noise_temp * loss_radar))^.25;

% Calculate number of pulses
%

non_coh_pulses = (Azimuth_bw/az_rate) * BW_dop;
coh_pulses = PRF/BW_dop;

% If target length is greater than zero,computes number of
% independent  pulses for FA radars
%
if target_length > 0
  N_exp = 1 + (BW_fa/(150 / target_length));
  else N_exp = non_coh_pulses;
end

% Determine detectability factors
%
det_fact;

% Calculate ranges for the different target types, clear, no
% atmospheric or weather attenuation accounted for here.
%

swrlg_case= [' 0     1        2       3      4      1fa   3fa'];
ranges_clear = range_index * 10.^((-det_facts)/40);

% echo on
% Now, the atmospheric attenuation and rain attenuation is
% calcuated.  First, the Swerling cases being calculated are
% listed.
% Then under ar_atten_ans, row 1 is the atmospheric attenuation
% (dB) and row 2 is the rain attenuation (dB).  The 3rd row
% contains the  affected ranges (NMI).  Under after_jam_ans, the
% meaning of the rows are the same as for the attenuation matrix.
```

```
% echo off
ar_atten;

% Effects of Jamming...
%
if SSJ==1
  pwrj_density = SSJ_pwr_jam/SSJ_bw;
  SSJ_gain = db_ratio(SSJ_gain_dB);
  SSJ_loss = db_ratio(SSJ_loss_dB);

% Range with just self-screening jamming
%
rng_ssj_only = 0.0048116 * sqrt(trans_pwr_radar * GT_radar * RCS
* pulse_width ./ ...
(pwrj_density .* SSJ_gain .* (loss_radar/SSJ_loss)));
  if SOJ==1
    soj_eff;
  end
  jam_eff;
end
if (SOJ==1)&(SSJ==0)
  soj_eff;
  rng_ssj_only = rs00;
  jam_eff;
end
```

» rj

Radar Parameters

Pt(kW)	f(MHz)	PW(1e-6*s)	PRF(pps)	Gt(dB)	Gr(dB)	NF(dB)
4000.0	3000.0	6.5	250.0	36.0	40.0	4.5

Lr(dB)	Dop Bw(Hz)	FA Bw(MHz)	Az Bw(deg)	Az Rate(deg/s)	Pd	Pfa
12.00	250.00	200.00	1.10	36.00	0.9	le-006

Detection Range with Atmospheric and Weather Attenuation

Swerling Case	0	1	2	3	4	1FA	3FA
Atmospheric Atten (dB)	2.30	1.79	2.23	2.04	2.27	2.27	2.29
Weather Atten (dB)	0.00	0.00	0.00	0.00	0.00	0.00	0.00
Detection Range (n mi)	146.48	94.31	138.26	117.53	142.31	143.43	144.95

Detection Range with Jamming

Swerling Case	0	1	2	3	4	1FA	3FA
Atmospheric Atten (dB)	1.01	0.67	0.96	0.83	0.98	0.99	1.00
Weather Atten (dB)	0.00	0.00	0.00	0.00	0.00	0.00	0.00
Jamming Range (n mi)	43.50	27.20	40.91	34.40	42.19	42.54	43.02

»

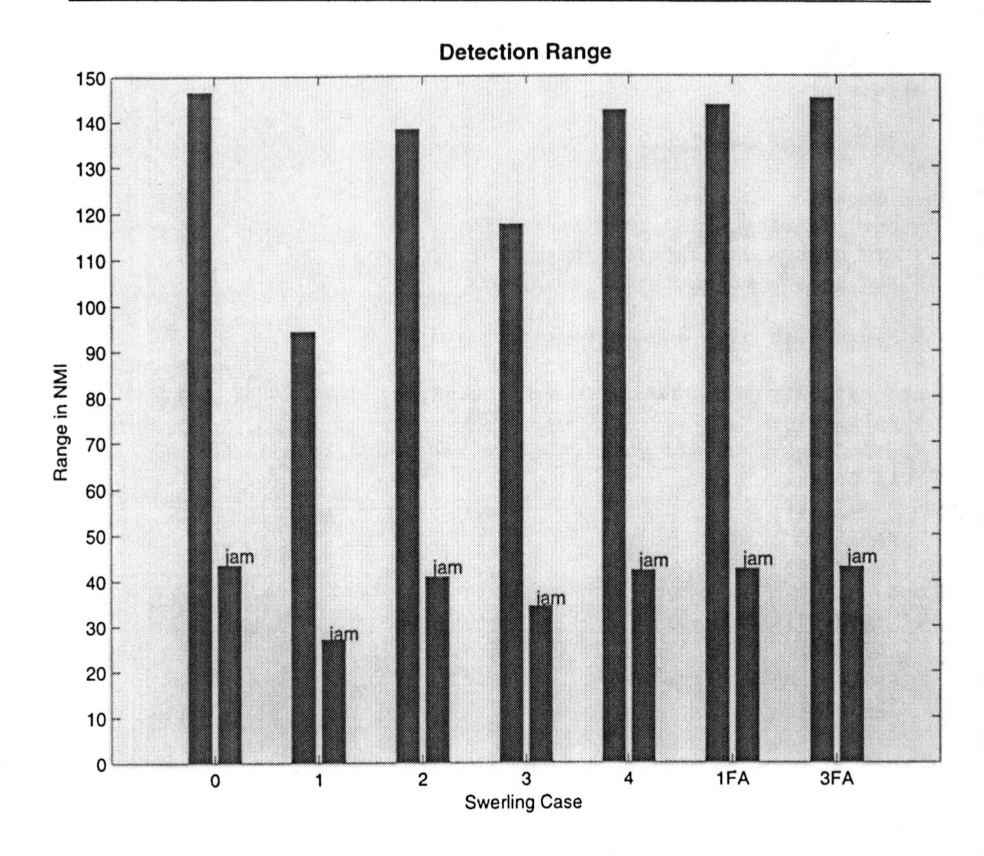

▶ *Example A.2*

The TPS-43 radar is modified by the addition of an ultralow sidelobe antenna (the radar is renamed the TPS-70). The sidelobe is advertised at 45 dB down from the peak antenna main-lobe response or −5 dBi. Determine the detection range under the stand-off conditions listed in Example A.1. Use the rjx.m version of RGJMAT for expanded parameter print out.

```
% Range Calculations in a Jamming Environment
% -------------------------------------------
% rjx.m
%
% This program calculates radar ranges in a jamming environment.
% It works with both Stand-off jamming and self-screening jamming
% for steady and Swerling type targets with frequency agility,
% coherent integration and standard atmosphere/rain attenuation
%
% Code translated from BASIC "RGJMAT.BAS"
%
```

```
% Last edited: 28 August 1997.

%%%%%%%%%%%%%%%%%%%%%%%%%%%%%%%%%%%%%%%%%%%%%%%%%%%%%%%%%%%%%%%
%%%%%%%%%%%%%%%%%%%%%%%%%
%
clear
clc
echo off
%
% Set user variables.  The user should change the following
% variables as needed for the simulation they are running.
% Descriptions of the variables follow.
%
% Radar related:
%
trans_pwr_radar = 4000;  % Transmitter power of radar - kw
freq_radar = 3000;       % Radar's frequency - MHz
pulse_width = 6.5;       % Radar's pulse width - microsecs
GTDB_radar = 36;         % Transmitter gain of radar - dB
GRDB_radar = 40;         % Receiving gain for radar - dB
GRDB_sl_radar =-5;       % Side-lobe of radar - dBi
noise_fig_radar = 4.5;   % Noise figure of receiver - dB
loss_radar_dB = 12;      % Radar losses - dB
PRF = 250;               % Pulse repetition freq - pps
prob_det = 0.90;         % Probability of detection  < 1
prob_false_alarm = 1e-6; % Probability of false alarm < 1
BW_dop = 250;            % Doppler filter BW - Hz
BW_fa = 200;             % Frequency agility BW  - MHz
Azimuth_bw = 1.5;        % Azimuth beamwidth - degrees
az_rate = 36;            % Azimuth scan rate - degrees/sec

% Target related:
%
RCS=1;                   % Radar cross section - m^2
el_tgt_deg = 0.4;        % Elevation of target - degrees
target_length = 10;      % Length of target - m

% Select jamming type (enter 1 for yes, 0 for no; can select
% none, one, or both):
%
SSJ = 0;                 % Selfscreening Jamming? - boolean
SOJ = 1;                 % Stand-off Jamming? - boolean

% Enter characteristics for selfscreening Jamming.  If not
% selected, values won't be used.
%
SSJ_pwr_jam = 10;        % SSJ power of jammer - w
SSJ_gain_dB = 0;         % Gain of SS jammer in dB - dB
SSJ_bw = 20;             % Bandwidth of SSJ - MHz
SSJ_loss_dB = 7;         % Losses with SSJ - dB
```

```
% Enter characteristics for stand-off Jamming.  If not, selected
% values won't be used.
%
% Each array must be equal and there must
% be a value for each stand-off jammer you
% wish to simulate
%
SOJ_pwr_jam = [2000 2000 2000 2000 2000];      % SOJ power - W
SOJ_gain_dB = [5 5 5 5 5];                     % SOJ Gain - dB
SOJ_bw = [200 200 200 200 200];                % SOJ Bandwidth - MHz
jam_range = [50 50 50 50 50];                  % Range of jammer - Nmi
jam_height = [20000 20000 20000 20000 20000]; % Height of jammer - ft
SOJ_loss_dB = [7 7 7 7 7];                     % SOJ Losses  - dB

% Weather
% rain_fall = 0;          % Rain fall rate - mm/hr

% Print radar, target , jamming and weather parameters

fprintf('Radar Parameters \n\n');
par1=[trans_pwr_radar;freq_radar;pulse_width;PRF;GTDB_radar;
RDB_radar;noise_fig_radar; ...
     GRDB_sl_radar];
fprintf(' Pt(kw)          f(Mhz)      PW(1e-6*s)   PRF(pps)     Gt(dB)
Gr(dB)      NF(dB)');
fprintf('        Gsl(dB)   \n ');
fprintf('%-14.1f',par1);fprintf('\n\n');
par2=[loss_radar_dB;BW_dop;BW_fa;Azimuth_bw;az_rate];par3=
[prob_det;prob_false_alarm];
fprintf('Lr(dB)    Dop Bw(Hz)  FA Bw(MHz)  Az Bw(deg)  Az Rate(deg/s)   Pd
Pfa\n');
fprintf('%-17.2f',par2);fprintf('%-14.1g',par3);fprintf('\n\n');
fprintf('Target and Weather Parameters\n\n');
par4=[RCS;el_tgt_deg;target_length;rain_fall];
fprintf(' RCS(m^2)  tar el(deg)  tar lgth(m)  rain(mm/hr)\n');
fprintf('%-18.1f',par4); fprintf('\n\n');
if SSJ==1;fprintf('Self Screening Jamming Parameters\n\n');
   par5=[SSJ_pwr_jam;SSJ_gain_dB;SSJ_bw;SSJ_loss_dB];
   fprintf(' Pj(w)       Gj(dB)     Bwj(MHz)    Lj(dB)\n');
   fprintf('%-14.1f',par5);fprintf('\n\n');
end;
if SOJ==1;fprintf('Stand-Off Jamming
Parameters\n\n');num=length(SOJ_pwr_jam);
   fprintf('Number of Jammers = %4.0f\n',num);
   fprintf('Pj(w) =          ');fprintf('%-
10.0f',SOJ_pwr_jam);fprintf('\n');
   fprintf('Gj(dB) =       ');fprintf('%-
12.1f',SOJ_gain_dB);fprintf('\n');
   fprintf('BWj(Mhz) = ');fprintf('%-
11.0f',SOJ_bw);fprintf('\n');
   fprintf('Rj(n mi) =    ');fprintf('%-
```

```
11.1f',jam_range);fprintf('\n');
   fprintf('Hj(ft) =       ');fprintf('%-
9.0f',jam_height);fprintf('\n');
   fprintf('Lj(dB) =      ');fprintf('%-
12.1f',SOJ_loss_dB);fprintf('\n\n');
end;

%%%%%%%%%%%%%%%%%%%%%%%%%%%%%%%%%%%%%%%%%%%%%%%%%%%%%%%%%%%%%%%%%
%%%%%%%%%%%%%%%%%%%%%%
% STOP EDITING.  ONLY PROGRAM FOLLOWS
%%%%%%%%%%%%%%%%%%%%%%%%%%%%%%%%%%%%%%%%%%%%%%%%%%%%%%

%%%%%%%%%%%%%%%%%%%%%%%%%%%
%
% Begin Main
%

% Convert radar gain, loss, noise to ratios
%
loss_radar = db_ratio(loss_radar_dB);
GT_radar = db_ratio(GTDB_radar);
GR_radar = db_ratio(GRDB_radar);
noise_factor_radar = db_ratio(noise_fig_radar);

% Calculate system input noise temp
%
sys_noise_temp = 290 * (noise_factor_radar - 1) + 150;  % in K

% Calculate range without jamming, FA, etc (S/N = 1)
%
range_index = 129.2 * ((trans_pwr_radar * pulse_width * GT_radar
* GR_radar * RCS) / ...
(freq_radar^2 * sys_noise_temp * loss_radar))^.25;

% Calculate number of pulses
%

non_coh_pulses = (Azimuth_bw/az_rate) * BW_dop;
coh_pulses = PRF/BW_dop;

% If target length is greater than zero,computes number of
% independent pulses for FA radars
%
if target_length > 0
  N_exp = 1 + (BW_fa/(150 / target_length));
  else N_exp = non_coh_pulses;
end

% Determine detectability factors
%
det_fact;
```

```
% Calculate ranges for the different target types, clear, no
% atmospheric or weather attenuation accounted for here.
%

swrlg_case= ['    0        1         2          3          4       1fa
3fa'];

ranges_clear = range_index * 10.^((-det_facts)/40);

% echo on
% Now, the atmospheric attenuation and rain attenuation is
% calcuated.  First, the Swerling cases being calculated are
% listed. Then under ar_atten_ans, row 1 is the atmospheric
% attenuation (dB)and row 2 is the rain attenuation (dB).  The
% 3rd row contains the  affected ranges (NMI).Underafter_jam_ans,
% the meaning of the rows are the same as for the attenuation
% matrix.
% echo off
ar_atten;

% Effects of Jamming...
%
if SSJ==1
  pwrj_density = SSJ_pwr_jam/SSJ_bw;
  SSJ_gain = db_ratio(SSJ_gain_dB);
  SSJ_loss = db_ratio(SSJ_loss_dB);

% Range with just self-screening jamming
%
rng_ssj_only = 0.0048116 * sqrt(trans_pwr_radar * GT_radar * RCS
* pulse_width ./ ...
(pwrj_density .* SSJ_gain .* (loss_radar/SSJ_loss)));
  if SOJ==1
     soj_eff;
  end
  jam_eff;
end
if (SOJ==1)&(SSJ==0)
  soj_eff;
  rng_ssj_only = rs00;
  jam_eff;
end
```

```
» rjx
Radar Parameters

 Pt(kW)   f(MHz)   PW(1e-6*s)  PRF(pps)  Gt(dB)   Gr(dB)   NF(dB)   Gsl(dB)
 4000.0   3000.0       6.5        250.0     36.0     40.0     4.5      -5.0

Lr(dB)  Dop Bw(Hz)  FA Bw(MHz)  Az Bw(deg)  Az Rate(deg/s)   Pd      Pfa
 12.00    250.00      200.00       1.50         36.00        0.9   1e-006

Target and Weather Parameters

 RCS(m^2)  tar el(deg)  tar lgth(m)  rain(mm/hr)
    1.0        0.4         10.0          0.0

Stand-Off Jamming Parameters

Number of Jammers = 5
Pj(w) =         2000    2000    2000    2000    2000
Gj(dB) =         5.0     5.0     5.0     5.0     5.0
BWj(Mhz) =       200     200     200     200     200
Rj(n mi) =      50.0    50.0    50.0    50.0    50.0
Hj(ft) =       20000   20000   20000   20000   20000
Lj(dB) =         7.0     7.0     7.0     7.0     7.0

Detection Range with Atmospheric and Weather Attenuation

Swerling Case               0       1       2       3       4      1FA     3FA
Atmospheric Atten (dB)    2.36    1.85    2.31    2.10    2.33    2.32    2.34
Weather Atten (dB)        0.00    0.00    0.00    0.00    0.00    0.00    0.00
Detection Range (n mi)  154.29   99.16  147.86  123.67  151.04  149.59  151.92

Detection Range with Jamming

Swerling Case               0       1       2       3       4      1FA     3FA
Atmospheric Atten (dB)    1.90    1.38    1.85    1.63    1.87    1.86    1.88
Weather Atten (dB)        0.00    0.00    0.00    0.00    0.00    0.00    0.00
Jamming Range (n mi)    103.75   65.10   99.21   82.22  101.45  100.43  102.07
»
```

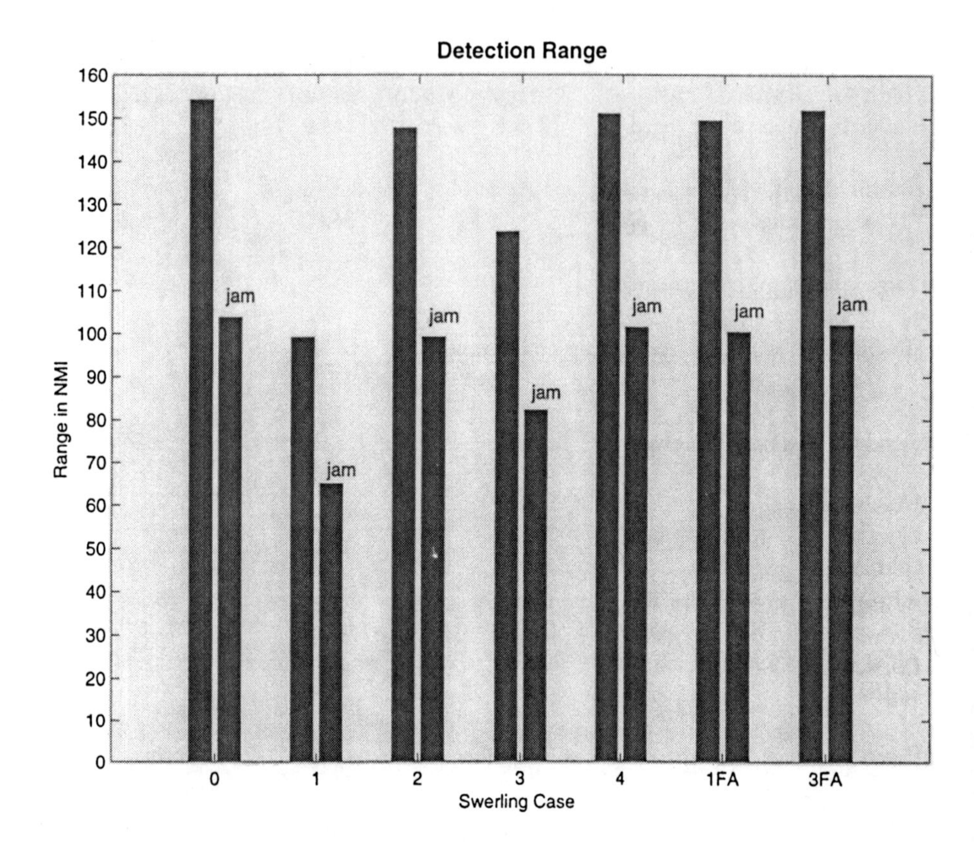
Detection Range
Range in NMI
0
10
20
30
40
50
60
70
80
90
100
110
120
130
140
150
160
jam
jam
jam
jam
jam
jam
jam
0
1
2
3
4
1FA
3FA
Swerling Case

▶ *Example A.3*

The LPD-20 PD radar is used as an alerting radar for the ADATS missile system. Its parameters follow. Determine the detection range of the LPD-20 under clear weather conditions and when subjected to the ensuing stand-off jamming. Use the rjv2.m version of PGJMAT.

Radar Parameters	**Jammer Parameters**
P_t = 8 kW	Stand-off
f = 10,000 MHz	P_j = 1000W
PRF = 7950 pps	G_j = 13 dB
G_t = 29 dB	B_j = 100 MHz
G_r = 29 dB	R_j = 20 nmi
θ = 1.4 degrees	H_j = 20,000 ft
θ = 360 degrees/s	Number = 1
NF = 6 dB	G_{sl} = 3 dBi
L = 15 dB	
ϕ_t = 0 degrees	
RCS = 1 m^2	
B_{dop} = 1000 Hz	
PW = 6.67 μs	

```
% Range Calculations in a Jamming Environment
% -------------------------------------------
% rjv2.m
%
% This program calculates radar ranges in a jamming environment.
% It works with both Stand-off jamming and self-screening jamming
% for steady and Swerling type targets with frequency agility,
% coherent integration and standard atmosphere/rain attenuation
%
% Code translated from BASIC "RGJMAT.BAS"
%
% Last edited: 28 August 1997.

%%%%%%%%%%%%%%%%%%%%%%%%%%%%%%%%%%%%%%%%%%%%%%%%%%%%%%%%%%%%%%%%%%%%%%%%%%%%%%%%%%%%%%%%%%%%
%
clear
clc
echo off
%
% Set user variables.  The user should change the following
```

```
% variables as needed for the simulation they are running.
% Descriptions of the variables follow.
%
% Radar related:
% trans_pwr_radar = 8;    % Transmitter power of radar - kW
freq_radar = 10000;       % Radar's frequency - MHz
pulse_width = 6.67;       % Radar's pulse width - microsecs
GTDB_radar = 29;          % Transmitter gain of radar - dB
GRDB_radar = 29;          % Receiving gain for radar - dB
GRDB_sl_radar = 3;        % Side-lobe of radar - dBi
noise_fig_radar = 6;      % Noise figure of receiver - dB
loss_radar_dB = 15;       % Radar losses - dB
PRF = 7950;               % Pulse repetition freq - pps
prob_det = 0.90;          % Probability of detection < 1
prob_false_alarm = 1e-6;  % Probability of false alarm < 1
BW_dop = 1000;            % Doppler filter BW - Hz
BW_fa = 0;                % Frequency agility BW - MHz
Azimuth_bw = 1.4;         % Azimuth beamwidth - degrees
az_rate = 360;            % Azimuth scan rate - degrees/sec

% MENU Driven Radar Parameters
% Use a dialog box to get input
title = 'Radar Parameters Page 1';
prompt = {'Transmitter power (kW)', 'frequency (MHz)',...
      'pulse width (microsecs)', ...
     'Transmit antenna gain (dB)', 'Receive antenna gain (dB)'...
      'Receive antenna sidelobe (dBi)', 'Noise Figure (dB)', ...
      'Losses (dB)', 'PRF (Hz)'};
default = {trans_pwr_radar, freq_radar, pulse_width,
GTDB_radar,...
     GRDB_radar, GRDB_sl_radar, noise_fig_radar, loss_radar_dB, PRF,};
response = inputdlg(prompt, title, 1, default);
fields =
{'trans_pwr_radar','freq_radar','pulse_width','GTDB_radar',...

'GRDB_radar','GRDB_sl_radar','noise_fig_radar','loss_radar_dB','PRF'};
input = cell2struct(response,fields,1);

% Convert cell structure created by dialog box back to numbers

trans_pwr_radar = str2num(input.trans_pwr_radar);
freq_radar = str2num(input.freq_radar);
pulse_width = str2num(input.pulse_width);
GTDB_radar = str2num(input.GTDB_radar);
GRDB_radar = str2num(input.GRDB_radar);
GRDB_sl_radar = str2num(input.GRDB_sl_radar);
noise_fig_radar = str2num(input.noise_fig_radar);
loss_radar_dB = str2num(input.loss_radar_dB);
PRF = str2num(input.PRF);

clear input response fields default prompt title;
```

```
title = 'Radar Parameters Page 2';
prompt = {'Probability of detection (<1)', ...
      'Probability of false alarm (<1)', 'Doppler filter
bandwidth (Hz)',...
      'Frequency agility bandwidth (MHz)', 'Azimuth beamwidth
(degrees)',...
      'Azimuth Scan rate (degrees/sec)'};
default = {prob_det, prob_false_alarm, BW_dop, BW_fa, Azimuth_bw,
az_rate};
response = inputdlg(prompt, title, 1, default);
fields =
{'prob_det','prob_false_alarm','BW_dop','BW_fa','Azimuth_bw','az_rate'};
input = cell2struct(response,fields,1);

% Convert cell structure created by dialog box back to numbers

prob_det = str2num(input.prob_det);
prob_false_alarm = str2num(input.prob_false_alarm);
BW_dop = str2num(input.BW_dop);
BW_fa = str2num(input.BW_fa);
Azimuth_bw = str2num(input.Azimuth_bw);
az_rate = str2num(input.az_rate);

clear input response fields default prompt title;

% End Menu for Radar Parameters

% Target related:
%
RCS = 1;              % Radar cross section - m^2
el_tgt_deg = 0.0;     % Elevation of target - degrees
target_length = 15;   % Length of target -     m

% Other
%
rain_fall = 0;        % Rain fall rate -        mm/hr

% MENU for Target and Atmosphere information

title = 'Target and Atmosphere information';
prompt = {'RCS (m^2)', 'Elevation of target (degrees)',...
      'Length of target (m)', 'Rain fall rate (mm/hr)'};
default = {RCS, el_tgt_deg, target_length, rain_fall};
response = inputdlg(prompt, title, 1, default);
fields = {'RCS', 'el_tgt_deg', 'target_length', 'rain_fall'};
input = cell2struct(response,fields,1);

% Convert cell structure created by dialog box back to numbers

RCS = str2num(input.RCS);
el_tgt_deg = str2num(input.el_tgt_deg);
```

```
target_length = str2num(input.target_length);
rain_fall = str2num(input.rain_fall);

clear input response fields default prompt title;
% End menu for Target and Atmosphere Information

% Select jamming type (enter 1 for yes, 0 for no; can select
% none, one, or both):
%
jamming_type = menu('Select jamming type', 'Stand off',...
   'Selfscreening', 'None', 'Both');
switch jamming_type
case 1
   % Stand off
   SOJ = 1;       % Stand-off boolean
   SSJ = 0;       % Selfscreening boolean
case 2
   % Selfscreening
   SSJ = 1;       % Selfscreening boolean
   SOJ = 0;       % Stand-off boolean
case 3
   % No Jamming
   SSJ = 0;       % Selfscreening boolean
   SOJ = 0;       % Stand-off boolean
case 4
   % Both
   SOJ = 1;
   SSJ = 1;
end

% Enter characteristics for selfscreening Jamming.  If not,
selected, values won't be used.
%
SSJ_pwr_jam = 10; % SSJ power of jammer  W
SSJ_gain_dB = 0;  % Gain of SS jammer in dB dB
SSJ_bw = 20;      % Bandwidth of SSJ  MHz
SSJ_loss_dB = 7;  % Losses with SSJ  dB

if SSJ == 1

   % MENU for selfscreening Jamming chars

   title = 'Selfscreening Jamming information';
   prompt = {'Power (W)', 'Antenna Gain (dB)',...
         'Bandwidth (MHz)', 'Losses (dB)'};
   default = {SSJ_pwr_jam,SSJ_gain_dB,SSJ_bw,SSJ_loss_dB};
   response = inputdlg(prompt, title, 1, default);
         fields   =   {'SSJ_pwr_jam',   'SSJ_gain_dB',   'SSJ_bw',
'SSJ_loss_dB'};
   input = cell2struct(response,fields,1);
```

```
   % Convert cell structure created by dialog box back to numbers

   SSJ_pwr_jam = str2num(input.SSJ_pwr_jam);
   SSJ_gain_dB = str2num(input.SSJ_gain_dB);
   SSJ_bw = str2num(input.SSJ_bw);
   SSJ_loss_dB = str2num(input.SSJ_loss_dB);

   clear input response fields default prompt title;

   % End MENU for selfscreening Jamming chars

end

% Enter characteristics for stand-off Jamming.  If not, selected,
% values won't be used.
%
% Each array must be equal and there must
% be a value for each stand-off jammer you
% wish to simulate
%
SOJ_pwr_jam = [1000 1000 1000 1000 1000];
          % SOJ power of jammers  W
SOJ_gain_dB = [13 13 13 13 13];
          % Gain of SO jammer in dB dB
SOJ_bw = [100 100 100 100 100];
          % Bandwidth of SOJ's  Mhz
jam_range = [20 20 20 20 20];
          % Range of jammer  NMI

jam_height = [20000 20000 20000 20000 20000];
          % Height of jammer  ft
SOJ_loss_dB = [7 7 7 7 7];
          % Losses with SOJ's   dB
% MENU for Stand off jamming input

if SOJ==1
   num_jams = length(SOJ_pwr_jam);
   title = 'Stand off Jamming information';
   prompt = {'Number of Jammers'};
   default = {num_jams};
   response = inputdlg(prompt, title, 1, default);
   fields = {'num_jams'};
   input = cell2struct(response,fields,1);
   num_jams = str2num(input.num_jams);
   clear input response fields default prompt title;

   if num_jams ~= length(SOJ_pwr_jam)
      SOJ_pwr_jam = ones(1,num_jams).*SOJ_pwr_jam(1);
      SOJ_gain_dB = ones(1,num_jams).*SOJ_gain_dB(1);
      SOJ_bw = ones(1,num_jams).*SOJ_bw(1);
      jam_range = ones(1,num_jams).*jam_range(1);
```

```
      jam_height = ones(1,num_jams).*jam_height(1);
      SOJ_loss_dB = ones(1,num_jams).*SOJ_loss_dB(1);
   end % if num_jams

   for loop = 1:num_jams
      title = ['Standoff Jammer #' num2str(loop)];
      prompt = {'Power (W)', 'Gain (dB)', 'Bandwidth (MHz)',...
            'Range (nm)', 'Height (ft)', 'Loss (dB)'};
      default = {SOJ_pwr_jam(loop), SOJ_gain_dB(loop),...
            SOJ_bw(loop), jam_range(loop), jam_height(loop),...
            SOJ_loss_dB(loop)};
      response = inputdlg(prompt, title, 1, default);
      fields =
{'SOJ_pwr_jam','SOJ_gain_dB','SOJ_bw','jam_range',...
            'jam_height','SOJ_loss_dB'};
      input = cell2struct(response,fields,1);

      % Convert cell structure created by dialog box back to numbers
      SOJ_pwr_jam(loop) = str2num(input.SOJ_pwr_jam);
      SOJ_gain_dB(loop) = str2num(input.SOJ_gain_dB);
      SOJ_bw(loop) = str2num(input.SOJ_bw);
      jam_range(loop) = str2num(input.jam_range);
      jam_height(loop) = str2num(input.jam_height);
      SOJ_loss_dB(loop) = str2num(input.SOJ_loss_dB);

      clear input response fields default prompt title;

    end % for loop
end % if SOJ

    % End MENU for Stand off Jamming chars

%%%%%%%%%%%%%%%%%%%%%%%%%%%%%%%%%%%%%%%%%%%%%%%%%%%%%%%%%%%%%%%%
%%%%%%%%%%%%%%%%
% STOP EDITING.  ONLY PROGRAM FOLLOWS
%%%%%%%%%%%%%%%%%%%%%%%%%%%%%%%%%%%%%%%%%%%%%%%%%%%%%%

%%%%%%%%%%%%%%%%%%%%%%%%%%%
%
% Begin Main
%

% Convert radar gain, loss, noise to ratios
%
loss_radar = db_ratio(loss_radar_dB);
GT_radar = db_ratio(GTDB_radar);
GR_radar = db_ratio(GRDB_radar);
noise_factor_radar = db_ratio(noise_fig_radar);

% Calculate system input noise temp
%
```

```
sys_noise_temp = 290 * (noise_factor_radar - 1) + 150;  % in K

% Calculate range without jamming, FA, etc (S/N = 1)
%
range_index = 129.2 * ((trans_pwr_radar * pulse_width * GT_radar
* GR_radar * RCS) / ...
       (freq_radar^2 * sys_noise_temp * loss_radar))^.25;

% Calculate number of pulses
%
non_coh_pulses = (Azimuth_bw/az_rate) * BW_dop;
coh_pulses = PRF/BW_dop;

% If target length is greater than zero,computes number of independent
% pulses for FA radars
%
if target_length > 0
  N_exp = 1 + (BW_fa/(150 / target_length));
  else N_exp = non_coh_pulses;
end

% Determine detectability factors
%
det_fact;

% Calculate ranges for the different target types, clear, no
% atmospheric or weather attenuation accounted for here.
%

swrlg_case= ['     0        1          2         3        4
1fa        3fa'];

ranges_clear = range_index * 10.^((-det_facts)/40);

% echo on
% Now, the atmospheric attenuation and rain attenuation is
% calculated.  First, the Swerling cases being calculated are
% listed. Then under ar_atten_ans, row 1 is the atmospheric
% attenuation (dB) and row 2 is the rain attenuation (dB).  The
% 3rd row contains the affected ranges (NMI). Underafter_jam_ans,
% the meaning of the rows
% are the same as for the attenuation matrix.
% echo off

ar_atten;

% Effects of Jamming...
%
if SSJ==1
   pwrj_density = SSJ_pwr_jam/SSJ_bw;
   SSJ_gain = db_ratio(SSJ_gain_dB);
```

```
   SSJ_loss = db_ratio(SSJ_loss_dB);

% Range with just self-screening jamming
%
  rng_ssj_only = 0.0048116 * sqrt(trans_pwr_radar * GT_radar *
RCS * pulse_width ./ ...
             (pwrj_density .* SSJ_gain .*
(loss_radar/SSJ_loss)));
   if SOJ==1
     soj_eff;
   end
   jam_eff;
end
if (SOJ==1)&(SSJ==0)
   soj_eff;
   rng_ssj_only = rs00;
   jam_eff;
end
```

» rjv2

Detection Range with Atmospheric and Weather Attenuation

Swerling Case	0	1	2	3	4	1FA	3FA
Atmospheric Atten (dB)	0.33	0.21	0.30	0.27	0.31	0.21	0.27
Weather Atten (dB)	0.00	0.00	0.00	0.00	0.00	0.00	
Detection Range (n mi)	7.63	4.83	6.79	6.07	7.20	4.83	6.07

Detection Range with Jamming

Swerling Case	0	1	2	3	4	1FA	3FA
Atmospheric Atten (dB)	0.16	0.10	0.14	0.13	0.15	0.10	0.13
Weather Atten (dB)	0.00	0.00	0.00	0.00	0.00	0.00	0.00
Jamming Range (n mi)	3.60	2.26	3.20	2.85	3.39	2.26	2.85

»

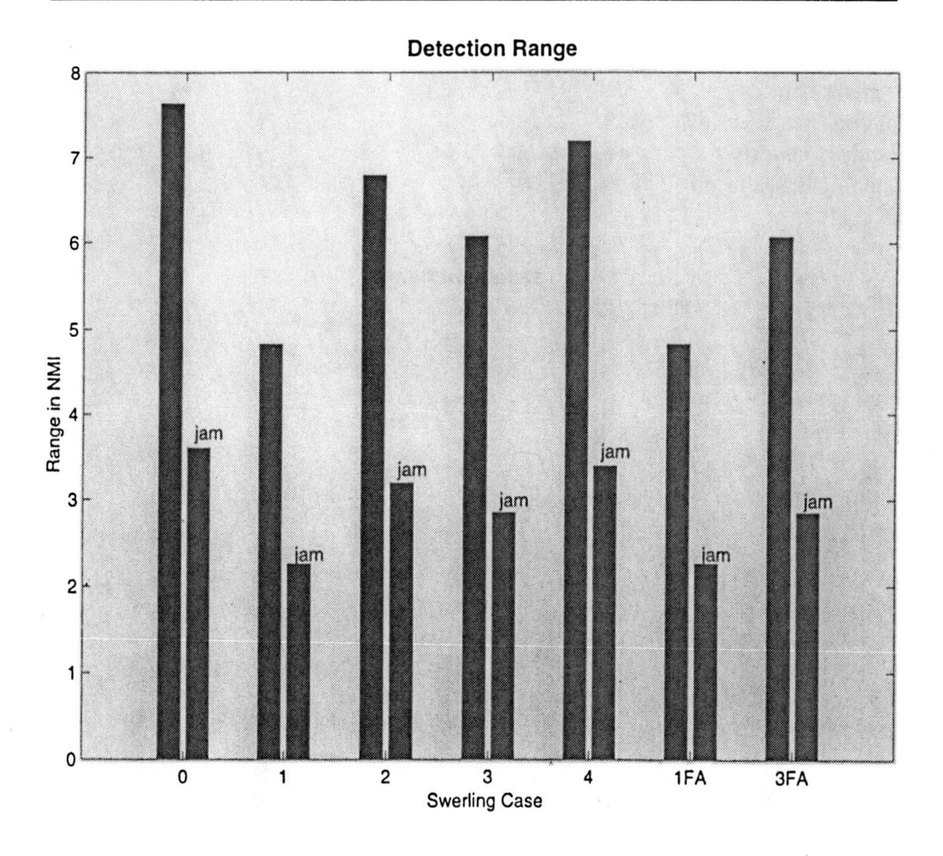

» rjv2

▶ *Example A.4*

For the LPD-20 radar, determine the detection performance in 4 mm/hr of rain for the radar and jamming parameters given in Example A.3.

Detection Range with Atmospheric and Weather Attenuation

Swerling Case	0	1	2	3	4	1FA	3FA
Atmospheric Atten (dB)	0.30	0.20	0.27	0.24	0.28	0.20	0.24
Weather Atten (dB)	1.88	1.24	1.69	1.53	1.79	1.24	1.53
Detection Range (n mi)	6.86	4.50	6.17	5.57	6.51	4.50	5.57

Detection Range with Jamming

Swerling Case	0	1	2	3	4	1FA	3FA
Atmospheric Atten (dB)	0.19	0.12	0.17	0.15	0.18	0.12	0.15
Weather Atten (dB)	1.18	0.74	1.04	0.93	1.11	0.74	0.93
Jamming Range (n mi)	4.29	2.69	3.80	3.40	4.04	2.69	3.40

»

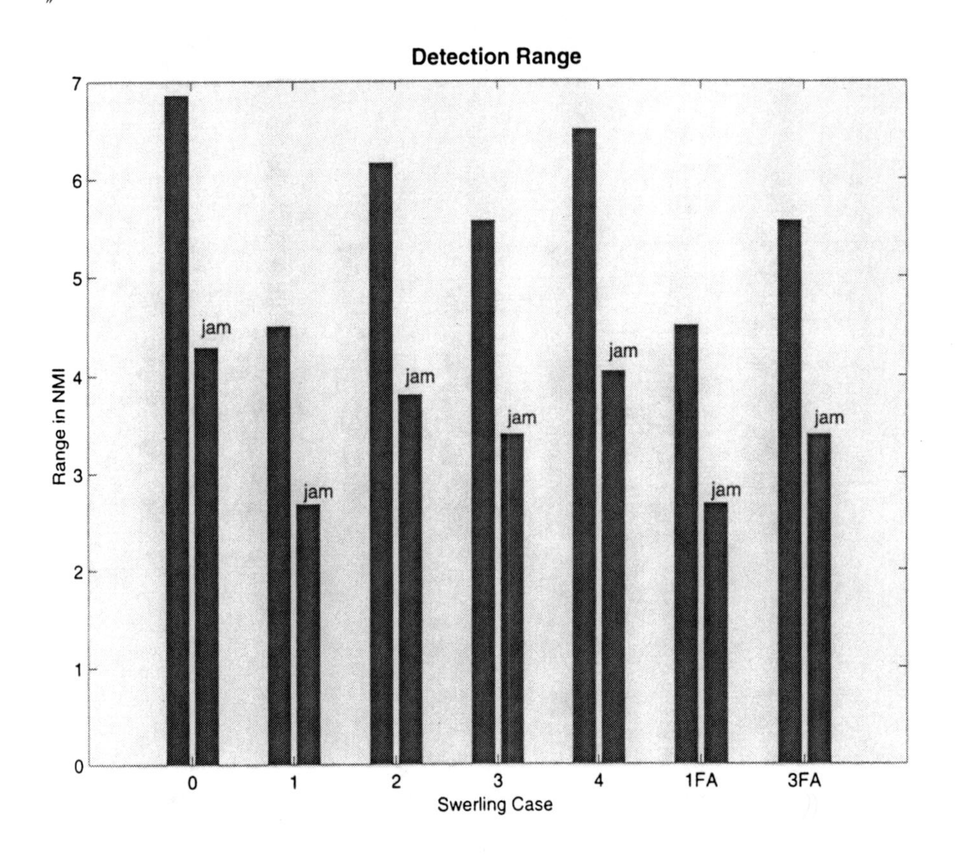

▶ *Example A.5*

The ASR-9 2-D search radar uses an eight-point FFT processor. The Doppler bandwidth is found to be $BW_{dop} = PRF/8 = 1200/8 = 150$ Hz. Determine the detection range for this radar under clear weather conditions and in the presence of five stand-off jammers. Use the rjx.m version. The parameters for the radar are given in the associated print out.

```
» rjxx
Radar Parameters

 Pt(kW)   f(MHz)   PW(1e-6*s)  PRF(pps)  Gt(dB)   Gr(dB)  NF(dB)  Gsl(dB)
 1200.0   2900.0      1.0       1200.0    33.5     33.5    5.0      0.0

 Lr(dB)  Dop Bw(Hz)  FA Bw(MHz)  Az Bw(deg)  Az Rate(deg/s)   Pd     Pfa
 12.00     150.00       0.00        1.30          75.00       0.9   1e-006

Target and Weather Parameters

 RCS(m^2)  tar el(deg)  tar lgth(m)  rain(mm/hr)
   1.0        0.4          10.0          0.0

Stand-Off Jamming Parameters

Number of Jammers = 5
Pj(w) =        2000    2000    2000    2000    2000
Gj(dB) =        5.0     5.0     5.0     5.0     5.0
BWj(Mhz) =      200     200     200     200     200
Rj(n mi) =     50.0    50.0    50.0    50.0    50.0
Hj(ft) =      20000   20000   20000   20000   20000
Lj(dB) =        7.0     7.0     7.0     7.0     7.0

Detection Range with Atmospheric and Weather Attenuation

Swerling Case             0      1      2      3      4     1FA    3FA
Atmospheric Atten (dB)  1.26   0.88   1.10   1.06   1.18   0.88   1.06
Weather Atten (dB)      0.00   0.00   0.00   0.00   0.00   0.00   0.00
Detection Range (n mi) 58.27  37.47  49.20  46.75  53.54  37.47  46.75

Detection Range with Jamming

Swerling Case             0      1      2      3      4     1FA    3FA
Atmospheric Atten (dB)  0.71   0.47   0.61   0.58   0.66   0.47   0.58
Weather Atten (dB)      0.00   0.00   0.00   0.00   0.00   0.00   0.00
Jamming Range (n mi)   29.15  18.39  24.45  23.17  26.72  18.39  23.17
»
```

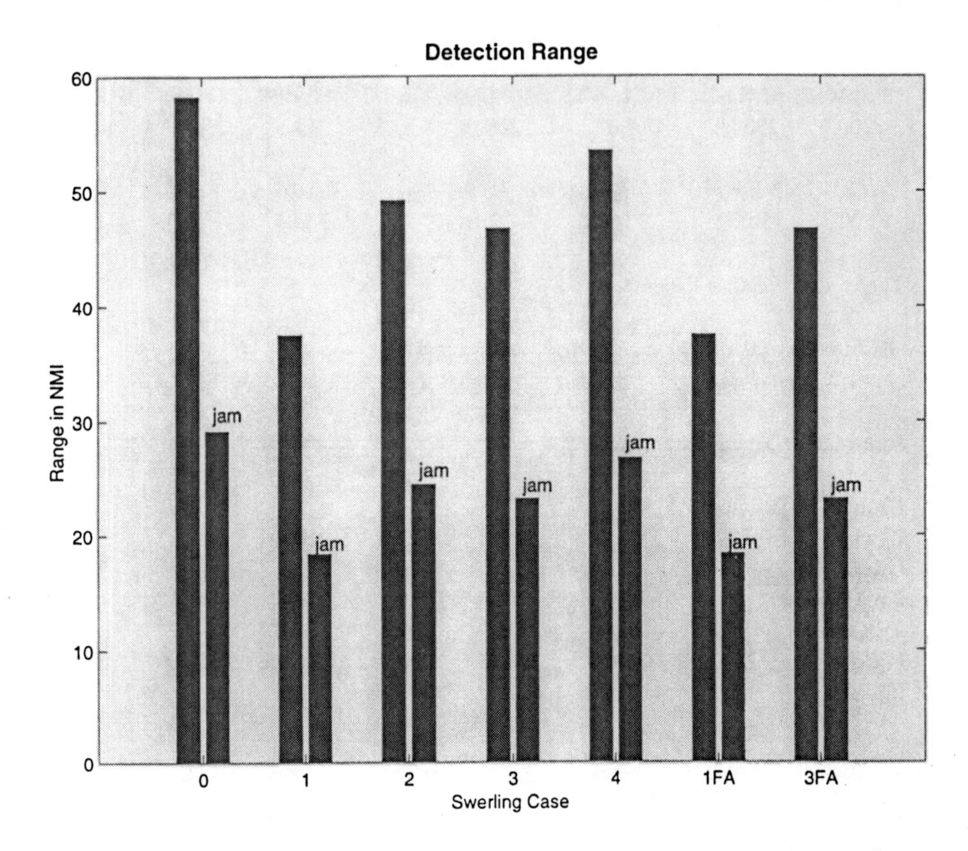
Detection Range
Range in NMI
60
50
40
30
20
10
0
jam
jam
jam
jam
jam
jam
jam
0
1
2
3
4
1FA
3FA
Swerling Case

▶ Example A.6

The Pilot FM-CW radar transmits only 1W of power. It employs a 50-MHz sweep over a repetition time of 1 ms. The parameters of this radar follow. Use the rjx.m version to provide extended output data.

Radar Parameters	**Stand-off Jamming Parameters**
P_t = .001 kW	P_j = 2000W
f = 9375 MHz	G_j = 13 dB
PRF = 1000 Hz	B_j = 100 MHz
PW = 1000 μs	R_j = 50 nmi
G_t = 30 dB	L_j = 7 dB
G_r = 30 dB	H_j = 5000 ft
θ = 1.2 degree	G_{SL} = −3 dBi
NF = 3 dB	
L = 10 dB	
ϕ_t = 0 degree	
RCS = 1000 m^2	
B_{DOP} = 1000 Hz	

» rjxpilot
Radar Parameters

Pt(kW)	f(MHz)	PW(1e-6*s)	PRF(pps)	Gt(dB)	Gr(dB)	NF(dB)	Gsl(dB)
0.001	9375.000	1000.000	1000.000	30.000	30.000	3.000	−3.000

Lr(dB)	Dop Bw(Hz)	FA Bw(MHz)	Az Bw(deg)	Az Rate(deg/s)	Pd	Pfa
10.00	1000.00	0.00	1.20	288.00	0.9	le-006

Target and Weather Parameters

RCS(m^2)	tar el(deg)	tar lgth(m)	rain(mm/hr)
1000.0	0.0	100.0	0.0

Stand-Off Jamming Parameters

Number of Jammers = 1
Pj(w) = 2000
Gj(dB)= 13.0
BWj(MHz)= 100

Rj(n mi)= 50.0
Hj(ft)= 5000
Lj(dB)= 7.0

Detection Range with Atmospheric and Weather Attenuation

Swerling Case	0	1	2	3	4	1FA	3FA
Atmospheric Atten (dB)	0.73	0.47	0.66	0.59	0.69	0.47	0.59
Weather Atten (dB)	0.00	0.00	0.00	0.00	0.00	0.00	0.00
Detection Range (n mi)	17.85	11.37	16.02	14.25	16.91	11.37	14.25

Detection Range with Jamming

Swerling Case	0	1	2	3	4	1FA	3FA
Atmospheric Atten (dB)	0.55	0.35	0.50	0.44	0.52	0.35	0.44
Weather Atten (dB)	0.00	0.00	0.00	0.00	0.00	0.00	0.00
Jamming Range (n mi)	13.34	8.43	11.93	10.63	12.62	8.43	10.63

»

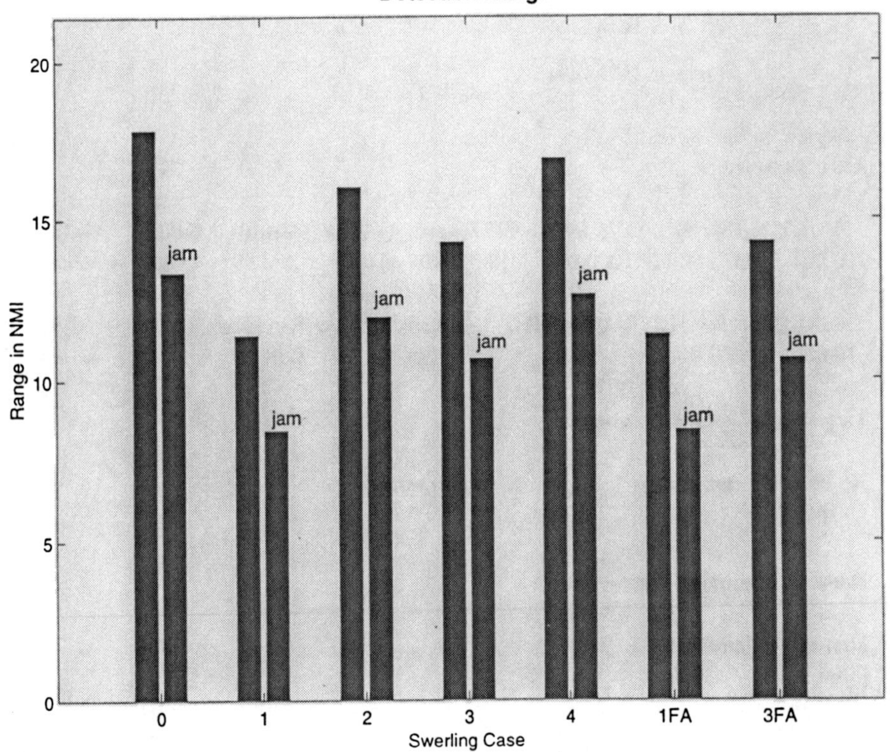

▶ *Example A.7*

RGJMAT can be used to determine the detection range of a CW radar. For this case, the average power of the radar is entered as the peak power. The PW is calculated as 1.0 divided by Doppler bandwidth. The parameters for the CW radar are given. Determine its detection performance under clear weather conditions and its performance against a stand-off jammer.

Radar Parameters	**Stand-off Jamming Parameters**
P_t = 1 kW	P_j = 2000W
f = 2850 MHz	G_j = 5 dB
PRF = 100Hz	B_j = 5 MHz
PW = 10,000 μs	R_j = 100 nmi
G_t = 34 dB	L_j = 7 dB
G_r = 34 dB	H_j = 10,000 ft
θ = 2 degrees	G_{SL} = 0 dB
NF = 3 dB	Number = 1
L = 15 dB	
ϕ_t = 0 degrees	
RCS = 1 m^2	
B_{DOP} = 100 Hz	

```
» rjxcw
Radar Parameters

 Pt(kW)   f(MHz)   PW(1e-6*s)  PRF(pps)  Gt(dB)   Gr(dB)   NF(dB)  Gsl(dB)
  1.0     2850.0    10000.0     100.0     34.0     34.0     3.0      0.0

Lr(dB)  Dop Bw(Hz)  FA Bw(MHz)  Az Bw(deg)  Az Rate(deg/s)   Pd     Pfa
15.00    100.00       0.00        2.00         200.00       0.9   le-006

Target and Weather Parameters

 RCS(m^2)  tar el(deg)  tar lgth(m)  rain(mm/hr)
   1.0        0.0          10.0         0.0

Stand-Off Jamming Parameters

Number of Jammers = 1
Pj(w) =         2000
```

Gj(dB) = 5.0
BWj(Mhz) = 200
Rj(n mi) = 50.0
Hj(ft) = 10000
Lj(dB) = 7.0

Detection Range with Atmospheric and Weather Attenuation

Swerling Case	0	1	2	3	4	1FA	3FA
Atmospheric Atten (dB)	1.18	0.80	0.80	0.98	0.98	0.80	0.98
Weather Atten (dB)	0.00	0.00	0.00	0.00	0.00	0.00	0.00
Detection Range (n mi)	49.21	31.83	31.83	39.62	39.62	31.83	39.62

Detection Range with Jamming

Swerling Case	0	1	2	3	4	1FA	3FA
Atmospheric Atten (dB)	0.79	0.52	0.52	0.64	0.64	0.52	0.64
Weather Atten (dB)	0.00	0.00	0.00	0.00	0.00	0.00	0.00
Jamming Range (n mi)	31.54	20.04	20.04	25.17	25.17	20.04	25.17

»

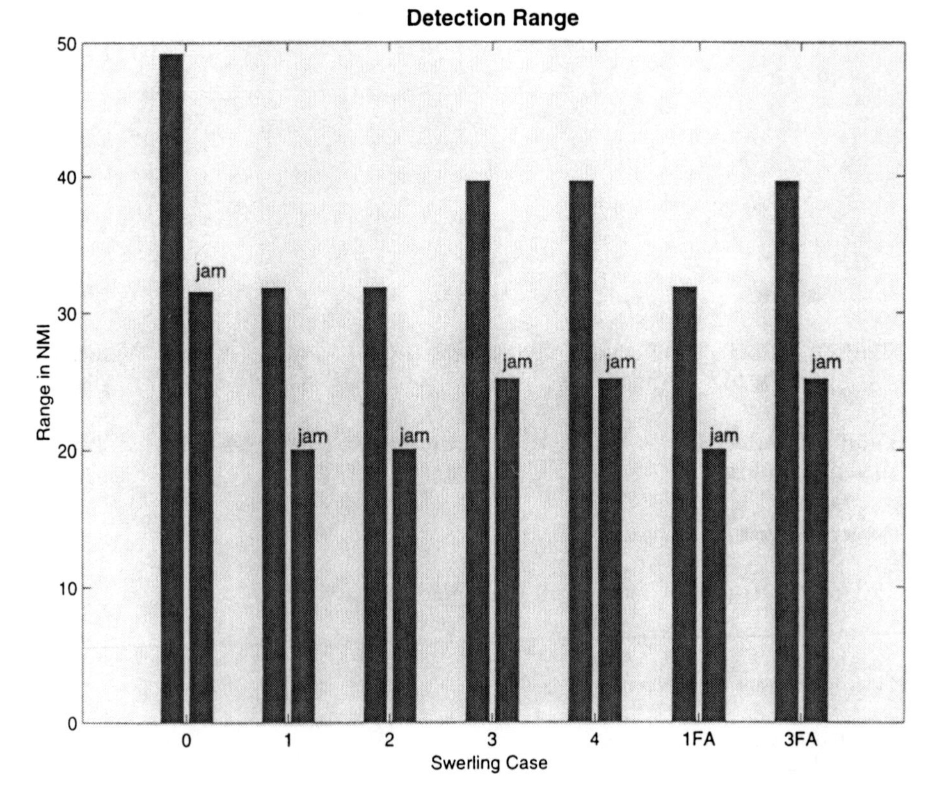

RGJMAT Supporting Programs

The print outs of MATLAB files ar_atten.m, dB_ratio.m, det_fact.m, jam_eff.m, and sof_eff.m follow. The print command for nonjamming detection performance is contained in ar_atten.m, while the jamming detection performance is in jam_eff.m. The bar graph command is contained in jam_eff.m.

```
% Calculates Atmospheric and Rain Attenuation
% ------------------------------------------
% ar_atten.m
%
% Last edited: 15 Mar 1996
%

% Attenuation parameters.  First two values in each row are
% the frequency range the attenuation parameters are valid.
%               fl     fh    sl       sh       wl       wh
%                1      2    3        4        5        6
Att_param = [0.4    1.3   0.01     0.012    1.0e-6   0.0003;
             1.3    3     0.012    0.015    0.0003   0.0013;
             3      5.5   0.015    0.017    0.0013   8.000001e-3;
             5.5    10    0.017    0.024    8.000001e-3 0.037;
             10     15    0.024    0.055    0.037    0.083;
             15     22    0.055    0.3      0.083    0.23;
             22     35    0.3      0.14     0.23     0.57;
             35     60    0.14     35       0.57     1.3;
             60     95    35       0.8      1.3      2;
             95     140    0.8     1        2        2.3;
             140    240    1       14       2.3      2.2];

% Shift frequency to GHz scale and test to see which parameters
% to use for subsequent calculations.  Program will still crash
% if freq_radar outside of 400MHz - 240GHz range.
%
freq_ghz = freq_radar /1000;
i = find(freq_ghz > Att_param(:,1) & freq_ghz <= Att_param(:,2));

% Use working variables to calculate the resulting attenuation
%
ax = log (Att_param(i,4)/Att_param(i,3)) / log
(Att_param(i,2)/Att_param(i,1));
sx = Att_param(i,3)*((freq_ghz/Att_param(i,1))^ax);
bx = log(Att_param(i,6)/Att_param(i,5)) / log
(Att_param(i,2)/Att_param(i,1));
wx = rain_fall * Att_param(i,5) * ((freq_ghz/Att_param(i,1))^bx);

el_tgt = el_tgt_deg * pi/180;
re = 3/sin(el_tgt + 0.00025/(el_tgt+0.028));
```

```
% Convert earlier calculated clear ranges to kilometers
%
rng_km = ranges_clear.* 1.854;
rx = rng_km;

rdel = 0;
done = 0;
while done==0
  rx = rx - rdel;
  lea = sx .* re *(1-exp(-rx ./ re));
  ler = wx .* rx;
  fr = 0.4343 .* log(rx ./ rng_km) + (lea + ler)/40;
  dfr = (1./rx) + wx/40 + sx.*(exp(-rx./re))/40;
  rdel = fr./dfr;
       a = find(rdel <= 0.01);
  b = length(a);
  if b > 0
     for i = 1:b,
       rdel(a(i)) = 0;
     end
  end
  if rdel <= 0.01
     done = 1;
  end
end

% swrlg_case

ar_atten_ans = [lea; ler; (rx./1.854)];

% Print Detection with Atmospheric and Weather Attenuation

fprintf('Detection Range with Atmospheric and Weather
Attenuation\n\n');
fprintf('Swerling Case                    0          1        ');
fprintf(' 2          3            4        1FA         3FA\n');

fprintf('Atmospheric Atten (dB)       ');fprintf('%-
10.2f',lea);fprintf('\n');
fprintf('Weather Atten (dB)              ');fprintf('%-
10.2f',ler);fprintf('\n');
if rx<50;
   fprintf('Detection Range (n mi)       ');fprintf('%-
10.2f',rx/1.854);
   else,
      fprintf('Detection Range (n mi)       ');fprintf('%-
9.2f',rx/1.854);
end;
fprintf('\n\n\n');
```

```
% Function: dB to ratio conversion
% --------------------------------
% db_ratio.m

%
% Converts a given dB (Y) to a ratio and returns value
% to calling program

function X = dB_ratio(Y)
X = 10.^(Y./10);

% Determines Detectability Factors
% --------------------------------
% det_fact.m
%
% Last edited:  8 Mar 1996
%%%%%%%%%%%%%%%%%%%%%%%%%%%%%%%%%%%%%%%%

P(1) = prob_false_alarm;
P(2) = 1 - prob_det;
T = sqrt(log(1./(P.^2)));
B = 2.30752 + .27061 * T;
C = 1 + 0.99229 * T + 0.04481 * T.^2;
YB = T - (B./C);
YMS = sum(YB);
snr_coh_det = YMS^2 /2;
AA = snr_coh_det / (2*non_coh_pulses*coh_pulses);
BB = 1 + sqrt(1+9.2*non_coh_pulses./snr_coh_det);
snr_non_coh_det = 4.343 * log((AA) * BB);

Z1 = log(prob_false_alarm) / log(prob_det) - 1;
det_fact_1_1 = 4.343 * log(Z1);
Z2 = ((T(1) + YB(2))^2 - 1)/2;
det_fact_0_1 = 4.343 * log(Z2);
W = 1 + 0.03 * 0.4343 * log(non_coh_pulses);
fluct_loss = det_fact_1_1 - det_fact_0_1;

det_facts(1) = snr_non_coh_det;
det_facts(2) = snr_non_coh_det + W * fluct_loss;
det_facts(3) = snr_non_coh_det + W * fluct_loss / non_coh_pulses;
det_facts(4) = snr_non_coh_det + W * fluct_loss / 2;
det_facts(5) = snr_non_coh_det + W * fluct_loss / (2*non_coh_pulses);
det_facts(6) = snr_non_coh_det + W * fluct_loss / N_exp;
det_facts(7) = snr_non_coh_det + W * fluct_loss / (2*N_exp);

% Calculates the Effect of Jamming
% Takes inputs from both the self-screening and
% stand-off jamming processes
% --------------------------------------------------------
% jam_eff.m
%
```

```
%

% If SSJ, account for that with aax factor
%
if SSJ==1
   aax = (range_index./rng_ssj_only).^4;
else
   aax = 0;
end;

% If stand-off jamming as well, figure in those effects
% or, if SOJ only, then account for that
%

if SOJ==0
   bbx = 0;
else
   bbx = (range_index./rs00).^4;
end

rc = ranges_clear .* 1.854;
rss = (rng_ssj_only .* (10.^((-det_facts)./40))) .* 1.854;

rxx = rss;

rxdel = 0;
done = 0;
while done==0
  rxx = rxx - rxdel;
  lax = sx .* re *(1-exp(-rxx ./ re));
  lwx = wx .* rxx;
  lxx = 10.^((lax+lwx)./20);
  lyy = 10.^((lax+lwx)./10);
  xfrx = .4343 .* log(rxx ./ rc) + .4343 * (log(lyy + aax * lxx +
bbx)) ./ 4;
  dlxx = 2.3026 .* lxx .* ((1 ./ rxx) + wx / 20 + (sx / 20) .*
exp(-rxx ./ re));
  dlyy = 2.3026 .* lyy .* ((1 ./ rxx) + wx / 10 + (sx / 10) .*
exp(-rxx ./ re));
   dfrx = (1./ rxx) + (dlyy + aax * dlxx) ./ (4 * (lyy + lxx *
aax + bbx));
        rxdel = xfrx./dfrx;
          a = find(abs(rxdel) <= 0.1);
  b = length(a);
  if b > 0
     for i = 1:b,
        rxdel(a(i)) = 0;
     end
  end
  if rxdel <= 0.1
     done = 1;
```

```
  end
end

% Print out ranges after jamming accounted for
%
after_jam_ans = [lax; lwx; (rxx./1.854)];

fprintf('Detection Range with Jamming\n\n');
fprintf('Swerling Case                    0          1        '); ...
fprintf('    2           3          4          1FA       3FA\n');
fprintf('Atmospheric Atten (dB)       ');fprintf('%-
10.2f',lax);fprintf('\n');
fprintf('Weather Atten (dB)               ');fprintf('%-
10.2f',lwx);fprintf('\n');
if max(rxx<50);
fprintf('Jamming Range  (n mi)        ');fprintf('%-
10.2f',rxx/1.854);fprintf('\n');
else,
   fprintf('Jamming Range  (n mi)        ');fprintf('%-
9.2f',rxx/1.854);fprintf('\n');
end;

rxx = (rxx / 1.854);
lea = lax;
ler = lwx;

% Set Figure Parameters

ra=rx/1.854;
yz=[ra' rxx'];
z=0:6;
rmx=1.2*max(ra);

figure
bar(z,yz);
if rmx<50;set(gca,'YTick',[0:5:rmx]);
else,set(gca,'YTick',[0:10:rmx]);end;
if rmx<10;set(gca,'YTick',[0:1:rmx]);end;
set(gca,'XTicklabel','0|1|2|3|4|1FA|3FA');
xlabel('Swerling Case ')
ylabel('Range in NMI')
% axis([-1 7 0 rmx]);
title('Detection Range','Fontsize',12,'Fontweight','bold');
for ii=1:7;
   if ii==1;
      text(ii-.9,1.05*rxx(ii),'jam');
      %text(ii-1.5,1.03*ra(ii),'w/o jam');
   else
      text(ii-.9,1.05*rxx(ii),'jam');
      %text(ii-1.5,1.03*ra(ii),'w/o jam');
   end;
```

```
end;

% Stand-off Jamming Effects
% -------------------------
% soj_eff.m
%

slx = GRDB_radar - GRDB_sl_radar;

sl = db_ratio(slx);

% Jammer height's and ranges are converted to meters
%
jam_height = 0.3048 .* jam_height;
jxr = jam_range .* 1854;

%
%
xrx = jxr ./ 1.7E+07;
xsx = jam_height ./ jxr;
elx = (xsx - xrx) + ((xsx - xrx) .^ 3) ./ 6;
effj = elx + .00025 ./ (elx + .028);

rej = 3 ./ sin(effj);
leaj = sx * rej * (1 - exp(-jxr / rej));
lerj = wx * jxr ./ 1000;
jlj = db_ratio(SOJ_loss_dB);
jlj = jlj + leaj + lerj;
jpjj = SOJ_pwr_jam ./ SOJ_bw;

jg = db_ratio(SOJ_gain_dB);

rng_soj_only = 0.0048116 * sqrt(trans_pwr_radar * GT_radar * RCS
* pulse_width ./ (jpjj .* jg .* ( loss_radar./jlj)));

rs0 = sqrt(rng_soj_only .* jam_range) * (sl) ^ .25;

s = 1./((rs0).^4);

rs00 = 1 / (sum(s))^.25;
```

References

[1] Schleher, D. C., *Introduction to Electronic Warfare,* Norwood, MA: Artech House, 1986.

Acronym List

AAM	air-to-air missile
ABCCC	airborne battlefield command and control center
ACF	autocorrelation function
ACINT	acoustical intelligence
A/D	analog-to-digital
ADATS	air defense anti-tank system
AED	active electronic decoy
AEW	airborne early warning
AFSATCOM	Air Force Satellite Communications Network
AGC	automatic gain control
AI	airborne interceptors
ALCM	air-launched cruise missiles
AM/PM	amplitude-to-phase modulation
AO	acoustic optic
AOA	angle of arrival
AOSA	acousto-optic spectrum analyzer
ARM	antiradiation missile
ASAR	advanced synthetic aperture radar
ASCM	anti-shipping cruise missiles
ASIC	application-specific integrated circuit
ASPJ	advanced self-protection jammer
ASMD	antiship missile defense
ASW	antisubmarine warfare operations

ATIRCM	advanced threat infrared countermeasures
AWACS	airborne warning and control system
BCH	Bose-Chaudhuri Hocquenghem
BDA	battle damage assessment
BER	bit error rate
BWO	backward wave oscillator
C2W	command and control warfare
C2	command and control
C3	command, control, and communication
CDMA	code division multiple access
CEC	cooperative engagement capability
CFAR	constant false-alarm rate
CHAALS	Communication High-Accuracy Airborne Location System
CI	counter intelligence
CNR	combat net radio
COMINT	communications intelligence
COIL	chemical oxygen iodine lasers
COMSEC	communications security
CPB	changed particle beams
CPI	coherent processing interval
CW	continuous wave
CSLC	coherent sidelobe canceiler
D/A	digital-to-analog converter
DBS	Doppler beam sharpening
dBW	decibels referenced to one watt
DD	differential Doppler
DDS	direct digital synthesizers
DE	directed-energy
DECM	deceptive electronic countermeasures
DEW	DE weapons
DF	direction finding
DIFM	digital IFM
DINA	direct noise amplification
DME	distance-measuring equipment
DOA	direction of arrival
DPCA	displaced phase center antenna

DRFM	digital RF memory
DSCS	Defense Satellite Communications System
DTDMA	distributed time-division multiple access
EA	electronic attack
ECL	emitter-coupled logic
ECM	electronic countermeasures
ECCM	electronic counter-countermeasures
EHF	extremely high frequency
EIRP	effective radiated power with respect to an isotropic radiator
ELINT	electronic intelligence
ELS	emitter location system
EM	electromagnetic
EMCON	electromagnetic control
EMP	electromagnetic pulse
EO	electro-optical
EOB	electronic order of battle
EP	electronic protection
EPL	ELINT parameter limits
ERP	effective radiated power
erf	error function
ES	electronic warfare support
ESM	ES measures
EWIR	EW Integrated Reprogramming Database
EW	electronic warfare
FDOA	frequency difference of arrival
FEL	free-election lasers
FFT	fast Fourier transform
FISINT	foreign instrumentation signals intelligence
FLIR	forward-looking infrared
FMOP	frequency modulation on-the-pulse
FO	fiber optic
FOTD	FO towed decoy
FOV	field of view
FPA	focal point array
FSK	frequency shift keying
FLTSATCOM	Fleet satellite communications

GBS	Global Broadcast Service
GCI	ground control intercept
GDOP	geometric dilution of precision
GEO	geostationary
GII	Global Information Infrastructure
GMTI	ground moving target indication
GPS	Global Positioning System
GRCS	Guardrail common sensor
HALE	high-altitude long-endurance
HARM	high-speed ARM
HEL	high-energy lasers
HERF	high-energy radio frequency
HF	high frequency
HPM	high-powered microwave
HUMINT	human intelligence
IDECM	integrated defensive electronic countermeasures
IFF	identification friend from foe
IFM	instantaneous frequency measurement
IMINT	imagery intelligence
IR	infrared
IRCCM	IR counter countermeasures
IRINT	infrared intelligence
IRST	infrared search and track
ISAR	inverse synthetic aperture radar
IW	information warfare
JEM	jet engine modulation
J/S	jam-to-signal ratio
JTAEWS	Joint Tactical Air Electronic Warfare Study
JTIDS	Joint Tactical Information Distribution System
LASINT	laser intelligence
LEO	low Earth orbit
LNA	low-noise amplifier
LORO	lobe-on-receive only
LPI	low probability of intercept
MASINT	measurement and signature intelligence
MAW	missile approach warning
MCM	mine countermeasures

MILSATCOM	military satellite communication
MILSTAR	military strategic and tactical relay
MMIC	monolithic microwave integrated circuits
MSE	mobile subscriber equipment
MTD	moving target detection
MTI	moving target indication
NPB	neutral particle beams
NUCINT	nuclear intelligence
OPINT	optical intelligence
OPSEC	operation security
OSINT	open-source intelligence
OTH	over-the-horizon
PA	pulse amplitude
PC	pulse compression
PD	pulsed Doppler
PDW	pulse description words
PHOTINT	photo intelligence
POI	probability of intercept
PPI	plan position indicator
PRF	pulse repetition frequency
PRI	pulse repetition interval
PSYOP	psychological operations
PTD	post-tuning drift
PTI	processing time interval
PW	pulse width
RADINT	radar intelligence
RAM	radar absorbent material
RANT	unintentional radiation intelligence
RCS	radar cross section
RGPO	range deception
RGWO	range gate walk off
RF	radio frequency
ROM	read-only-memory
RPV	remotely piloted vehicle
RSTA	reconnaissance, surveillance, target acquisition
RWR	radar warning receiver
SAM	Surface-to-Air Missile

SAR	synthetic aperture radar
SAW	surface acoustic wave
SEAD	suppression of an enemy air defense
SHF	super high frequency
SHORAD	short-range air defense
SIGINT	signal intelligence
SINCGARS	Single Channel Ground and Air Radio System
SLB	sidelobe blanker
SLC	sidelobe cancellation
SNAP	steerable null antenna processor
SSR	secondary surveillance radar
STAP	space-time adaptive processing
STC	sensitivity-time-control
TACCAR	time average coherent airborne radar
TDOA	time difference of arrival
TDMA	time division multiple access
TECHINT	technical intelligence
TELINT	telemetry intelligence
TERCOM	terrain contour matching
TEWS	tactical EW system
TG	techniques generator
TIBS	Tactical Information Broadcast System
TOA	time of arrival
T-R	transmit-receive
TWS	track-while-scan
TWT	traveling wave tube
UAV	unmanned airborne vehicle
UFMOP	unintentional frequency modulation on pulse
ULSA	ultralow sidelobe antenna
UV	ultraviolet
UWB	ultrawideband
VAB	Van Allen radiation belts
VCO	voltage controlled oscillator
VGPO	velocity gate pull-off
VSWR	voltage standing wave radio
WARM	wartime reserve mode

Index

Recent Titles in the Artech House Radar Library

David K. Barton, Series Editor

Airborne Pulsed Doppler Radar, Second Edition, Guy V. Morris and Linda Harkness, editors

Design and Analysis of Modern Tracking Systems, Samuel Blackman and Robert Popoli

Digital Techniques for Wideband Receivers, James Tsui

Electronic Intelligence: The Analysis of Radar Signals, Second Edition, Richard G. Wiley

Electronic Warfare in the Information Age, D. Curtis Schleher

High-Resolution Radar, Second Edition, Donald R. Wehner

Introduction to Electronic Warfare, D. Curtis Schleher

Introduction to Multisensor Data Fusion: Multimedia Software and User's Guide, TECH REACH, Inc.

Millimeter-Wave and Infared Multisensor Design and Signal Processing, Lawrence A. Klein

Modern Radar System Analysis, David K. Barton

Modern Radar System Analysis Software and User's Manual, David K. Barton and William F. Barton

Principles of High-Resolution Radar, August W. Rihaczek

Radar Cross Section, Second Edition, Eugene F. Knott, *et al.*

Radar Evalution Handbook, David K. Barton, *et al.*

Radar Meteorolgy, Henri Sauvageot

Radar Signal Processing and Adaptive Systems, Ramon Nitzberg

Radar Technology Encyclopedia, David K. Barton and Sergey A. Leonov, editors

For further information on these and other Artech House titles, including previously considered out-of-print books now available through our In-Print-Forever® (IPF®) program, contact:

Artech House
685 Canton Street
Norwood, MA 02062
Phone: 781-769-9750
Fax: 781-769-6334
e-mail: artech@artechhouse.com

Artech House
46 Gillingham Street
London SW1V 1AH UK
Phone: +44 (0)171-973-8077
Fax: +44 (0)171-630-0166
e-mail: artech-uk@artechhouse.com

Find us on the World Wide Web at:
www.artechhouse.com

CPSIA information can be obtained at www.ICGtesting.com
Printed in the USA
LVOW101101160113

315932LV00001B/43/P